FOOD-BORNE INFECTIONS
AND INTOXICATIONS

FOOD SCIENCE AND TECHNOLOGY

A SERIES OF MONOGRAPHS

Deceased, September, 1968.

FOOD-BORNE INFECTIONS

AND INTOXICATIONS

Edited by Hans Riemann

DEPARTMENT OF EPIDEMIOLOGY
AND PREVENTIVE MEDICINE
SCHOOL OF VETERINARY MEDICINE
UNIVERSITY OF CALIFORNIA
DAVIS, CALIFORNIA

ACADEMIC PRESS   New York and London   1969

ACADEMIC PRESS, INC.
111 Fifth Avenue, New York, New York 10003

United Kingdom Edition published by
ACADEMIC PRESS, INC. (LONDON) LTD.
Berkeley Square House, London W.1

LIBRARY OF CONGRESS CATALOG CARD NUMBER: 68-28896

PRINTED IN THE UNITED STATES OF AMERICA

LIST OF CONTRIBUTORS

Numbers in parenthesis indicate the pages on which the authors' contributions begin.

ROBERT ANGELOTTI, National Center for Urban and Industrial Health, Bureau of Disease Prevention and Environmental Control, Public Health Service, Department of Health, Education, and Welfare, Cincinnati, Ohio (359)

FRANK L. BRYAN, Department of Health, Education, and Welfare, Public Health Service, Bureau of Disease Prevention and Environmental Control, National Communicable Disease Center, Atlanta, Georgia (223)

DEAN O. CLIVER, Food Research Institute and Department of Bacteriology, University of Wisconsin, Madison, Wisconsin (73)

NEVA N. GLEASON, Department of Health, Education, and Welfare, Public Health Service, Bureau of Disease Prevention and Environmental Control, National Communicable Disease Center, Atlanta, Georgia (175)

RICHARD A. GREENBERG, Research and Development Center, Swift and Company, Chicago, Illinois (455)

GEORGE R. HEALY, Department of Health, Education, and Welfare, Public Health Service, Bureau of Disease Prevention and Environmental Control, National Communicable Disease Center, Atlanta, Georgia (175)

BETTY C. HOBBS, Food Hygiene Laboratory, Central Public Health Laboratory, Public Health Laboratory Service, London. N.W.9., England (131)

J. H. McCOY, Public Health Laboratory, Hull Royal Infirmary, Anlaby Road, Hull, Yorkshire, England (3)

HANS RIEMANN, Department of Epidemiology and Preventive Medicine, School of Veterinary Medicine, University of California, Davis, California (291)

GENJI SAKAGUCHI, Department of Food Research, National Institute of Health, Shinagawa-ku, Tokyo, Japan (329)

RIICHI SAKAZAKI, National Institute of Health, Tokyo, Japan (115)

JOHN H. SILLIKER, Silliker Laboratories, Chicago Heights, Illinois (455)

HAROLD GEORGE SCOTT, U.S. Public Health Service Hospital, New Orleans, Louisiana (543)

JOAN TAYLOR, Salmonella Reference Laboratory, Central Public Health Laboratory, London, England (3)

GERALD N. WOGAN, Department of Nutriton and Food Science, Massachusetts Institute of Technology, Cambridge, Massachusetts (395)

FOREWORD

By any standards one might apply "Food-Borne Infections and Intoxications" is a monumental work. It is certainly in a class by itself in this specific field — comprehensive in subject-matter coverage, authoritative in terms of the stature and competence of the several authors and the editor involved, and critical and integrative in its approach and style of presentation.

This treatise covers a subject of great current interest in food science and technology, and while other works such as Dack's "Food Poisoning" and Frazier's "Food Microbiology" are both valuable and useful in their own way, they can hardly be compared to the present treatise in terms of comprehensiveness of subject-matter coverage, and in providing a critical evaluation of the enormous and widely scattered literature. There is little doubt in my mind that this work will very quickly find its place in the world's scientific libraries and on the shelves of scientists' offices as *the* standard reference work in the field.

This work has been organized in three sections: (1) Food-Borne Infections, (2) Intoxications of Microbial Origin, and (3) Miscellaneous, including Poisonous Plants and Animals, Food Processing and Preservation (as factors influencing these infections and intoxications), and Laboratory Methods (for the examination of foods for causative organisms). Certain aspects of the epidemiological characteristics of these infections and intoxications and of environmental sanitation are covered.

FOOD-BORNE INFECTIONS

Infections by Salmonella and Related Organisms. This chapter is authored by two key U. K. scientists, Taylor and McCoy, who have been closely associated with the intense research and regulatory efforts carried out on these microorganisms over the past twenty-five years. After an introduction and a short section on classification, the authors go into a comprehensive and detailed discussion of human salmonellosis and the related food and nonfood animal infections. A key section of this chapter deals with salmonellae in animal feedstuffs and means for prevention or control, including a thorough and critical review of the literature.

Another important section relates to the transfer of the salmonellae from animals to man, including consideration of the vehicles such as eggs and egg products, milk and dairy products, and red meat and poultry products.

Viral Infections. This chapter, by a talented young scientist, appears to be the first comprehensive and critical review of the literature on food-borne viral infections. This was a difficult area in which to obtain a satisfactory contribution, both in terms of finding a qualified author and in view of the paucity of literature. Dr. Cliver of the University of Wisconsin has proved to be an excellent choice because he is devoting his entire research to the food-borne viruses, and has demonstrated competence.

The chapter appropriately begins with a discussion of the types and properties of the known animal viruses. Many in food science and technology have little background training or experience with these disease agents, and will very much appreciate this orientation. Included in the discussion are the nonviral agents, bedsonia and rickettsia, since they are similar to viruses and also because they are not discussed elsewhere.

Next, the author considers epidemiological aspects, including mode of transmission, population immunities, and the food vehicles. This is followed by a detailed coverage of the viruses associated with foods. It is recognized that, while some of these are not strictly food-borne, they are definite occupational hazards.

The author next addresses himself to the question of how viruses get into food. Primary contamination, fecal pollution, and contamination from human carriers are discussed. Processing and preservation methods as they affect virus inactivation in foods are considered next. A final section is devoted to the detection and identification of viruses in foods. It becomes obvious that there is an urgent and serious need for more research.

Halophilic Vibrio Infections. Sakazaki of Japan's National Institute of Health brings to us a thorough account of the research which has been done on this most interesting disease. Many food scientists are completely unaware of this particular infection, the causative organism of which was identified only seventeen years ago.

The author first presents a historical introduction, then immediately goes into a detailed discussion of the bacteriology of the causative agents — morphology, biochemistry, physiology, and serology. This is followed by a commentary on its pathogenicity for animals and man and the geographical distribution of the organism and infection. Epidemiological aspects are thoroughly covered in the following section. Then the author concludes by providing an excellent review of the methodology for the isolation and identification of the organism.

To my knowledge this is the first up-to-date status report in English on this food-borne infection. As a matter of fact, perusal of the author's bibliography reveals only five articles published on this subject outside of Japan, showing how little this disease has been studied internationally.

Clostridium perfringins and Bacillus cereus Infections. Hobbs, one of U. K.'s outstanding authorities, provides us with a critical and thorough review of the literature on these important food-borne infections. This is fortunate, indeed, because of the considerable controversy that has raged over whether these organisms are actually capable of producing illness in man.

The author has organized her chapter into four broad sections — a historical introduction, characteristics of the organisms, disease outbreaks, and disease prevention. A great deal of attention is paid to the biochemical, physiological, and serological characteristics of the organisms. The evidence as to whether these organisms are capable of causing illness in man is carefully and systematically reviewed and evaluated. A final section deals with ways and means for controlling in humans infections caused by these organisms.

I am sure that this chapter will be of immense help to all who are interested or must be concerned with these food-borne infections.

Parasitic Infections. Frequently the food technologist is not even aware or appreciative of the fact that there are infections due to parasites found in many raw food materials. Healy and Gleason of the U. S. National Communicable Disease Center have done a notable job of reviewing and summarizing the widely scattered and extensive literature on this subject. Each of a number of parasites — nematodes, cestodes, trematodes, *Toxoplasma gondii,* and the protozoa and helminths — are treated in detail, In each case the life cycle for the parasite is presented, including its manifestation in the human. Also treated are the means for control, where this is known. Methods of detection are also given in some detail, all of which should prove to be of value to those who are concerned with quality control in fresh and processed foods likely to be contaminated with these organisms.

Miscellaneous Infections. Bryan of the U. S. Public Health Service has done a masterful job of bringing together and digesting the enormous literature on the other infections and intoxications not dealt with in other chapters of this treatise. More than 300 literature citations are listed in the bibliography for this chapter, attesting to the enormity of the task the author undertook.

The author has organized his chapter into four sections: (1) organisms for which proof of pathogenicity or toxigenicity is lacking, (2) certain milk-borne organisms (e.g., *Brucella* and *Corynebacteria*), (3) organisms usually transmitted to man by means other than food, and (4) organisms with unknown food-borne relationship. A short, final section deals with contact infections.

The food scientist will be very appreciative of the author's critical and exhaustive treatment of the widely scattered literature on these numerous organisms that have been implicated at one time or another in causing illness in man and in possibly being food-borne.

INTOXICATIONS OF MICROBIAL ORIGIN

Botulism. The chapters on botulism have been prepared by two prominent researchers; one is on types A, B, and F by Reimann of the University of California at Davis, and the other on type E by Sakaguchi of the Japanese Institute of Health. Both chapters are organized along similar lines—after a brief historical introduction a review is presented of the incidence of outbreak, foods involved, and clinical manifestations. This is followed by a discussion on the ecology of the causative organisms.

Considerable discussion is devoted to the morphology, biochemistry, and physiology of the organisms, including toxin formation and sporogenesis. Factors affecting spore germination are also discussed extensively as is also the resistance of the spores to inactivation, especially by heat and radiation. Critical and exhaustive review is provided of the chemistry and biochemistry and toxicology of the several botulinum toxins, including the toxicity to various wild and laboratory species of animals.

While the early literature is well covered, of special note is the large number of references to the recent literature on botulism. This is particularly true for the chapter dealing with type E, on which a large amount of research has been carried out in recent years. Also of significance is the international flavor of the studies referred to in these chapters. Particularly notable is the large concentration of research effort in Japan, North America, and Northern Europe.

Staphlococcal Enterotoxoses. This chapter has been written by a well-qualified scientist—Dr. Angelotti of the U.S. Public Health Service. This in-depth, critical account updates our knowledge of this most important food-borne intoxication in first-class fashion.

After a short introduction and historical review, he plunges into a discussion of the morphological, biochemical, and physiological characteristics of the toxigenic staphylococci, including phage typing and toxigenesis. This is followed by a comprehensive discussion of the entero-toxoses, including clinical symptoms, incidence, and species suscepti-bility. A special section is devoted to the toxic compounds — types, labora-tory methods for their production, isolation, purification, identification, and characterization. A final section is devoted to methods for the isola-tion, identification, and enumeration of the organisms and their toxins in foods.

More than a hundred and ten references are cited. A substantial pro-portion of these articles were published in the last fifteen years.

Mycotoxicoses. This is an up-to-date review of the literature by one of the leading researchers in the field, Wogan of Massachusetts Institute of Technology. After a brief historical introduction he proceeds immediately to a detailed discussion of the best-known fungal intoxications: aflatoxico-sis, toxic aleukia, and the intoxication caused by moldy rice. Others are briefly mentioned. In the case of each toxicosis the author systematically provides a critical and exhaustive review of the following aspects: micro-biological, chemical, biochemical, and toxicological.

Although certain mycotoxicoses have been recognized medically for centuries (e.g., ergotism), only in the past decade has there been any extensive research on the microbiology, chemistry, biochemistry, and toxicology. Less than ten of the articles listed in the very extensive bibliography in this chapter predate 1956. It seems likely that myco-toxicoses will be a very active field for research in the next decade or two. Certainly a great deal of research is urgently needed.

MISCELLANEOUS SECTIONS

Laboratory Methods for Microbiological Contamination. Silliker, a well-known U.S. public health scientist, and Greenberg, a U.S. food in-dustry microbiologist with broad experience, have teamed up to produce a chapter on the detection and enumeration of organisms from food-borne outbreaks. Special emphasis is placed on the problems of the specific foods involved, proper sampling, and, in particular, the laboratory methods for the identification and enumeration of the pathogenic and toxigenic organisms. Selected for emphasis in laboratory methodology are those for the salmonellae, staphylococci, and clostridias. Methods for the examination of foods and other elements of the environment for the other infections and intoxications covered in this work are not dis-cussed in this chapter. However, considerable information on their de-tection and enumeration and toxicity tests are given in the chapters deal-ing with specific infections or intoxications.

Food Processing and Preservation. Riemann of the University of California, Davis has produced an excellent review of the effects of processing and preservation methods on the food-borne pathogenic and toxigenic organisms. He has approached the problem by thoroughly and critically examining the literature relative to the effects of the various processing and preservation methods: heat treatment, radiation, refrigeration, chemical preservation, and dehydration. The effects of packaging are also discussed in detail.

The author concludes his presentation by pointing out the serious information gaps in this area and the tremendous need for a greater research effort in this aspect of the food-borne infections and intoxications.

Poisonous Plants and Animals. A very extensive chapter on the highly toxic plant and animals species has been provided by Scott of the U.S. Public Health Service. Not covered are the species which are only mildly toxic to man. For a discussion of the latter the reader is referred to "Toxicants Occurring Naturally in Foods" published by the National Academy of Sciences/National Research Council, Washington, D. C.

In the early sections of his chapter the author deals with historical aspects and epidemiological studies. He then proceeds immediately to a lengthy treatment of the specific intoxications. Each of a large number of genera is specifically discussed: its geological locale, chemical entity (if known), and symptoms. Examples of the toxic plants discussed are *Amanita, Malvella, Psilocybe* (all three commonly known as mushrooms), *Dryoteria* (ferns), *Equisetum* (horsetails), *Taxus* (yew), *Ricimus* (castor bean), *Phytococca* (American nightshade), *Salanum* (common nightshade), *Cicuta* (wild carrot, dill, celery, etc.), *Conium* (California fern and others), *Aesculus* (buckeye), *Robinia* (black locus), *Ilex* (holly), and *Berberis* (barberry).

Poisonous animals are treated similarly, including certain of the fishes, mussels and oysters, clams, eels, and turtles. Finally, a number of toxic mammals are mentioned: certain dogs, seals, and polar bears.

This chapter should prove of great value to food scientists and technologists because they rarely have any background or experience with these toxicants. Over a hundred references are cited; this bibliography will be especially valuable for those wishing to delve into the subject in greater detail.

January, 1969   GEORGE F. STEWART
                                        Food Protection and Toxicology Center,
                                        University of California,
                                        Davis, California

PREFACE

The broadness of the subject of this volume has made selectivity a necessity. Several types of food poisoning have not been included, such as those caused by toxic chemicals intentionally or unintentionally added by man. Among the naturally occurring agents only those considered most important have been included.

The content of this work could have been organized in several ways. We have chosen to group together those agents which must be present in food in a viable state in order to cause disease, although the disease which is provoked is often termed food poisoning (e.g., Salmonella food poisoning, perfringens food poisoning). The subdivision of the chapter on botulism into one on type E and another on types A, B, and F was motivated by the higher incidence of type E botulism in recent years and the accumulating literature dealing specifically with type E.

The purpose of the chapters on laboratory methods and food processing and preservation is to make the treatise more useful to readers who have special interests in these aspects of food-borne infections and intoxications. Since these chapters deal with a variety of organisms which are discussed in other chapters some repetition and scattering of information have resulted.

I am indebted to my co-authors for their willingness to make contributions without which this work could not have materialized. I also gratefully acknowledge their advice, understanding, and patience. I am indebted to Dr. G. M. Dack for his encouragement and advice in connection with the planning of this volume. Special thanks are due to Mrs. Patricia Akrabawi for excellent editorial assistance and to the staff of Academic Press for their cooperation.

Davis, California   HANS RIEMANN
January, 1969

CONTENTS

Chapter IV.   *Clostridium perfringens*
               and *Bacillus cereus* Infections

               Betty C. Hobbs

Chapter V.   **Parasitic Infections**

               George R. Healy and Neva N. Gleason

Chapter VI.   **Infections Due to Miscellaneous Microorganisms**

               Frank L. Bryan

PART II.   **INTOXICATIONS OF MICROBIAL ORIGIN**

Chapter VII.   **Botulism — Types A, B, and F**

                Hans Riemann

Chapter VIII. Botulism — Type E

Genji Sakaguchi

Chapter IX. Staphylococcal Intoxications

Robert Angelotti

Chapter X. Alimentary Mycotoxicoses

Gerald N. Wogan

PART III.   MISCELLANEOUS

Chapter XI.   Laboratory Methods

John H. Silliker and Richard A. Greenberg

Chapter XII.   Food Processing and Preservation Effects

Hans Riemann

Chapter XIII.   Poisonous Plants and Animals

Harold George Scott

GLOSSARY*

Abattoir — A slaughterhouse

Abscess — A circumscribed collection of pus

Adenoma — An ordinary benign neoplasm (any new growth of cells or tissues) in which the tumor cells form glands or glandlike structures

Adrenergic — Relating to sympathetic nerve system

Afebrile — Without fever

Aflatoxin — Toxic substance produced by the fungus *Aspergillus flavus*

Agranulocytosis — Acute condition characterized by pronounced reduction in white cells

Albuminuria — The presence of protein in urine

Aleukia — (a) Absence or extremely decreased number of leukocytes in circulating blood; (b) Absence or extremely decreased number of blood platelets

Alimentary toxic aleukia — A form of poisoning of the bone marrow

Alkaloid — A basic usually bitter substance found in the leaves, bark, seeds, and other parts of plants. Some are highly poisonous.

Alzheimer, or Picks Disease — A presenile mental deterioration occurring usually in persons under fifty years of age

Amnion — The inner of the membranes composing the sac that envelops the fetus in the uterus

Anemia, aplastic — Characterized by a greatly decreased formation of erythrocytes and hemoglobin, and usually associated with low numbers of granular leukocytes and platelets in the circulating blood

Anorexia — Loss of appetite

Antimetabolite — A substance that resembles chemically a particular metabolite, but competes with, replaces, or antagonizes it

Anuria — Total suppression of urine

Aphonia — Loss of the voice due to disease or injury of the organ of speech

Ascarid — A nematode worm; intestinal parasite

Ascites — An accumulation of serous fluid in the peritoneal cavity

Asymptomatic — Without symptoms, or producing no symptoms

Atabrine — Trade name for quinacrine hydrochloride

Atrophy — A wasting of tissues, organs, or the entire body

Autochthonous — Originating in the place where found; said of a disease originating in the part of the body where found, or of a disease acquired in the place where the patient is

Autochthonous parasite — A parasite descended from the tissues of the host

---

Stedman's Medical Dictionary, Twentieth Edition, The Williams and Wilkins Company, Baltimore, 1961.

Webster's New World Dictionary of the American Language, The World Publishing Company, Cleveland and New York, 1962.

Avitaminosis — A deficiency disease state resulting from an inadequate supply of one or more vitamins in the diet

Bacteriocins — Antibacterial compounds produced by related organisms

Biliary — Relating to bile

Biopsy — Gross and microscopic examination of tissues or cells removed from a living patient, for the purpose of diagnosis or prognosis of disease

Bithionol — A therapeutic agent for pulmonary paragonimiases

Bladderworms — The tapeworm larva at the stage in which its head and neck are enclosed in a bladderlike sac

Blanch — In cookery, scalding foods by dipping in hot water

Blepharoptosis — Drooping of the upper eyelid

Buccal — The mouth or mouth cavity

Caltrops — Any of a number of plants with pointed flowers or fruits, as the star thistle, water chestnut, etc.

Carbuncles — An inflammation of the skin and subcutaneous tissues, e.g., caused by the anthrax bacillus

Carcinogenic — Cancerigenic; causing cancer

Carcinogens — Any cancer-producing substance

Casamino acid — Trade name for a hydrolyzate of casein

Cathartic — An agent causing active movement of the bowels

Cecal area — Ending blindly or in a *cul-de-sac*

Centrelobular — At or near the center of a lobe

Cervical — Relating to a neck or cervix

Cestode — A tapeworm

Chlorpromazine — A drug with a depressing effect upon the liberation of adrenaline

Cholangitis — Inflammation of a bile duct

Choleglobin — A pigmented compound of globulin and iron prophyrin occurring as an intermediary metabolite during the formation of bile pigment

Chorioallantoic cavity — In bird embryos, the cavity surrounded by an extraembryonic membrane

Chorioretinitis — Inflammation of the middle coat of the eyeball and retina

Cirrhosis, biliary — A chronic inflammation of a bile duct resulting in jaundice, attacks of abdominal pain, and enlargement of the liver and spleen

Clonorchiasis — Clonorchiosis. A condition marked by invasion of the bile ducts by *Clonorchis endemicus* or *C. sinensis*

Conidia — Pleural of condidium, a nonsexual spore of certain fungi

Conjunctivitis — Inflammation of the mucous membrane covering the anterior surface of the eyeball

Convulsion — A violent involuntary muscular contraction

Coprophagous — That which eats human or animal excrement

Corneal opacity — A lack of transparency of the cornea

Corticosteroids — A steroid hormone produced by the adrenal cortex

Cotyledon — Earliest leaf or one of the earliest leaves growing out of a seed

Cutaneous nodule — A small knot on the skin

Cyanogen — A compound of two carbon and two nitrogen atoms $(CN)_2$

Cyanogenic — Containing cyanide

Cyanosis — A dark bluish or purplish coloration of the skin and mucous membrane due to deficient oxygenation of the blood

Cystic — (a) Relating to the urinary bladder or gall bladder;
(b) Relating to a cyst

Cysticerci—Cysticercus: The encysted larvae of various tapeworms

Cytolytic—Pertaining to cytolysis; possessing a solvent or destructive action on cells

Cytolysis—The dissolution of a cell

Cytotoxic—Cytolytic; destructive to cells

D value—The number of minutes at a defined temperature required to destroy 90% of the organisms present

Deciduous—Falling off at a certain season or stage of growth. Shedding leaves annually

Deltoid muscle—A muscle which controls the rotation, flexion and extension of the arm

Dermatitis—Inflammation of the skin

Dermonecrotic—Causing the pathological death of skin

Diathesis, hemorrhagic—Any of the many syndromes showing a tendency to spontaneous hemorrhage, resulting from weakness of blood vessels or a clotting defect, or both

Dilatation—Enlargement of a cavity, canal, blood vessel, or opening

Diuretic—An agent that increases the amount of urine

Duodenum—The first division of the small intestine

Dyscrasia—A morbid general state resulting from the presence of toxic matters in the blood; an old term to indicate an abnormal, depraved condition

Dysentery, amebic—Diarrhea resulting from ulcerative inflammation of the colon, caused by *Entamoeba dysenteriae*

Dyspnea—Difficulty or distress in breathing, frequent rapid breathing, usually associated with serious disease of the heart or lungs

Ectopic—Out of place

Eczema, facial—An inflammation of the skin of the face presenting multiform lesions, and often accompanied by itching and burning

Edema—A perceptible accumulation of excessive clear watery fluid in the tissues

Embolus—A plug, composed of a detached clot, mass of bacteria, or other foreign body, occluding a blood vessel

Emesis—Vomiting

Emetic—An agent that causes vomiting

Emulsifying agents—An agent, such as gum arabic or the yolk of an egg, used to make an emulsion of an oil

Encapsulated—Enclosed in a sheath or capsule

Endocarditis—Inflammation of the lining membrane of the heart. It may involve only the membrane covering the valves or the general lining of the chambers of the heart

Endoplasm—The cytoplasm, a mass of cell substance

Enteric—Relating to the intestine

Enteropathogenic—Causing an intestinal disease

Enterotoxemia—A disease caused by toxins produced in the intestines: microorganisms

Enterotoxin—A toxin specific for the cells of the mucous membrane of the intestine, e.g., *Staphylococcus enterotoxin*

Enzootic—Denoting a disease of animals which is indigenous to a certain locality, analogous to an endemic disease among men

Eosinophil—A cell or other element, especially a leukocyte, that has an affinity for eosin dye

Eosinophilia—A disease with increase in the number of eosinophilic leukocytes

Epidemic—The extensive prevalence in a community of a disease brought from without, or a temporary increase in number of cases of an endemic disease

Epigastric—Relating to the pit of the stomach

Epistaxis—Nosebleed

Epithelium—The cellular layer covering all the free surfaces of the body

Epizootic—The prevalence of a disease among animals, the same as epidemic in men. A disease attacking a large number of animals simultaneously

Ergot—The hardened weblike mass of threads of *Claviceps purpurea* which replaces the grain of rye attacked by this fungus

Ergotism—Poisoning by a toxic substance contained in the fungus *Claviceps purpurea*

Erysipeloid—A specific of the skin usually on the hand, caused by *Erysipelothrix rhusiopathiae*

Erythema arthriticum epidemicum—An infection that occurs in epidemic form, marked by multiple joint pains, a blotchy eruption, and general toxic symptome with fever; Haverhill fever

Erythrogenic—(a) Producing red, as causing an eruption or a red color sensation; (b) Pertaining to the formation of red blood cells

Erythema—Redness of the skin; inflammation

Erythematous—Relating to or marked by redness of the skin; inflammation

Erythrocyte—A mature red blood cell

Esophagus—The swallow; the portion of the digestive canal between the throat and the stomach

Estrogenic—Causing estrus (heat) in animals. Having the action of the female sex hormone, estradiol

Etiological agent—The agent that causes a disease

Etiology—The cause of disease

Euploid—Having a chromosome number that is an exact multiple of a single genome

Exanthema—A general disease accompanied by an eruption on the skin such as measles; an eruptive fever

Exotoxin—A toxin recovered by filtration from a culture

Exudate—A liquid or semisolid that is being discharged or has been discharged through the tissue to the surface or into a cavity

F value—The number of minutes at a specified temperature that destroy a microorganism or a given proportion of the microbial cells

Febrile—Relating to fever, feverish

Fever, hemorrhagic—A condition characterized by an acute onset of headache, chills, and high fever, thirst, cough, abdominal pain with nausea and vomiting

Fibrinolysin—An enzyme that causes the destruction of fibrin; causes coagulated blood to become fluid again

Fibroblast—An elongated, flattened cell present in connective tissue

Fibrosarcosomas—A malignant growth of tissue derived from fibrous connective tissue

Fibrosis—The formation of fibrous tissue

Flatulence—The presence of an excessive amount of gas in the stomach and intestines

Flatworms—A group of worms with a flattened body, as the liver fluke and the tapeworm

Fluke—The common name for any species of the class Trematoda; tremetodes, or flukes, are exclusively parasitic forms in man and animals

Gastric lavage—The washing out of the stomach by copious injections and rejections of water

Gastritis—Inflammation of the stomach

Gastrocnemius—A muscle in the leg

Gastroenteritis—Inflammation of the mucous membrane of both stomach and intestine

Gibberellic acid — A plant growth hormone

Globinuria — The excretion of a protein globulin in the urine

Glottis — The vocal apparatus of the larynx

Gnathostomiasis — Infection with a species of *Gnathostoma.*

Granuloma — Any one of a large group of focal lesions that (a) are formed as a result of inflammatory reactions; (b) are granule-like or nodular, i.e., fairly well circumscribed and firmer than the uninvolved adjacent tissue; (c) are usually characterized by the accumulation of phagocytes; and (d) ordinarily persist in the tissue as "slowly smoldering" inflammations

Gravid — Pregnant

Halophilic — Pertaining to microorganisms that are resistant to or require high concentrations of salt

HeLa cells — Ovarian cancer cells widely used in tissue culture work; one of the original cell lines established

Helminth — An intestinal worm

Hematopoietic — Hemopoietic; pertaining to or related to the formation of blood cells

Hemolysis — The alteration, dissolution, or destruction of red blood cells

Hemolytic — Destructive to blood cells, resulting in liberation of hemoglobin

Hemoptysis — The spitting of blood derived from the lungs or bronchial tubes

Hemorrhage — Bleeding; a flow of blood, especially if it is very profuse

Hepatic — Relating to the liver

Hepatitis — Inflammation of the liver

Hepatogenous — Of hepatic origin; formed in the liver

Hepatoma — Liver tumor

Hepatotoxic — Relating to an agent that damages the liver or pertaining to any such action

Hepatotoxin — A toxin that is destructive to liver cells

Hermaphroditic — An individual whose genital organs have the characters of both male and female in greater or less degree

Host — The organism at the expense of which a parasite lives

Hydrocephalus — A condition marked by an excessive accumulation of fluid in the cerebral ventricles

Hyperemia — The presence of an increased amount of blood in a part; congestion

Hyperkeratosis — Thickening of the horny layer of the epidermis

Hyperplasia — An increase in number of the individual tissue elements, excluding tumor formation, whereby the bulk of an organ is increased

Hypertrophy — Abnormal increase in the size of an organ or tissue caused by enlargement of its cellular components, not due to tumor formation

Hypoglycemia — An abnormally small concentration of glucose in the circulating blood

Ichthyism — Poisoning by eating stale or otherwise unfit fish

Ichthyosarcotoxism — Any condition caused by eating poisonous fish

Icterus — Jaundice

Immunity, active — A form of immunity in which the hosts tissues are stimulated by an antigen that leads to the formation of a specific antibody and increased resistance

Immunity, passive — A form of immunity acquired as the result of transfer of preformed protective substances, either naturally (from mother to fetus *in utero*) or artificially as in the injection of immune serum

Intermediate host — The host in which the asexual cycle of a parasite is completed

Intestinal mucosa — Mucous membrane of the intestine

Intraperitoneal —Within the peritonial (abdominal) cavity

Jaundice —A yellowish staining of the skin and deeper tissues and the excretions with bile pigments

Jejunum —The portion of the small intestine between the duodenum and the ileum

Lacrymation —The secretion of tears, especially in excess

Lairage —A place where animals are kept, e.g., before slaughter

Lamziekte —A disease of cattle in South Africa due to the toxin of type D *Clostridium botulinum*

Larvae —The wormlike form of an insect on issuing from the egg

Lassitude —A sense of weariness

LD_{50} —Lethal Dose 50%, a measure of toxicity, i.e., the quantity of drug, toxin, etc., which, when administered to experimental animals, will be 50% fatal

Lesion —(a) A wound or injury; (b) A more or less circumscribed pathologic change in the tissues; (c) One of the individual points or patches of a disease of the skin

Leukemia —Any one of a complex group of diseases of the reticuloendothelial system characterized by a great degree of uncontrolled proliferation of one of the types of white blood cells

Leukocyte —Any one of the white blood cells

Leukocytic —Pertaining to or characterized by any one of the white blood cells

Leukopenia —Any situation in which the total number of leukocytes in the circulating blood is less than normal

Leukosis —Abnormal proliferation of one or more of the tissues that produce white blood cells

Leukosis, avian — A group of conditions that occur chiefly in chickens and are characterized by an abnormal proliferation of lymphoid tissue

Lung fluke disease —Disease caused by *Paragonimus westermanii*

Lymphadenitis —Inflammation of a lymph node or lymph nodes

Lymphocytic choriomeningitis —A mild, often unrecognized, virus disease, prevalent in rodents and other animals including man

Lymphomatosis —Any condition characterized by the occurrence of multiple, widely distributed sites of involvement with diseases of the lymphoid tissues

Lymphosarcoma —A malignant disease of the lymphoid tissues which results in enlarged lymph nodes and infiltration of various tissues

Lysogenic — Causing or having the power to cause lysis, indicating the action of certain antibodies and chemical substances

Maculopapule —A spot on the skin

Mastitis —Inflammation of the mammary gland

Meningeal hydrops —Abnormal accumulatin of fluid in the membranous envelope of the brain and spinal cord

Meningoencephalitis —An inflammation of the brain and its membranes

Mesentery —A double layer of peritoneum attached to the abdominal wall

Metacercaria —The stage in the life history of a fluke in which the cercariae lose their tails and become encysted

Metamorphosis —A change in form, structure, or function

Methemoglobinemia —The presence in the circulating blood, of a nonfunctional transformation product of oxyhemoglobin

Miracidium —The ciliated larva of a trematode worm

Mitochondria —Granules in the cytoplasm of cells essential for life processes

M.L.D. —Abbreviation for the minimal lethal dose of a toxic substance, as assayed in experimental animals

Monocytosis—An abnormal increase in the number of monocytes (large leukocytes) in the circulating blood

Mononucleosis—The presence of abnormally large numbers of mononuclear leukocytes in the circulating blood, especially with reference to forms that are not normal

Morbidity—A deceased state

Mucin—One of a number of glycoproteins secreted by cells of the mucous glands and present also in connective tissue and in the umbilical cord

Mycelium—The mat or complex group of protoplasmic units, or the entangled mass of tube-like or filamentous structures, hyphae, that represents the "body" of fungi

Mycoplasma—Type genus of the microorganism *Mycoplasmataceae;* mostly parasitic in humans and animals

Mycotoxicoses—Toxicity syndromes resulting from ingestion of food stuffs contaminated with mycotoxins

Mycotoxins—Toxic compounds produced by fungal contaminants of foods and food raw materials

Mydriatic—Causing delation or mydriasis of the pupil

Mydriasis—Dilation of the pupil

Myocarditis—Inflammation of the muscular walls of the heart

Myocardium—The muscular substance of the heart

Narcosis—Stupor or general anesthesia produced by some narcotic drug

Nausea—Sickness at the stomach; an inclination to vomit

Necropsy—Autopsy

Necrosis—The pathologic death of one or more cells, or of a portion of tissue or organ

Necrotic angina—A severe type of sore throat occurring usually as a complication of scarlet fever and more rarely of diphtheria

Nematode—A common name for any parasitic worm of the class *Nematoda*

Nystagmus—Rhythmical oscillation of the eyeballs, either horizontal, rotary, or vertical

Oliguria—Scanty urination

Onchospheres—Little spherical embryos developed in the eggs of Cyclophyllideau tape-worms, characterized by the presence of three pairs of clawlike hooks

Oncogenic—Causing tumors

Operculum—The lid or caplike cover of a helminth egg

Ornithosis—A disease in birds caused by virus; this disease is contracted by man by contact with infected birds

Ossein—The organic constituent of bone; bone collagen

Oxytocic—Hastening childbirth; an agent that promotes the rapidity of labor

Panmyelopathy—Disease of the bone marrow

Panmyelotoxicosis—Systemic poisoning of the bone marrow

Paragonimiasis—Infestation with a worm of the genus *Paragonimus*

Parasite—An organism that lives on or in another and draws its nourishment therefrom

Parenchymal—Relating to the essential tissue of an organ as distinguished from its supportive framework

Parenchymatous—Relating to the tissue of a gland or organ contained in the connective tissue framework

Paresthesia—An abnormal spontaneous sensation, such as of burning, pricking, numbness, etc.

Pasteurization—The destruction of disease-producing organisms by heating

Pathogenic—Causing disease

Periorbital edema—A perceptible accumulation of excessive watery fluid in the tissues around the eyes

Periportal — Surrounding the portal vein entering the liver

Peritoneum — The serous sac lining the abdominal cavity and covering most of the viscera therein contained

Peritonitis — Inflammation of the membrane lining the abdominal cavity

Peritrichate — Denoting microorganisms having cilia or flagella projecting from all sides and not at the poles alone

Petechial — Relating to or accompanied by minute hemorrhagic spots, of pinpoint to pinhead size, in the skin

Pharyngeal erythema — Redding or inflammation of the throat

Pharyngitis — Inflammation of the mucous membrane and underlying parts of the throat

Photochromogenic — Capable of producing color in the light

Pica — A depraved appetite; a hunger for substances not fit for food

Planarian — Small, flat-bodied worm moving by means of cilia

Pleomorphic — Occurrence in more than one form; the existence of more than one morphological type in the same species

Plerocercoid — A stage in the development of a tapeworm following the procercord stage

Pleura — The serous membrane enveloping the lungs and lining the walls of the thoracic cavity

Pleural effusions — The escape of fluid from the membrane enveloping the lungs and lining the walls of the thoracic cavity

Pneumonitis, pneumonia — Inflammation of the lungs

ppb — Parts per billion

ppm — Parts per million

Primary host — The one in which the mature parasite resides when it has two or more stages or existence in different animals

Proglottids — One of the segments of a tapeworm

Prophylactic — Preventing disease

Prostration — A marked loss of strength; exhaustion

Protozoa — That part of the animal kingdom including all of the unicellular forms

Pruriginous — Relating to or suffering from a chronic disease of the skin that itch intensely

Pruritus — Itching

Psittacosis — A viral disease of birds, especially parrots, sometimes transmitted to man

Pulse — The rhythmical dilation of a blood vessel produced by contraction of the heart

Pulses — The edible seeds of peas, beans, lentils, and similar plants having pods

Purulent — Containing or forming pus

Pustule — A small circumscribed elevation on the skin, containing pus

Pyrexia — Fever

Pyrimethamine — An antimalarial drug

Quinacrine hydrochoride — An antimalarial drug

Rectal prolapse — The falling down of the rectum

Reservoir host — An animal that by harboring a species of parasite is the source of human infection

Reticulum — A fine network

Rigor mortis — Stiffening of the body after death

Saccharolytic — Something which ferments sugar

Saponin — Substances found in many plants, possess the common property of foaming or making suds

Saprophytes — A plant or microorganism that grows on dead decaying matter

Sarcoma — A tumor, usually highly malignant

Scabrous grain —Grain rough with small points or knobs, like a file; scaly; scabby

Scombroid —Fish of the mackeral family or group

Scotochromogenic —Capable of producing color in the dark

Sepsis —The presence of various pus-forming and other pathogenic organisms, or their toxins, in the blood or tissues

Septic —Relating to the presence of pus-forming and other pathogenic organisms, or their toxins, in the blood or tissues

Septicemia —A disease caused by the presence of microorganisms or their toxins in the circulating blood

Serotype —A subdivision of a species or subspecies distinguishable from other strains on the basis of antigenic character

Serum sickness —Allergic reaction to injected serum

Simian —Of or like an ape or monkey

Sinoatrial —Relating to chambers of the heart

Sparganum —The plerocercoid larva of *Diphyllobothrium* species

Steatorrhea —Fatty stools; the passage of fat in large amounts in the feces

Stenosis —A narrowing of any canal; a stricture; a narrowing of one of the valves of the heart

Sterilization —The destruction of microorganisms

Stomatitis —Inflammation of the mucous membrane of the mouth

Striate muscle —Skeletal muscles; voluntary muscles

Strobila —A number of consecutive tapeworm segments

Stupor —Lethargy; torpor; unconsciousness

Subclinical —Lack of appearance of manifest symptoms in a disease

Subcutaneous tissues —Tissues beneath the skin

Submaxillary —Beneath the lower jaw

Sulfadiazine —A drug, sulfa derivative, against infections

Syndrome —The aggregate of symptoms constituting together the picture of a disease

Synergistic —Synergetic; working together

Tachycardia —Excessively rapid action of the heart

Tapeworm —Taenia; an intestinal parasite

Tapeworm anemia —Anemia sometimes observed in persons infected with the fish tapeworm, *Diphyllobothrium latum*

Tenesmus —A painful spasm of the anal or vesical sphincter with an urgent desire to evacuate the bowel or bladder, involuntary straining, and the passage of but little fecal matter or urine

Tetany —A disorder marked by intermittent tonic muscular contractions, accompanied by fibrillary tremors, and muscular pains

Thrombocytopenia —A condition in which there is an abnormally small number of platelets in the circulating blood

Toxemia —The presence of an endotoxin, exotoxin, or other noxious substances in the circulating blood. The clinical syndrome caused by the toxic substances in the blood

Toxoid —A toxin that has been treated so as to destroy its toxic property while preserving its capability of producing active immunity

Trachea —Windpipe

Trematode —Parasitic worm, a fluke

Trichinoscope —A magnifying glass used in the examination of meat suspected of being trichinous

Trichinosis—The disease resulting from ingestion of raw or inadequately cooked meat (especially pork) that contains encysted viable larvae of *Trichinella spiralis*

Trismus—Lockjaw; a firm closing of the jaw due to tonic spasm of the muscles of mastication

Trophic—Relating to or dependent upon nutrition

Trophozoite—Schizont: An adult protozoan of the asexual cycle which produces only spores developing into adult forms without conjugation

Tropism—The phenomenon observed in living organisms of moving toward or away from a focus of light, heat, or other stimulus

Tularemia—A disease caused by the bacteria *Pasteurella tularensis*

Tumor—Neoplasm; a circumscribed growth, not inflammatory in character, arising from pre-existing tissue

"Undulant fever"—Brucellosis; an infectious disease caused by *Brucella abortis, melitensis,* or *suis*

Urticaria—Hives; nettle-rash; an eruption of itching wheals

Uterotrophic—Hormone-like effect on the uterus

Uveitis—Inflammation of the middle coat of the eye

Vermifuge—An agent that causes the expulsion of intestinal worms

Vertebrate—An animal having vertebrae, segmented spinal column

Vertigo—Dizziness; giddiness; a sensation of irregular or whirling motion

Vesicular exanthema—A viral disease of swine that resembles foot and mouth disease

Virulence—The quality of being poisonous; the disease-producing power of a microorganism

Vitreous chamber of the eye—The cavity of the eyeball behind the lens

Vulvar hypertrophy—An abnormal increase in the size of the external genital organs of the female

Z value—The number of degrees Fahrenheit (or Centigrade) required to produce a 10-fold change in the time required for destruction of an organism by heat

Zoonosis—Communicable disease common to man and lower animals, e.g., salmonellosis

PART I **FOOD-BORNE INFECTIONS**

CHAPTER I | SALMONELLA AND

ARIZONA INFECTIONS

Joan Taylor and J. H. McCoy

I. INTRODUCTION

A. Classification

The family Enterobacteriaceae in general is composed of bacteria living in the intestine of vertebrates either commonly as commensals or occasionally as pathogens, which present the following characteristics: nonsporing gram-negative rods, motile with peritrichous flagella or non-motile, which grow on ordinary media, reduce nitrates to nitrites, give a negative oxidase reaction, and break down carbohydrates by a fermentative rather than an oxidative reaction. All members ferment glucose, with the formation of acid or acid and gas both aerobically and anaerobically. The family is thus composed of a series of related bacteria which can be segregated into groups giving the same biochemical reactions and showing close serological relationships. Many transitional forms are found between groups. Typical members from various groups may possess common serological relationships. The following groups are recognized

3

within the family: *Shigella, Escherichia coli, Salmonella, Arizona, Citro-bacter, Klebsiella-Aerobacter-Serratia,* and *Proteus-Providence* (Edwards and Ewing, 1962). The classification of groups within the family presented difficulties, as attention has always been centered mainly on those groups *(Shigella, Salmonella)* causing epidemic human disease. As the recognition of members of the *Shigella* and *Salmonella* groups in primary cultures is based largely on their lack of ability to ferment lactose, members of other groups within the family showing this characteristic have at one time or another been regarded as pathogenic. There is now general agreement that, although isolated individual human or animal infections may be caused by members of groups other than *Shigella* or *Salmonella,* evidence of epidemic spread is lacking. A classification based on the lack of ability to ferment lactose excludes, of course, the *Escherichia coli* group, which includes the bacteria most commonly present in the intestine of normal animals, of which many serotypes cause epidemic disease in babies and young animals. The heterogeneous collection of bacterial groups which comprise the Enterobacteriaceae thus presents problems to the taxonomist which are still far from solved. Agreement has been reached that the biochemically defined groups of the Enterobacteriaceae can be regarded as genera (Carpenter, 1963). For this paper it is therefore proposed to confine discussion to groups of the family causing epidemic disease in man and animals through the medium of food and water.

B.   Food-Borne Enterobacteriaceae

In the normal individual, the bulk of the feces is composed of bacteria living normally in the intestine. *Escherichia coli* is the most abundant of these organisms. In individuals in good health the mean number of organisms normally present in the feces is approximately 10^9 per gram (Thomson, 1954; Smith and Crabb, 1961; Williams Smith, 1961). In acute infections with Enterobacteriaceae and in the carrier state, pathogenic bacteria in the feces may equal or surpass in number the normal flora (Thomson, 1955). Whereas the duration of excretion in an infected individual varies from a few days to a lifetime, the period of excretion of a group of individuals infected at the same time through the same vehicle follows a more definite pattern. In salmonella infection, excretion continues undiminished for some weeks, after which a more or less constant proportion of individuals ceases to excrete in each week until the persistent excretors are revealed (George *et al.,* 1953; Kwantes, 1952; Lennox *et al.,* 1954). In infection with the Enterobacteriaceae, therefore, the reservoir of infection is the intestine of the infected human or animal; the source of infection is the feces of the excreting individual, either acutely infected, convalescent from infection, a chronic carrier, or a symptomless excretor. The method of infection is direct through contact with the feces of an excretor, or indirect through a vehicle, food or water, contaminated with the feces of an excretor either directly or indirectly. In general, in salmonella infection the reservoir of infection is the vertebrate intestine. Direct infection through contact is relatively uncommon in human disease except in hospitals where, either through age or through illness, the host is peculiarly susceptible. Direct infection is more common in domestic food animals, owing to intensive farming methods. Infection through contaminated water is usually limited to a few host-specific serotypes such as *S. typhi* for man and *S. dublin* for cattle. In infection with serological types of *E. coli,* the reservoir is the human child. Infection is confined to the young human or animal, and a considerable degree of host specificity is shown. The majority of infections appear to result from direct contact. Infection via water or food is rare. Infections caused by members of the Enterobacteriaceae other than the *Salmonella, Shigella,* and *E. coli* groups rarely appear in epidemic form. Epidemics of human infection by members of the *Arizona* and *Proteus-Providence* groups are rare, though infection of individuals occurs. It is usual to isolate members of the Enterobacteriaceae, other than those named, from foods associated with food poisoning; these organisms are usually present in large numbers together with other fecal organisms such as fecal streptococci. In these instances no conclusion can be reached as to the part played by the Enterobacteriaceae. The Enterobacteriaceae most commonly food-borne and causing disease are therefore members of the *Salmonella* group.

II. SALMONELLAE

A. Definition and Classification

In addition to the general characteristics already mentioned, salmonellae in general conform to the following definition of the Enterobacteriaceae Subcommittee of the International Association of Microbiological Societies (Carpenter, 1963): Ferment mannitol, and usually dulcitol. May ferment inositol. Do not ferment adonitol, lactose, salicin, sucrose. Indole negative. H_2S usually positive, no growth in KCN medium. M.R. positive. V-P negative. Citrate usually utilized. Decarboxylases, lysine positive, arginine positive usually delayed, ornithine positive. Phenylalanine deaminase negative. Gelatin rarely liquefied. Techniques for the detection of salmonellae are discussed in Chapter XI, Laboratory Methods, Section III, A.

Individual members of the group (serotypes) are identified by serological recognition of their main antigens and, after international approval as a member of the group, are given a specific name, now usually that of the laboratory or area in which they were first isolated. Approximately 1000 serotypes have been granted specific names. This proliferation of serotypes is of immense value to the epidemiologist in aiding the tracing of infection through the vehicle to the source. In most countries, however, the serotypes isolated from the majority of human infections belong to a few common types. Phage typing provides a means of distinguishing between strains of the same serotype and has been applied to *S. typhi* (Craigie and Yen, 1938*a, b*; Felix, 1955), *S. paratyphi* B (Felix and Callow, 1943, 1951), and *S. typhimurium* (Callow, 1959; Anderson, 1962).

In addition to these laboratory characteristics, salmonellae may be classed on the basis of the degree of association with a particular host (Table I) and on the type of disease produced. A few types are host-specific, producing disease in only one animal species and never or rarely in others. In the definitive host the disease is produced by the ingestion

TABLE I

SALMONELLA TYPES ISOLATED FROM MAN, ANIMALS, AND FOWL — ENGLAND AND WALES
(Salmonella Reference Laboratory, 1956–1961)

Type	Human[a]	Bovine[b]	Porcine	Fowls (hens, ducks, turkeys)
S. typhi[c]	707 (3)[d]			
S. paratyphi B[c]	1,838 (8)			1
S. dublin	127	1,375 (65)	20 (5)	5
S. cholerae-suis	53	2	32 (7)	1
S. thompson	563 (2)	9	12	236 (16)
S. menston	103	3	5	139 (9)
S. typhimurium	13,998 (60)	560 (26)	80 (18)	927 (61)
Other types	6,036 (26)	167 (8)	287 (66)	206 (14)
All types	23,425	2,116	436	1,515
Number of other types	92	36	52	43

[a]Types appearing only once in the period omitted.

[b] Compiled from Table 3, Report (1965).

[c] Figures kindly supplied by Dr. E. S. Anderson, Enteric Reference Laboratory and Bureau.

[d] Percentages in parentheses.

of a small number of organisms and is characterized by a prolonged septicemic stage with massive invasion of the tissues via the bloodstream by salmonellae from the bowel.

B. Factors Influencing Salmonella Infection in Man and Animals

1. POPULATION INCREASE

The most important single factor in the increase of salmonella infection in England and Wales over the past 30 years would appear to be the increase in population (Taylor, 1965a). England is now the second most densely populated country in the world, with the greater part of the population housed in ever-expanding urban areas and conurbations. This herding of population has led to changes in feeding in that meals are no longer all prepared and eaten in the home. A large proportion of the working population now has at least one meal daily away from home in schools, canteens, cafes, or hotels. Most of the food consumed is now mass-produced on farms, mass-prepared in factories, and then distributed both locally and nationally. Since the introduction of freezing, chilling, and canning of food, trade in food is international. An error in hygiene in the preparation of a food distributed internationally may now result in an epidemic of salmonella infection among consumers some thousands of miles distant from the source of contamination, and some months or years after the contamination has occurred (Report, 1964a).

2. RESERVOIRS OF INFECTION

The primary reservoir of salmonellae is the vertebrate intestine. The only vertebrates from which salmonellae have not been isolated under natural conditions are aquatic vertebrates living in unpolluted waters. In the infected individual the number of salmonellae present in the feces is approximately 10^9 per gram (Section I, B). This level of excretion is maintained for several weeks and then falls gradually until the individual no longer excretes. There is, however, little correlation between clinical recovery and the duration of excretion, which in general greatly exceeds the duration of clinical signs and symptoms. In all outbreaks of salmonella infection a number of symptomless excretors are invariably found.

3. THE INFECTING DOSE

McCullough and Eisele (1951a, b, c) conducted human volunteer feeding trials on adults at various levels of dosage of several types of salmonella found frequently in spray-dried egg products. Table II presents a summary of their findings. The size of the infecting dose necessary to

TABLE II
INFLUENCE OF SEROTYPE AND INFECTING DOSE ON CLINICAL DISEASE[a]

Type	Infecting dose (millions)	Volunteers				Days of excretion			
		Number exposed	Clinical illness	Symptomless excretion	No effect	Clinical illness		Symptomless excretion	
						Range	Median	Range	Median
S. meleagridis	— 0.16	24	—	13	11	—	—	2–21	4
	— 5.50	53	—	44	9	—	—	2–128	9
	—50.00	41	16	23	2	2–72	14	2–89	8
S. anatum	— 0.16	41	—	20	21	—	—	2–15	9
	— 5.50	48	11	30	7	2–37	18	2–28	10
	—67.00	24	5	19	—	4–14	6	2–107	7
S. newport	— 0.15	6	1	2	3	51		9–14	
	— 0.38	8	1	5	2	2		4–28	
	— 1.30	6	3	3	—	9–16		7–16	
S. derby	— 0.13	6	—	3	3	—		1–36	
	— 5.00	12	—	8	4	—		2–11	
	—15.00	12	3	4	5	4–11		2–44	
S. bareilly	— 0.12	6	1	4	1	6		1–3	
	— 1.70	12	6	5	1	2–11		2–9	
S. pullorum	1,300	12	5	1	6	0–2		1	
	7,000	11	10	1	—	1–2		—	
	16,000	12	12	—	—	0–2		—	

[a] Summarized from McCullough and Eisele (1951a, b, c).

produce clinical illness varied according to the type and particular strain of salmonella fed. *Salmonella newport* produced clinical illness at levels of approximately one-tenth of a million organisms, *S. anatum* at 5 millions, *S. meleagridis* at 50 millions, and *S. pullorum,* a host-specific type in fowl, at 1300 millions. Dosage at all levels, however, produced symptomless excretion in a proportion of subjects which increased as the dose was increased; at the highest dose level almost all the subjects showed either clinical illness or symptomless excretion. The mean duration of excretion in clinical illness and in symptomless excretion was approximately the same. During these trials there was apparently no transmission of infection from symptomless excretors. Infection, both clinical and subclinical (symptomless excretion), in adults is thus seen to depend mainly on the size of the infecting dose. The type of salmonella involved would appear to be of much lesser importance, as a few hours of multiplication in food at a favorable temperature easily converts an innocuous number of salmonellae into many millions. In man there is much indirect evidence to support the view that the infecting dose of *S. typhi,* a host-specific type in man, is very small and may indeed be numbered in units rather than in tens or hundreds. Most of the evidence depends on the long-recognized association of typhoid fever with sewage-polluted water supplies where the dilution is often very great. Water, of course, permits survival of pathogenic bacteria in ever-diminishing numbers but does not offer an environment suitable for multiplication. It is highly probable that the infecting dose of other host-specific serotypes for their host and possibly even some host-associated serotypes is small.

4. HUMAN DISEASE

In man, food poisoning by salmonellae is due to the ingestion of these organisms which then multiply within the small intestine. The incubation period during which this takes place is about 24 hours to 48 hours. The patient usually complains of headache, vomits, and feels unwell. Very often there is abdominal pain, which starts in the upper part of the abdomen. As the process is infective, the patient commonly has a rise in temperature to about 100°F, but it may be higher. Although the acute stage may last only for about 2 days, the patient rarely feels completely well until the seventh day. Prostration is uncommon in the acute stages but may occur after some days of illness and is due to dehydration, as the result of diarrhea and vomiting, and the toxic effect of the salmonella infection. The three main causes of food poisoning—*Staphylococcus, Clostridium welchii,* and *Salmonella*—present somewhat different clinical pictures which may give some indication as to the cause of the condition

(Table III); nevertheless, the correct diagnosis can be made only by the isolation and identification of the organism concerned.

TABLE III

INCUBATION PERIODS AND MAIN SYMPTOMS OF BACTERIAL FOOD POISONING[a]

Symptom	Staphylococcus (2–4 hours)[b]	Clostridium welchii (12–16 hours)[b]	Salmonella (24–48 hours)[b]
Vomiting	Extremely common	Very rare	Common
Diarrhea	Common	Extremely common	Very common
Pyrexia	Absent	Absent	Very common
Prostration	Common	Absent	Rarely in early stage

[a] Reproduced, by permission, from Taylor (1965b).
[b] Incubation period.

In young children and in patients over 60 years of age, the illness tends to be more severe and to last for a longer period. The most important material for the isolation of the organism is the patient's feces, which must be cultured before antibiotic treatment has commenced. It is sometimes possible to isolate the organism from the patient's blood, cerebrospinal fluid, or urine in those rare cases who have prolonged pyrexia. Very occasionally it is possible to isolate the organism from the vomit.

Enteric fever caused by host-specific types such as S. typhi and S. paratyphi A, B, and C usually shows a different clinical picture. In the early stages the patient has a high temperature which may last for about 2 weeks; after about 7 to 10 days there may be diarrhea. In these infections it is usual to isolate the organism from the patient's blood and from the feces.

In enteric fever after about the tenth day of the disease it is possible to show the presence in the patient's serum of agglutinins to the infecting organism; this is rarely possible in salmonella infection of the food-poisoning type. We have never been able to obtain an indication of the salmonella serotype in an outbreak or sporadic case of food poisoning by the examination of the patient's serum alone. In an outbreak where the serotype is known, then it is occasionally possible in some patients to show a raised titer to the causal serotype, but the reaction is too variable in the individual to be of any general use in outbreaks. The reaction is useless in the investigation of the sporadic case.

5. THE ROUTE OF INFECTION

The most common route of infection with salmonellae in man and animals is by the alimentary tract. Moore (1957), however, experimentally infected guinea pigs through the conjunctivae and, by masking the conjunctivae of animals exposed to dust-borne infection, terminated an epidemic occurring naturally in an animal house. Pritulin (1959) used face masks to infect sheep and calves through the respiratory tract with aerosols of salmonella cultures and reported that the minimum infective dose was lower, by a factor of 3 to 4, than that necessary when given by mouth. In hatchery practice, lung infection of healthy chicks from the inhalation of contaminated fluff from an infected chick may be extensive (Hinshaw *et al.*, 1926). Gibson (1965) noted that in cattle salmonellosis the spread of infection within an infected building was much wider than indicated by visible soiling, suggesting that the animals in such a building have been exposed to aerosols of the organisms. Taylor (1965a) mentions a hospital outbreak in which contaminated ward dust appeared to be responsible for an outbreak lasting for almost a year. Datta *et al.* (1960) described a hospital outbreak in which every ambulant patient using the water closet became infected although the causal salmonella was never isolated from door or chain handles or from the closet seat. It is suggested that the flush of water produced an aerosol from contaminated feces.

6. The Age of Infection

The ability to respond to infection by the development of active immunity is acquired only after some months of life. The newborn animal possesses only transient passive immunity derived from the maternal bloodstream, colostrum, or yolk sac. This lack of a specific defense mechanism renders the newborn animal the most susceptible to infection. Age susceptibility to infection is observed most frequently in poultry, where severe epidemics may involve the death of 90 to 100% newly hatched chicks (Buxton, 1957). In calves Gibson (1961) found salmonella infection to be more prevalent after the calves were 1 week old than before. Gitter (1959) isolated *S. cholerae-suis* from 15% of pigs aged 2 to 4 months but not from four times as many pigs of other ages examined at the same time. In a study of *S. typhimurium* infection in man, Taylor (1960) found that 37.9% of cases were under 5 years of age and 12% were less than 1 year.

7. The Transfer of Infection

As the numbers of salmonellae in the feces of an excretor may approach 10^9 organisms per gram, an infecting dose may be contained in a very small quantity of feces. Infection may be transferred directly by ingestion of excreta or indirectly through the medium of food or water contaminated with excreta. Such transfer constitutes the cycle of infection. Several possible cycles of infection exist within species and between species, in order of frequency animal to animal, animal to man, man to man, man to animal, in which the relative frequency of direct and indirect transmission of infection varies greatly. In general, the types of salmonellae involved in the man-to-man and animal-to-animal cycles are the host-specific and host-associated serotypes. In man the transfer of infection with *S. typhi* and *S. paratyphi* B is mainly indirect, through the medium of contaminated food or water. Direct transmission is rare. One of us investigated two sporadic cases of typhoid infection with different phage types of *S. typhi* which occurred within a few days at a refugee camp in 1947. Individual stool examination of all the members of the camp finally detected two chronic carriers of the phage types concerned. Contact occurred between the carriers and cases only during work in a remote fen. The method of infection appeared to have been the communal smoking of cigarettes passed from hand to hand. Fry (1951) investigated an outbreak of enteric fever caused apparently by an ambulant case, an insurance collector. The victims were all inmates of households visited weekly. The method of transfer appeared to be boiled sweets handed by the excretor to his clients.

Although the human excretor of *S. typhi* was important in these instances, one should not overemphasize this point. The intelligent adult or child over the age of 5 years who is excreting salmonellae (excluding *S. typhi*) should be allowed to return to duty provided that he has been given instructions in personal hygiene and is not allowed to handle food. For many years now we have advised the return to offices or schools of such excretors and have not encountered any return cases.

The situation is entirely different with the healthy baby or toddler who is a salmonella carrier. Such patients commonly pass fluid or semiformed stools which cause much environmental contamination. These children should not be admitted to open wards or to residential or day nurseries. On ethical grounds these children should not be subjected to institutional treatment but should stay with their families who should be warned of the danger of the spread of infection. By the use of ordinary hygienic methods in the home it is unlikely, in our experience, that any member of the family will ingest a dose of salmonella sufficient to cause symptoms, even though he may become a temporary healthy excretor.

Healthy patients, irrespective of age, who are excreting salmonellae should never be treated with antibiotics, for it is the normal bowel flora which displaces the salmonellae. Antibiotic treatment, because it changes the bowel flora and reduces the numbers and types of the normal bacteria, tends to prolong the period of salmonella excretion (Dixon, 1965).

In animals direct transmission of infection is much more frequent than in man. Transmission of *S. pullorum* infection from the adult hen to the chick via the infected egg was recognized more than 50 years ago (Rittger and Stonebrun, 1909). In the United Kingdom *S. thompson, S. typhimurium,* and *S. menston* are now known to be transmitted by eggs. In salmonella infection of young calves, Taylor (1965a) has drawn attention to the importance of diarrhea in association with the tendency of young calves to lick each other, in the transmission of infection. In adult animals aerosols resulting from the splashing of feces of an infected animal may initiate direct infection.

The man-to-animal cycle of infection is of least importance. Although salmonellae are constantly present in urban sewage and sewage-polluted water (McCoy, 1962), there is little evidence to suggest that the use of such water is of importance in initiating infection in animals (Gibson, 1965).

C. Salmonellosis in Food Animals

The intensive rearing of food animals is characterized by a pattern of concentration of animals for production, dispersal of young animals for rearing, and reconcentration of reared animals for slaughter.

1. CATTLE

a. Movements. In England and Wales the primary purpose of the national herd is milk production. About three-quarters of home-produced beef originates from dairy herds and consists of cow beef and beef produced from bull calves and from heifer calves surplus to replacements for the dairy herd. The annual calf crop numbers about 3.5 million calves, of which half a million heifer calves form replacements for the dairy herd. About 2 million are reared for beef production and 1 million are slaughtered (Report, 1965). Since the introduction of early weaning systems of calf rearing in the late 1950's, the number of calves slaughtered has fallen to about half a million (Intelligence Bulletin, 1965a). Most of the calves from pure dairy herds are slaughtered within a few days of birth. The proportion of calves reared for veal and slaughtered at 3 months of age is not known but is thought to be increasing. Surplus calves from individual farms in the dairying areas of the south and west are concentrated soon after birth in markets or dealers' premises and then dispatched either to abattoirs for immediate slaughter or to individual farms in the east and northeast for rearing. After rearing, cattle are again concentrated in markets, then in abattoir lairages for slaughter. In England and Wales, as the result of the preponderance of dairy cattle, the adult cow would appear to constitute the main reservoir of salmonella infection. The concentration of susceptible calves sold soon after birth encourages the transmission of infection. The dispersal of infected calves to rearing farms distributes infection widely.

b.   *Serotypes.*   Since the definition of the genus *Salmonella* and the adoption of serological classification of types (Report, 1934), a clear picture has emerged of the serotypes causing clinical disease in cattle. Two serotypes, *S. dublin* and *S. typhimurium,* predominate both in calves and in older cattle, other serotypes rarely cause clinical disease. *Salmonella dublin,* which may be regarded as host-specific, is widely distributed and presents a major problem in many countries. *Salmonella typhimurium,* some phage types of which may be regarded as host-associated (Anderson, 1962), is usually also present but with lower incidence in these countries. In the absence of *S. dublin, S. typhimurium* usually constitutes the chief salmonella of cattle (Gibson, 1965). *Salmonella dublin* has been shown to be the predominant serotype of cattle in Italy (Strozzi, 1934), Germany (Bartel, 1938), Holland (Clarenburg and Vink, 1949), and Denmark (Muller, 1954), and in certain areas of the British Isles, South Wales (Field, 1948; Gibson, 1958), Southwest England (Grunsell and Osborne, 1948), Northern Ireland (Murdock and Gordon, 1953), and Eire (Mullaney, 1949). It should be noted that the areas in the British Isles in which *S. dublin* is the predominant serotype are those with the highest proportion of dairy cattle. A high incidence of *S. dublin* has been reported from South Africa (Henning, 1939), Venezuela (Galto, 1939), and Brazil (Penha and D'Apice, 1946); a low incidence from the United States (Edwards *et al.,* 1948) and New Zealand (Salisbury, 1958). The low incidence from the United States may be associated with the low proportion of dairy cattle in its national herd: Less than one quarter of beef produced in the United States is derived from dairy cattle (Intelligence Bulletin, 1965*b*).

c.   *Pattern of Salmonellosis.*   Gibson (1961) described sporadic infection in calves in areas where salmonella infection was endemic in yearling and adult cattle. The epidemic form of the disease occurred typically among large batches of bought calves often on farms with no older cattle. Field (1948) in South Wales estimated that approximately 1% of adult cattle on the farms surveyed were excreting salmonellae. Calves born on farms harboring excreting adults are exposed to infection from birth, and a proportion of these calves might be expected to acquire infection after birth. Gibson (1961) examined calves arriving at rearing farms from dealers and isolated salmonellae from 7 out of 50 batches (14%). In all, salmonellae were isolated from 12 out of 2518 calves (0.5%). Anderson *et al.* (1961) isolated salmonellae from 1 out of 164 calves (0.6%) held overnight at a collecting premises, but from 73 out of 205 calves (36%) held from 2 to 5 days. At the dealer's premises, salmonellae were not isolated from food or water, or from bedding, suggesting that the salmonellae were introduced repeatedly by infected calves and spread among the animals herded together for several days (Galbraith, 1961).

The transfer of infection on a calf-rearing farm following the receipt of infected animals has been demonstrated (Wormald and Greasby, 1965). The herd consisted of approximately 60 suckling calves bought in batches from different sources (in one batch alone calves from six widely separated sources were present) and reared indoors in pens on a proprietary milk-substitute diet. Fecal material from the pens was examined several times monthly. The last batch of calves to arrive started to scour a week after arrival, and two were found to be excreting *S. typhimurium*. Routine sampling from the pens 2 days later showed 4 out of 9 contaminated. Three days later all pens were found to contain infected calves. Salmonellae were isolated from 29 out of 62 calves (46%). Eleven days later 6 more calves were found to be infected. Thirty-eight days after onset no excretors were found. During the period of excretion most calves were symptomless. No evidence of contamination of food or water was found.

The general picture of cattle salmonellosis in England and Wales is now clear. The adult cow population forms the main reservoir, from which infection is transferred to a proportion of calves born. Transfer of calves for slaughter or rearing spreads the infection by direct contact to numbers of susceptible animals. As the dairy herd in England and Wales supplies three-quarters of home-produced beef, cattle must be regarded as the largest and most important animal reservoir of *Salmonella typhimurium* in the United Kingdom. The influence of the introduction of calf-rearing establishments on cattle salmonellosis is shown by the results of surveys. Withers (1952, 1953) in 1946–1948 encountered only one outbreak of salmonellosis, and from this *S. dublin* was isolated. A survey (Report, 1964b) of the cause of death in more than 2000 calves up to 6 months of age, in 1959–1961, found salmonella infection with *S. dublin* and *S. typhimurium* responsible for the deaths of 23% of calves, second only to *Escherichia coli* infection (45%).

2. PIGS

a. Movements. Pig movements in England and Wales are less frequent and less extensive than cattle movements. Pig production is based
on the feeding of home-grown barley supplemented with animal and
vegetable proteins, minerals, and vitamins. Two units are distinguished,
breeding units and fattening units. Breeding units produce piglets and
retain them until they are weaned. Fattening units buy in weaned pigs
from breeding units and retain them until pork or bacon weight is reached.
There is thus little movement of live pigs, comparable to the movements
of calves through dealers, markets, and farm premises. There is, however,
considerable movement of pork carcasses, portions of carcasses, and
offals from bacon factories after slaughter.

b. Serotypes. Sojka and Gitter (1961) have listed the serotypes
isolated from diseased pigs by Veterinary Investigating Centres in
England and Wales between 1953 and 1959. Of 351 isolations, *S.
cholerae-suis* accounted for 283 (81%), *S. typhimurium* for 32 (9%),
and 14 other serotypes for 36 isolations (10%). A similar preponderance
of *S. cholerae-suis* has been recorded in New Zealand (Salisbury, 1958),
Australia (Simmons and Sutherland, 1950), Germany (Lutje, 1938),
Argentina (Quiroqa and Monteverde, 1945), and the United States
(Levine *et al.,* 1945).

From the mesenteric glands of healthy pigs at slaughter, *S. cholerae-
suis* has been isolated less frequently. Hormaeche and Salsamendi (1936,
1939) isolated salmonellae from 54 out of 83 batches (65%) of mesenteric
glands from 20 pigs each; from 8 out of 14 batches (57%) of glands from
2 to 5 pigs each; and from 16 out of 25 (64%) batches of glands from
single pigs. One strain of *S. cholerae-suis* was isolated. Fourteen other
types were isolated. The infrequency of *S. cholerae-suis* and the frequency of isolation of other serotypes from mesenteric glands have been
reported in Mexico (Varela and Zozaya, 1941), the United States (Edwards *et al.,* 1948), and Yugoslavia (Zakula, 1956).

From the feces of healthy pigs, *S. cholerae-suis* is rarely isolated.
Galton *et al.* (1954) isolated *S. cholerae-suis* once from almost 3000
samples of pig feces from an abattoir. Salmonellae belonging to 35 types
were, however, isolated from 50% of fecal samples.

c. The Transfer of Infection in Pigs. In intensive pig production
in England and Wales, pigs after weaning are fattened in pens and sent
for slaughter when the necessary live weight has been reached. Toward
the end of the fattening period pigs are weighed weekly, and those reaching the necessary weight are sent for slaughter. There is a general move-

ment of pigs from pen to pen during this process. Culling of unthrifty pigs is not generally practiced. Opportunity for the transfer of infection occurs within pens by fecal transfer either directly or indirectly by the contamination of food and water; transfer between pens occurs by the intake of pigs from infected pens to replace those sent for slaughter and the introduction of fresh pigs for fattening. The situation is thus the classic one in which susceptible animals are added at regular intervals to an infected herd, so prolonging the outbreak. As the attainment of slaughter weight is the sole criterion for removal, unthrifty pigs may remain in the herd as a source of infection for many months. At slaughter in bacon factories pigs from many herds come into close contact in lairages: at abattoirs pigs come into contact with animals of other species. At bacon factories the day's intake of pigs is usually slaughtered within a few hours: at abattoirs animals may be held for many days in lairages, and opportunity occurs for the transfer of infection within and between species.

3. POULTRY

a. Movements. The poultry industry in England and Wales was reestablished after World War II, and intensive production of hens, turkeys, and ducks is now common. Hens are reared for both egg and meat production, turkeys and ducks mainly for meat production. In these species, egg production is limited to that necessary to maintain breeding flocks. In general the pattern of movement of poultry is more limited than movements of cattle and pigs. Chicks are produced in a very small number of hatcheries and sent to some thousands of farms for rearing and egg production. The limitation of hatcheries to a small number is considered essential for the maintenance of the desired genetic characteristics of the stock. In egg-laying strains, early maturity and long laying are the desired characteristics; in table poultry, food utilization and carcass conformation are desirable. As egg transmission in fowl salmonellosis is common, infection of hatchery stocks is followed by a wide distribution of infection to producing farms. In 1962–1963 (Intelligence Bulletin, 1964*a*) in England and Wales, estimated placings of chicks by hatcheries for breeding and other purposes averaged 5 million, for egg laying 50 million, and for table poultry 125 million.

The structure of the industry is as follows.

i. Egg production. At central hatcheries a master breeding flock is maintained in which the desired genetic character of the stock is fixed. The master flock supplies stock to multiplying farms which return fertile eggs for setting to the central hatchery. Day-old chicks from the central hatchery are supplied to individual egg producers. In terms of numbers, master breeding stocks probably amount to some tens of thousands of birds, multiplying stock to some hundreds of thousands, and egg-producing stock to 50 million. Infection in the master breeding flock is transmitted by egg infection and incubator infection to multiplying stock, and from multiplying stock by the same routes to individual farms receiving day-old chicks. We have found that the time lag between introduction of infection to the master breeding stock and its dissemination through day-old chicks to individual farms is approximately one year. Pullets come into lay at 5 to 6 months of age. An interval of 6 months therefore elapses between the distribution of day-old chicks from hatcheries to egg-producing farms, and the appearance on the market of eggs laid by these birds.

Table IV shows the time relation between placings of day-old chicks from hatcheries in England and Wales, salmonella outbreaks in chicks *(S. menston, S. thompson, S. typhimurium),* and the presence of these serotypes in liquid egg sampled daily during production. The incidence of salmonellae in liquid egg is seen to be clearly related to the incidence of salmonella infection in chicks 5 to 6 months earlier.

ii. Meat production. The broiler industry is arranged in groups. A typical group consists of a master breeding flock in which the desired genetic characters are carefully fixed, and multiplying farms which supply fertile eggs to the central hatchery which sends day-old chicks to rearing farms. On rearing farms the chicks are housed in batches of 10,000 to 15,000 birds, reared to the desired weight, and then sent to a central plant for processing into food. The time taken to arrive at killing weight is usually 10 weeks. Each broiler house thus supplies four crops of broilers annually, with a resting period of about 3 weeks between crops during which time the house is cleaned and fumigated. Opportunities for the transmission of infection occur in the master breeding and multiplying flocks by egg and incubator contamination; on rearing units by environmental contamination; and at the processing plant by the processing of flocks from different farms.

iii. Egg disposal. Annual egg production in the United Kingdom in 1960–1963 exceeded 1000 million dozen eggs (Intelligence Bulletin, 1964b). Eggs produced are graded at packing stations, and first-quality eggs are distributed for retail. Eggs rejected for the retail trade (oversize, undersize, chipped, or cracked) are processed to liquid egg, as are unsold eggs from the retailers. The multiplying farms of the day-old chick and the broiler also contribute to the supply of liquid egg, as eggs unsuitable for setting and those found infertile on incubation are also

TABLE IV

SEASONAL INCIDENCE OF SALMONELLAE IN CHICKS AND LIQUID EGG

Month	Chick placings for egg laying (millions, 1961–1964)[a]	Salmonella outbreaks (S. menston, S. thompson, S. typhimurium) in chicks (1960–1964)	Salmonella isolations (S. menston, S. thompson, S. typhimurium) from liquid egg during production (1958–1962)[b]
January	17.3 (11.4)[c]	74 (9.7)	—
February	17.4 (11.4)	78 (10.3)	1 (0.8)
March	20.3 (13.3)	90 (11.8)	2 (1.7)
April	20.2 (13.3)	100 (13.2)	11 (9.3)
May	12.5 (8.2)	74 (9.7)	10 (8.5)
June	9.6 (6.3)	50 (6.6)	16 (13.6)
July	9.0 (5.9)	55 (7.3)	23 (19.5)
August	7.3 (4.8)	36 (4.7)	3 (2.5)
September	7.9 (5.2)	48 (6.3)	21 (17.8)
October	11.2 (7.4)	55 (7.3)	16 (13.6)
November	10.3 (6.8)	56 (7.3)	10 (8.5)
December	9.2 (6.0)	44 (5.8)	5 (4.2)
Total	152.2	760	118

[a] Intelligence Bulletin (1964a, 1965a).
[b] Unpublished reports: L. A. Little (Wakefield) and J. H. McCoy (Hull).
[c] Percentages in parentheses.

processed for human consumption. Egg production is seasonal, being characterized by the "spring flush." The processing of surplus eggs forms an important part of the egg trade both nationally and internationally, and most continents export both whole eggs and egg products. Egg products consist of whole egg — liquid, frozen, or dried; egg albumen — liquid, frozen, crystalline, or powdered; and egg yolk — liquid, frozen, or dried. Egg products can be transported more easily than shell egg and can be stored tor long periods after production. Their main use is in large scale catering.

iv. Serotypes of salmonellae in fowl and egg products. In distinction to the United States, where many serotypes have been isolated from hens (Hinshaw *et al.*, 1944; Edwards *et al.*, 1948) and from egg products (Solowey *et al.*, 1947), the greater part of avian salmonellosis and salmonellae in liquid egg in England and Wales has been produced by only 5 types (Table V). The sudden appearance of *S. thompson* in 1943 and *S. menston* in 1961 may be attributed to infection of master breeding stocks at hatcheries.

v. Summary. The pattern of infection in the fowl industry in England and Wales is dictated by the structure of the industry. Infection in master breeding flocks is egg-transmitted to offspring and widely distributed to rearing farms. Salmonellae from infected laying stocks are transmitted in liquid egg.

4. ANIMALS OTHER THAN FOOD ANIMALS

Edwards (1958) has reviewed the isolations of salmonellae from vertebrates other than food animals and insects up to 1957. Salmonellae have been isolated from cats and dogs, and in some instances direct transmission of infection from dogs has been observed. The role of contaminated meat in salmonellosis of cats and dogs will be discussed later. Rodents have long been known to carry *S. enteritidis* and *S. typhimurium*. For many years it was thought that rats and mice carried these types exclusively, but it is now known that types carried are those found in the environment of the animals, and many types have been isolated. Tortoises, lizards, and snakes have been shown to harbor salmonellae. The source of infection of tortoises is obscure. Lizards probably become infected by eating flies which have fed on contaminated human or animal excreta; snakes become infected by the consumption of infected lizards. Infection in these animals is thus a response to environment.

Flies carry organisms on the surface of their bodies and for short periods in their intestine and act as mechanical transmitters of infection from excreta to food (Greenberg, 1964).

TABLE V

AVIAN SALMONELLOSIS—ENGLAND AND WALES, 1933–1944,[a] 1948–1956[b]
SALMONELLAE IN LIQUID EGG, 1959–1963[c]

Types	Avian salmonellosis[d]			Liquid egg[d]	
	1933–1942	1943–1944	1948–1956	1959–1960	1961–1963
S. pullorum/gallinarum	1252 (91)	464 (84)	2417 (68)	28 (3.4)	18 (1.7)
S. typhimurium	104 (7.5)	18 (3)	704 (20)	12 (1.5)	11 (1.0)
S. thompson		50 (9)	295 (8.3)		
S. menston					122 (11.7)
All outbreaks	1380	553	3541	817	1039

[a] Abstracted from Gordon and Buxton (1946).
[b] Abstracted from Blaxland et al. (1958).
[c] Unpublished data: Dr. L. A. Little (Wakefield) and J. H. McCoy (Hull).
[d] Percentages in parentheses.

Salmonellae have often been isolated from sewage-polluted natural waters (Report, 1959; McCoy, 1962, 1963) but rarely from fish caught in such waters. Shellfish from beds in contact with sewage-polluted sea water have long been known as a source of typhoid infection (Gay, 1918).

D. Salmonellae in Animal Foods

1. RAW MATERIALS OF ANIMAL FOODS

Essential animal foods (roughage, carbohydrates, fats, protein, minerals, and vitamins) are derived from coarse and green fodders, roots and tubers, and concentrated foods. The fodders are composed of whole plants, such as grass, lucerne, clover, and kale, or parts of plants, such as hay, straw, and sugar beet tops; they are distinguished by their bulk and high content of fiber.

Concentrated foods are characterized by a high content of essential foods in low bulk and supply carbohydrate, protein, or a mixture of protein and carbohydrate, some mineral element, or a vitamin.

Protein concentrates are derived from plants and animals.

Plant concentrates contain a high concentration of protein as the result of the removal of either oils or carbohydrates, whereas animal protein supplements contain little carbohydrate and are, in general, by-products of other industries. The main animal protein supplements are meat meal, meat and bone meal, liver meal, dried blood, whale meat meal, whale bone meal, dried skim milk, "fish" meal, white fish meal, and herring meal. Dried blood possesses many undesirable qualities as a protein supplement and, except in an emergency when animal protein supplements are limited, finds little application in animal feeding.

Herbivorous animals—horses, cattle, and sheep—do not as a rule receive animal protein supplements in their diet, but pigs and poultry, with their simpler digestive tract, require protein supplements to supply essential amino acids.

a. Animal Diets. Animal diets are made up to supply the needs of groups of animals for the essential nutriments. In general, complete foods are prepared for the following groups of animals.

Cattle	Pigs	Sheep	Poultry	Turkeys
Calves	Early weaners	Nursing ewes	Chicks	Poults
Cows in milk	Breeding sows	Lambs	Growing birds	Growing birds
Breeding cows	Market pigs	Older fattening	Broilers	Table birds
Store cattle	All others	sheep	Layers	
Fattening cattle			Breeders	

These diets have a common basis of roughages, grain, and millers' by-products, with varying requirements of protein, mineral, and vitamin supplements. Cattle and sheep as a rule receive vegetable but no animal protein supplements, while hens receive both vegetable and animal protein supplements and mineral supplements, younger birds receiving proportionately greater amounts. The diets of turkeys, pheasants, and grouse are distinguished sharply by the high content of protein supplements required at all ages.

Where grain produced on the home farm forms the basis for animal rations, protein concentrate meals or grain balancer rations are usually added to supply the protein, minerals, and vitamins lacking in the grain. Such meals are prepared from a mixture of animal, marine, and vegetable proteins with added minerals. The grain balancer rations may be supplied as raw materials bought in quantity or as protein concentrates ready for farm mixing. Complete rations already mixed in the correct proportions may be bought from compound millers. Such rations are supplied as mashes (meals, supermixes, etc.) or in the form of dust-free crumbs, pellets, pencils, or cubes of convenient size. The compound miller obtains the various contents of his feeding stuffs already processed from the makers, and blends them in the correct proportion to produce a palatable product.

In the manufacture of mashes, the ground or powdered ingredients are mixed mechanically, molasses often being used as a binder to reduce dust and to add to the palatability of the ration. The only ingredient to which heat is applied by the compounder is the molasses, which is a very viscous liquid at atmospheric temperature and requires heating to about 90°F before it will flow as a liquid.

Pelleted rations are prepared from meals, heated only enough to facilitate pelleting, as temperatures above 140°F (60°C) denature proteins. The process lasts only a few minutes, so the ration is not cooked in the true sense of the word; temperatures rarely exceed 160°F (71°C) and may be as low as 90°F (32°C).

Low-grade samples of meat meal, meat and bone meal, blood meal, and fish meal are used for fertilizers; as the cost of these manurial grades is less than that of feeding grades, they may be attractive to the pig or poultry producer making up his own ration.

b. *Definitions* (Table VI). The control in the United Kingdom of the quality of fertilizers and feeding stuffs is effected by the Fertilizers and Feeding Stuffs Act (1926), The Fertilizers and Feeding Stuffs Regulations (1955), and The Fertilizers and Feeding Stuffs (Amendment) Regulations (1956).

TABLE VI

FERTILIZER AND FEEDING GRADES ANIMAL AND MARINE PRODUCTS

Name	Definition of fertilizer grade[a]	Name	Definition of feeding grade[a]
Dried blood	Blood which has been dried, to which no other matter has been added	Feeding dried blood	Blood which has been dried, to which no other matter has been added
Bone meal	Commercially pure bone, raw or degreased, which has been ground or crushed and which contains not less than 20% phosphoric acid	Feeding bone meal, ground bone	Commercially pure bone, raw or degreased, which has been ground or crushed
Bone meal, grade II	Commercially pure bone, raw or degreased, which has been ground or crushed and which contains less than 3.5% nitrogen or less than 20% phosphoric acid		
Steamed bone meal, steamed bone flour	Commercially pure bone from which nitrogen has been removed by steam	Feeding bone flour	The product obtained by grinding commercially pure steamed bone
Precipitated bone phosphate, dicalcium bone phosphate	An insoluble calcium phosphate prepared by treating commercially pure bone with acid and precipitation of phosphate from the solution		
Dissolved or vitrolized bone	Commercially pure bone which has been treated with sulfuric acid		

Meat meal, carcass meal, meat and bone tankage, meat and bone meal	The product of drying and grinding or otherwise treating bone, flesh, flesh fiber (including whale meat) and other slaughterhouse residues to which no other matter has been added
Feeding meat meal	The product, containing not less than 55% of albuminoids obtained by drying and grinding animal carcasses or portions thereof (excluding hoof and horn) and bone, to which no other matter has been added, but which may have been preliminarily treated for removal of fat
Feeding meat and bone meal	The product, containing not less than 40% of albuminoids, (protein) and not more than 4% salt, obtained by drying and grinding animal carcasses or portions thereof (excluding hoof and horn) and bone, to which no other matter has been added, but which may have been preliminarily treated for removal of fat
Fish, guano, fish manure	A product obtained by drying or grinding or otherwise treating fish or fish waste to which no other matter has been added
Fish meal, fish residue meal	A product obtained by drying and grinding or otherwise treating fish or fish waste, to which no other matter has been added
White fish meal	A product (containing not more than 6% of oil and not more than 4% of salt) obtained by drying or grinding or otherwise treating white fish waste or white fish, to which no other matter has been added

a"Commercially pure" implies that no other matter may be added.

The Act defines fertilizers and feeding grades of animal and marine protein and/or mineral products, as these products are produced in the same plant from common raw material.

The variation encountered in the composition of the raw material is so great and so uncontrollable that the consistency of the final product cannot be guaranteed. The definitions within the Act which follow show how some standardization has been attempted. The definitions of hoof, hoof and horn, and horn meal have been excluded from the table, as these substances are used only for fertilizer and their use in feeding grades of meat and meat and bone meal is expressly prohibited. *From the definitions it will be seen that fertilizer and feeding grades of many products are interchangeable.*

In the group of bone products the basic definition for both grades is the same, so that fertilizer grades of bone meals and flour may also be used as feeding stuffs. Feeding grades, however, may not conform chemically to the fertilizer grades. Precipitated bone phosphate and dicalcium bone phosphate, though defined in the Act only as fertilizer, are used also in feeding stuffs as a source of minerals. In the group of meat and meat and bone meal products, feeding grades are defined by their content of protein for which minimum amounts are specified and by the exclusion of hoofs and horns from the raw material. In the group of fish products, no distinction exists between fertilizer and feeding grades of fish meal, but white fish meal for feeding is distinguished by defining its raw material and by setting maximum limits to its content of oil and salt.

The definitions in the Act are not exhaustive, and many other animal and marine products such as greaves, whale liver meal, mammal liver meal, herring meals and flour, red fish meal, and crayfish meal are used for animal food.

c. Salmonellae in Raw Materials. Between 1948 and 1958 salmonellae were demonstrated in bone, animal, marine, and vegetable products used for animal foods in Norway, Sweden, the Federal Republic of Germany, the United Kingdom, and the United States. The sole constituents of animal feeding stuffs in which salmonellae were not demonstrated were millers' offals obtained from grain and by-products of the production of flour (Report, 1958a).

In the United Kingdom it is believed that salmonellae are present in home-produced animal offal used for animal foods. Ludlam (1954), between 1949 and 1954, isolated salmonellae from 38 out of 183 rats (20.8%) killed at a butchers' by-products factory processing offal from slaughterhouses, retail butchers' shops, and other premises to tallow, bone meal, dried blood fertilizer, and general fertilizer, but from only 23 out of 518 rats (4.4%) killed on other premises. From rats killed at the factory 46 strains of salmonellae belonging to 10 serotypes were isolated; from rats killed on other premises, 23 strains belonging to 2 serotypes. *Salmonella enteritidis* var. *dansyz* comprised 17 out of 46 strains (36.5%) isolated from the factory; 20 out of 23 strains (86.9%) isolated from other premises. It was concluded that the difference both in serotypes and in the percentage of rats found to harbor salmonellae could be explained by the presence of a large variety of salmonellae in the offal brought into the factory.

Muller (1956) reported the isolation in Denmark between 1948 and 1954 of 49 strains of salmonellae belonging to 21 serotypes from 4.5% of 1093 samples of imported meat meal, blood meal, and feeding and ferti-lizer grades of bone meal. In the same period 7 strains of salmonellae belonging to 6 serotypes were isolated from 10 samples of imported bones. With the exception of *S. typhimurium* about two-thirds of the types isolated had not been previously isolated in Denmark, and the re-mainder only sporadically.

Rhode and Bischoff (1956) reported the isolation of salmonella sero-types in the Federal German Republic from imported Indian bone meal, imported fish meal, imported whale meat meal, imported meat meal, and home-produced bone meal.

Walker (1957) in England isolated 34 serotypes of salmonellae from 50 out of 123 samples (40%) of organic fertilizers, including bone meal, dried blood, and meat and bone meal.

Swahn and Rutquist (1957) isolated 4 serotypes of salmonellae from 10 out of 28 samples (35.7%) of lucerne flour imported into Sweden.

Hauge and Borre (1958) isolated 17 serotypes of salmonellae from 42 out of 910 samples (4.6%) of imported vegetable protein concentrates (coconut cake, ground nut cake, linseed cake, sunflower cake, cotton seed cake meal, soya bean cake meal, extracted soya bean meal, and rape meal).

   d.   Salmonellae in Raw Materials and Manufactured Foods. Be-tween 1959 and 1963, several investigations of the salmonella content of the raw materials used in animal feeding stuffs and of complete foods manufactured from these raw materials were carried out in England (Report, 1959*a*, 1961*a*; Galbraith *et al.*, 1961, 1962*a*; Harvey and Price, 1962). The findings in these investigations, together with previously unpublished data (Cook, McCoy), are summarized in Table VII.

Of the raw materials, examined bone products proved to be the most

TABLE VII

Number of Salmonella Isolations from Raw Materials and From Animal Foods–United Kingdom, 1959–1963

Product	Samples positive	Percent positive	Serotypes					
Raw materials								
Bone products	216/531	40.7	adamstown	1	heidelberg	1	orion	2
			adelaide	2	huduvsta	2	panama	1
			alachua	4	hvittingfoss	9	paratyphi B	7
			anatum	34	infantis	1	pomona	1
			ardwick	1	jedburg	1	poona	11
			bareilly	2	jodphur	2	reading	13
			bere	3	kandla	2	richmond	8
			blockley	1	kapsovar	1	ruiru	1
			bovis-morbificans	6	karachi	1	rutgers	1
			brazzaville	1	kentucky	12	saint-paul	7
			bredeney	8	kirkee	1	san diego	4
			bronx	3	lexington	14	schwarzengrund	3
			butantan	8	livingston	4	seigberg	2
			cerro	10	london	5	senftenberg	49
			chailey	4	luke	1	shangani	1
			champaign	1	marylebone	1	takoradi	3
			chester	2	matopeni	1	taksony	12
			cubana	30	meleagridis	18	tel-hashomer	3
			derby	1	minnesota	10	tennessee	14
			dublin	2	missouri	1	treforest	2
			enteritidis var. jena	3	montevideo	21	typhimurium	26
			frintrop	1	muenchen	2	waycross	1
			gaminara	15	muenster	2	weltevreden	1
			give	16	newington	1	westhampton	13

Meat products — 237/827 — 28.6

Serotype	n	Serotype	n	Serotype	n
agama	1	enteritidis	1	newington	8
alachua	9	fresno	2	oranienburg	5
albany	2	halmstad	1	os	1
amager	1	havana	5	paratyphi B	3
anatum	14	hillsborough	1	ruiru	1
anfo	2	indiana	1	san diego	1
ardwick	1	infantis	5	schwarzengrund	4
bareilly	4	johannesburg	27	senftenberg	45
braenderup	1	kentucky	19	siegberg	1
bredeney	11	lexington	10	simsbury	1
binza	3	liverpool	3	tennessee	8
california	10	livingstone	42	taksony	5
cairo	4	madelia	1	thompson	1
cerro	27	manila	1	typhimurium	5
champaign	1	mbandaka	3	urbana	3
chester	6	meleagridis	19	westhampton	1
corvallis	1	menston	3	worthington	23
cubana	11	minnesota	3	unidentified	1
derby	9	montevideo	31		
godesberg	1	newport	25	worthington	6
grumpensis	3	onderstepoort	3	yalding	1
guildford	1	oranienburg	20	unidentified	3
havana	6				

Marine products — 77/955 — 8.1

Serotype	n	Serotype	n	Serotype	n
amager	7	give	7	montevideo	4
anatum	8	goerlitz	8	muenchen	1
binza	6	hamilton	6	newington	7
blockley	2	havana	2	newport	7
braenderup	3	indiana	3	oranienburg	7
bredeney	2	infantis	2	orion	4

TABLE VII (Continued)

NUMBER OF SALMONELLA ISOLATIONS FROM RAW MATERIALS AND FROM ANIMAL FOODS—UNITED KINGDOM, 1959–1963

Product	Samples positive	Percent positive		Serotypes				
			1	canoga	4	kentucky	1	paratyphi B
			2	cerro	1	kiambu	1	poona
			1	corvallis	1	kottbus	8	schwarzengrund
			1	chailey	2	lomita	5	senftenberg
			1	chester	1	meleagridis	1	stockholm
			2	cubana	1	menston	5	typhimurium
			1	derby	4	minnesota	1	worthington
Vegetable	14/387	3.6	4	cubana	1	oranienburg	1	tennessee
			1	kentucky	1	san diego	1	typhimurium
			3	meleagridis	4	senftenberg		
Manufactured foods								
Protein concentrate meals	94/893	10.5	1	alachua	1	fresno	1	orion
			1	anatum	1	gaminara	1	paratyphi B
			1	anfo	2	give	6	ruiru
			1	babelsberg	5	havana	1	rutgers
			1	binza	1	illinois	1	schwarzengrund
			1	braenderup	8	kentucky	24	senftenberg
			2	bredeney	4	livingstone	1	taksony
			1	cambridge	1	meleagridis	1	tennessee
			1	chailey	2	montevideo	5	typhimurium
			1	chester	1	newington	1	worthington
			4	cubana	11	oranienburg		

			Serotype		Serotype		Serotype	
Meals, mashes	64/1955	3.3	alachua	1	derby	2	newington	2
			anatum	4	fresno	1	oranienburg	1
			bareilly	1	gaminara	1	paratyphi B	1
			bere	1	johannesburg	1	ruiru	2
			binza	2	kentucky	3	senftenberg	10
			brancaster	3	lille	1	taksony	4
			bredeney	1	livingstone	2	tennessee	2
			california	1	manila	1	worthington	1
			champaign	1	meleagridis	3	unidentified	1
			cubana	9	montevideo	1		
Pelleted foods	8/1459	0.5	bere	1	mbandaka	1	tennessee	1
			brancaster	1	senftenberg	1	typhimurium	2
			california	1				

heavily contaminated (40% of samples), closely followed by meat products (29%). Marine products were less contaminated (8%), and vegetable products least (4%).

Of complete foods manufactured from these raw materials, protein concentrate meals, containing mixtures of animal and vegetable protein with added minerals, and intended to be diluted on the farm with grain products, proved the most heavily contaminated (10.5%). Meals and mashes showed a lower degree of contamination (3%) as might be expected, for the animal or marine protein in these products rarely exceeded 7.5%. Pelleted foods prepared from these meals or mashes showed the least contamination (0.5%).

The reduction of salmonella in the preparation of pelleted foods from meals or mashes can be attributed to the heat used in pelleting, as Swahn and Rutquist (1957) examined salmonella-contaminated lucerne flour before and after pelleting and were unable to isolate salmonellae from the pellets; also, Gray *et al.* (1958) pelleted mixtures of feeding stuffs containing up to 17% of bone meal heavily contaminated with salmonellae and failed to recover salmonellae from the pellets.

In Report (1961a), the effect of heat on salmonella in protein concentrate meals was investigated. Salmonellae were isolated from 41 out of 219 samples (18.7%) of unheated protein concentrate meal. The manufacturers then heated the animal ingredients before incorporation in the final concentrate meals; of these, 31 out of 627 samples (4.9%) were positive. The pelleting of animal foods is thus shown to effect a considerable destruction of salmonellae present in the raw materials of compound foods.

In the manufacture of complete foods from contaminated raw materials, two factors work together to effect a reduction of salmonellae in the complete food. In addition to the effect of the application of heat in the production of pelleted foods, the dilution of contaminated protein and mineral constituents with millers' offals and other carbohydrate constituents free from salmonellae brings about a considerable reduction of salmonellae in the meals or mashes from which the pelleted foods are prepared. Table VIII demonstrates this dilution effect.

In 19 out of 33 samples (57.6%) of bone products, in 8 out of 33 samples (24.2%) of meat products, and in 14 out of 28 samples (50%) of marine products containing salmonellae, the most probable number (M.P.N.) of salmonellae present exceeded 10 per 100 grams. In 26 samples of meals and mashes prepared from these products and found to contain salmonellae, the M.P.N. did not exceed 10 per 100 grams in any sample, and did not exceed 1 per 100 grams in 22 out of 26 samples (84.6%).

TABLE VIII
MOST PROBABLE NUMBER OF SALMONELLAE PER 100 GRAMS
IN RAW MATERIALS AND COMPLETE FOODS

Material	Samples examined	< 1	1	−10	−50	−100	−500	−1000	>1000
Bone products	33	4	5	5	5	6	3	4	1
Meat products	33	1	8	16	6	2	−	−	−
Marine products	28	6	−	8	3	3	5	−	3
Meals, mashes	26	3	19	4	−	−	−	−	−

e. Sources and Uses of Bone Products. The sources and methods of manufacture of bone products will now be discussed, as the numbers and serotypes of salmonella present in the finished products are related to the nature of the material processed and to the method of processing.

The agricultural use of bones and bone products has been practiced in the United Kingdom since the latter part of the eighteenth century (Fussell, 1962). Since about 1770, bones, crushed bones, bone waste, and the ash of calcined bone were used as dressings for dairy pastures and for root crops.

In the United Kingdom the main bone-using industries are the fertilizer industry, the animal fat and tallow and the glue and gelatin industries, the bone charcoal industry, the ceramics industry, and the animal feeding stuffs industry.

The main source of raw bone in the United Kingdom is native. The bones are derived from animals killed for food, from imported carcass meat, and from the oil and grease industry processing dead animals. Annual production is approximately 100,000 tons. Annual imports of bones amount to approximately 50,000 tons, of which approximately 20,000 tons are derived from slaughterhouses in the Argentine and in West and North Africa. Imports of sun-dried bones from India and Pakistan amount to approximately 30,000 tons annually.

Imported bone meal amounts to approximately 10,000 tons annually, roughly equal quantities being imported from Europe, South America, and the remainder of the world. Bone products imported from Europe under the Importation of Carcasses and Animal Products Order 1954, made by the Minister of Agriculture, Fisheries and Food, are required to be subjected to a sterilization process designed to destroy the virus of foot and mouth disease.

The fertilizer industry uses bone meal both home-produced and imported; the fat and tallow and the glue industries use raw bones both home-produced and imported; and the gelatin and bone charcoal industries use imported sun-dried bones, defatted and cleaned of organic matter by natural processes. Bone meal is also produced as a by-product in the preliminary selection of bone pieces in the manufacture of gelatin and bone charcoal. Bone used in the manufacture of bone china and steamed bone flour used by the animal feeding stuffs industry are produced as by-products in the glue industry. Dicalcium phosphate used as a mineral supplement in the animal feeding stuffs industry is produced as a by-product in the manufacture of gelatin from bone.

f. Bone Products and Animal Feeding Stuffs. In the United Kingdom the problem of the use of unsterilized bone meal relates essentially to fertilizers only. The risk of incorporation of raw bone meal in animal feeding stuffs is comparatively slight, although it could still occur when a farmer makes up his own feed. The bone products normally used in commercially produced animal feeding stuffs are dicalcium phosphate and steamed bone flour (Report, 1959b). Dicalcium phosphate is an inorganic product resulting from the chemical extraction of salts in the preparation of ossein from bone pieces in the manufacture of gelatin. Steamed bone flour is produced by grinding bone after steam extraction in the manufacture of glue. In the commercial compounding of animal foods in the United Kingdom, the essential minerals, trace elements, and vitamins are incorporated as a mineral mixture which may be supplied by specialist firms. Table IX contrasts the salmonella isolations in the United Kingdom from bone materials of fertilizer grade, and from samples obtained on the open market, with those from mineral mixtures and bone products used in the compounding of animal foods.

In all, salmonellae were isolated from 55% of bone products of fertilizer and unspecified grades, but from only 10% of feeding grades of bone meals, mineral–bone mixtures, and steamed bone flours. It will be remembered that in the preparation of animal foods in the form of meals or mashes, the concentration of salmonellae in the meal will be reduced by dilution with other ingredients, and further reduced by the heat used in the preparation of pelleted foods from these meals. Bone meal, however, is unique among animal foods in that it may be used in concentrated form as a major constituent of "salt licks"—a convenient method of supplementing the mineral requirement of dairy cows and cattle on grass pastures.

TABLE IX

SALMONELLA ISOLATIONS FROM BONE PRODUCTS—UNITED KINGDOM, 1957–1962

Reference	Grade	Product	Samples positive	Samples examined	Percent positive	Mean percent positive
Walker (1957)	Fertilizer	Bone meal	28	40	70.0	
Galbraith et al. (1962b)		Bone meal	8	30	26.7	
Harvey and Price (1962)	Unspecified	Indian crushed bone	56	57	98.2	55.3
Report (1959a)		Crushed bone	14	18	77.8	
		Bone meal	118	230	51.3	
		Bone flour	11	50	22.0	
Report (1961)	Feeding	Bone meal	1	8	12.5	
		Mineral–bone mixture	11	101	10.9	10.0
Galbraith et al. (1962a)		English bone flour	5	61	8.2	

Gray *et al.* (1958) describe a series of outbreaks of clinical salmonellosis in dairy cows on nine farms in Australia. Forty-nine out of 575 cows on these farms required veterinary attention. Eighteen cows died. The souce of infection on seven farms was native bone meal incorporated in a commercial cow meal, and on the remaining two farms a salt lick containing the same bone meal. Salmonellae were not isolated from the cow meal after the replacement of bone meal by rock phosphate as the mineral supplement. In this outbreak, 11 species of salmonellae were isolated from the bone meal. The occurrence of multiple salmonella types in bone products is common. Smith (1960) isolated 16 serotypes of salmonellae from a sample of Pakistan bone meal. Harvey and Price (1962) isolated 216 serotypes from 57 samples of Indian crushed bone, the maximum number of serotypes isolated from a single specimen being 17.

g. Sources and Uses of Meat Products. Meat meals and meat and bone meals are by-products of the industries extracting animal fats, oils, greases, and tallows, including the whaling industry. The raw materials include the whole carcasses of animals dying from accident, disease, or other natural causes. The carcasses include cattle, calves, sheep, pigs, and the carcasses of cats and dogs which have been destroyed; dressed carcasses of animals which have been condemned as unfit for human consumption and condemned meats, fresh, salted, pickled, and tinned; inedible offals from abattoirs, animal heads and feet, lungs, spleens, livers, and intestines; and butchers' scraps, bones, and rough fats. The list is not exhaustive but contains the materials most commonly met (Weaver, 1952). In addition to waste meats, any waste, surplus, or condemned food is processed to recover fat and protein. Whey, cheese, egg products, poultry carcasses, and desiccated coconut are known to have been processed. Chinese egg products and desiccated coconut are known to be contaminated with salmonellae, including *S. paratyphi B*. Again, the list is not exhaustive.

The processing plants may be sited at abattoirs and bacon factories, or they may be located in areas remote from the sources of supply, in which case raw materials are collected from a wide area and delivered to the plant by rail or road transport (Ludlam, 1954). Smaller amounts of meat meal and meat and bone meal are produced as a by-product in the manufacture of glue from animal bones.

h. Manufacture of Meat Products. Hoofs and horns are removed, and whole carcasses are skinned. The raw material is then reduced to suitable size by cutting it or by passing it through a crusher, and loaded into the agitator for dry rendering. The agitator consists of a cylindrical steam-jacketed cylinder traversed by a shaft carrying agitator arms. Processing may take place in vessels open to the atmosphere or under pressure; however, a temperature somewhat less than sterilizing temperature is required to assist in rupture of the fat cells. An internal pressure of 35 pounds ensures sterilization and disintegration of the material under treatment. The danger of recontamination of the final product during cooling and grinding is high, unless the layout of the plant and the hygienic precautions are strictly controlled (Albertsen, 1957). The classification of the final product (high-protein meat meal, meat meal, meat and bone meal, bone meal, liver meal, etc.) will depend on the nature of the raw material processed and will vary from charge to charge. The final classification will be determined by chemical analysis (Table VI).

In the United Kingdom little attempt has yet been made to process the by-products from broiler plants. These include heads and necks, feet, intestines, and feathers. One or two plants produce a "feather meal" from feathers keratinized by heating; a few plants produce a "feather, blood, and offal meal" which uses all the inedible portions of the fowl; the inedible products of most broiler plants are usually destroyed. Heads and necks of fowl caponized by hormone implants are unsuitable for conversion into feeding meals.

i.   World Production of Meat Products.   Meat meals, meat and bone meals, etc., are essentially a by-product of the slaughter of domestic animals (cattle, pigs, sheep, poultry) for human food and are produced domestically in almost every country in the world. In general, the main countries exporting meat meal are those that export carcass meat (Australia, New Zealand, Argentina) and produce meat meals in great excess of domestic requirements. The United States, which has a high per capita consumption of carcass and poultry meats, also produces meat meal and tankage in excess of her domestic requirements. The main importing countries are those in which intensive production of milk, cattle, pigs, fowl, and eggs are practiced (the United Kingdom, the Netherlands) and in which domestic production does not supply the demands for animal protein. Countries in which the production of farm animals, fowl, and their products is based on a farm economy rather than an intensive production are generally able to meet their demands for animal protein from domestic production. As animal proteins are entirely a by-product, variations in their supply depend almost entirely on the numbers of cattle, sheep, pigs, and poultry slaughtered. Though the supply available within any one country may show violent fluctuations over a period, the total annual production of all countries is reasonably stable (Allen *et al.,* 1961).

j. Salmonellae in Meat Products for Animal Foods.   Table X shows salmonella isolations from meat meals, meat and bone meals, etc., be-

TABLE X

SALMONELLAE IN ANIMAL PROTEIN

Reference	Product	Samples positive	Samples examined	Percent positive
Walker (1957)	Bone and meat meal	1	6	16.7
Report (1959a)	Meat products	28.	172	16.3
Kovacs (1959)	Meat meal, Australia	43	88	48.9
	Meat meal, New Zealand	3	4	75.0
Report (1961)	Meat and bone meal	54	287	18.8
Galbraith *et al.* (1961)	Meat meal, United States	93	101	92.1
Galbraith *et al.* (1962a)	Meat meal, United States	11	17	64.7
McCoy, 1961–1963 (unpublished)	Meat meal, home-blended	26	82	31.7

tween 1957 and 1963, in the United Kingdom and in Australia (Kovacs, 1959). During these years, domestic production of animal proteins in the United Kingdom averaged 60,000 tons annually (Petersen, 1961), and imports averaged 20,000 tons, of which 80% was imported from the United States, Australia, New Zealand, and the Argentine.

The majority of meat meals examined by Walker (1957) and by the Public Health Laboratory Service (Report, 1959a, 1961) were obtained through manufacturers or through compounding firms supplied by manufacturers and most probably represented a random selection of meals produced from native animals; the number of positive samples ranged from 16.3% to 18.8%. The meat meals examined in 1961–1963 (McCoy, unpublished), were described as "home-blended" and probably consisted of a mixture of home-produced and imported meals; salmonellae were isolated from 31.7% of samples.

The salmonellae isolated from animal protein, though fewer both in serotypes and in numbers of strains, correspond closely to those isolated from bone products (Table VII). The numbers of salmonellae in meat products are lower than in bone (Table VIII). These general findings are not unexpected, as the raw material from which both bone and meat products are produced is essentially the same. Only a small proportion of bone products, however, receive heat treatment in processing, whereas all meat products are heated. That complete sterilization of meat products is not effected during processing is probably the result of the combination of many factors, which include the nature of the raw material, bacterial multiplication between production and processing of raw material, incomplete heat penetration into the large masses of material treated, recontamination of processed material during cooling before grinding, and finally the dispersal of surviving organisms during grinding.

k. Use of Meat Products in Animal Foods. Animal protein products are used almost exclusively in feeding stuffs for pigs and poultry to supply the amino acids, methionine, lysine, and tryptophan which are distributed irregularly among vegetable proteins and are generally lacking in cereal grains. In this respect meat meals are generally regarded as interchangeable in the diets, though other considerations may limit their use. In the Federal German Republic, for example, imports of animal protein are insignificant partly because of salmonella regulations and partly because a maximum of 15% bone content is allowed (Petersen, 1961).

l. Marine Products. Production. Marine products used in animal foods include fish meals and whale meat meals. The methods used in production of whale meat meals and bone meals from the by-products of the whale oil industry differ in no essential respect from those used in the production of meat meals and bone meals. Whale products have been classed with fish products, however, as salmonellosis is not known to be a natural infection of fish or of marine animals.

Three arbitrary types of fish meal are recognized: white fish meal, meal prepared from oily fish, and sun-dried fish meal. White fish meal is prepared mainly from filleters' offals (heads, bone, and skins) of white fish (cod, haddock, etc.), from white fish condemned as unfit for human consumption, and from white fish which are not sold as the result of heavy landings.

The raw material of white fish meal is heated in a steam-jacketed cylinder with internal beating or stirring paddles. In batch-continuous driers, the preliminary heating of the raw material takes place under reduced pressure. Iceland and the United Kingdom are the main producers of white fish meal.

Oily fish meal is prepared from oily whole fish, waste fish, or offals, such as herring, menhaden, mackerel, sardine, and anchovy, from which as much oil as possible is extracted before the residue is dried for meal. Oil is recovered by cooking the fish with steam, and then pressing in continuous screw presses. North America, Norway, Peru, and South Africa are the main producers of fish meal from oily fish.

Sun-dried fish meal is produced by exposure of the raw material to the atmosphere, followed by grinding to meal. Angola is the largest producer of sun-dried fish meal.

Peru, Norway, the Union of South Africa (including South West Africa), Iceland, Denmark, Angola, and Canada are the most important exporting countries. The most important importing countries are the United States, the United Kingdom, the Federal Republic of Germany, and the Netherlands. The United Kingdom home production of white fish meal is used nationally.

m. The Bacteriology of Fish Meals. The general bacteriological picture of representative samples of white fish meal, meals prepared from oily fish, and other fish meals is shown in Table XI. The numbers of bacteria per gram of sample were least in meals from oily fish, slightly higher in white fish meals, and greatest in other fish meals. The highest counts were found in sun-dried fish meals.

In general, material leaving the concentrators for drying showed very low counts, comparable with those in meals from oily fish. Counts on the finished meals were somewhat higher, indicating some slight contamination during drying, cooling, and grinding. Bone meal and sun-dried fish meal known to contain salmonellae was added to several batches of raw offal in quantities up to 5% of the charge (4 cwt to 4 tons). In no case were salmonellae isolated from the finished meal, showing that the processing of contaminated white fish meal destroyed the salmonellae.

In addition, no salmonellae were isolated from 1400 individual cod sampled on landing, nor from any of the samples of offal before processing.

n. Salmonellae in Marine Products Used in Animal Foods. United Kingdom 1957–1963. Table XII presents the results of examination of marine products classed both by the nature of the product and by the country of origin so far as could be determined. Angola sun-dried fish meal proved to be the most heavily contaminated fish meal examined; 58.7% of samples contained salmonellae and generally contained multiple types, as many as 13 serotypes being isolated from a single sample. In this respect sun-dried Angola fish meal resembles Indian crushed bone. That this contamination of Angola sun-dried fish meal is not connected with the raw material of the industry is shown by the absence of salmonellae from South West African fish meal, machine-dried from the same type of fish used in the preparation of Angola sun-dried meals and caught in the same geographical area. In addition, salmonellae were not isolated from machine-dried Angola fish meal. Direct investigation into the method of contamination of fish with salmonellae during sun-drying has not proved possible, but the manner of contamination of bone products with

TABLE XI

General Bacteriological Examination of Representative Samples of Fish Meals[a]

Product	Area of origin	Number of samples	3 days, 22°C	2 days, 37°C	Type of organisms growing at 37°C[b]
White fish meals	Hull	1	800	200	95% presumptive coliforms, 5% A.S.B.
	Iceland	1	<200	10,400	100% A.S.B.
		1	7,400	15,000	95% presumptive coliforms, 5% A.S.B.
			20,000	10,600	100% A.S.B.
Herring meals	Yarmouth	4	<200	<200	100% A.S.B.
	Norway	3	200	<200	100% A.S.B.
	Denmark	1	16,000	<200	—
	Denmark	1	200	200	100% A.S.B.
Herring flour	Iceland	1	1,800	<200	—
	Norway	1	800	<200	—
		1	1,000	<200	—
		1	12,600	<200	—
		1	1,200	200	100% presumptive coliforms
		1	1,200	1,000	100% presumptive coliforms
Herring full meal	Norway	1	1,400	<200	—
		1	200	200	100% A.S.B.
Other meals					
Sea perch meal	Iceland	1	<200	200	100% A.S.B.
Crayfish meal	South Africa	1	600	<200	—
Fish meals	India	1	30,000	10,800	2% coliforms, 98% A.S.B.
	Pakistan	1	>60,000	Uncountable	Micrococci, coliforms, A.S.B.
	Angola	1	23,000	7,000	60% coliforms, 40% A.S.B.
	Sun-dried	5	—	$2.6\text{–}3.6 \times 10^6$	—

[a] Surface counts per gram on blood agar.

[b] A.S.B. = aerobic spore bearer.

TABLE XII

SALMONELLAE ISOLATED FROM MARINE PRODUCTS—UNITED KINGDOM, 1957–1963

Product	Country of origin	Report (1959)	Report (1961a)	McCoy (1961–1963)	Galbraith (1962a)	Total positive	Percent positive
White fish meals	United Kingdom	6/57					
	Iceland	0/11	0/17	0/8	0/13	7/114	6.1
	Canada	1/8					
Herring meals	United Kingdom	3/9					
	Denmark/Norway	0/20	1/161	0/11		4/201	2.0
Other oily fish meals	South West Africa	0/20	0/15	0/1			
	Chile	0/2	0/15	4/118	0/2	5/187	2.7
	Peru	1/14					
Fish meal	China	0/2					
	India	0/3				0/8	
	Parkistan	0/3					
Angola fish meal	Sun-dried	27/37					
	Machine-dried	0/9				27/46	58.7
"Fish meals"	Unspecified	9/120	16/226	4/9		29/355	8.2
Whale meat products (meals, etc.)	United Kingdom Argentine Chile	4/36		2/28		6/64	9.4

salmonellae in Israel (Hirsch and Sapiro-Hirsch, 1958) affords more than a hint as to the manner of contamination. Bones, after sterilization by steam at high pressure, were spread on the open, unfenced ground for sun-drying before being ground to bone meal. The bones were thus exposed to contamination by flies, jackals, dogs, rodents, and birds, whose excrements were collected and ground with the bones. The resultant product was not sterilized again and was sold as "steamed bone meal." Altogether 11 different salmonellae types were isolated from 27 samples of bone meal, probably all manufactured in Israel.

o. Sources of Contamination of Fish Meals. Of the fish meals examined (Report, 1959*a*) between 1957 and 1959 in the United Kingdom, the majority were collected from the ultimate user, except in the case of white fish meals manufactured in the United Kingdom, of which approximately half were sampled at the factory immediately after production.

Salmonellae were not isolated from 29 samples collected from the factory immediately after production; salmonellae were isolated from 6 out of 28 samples (21%) collected from retailers or from manufacturers of animal feeding stuffs. These results show that contamination with salmonellae occurred after processing.

The major source of contamination of fish meal after manufacture proved to be the addition to it of material recovered from empty sacks. In establishments receiving large quantities of products in sacks, the sacks after emptying are finally cleaned by exposure to a vacuum. The resultant dust is collected from the airstream and added to meals. As retailers of animal and marine products may deal in many different classes of material, a considerable degree of contamination of clean meals may result from the addition of contaminated extracted dust. Investigations showed that salmonellae were isolated from 9 out of 12 sacks which had held contaminated materials, after vacuum cleaning (Report, 1959*a*). Sacks in which contaminated products are imported are generally sold for other uses as second-hand sacks and may again result in contamination of a clean product.

Another minor method of contamination of imported fish meals is during importation. Fish meals are usually imported in sacks, and during shipment and transshipment a certain amount of spillage inevitably occurs. At the end of unloading, the spilt material is collected and re-bagged, resulting possibly in contamination of a clean material from the remains of previous cargoes.

A major source of contamination is the blending of imported and home-produced fish meals with meat meals to produce a product of standard protein or mineral content.

McCoy (1961–1963) found that 4 out of 9 samples (44%) of home-blended feeding fish meals contained salmonellae.

p. The Use of Marine Products in Animal Foods. In the Federal Republic of Germany, about 70% of all fish meal consumed is used for pig production and 30% for poultry production; in the United States the proportions are reversed, 70% of all fish meal consumed being used for poultry production. This might explain why in the Federal Republic of Germany many new salmonella serotypes were isolated from pigs whereas in the United States new serotypes were isolated from poultry. From 1956 to 1959 Germany and the United States used half of the entire production of Angola sun-dried fish meal (Meseck, 1961).

q. Feeding Practice. Two main types of feeding practice may be recognized as being associated with "farm" production and intensive production of pigs and poultry. In "farm" production, pigs and poultry are produced by a relatively large number of small producers. France, Italy, Spain, and to a lesser extent Denmark and the Federal Republic of Germany are mainly "farm" producers of pigs and poultry; the Netherlands and the United Kingdom are mainly intensive producers. The difference in the method of production is reflected in feeding practice. In the United Kingdom approximately 80% of the total quantity of animal food produced is in the form of heated compounded foods prepared by manufacturers. In the Federal Republic of Germany approximately only 30% of the total quantity of animal food used is compounded. German animal feeding practice is unique in that between 50 and 60% of the fish meal used is sold directly to farmers as fish meal, and only 40 to 50% is used by compounders in concentrates and mashes (Petersen, 1961). Feeding practice has an effect on salmonella in animals, as a progressive reduction has been demonstrated in the salmonella content of meals and mashes prepared from contaminated raw materials, and in pelleted foods prepared from these meals (Report, 1961). These investigators found salmonellae in 115 out of 1284 samples (9.0%) of raw materials, in 48 out of 1742 samples (2.8%) of meals and mashes prepared from these raw materials, and in 3 out of 1114 samples (0.27%) of pelleted foods prepared from these meals and mashes, a final reduction in salmonella content of 97%.

In general the risk of salmonella infection of domestic animals from feeding stuffs would appear to be greatest when meals mixed on the farm from raw materials bought by the farmer are used, and least when pelleted foods prepared commercially are used.

r. Pet Foods. The commercial production of foods for domestic pets, dogs, cats, and budgerigars, though small when set beside the production of foods for farm animals, nevertheless has increased considerably in the United Kingdom in recent years (Contango, 1962). In the United Kingdom the greater part of the food for dogs is furnished by raw meat including abattoir offals, lungs, livers, imported horse meat, and butchers' scraps: a lesser portion by canned products ranging from whole meat to complete foods balanced as to protein, fat, and carbohydrate; the least by dog biscuits and dog meals. The greater part of cat food is provided by raw fish or fish scraps, the remainder by canned foods, in a range comparable to canned dog foods but based on fish instead of meat.

Canned foods for dogs and cats is processed by the same methods as meat products for human consumption. Dog biscuits are composed of flour with added meat and are baked hard at a high temperature. Salmonellae have not been isolated from this material.

In the United States meats for dog foods are obtained from nationally known packing houses and rendering plants (Galton, *et al.,* 1955*a*). Raw meats for dogs include beef livers rejected for human consumption (McElrath *et al.,* 1952). Salmonellae were found in raw horsemeat, in other raw meat, and in cooked or dehydrated meats for animal feeding (Galbraith *et al.,* 1962*b*). Table XIII presents isolations of salmonellae from pet foods in the United States of America and the United Kingdom between 1952 and 1962. Raw and processed meats were the most heavily contaminated products.

2. RELATION TO ANIMAL SALMONELLOSIS

Clinical salmonellosis in food animals following the consumption of contaminated food would appear to be uncommon. In bovines, outbreaks of the type described by Gray *et al.* (1958) appear to be rare. The relation of salmonellae in feeding stuffs to bovine salmonellosis in England and Wales has recently been examined (Report, 1965). *Salmonella dublin* and *S. typhimurium* were the serotypes encountered most commonly in cattle but only infrequently in the raw materials of animal foods and in complete foods (Table VII). Bovine salmonellosis was highest in the last quarter of the year, the period of maximum calving. The high incidence of calf salmonellosis in this quarter was ascribed to calf sales and movements rather than to the consumption of contaminated food. In addition, it was quite clear that salmonellosis was spread into different regions of the country by calf movements.

In pigs symptomless excretion without evidence of clinical infection would appear to follow the ingestion of serotypes of salmonellae in con-

TABLE XIII

SALMONELLAE IN CAT AND DOG FOOD

Type of food	Galton et al. (1955a)	Report (1959)	Galbraith et al. (1962b)	Percent samples positive	Number of serotypes/strains isolated			
Biscuits, "bones," flakes, etc.	0/54	0/13	0/111	—				
Processed meats (cooked or dehydrated)	—	—	16/139	11.5	adelaide 1 agodi 1 anatum 4 bredeney 2	cerro 1 dublin 1 give 1 islington 1	meleagridis 2 oranienburg 1 reading 2 typhimurium 1	
Dog meals (containing meat)	26/98	—	—	26.5	anatum 5 bareilly 2 bredeney 1 cubana 2 derby 4 illinois 1	kentucky 1 lexington 1 minnesota 2 montevideo 2 newington 1 oranienburg 4	panama 1 senftenberg 4 tennessee 1 thompson 1 typhimurium 4	
Raw horsemeat	—	—	85/319	26.6	anatum 10 derby 10 enteritidis 1 finchley 1 give 1 havana 1	hvittingfoss 1 irumu 1 kentucky 4 meleagridis 27 minnesota 3 montevideo 2	ness-ziona 1 newport 2 oranienburg 25 san diego 1 typhimurium 4	
Other raw meat	—	—	9/56	16.0	brandenburg 1 dublin 3	give 1 heidelberg 1	oranienburg 2 typhimurium 1	

taminated foods (Report, 1947; Newell *et al.*, 1959; Smith, 1960). In the intensive production of pigs, characterized by the concentration of susceptible animals, infection once introduced is liable to spread widely and, through environmental contamination, to persist.

In poultry, infection may be introduced into the master breeding flock by contaminated food. If infection is allowed to continue uncontrolled, eventually ovarian infection of laying stock occurs and egg transmission is established. In the United Kingdom, infection in the master breeding or multiplying farms forms the main focus of infection from which infection is transmitted by the dispatch of newly hatched chicks to rearing farms. The introduction of infection through contaminated food at a rearing farm is likely to cause infection in that individual farm only, and then only for the life of the flock.

In England and Wales, cattle form the largest reservoir of *S. typhimurium* infection, followed by poultry (Table I). More serotypes of salmonellae have been isolated from pigs in England and Wales than from any other animal species. Host-specific strains of salmonellae play a minor part in pig salmonellosis, and there is as yet no evidence of the emergence of host-associated strains.

E.   Transfer of Salmonellae from Animals to Man

1.   MULTIPLYING PROCESSES

The multiplication of bacteria in a nutrient medium held at suitable temperature for sufficient time, which occurs, for example, in lightly contaminated food during storage at high ambient temperature or following reheating before consumption after storage, is too well known to merit further description.

In the biological sense under natural conditions, however, the most common process resulting in multiplication of salmonella is infection of the susceptible animal or human individual. In infection, the ingestion of a relatively small infecting dose is followed by the excretion in feces of up to 10^{11} salmonellae daily for periods of up to 3 weeks or longer. This multiplication in the individual leads to the infection of other individuals in close contact. Galton *et al.* (1954) isolated salmonellae from the feces of 7% of 374 individual hogs on farms; from the feces of 25% of 100 hogs held at an abattoir before slaughter for periods of up to 48 hours; and from the feces of 51% of 98 hogs after slaughter at the same abattoir. McDonagh and Smith (1958) isolated salmonellae from the feces of 3% of 171 healthy pigs arriving at an abattoir and from 14% of pigs held in lairs at the abattoir for up to 7 days. The abattoir thus serves as means of transfer of salmonellae between animals of the same species and between animals of different species (McDonagh and Smith, 1958).

The mechanisms whereby salmonella contamination is spread from carcass to carcass in abattoirs involves contaminated equipment (Galton et al., 1954) and wiping cloths (Camps, 1947).

2. DISPERSAL

The dispersal of salmonellae is brought about in the mass production of many foods. Bernstein (1952) failed to isolate salmonellae from 3648 individual hen eggs. McCoy and Little (Table V) isolated salmonellae from 191 out of 1856 batches of liquid egg in production. Each batch examined was made from more than 30,000 eggs. The contamination rate of more than 13,500 individual duck eggs was 0.15% (Report, 1954); that of batches of liquid duck egg, each batch containing between 4000 and 20,000 eggs, was 48% (Murdock, 1954). In the manufacture of sausages a similar distribution of contamination has been noted. Newell et al. (1959) isolated salmonellae from 3% of muscle samples from individual pigs but from 70% of batches of sausage meat prepared from these carcasses.

The largest recorded outbreak of human salmonellosis from animals (Lundbeck et al., 1955), 7717 notified cases of S. typhimurium infection with 90 deaths, resulted from a combination of transfer of infection among animals held at a central abattoir for up to 6 weeks; overtaxing of the abattoir's resources for sanitary slaughter when killing was resumed; insufficient cooling of meat after slaughter due to failure of refrigeration plant; and an unusually high ambient temperature which encouraged multiplication of organisms in meat during transport to retailers and before consumption in the home.

In the growing and processing of food, wide dispersal of contamination may also result from the use of sewage-polluted water. Growing vegetables may be contaminated or heat-treated foods contaminated after processing. Harmsen (1953, 1954) ascribed epidemics of typhoid fever at Luneberg and Vienna to the consumption of raw vegetables irrigated with domestic sewage applied as artificial rain. Canned meats have been contaminated with the typhoid bacillus from the use of sewage-polluted water for cooling after processing (Couper et al., 1956; Report, 1964a). Thomas et al. (1948) reported an outbreak of paratyphoid B fever spread by pasteurized milk, in which raw river water drawn from a sewage-polluted river was used for the final rinse of milk bottles before the bottles were filled with pasteurized milk; the water was chlorinated before use, but the time of contact was inadequate. The source of contamination of Chinese egg products and desiccated coconut from Ceylon with paratyphoid B bacilli, though not yet definitely ascertained, is most likely to be the use of sewage-contaminated water during processing.

3. VEHICLES OF TRANSMISSION

Table XIV shows the vehicle of infection in 733 outbreaks and family outbreaks of salmonella infection in England and Wales, 1949–1963, in which the vehicle was identified.

Meat served as the vehicle in almost half the outbreaks; egg products in one quarter; and sweetmeats and milk products in rather more than one fifth. Fish, fruit, and vegetables were the vehicles in the remaining outbreaks. Processed and made-up meat products were responsible for about 40% of outbreaks. *Salmonella typhimurium* was responsible for three-quarters of all outbreaks.

a. Meat. Human infection from the flesh of sick animals slaughtered as an emergency to save the value of the carcass as flesh meat is well known. The first isolations of *S. enteritidis* (Gärtner, 1888) were from the flesh of an emergency-slaughtered animal and from patients who had eaten the flesh. De Nobele (1898) isolated *S. typhimurium* from patients and from flesh of an animal suspected as the vehicle. Many such instances have been recorded subsequently. Healthy animals are slaughtered in abattoirs, however, whereas more clinically ill animals are slaughtered in knackeries. The presence of salmonellae in abattoirs slaughtering healthy animals has recently been investigated (Report, 1964c). Thirty-two abattoirs were studied. Salmonellae of 75 serotypes were isolated from 21% of 4496 sewer swabs.

In twenty abattoirs slaughtering cattle, pigs, and sheep, *S. typhimurium* was the dominant serotype (26% of strains), followed by *S. dublin* (16%) and *S. heidelberg* (12%). Six serotypes—*S. menston, S. senftenberg, S. livingstone, S. anatum, S. meleagridis,* and *S. give*—together accounted for a further 17% of strains.

In swabs from an abattoir slaughtering aged dairy cows, *S. typhimurium* made up 26% of strains. *Salmonella dublin* was not isolated. Four serotypes—*S. livingstone, S. menston, S. kiambu,* and *S. schwarzengrund*—accounted for a further 32% of strains.

In swabs from two bacon factories slaughtering pigs only, *S. typhimurium* was again the dominant serotype (23% of strains). Six serotypes —*S. bredeney, S. livingstone, S. thompson, S. menston, S. meleagridis,* and *S. anatum*—were responsible for 39% of strains isolated. *Salmonella heidelberg* accounted for only 2% of strains. The serotypes present in abattoir swabs and in animal tissues sampled at the abattoir were com-

TABLE XIV

Food Poisoning in England and Wales, 1949–1963

Vehicle	Description	Types		Outbreaks	
		S. typhimurium	Other types	Number	%
Meat	Fresh	1	1		
	Gravy, soup, stock, etc.	4	4		
	"Meat"	18	7	347	47.3
	Canned	19	11		
	Processed and made up	177	105		
Eggs	Dried egg	3	—		
	Hen egg	4	1	181	24.7
	Liquid and frozen	15	7		
	Duck	141	10		
Sweetmeats	Trifles, ice cream, custard, cream confectionery	97	22	119	16.2
Milk	Cheese	2	2		
	Canned	3	—	40	5.4
	Dried	4	1		
	Fresh	20	8		
Fish	Shellfish	5	—		
	Canned	5	2	27	3.7
	"Fish"	6	1		
	Processed	6	2		
Fruit	Fresh	8	1	10	1.4
	Canned	1	—		
Vegetables	Canned	2	—		
	Dried	2	1	9	1.2
	Fresh	4	—		
Total		547 (75)	186 (25)	733	

pared with serotypes isolated from human infections in the areas supplied with meat from the abattoirs. Of the 75 types of salmonellae isolated from abattoirs, 35, rather less than half, were isolated from human infections.

In the United Kingdom, however, 30% of the meat consumed is imported (Report, 1965). Salmonellae were isolated from 4% of chilled carcasses imported in 1956–1959 (Hobbs and Wilson, 1959) and from 16% of samples of frozen boneless imported meats in 1961–1963 (Hobbs, 1965). Frozen boneless New Zealand veal and South American meat were the most heavily contaminated imports. *Salmonella typhimurium* was the serotype most frequently isolated from New Zealand veal (95% of salmonellae isolated); *S. typhimurium* (27% of isolations), *S. newport* (25%), and *S. anatum* (19%) were the serotypes most frequently isolated from South American boneless beef.

Imported boneless beef and veal amount to approximately 5% of beef consumed in the United Kingdom. This type of meat is used mainly in manufactured made-up meat products, but it is also recommended for use as a cheap source of protein in school and hospital canteens.

b. Eggs. Table XV shows the types of salmonellae isolated from egg products in 1956–1962. Sixty-three serotypes were identified. In terms of serotypes, isolated Chinese egg products proved the most heavily contaminated, yielding 35 serotypes; United States and Australian egg followed (29 and 27 serotypes, respectively); Dutch egg yielded 14 serotypes; English egg, 7.

In products from each country a few serotypes predominated. In Chinese products, paratyphoid bacilli formed 10% of isolations, *S. thompson* 55%, and *S. aberdeen* 8%. In Australian egg, *S. typhimurium* formed 67% of isolations. In American egg, *S. oranienburg* formed 17% of isolations, *S. blockley* 13%, and *S. tennessee* 10%. In English egg, *S. typhimurium* formed 78% of isolations, *S. menston* 10%, and *S. thompson* 9%. In Dutch egg, *S. bareilly* formed 38% of isolations, *S. typhimurium* 18%, and *S. anatum* 13%. The differences in the frequency and types of salmonellae isolated from poultry in different countries are probably due to the establishment of certain serotypes in the master breeding flocks.

Chinese liquid egg products are unique in their content of paratyphoid bacilli, and in the high incidence of *S. thompson* and *S. aberdeen*. The import of Chinese egg and egg products into England is long-standing. *Salmonella aberdeen* and *S. thompson* were regarded as types indigenous to England and Wales, or rather types isolated up to 1939 (Report, 1947). From 1923 to 1939, *S. thompson* was the third most common serotype isolated from human infections in England and Wales, but

TABLE XV

Salmonellae in Egg Products, 1956–1962[a]

Source	Isolations	Serotypes[b]			
China	2126	thompson 55	muenchen 1	dahlem	oranienburg
		paratyphi B 10	senftenberg 1	derby	paratyphi C
		aberdeen 8	wandsworth 1	dublin	reading
		potsdam	bareilly	enteritidis	schwarzengrund
		typhimurium	blegdam	heidelberg	senegal
		meleagridis	blockley	infantis	stanley
		anatum 2	chincol	javiana	sundsvall
		newport 2	chinovum	kisarawe	tennessee
		bovis morbificans 1	cholerae-suis	mikawasima	
America	401	oranienburg 17	thompson 1	heidelberg 3	meleagridis
		blockley 13	alachua	cerro 2	menston
		tennessee 10	braenderup	give 2	new brunswick
		infantis 9	bredeney	kentucky 2	newington
		montevideo 9	chester	muenchen 2	san diego
		anatum 6	cubana	senftenberg 2	schwarzengrund
		bareilly 5	derby	newport 1	worthington
		typhimurium			
Australia	1766	typhimurium 67	potsdam 2	cambridge	menston
		hessarek 6	bredeney 1	chester	muenchen
		oranienburg 5	thompson 1	cholerae-suis	new brunswick
		anatum 4	worthington 1	derby	newport
		meleagridis 3	aberdeen	enteritidis	orion
		tennessee 3	adelaide	give	senftenberg
		bareilly 2	bovis morbificans	lexington	

Holland	137	bareilly 38 typhimurium 18 anatum 13 montevideo 8	oranienburg 6 tennessee 6 newington 5 give 4	binza 3 infantis 2 litchfield 2	thompson 2 cubana heidelberg
England	559	typhimurium 78 menston 10	thompson 9 derby 1	oranienburg 1 anatum	worthington
Argentina, Canada, Denmark, New Zealand, South Africa, Sweden, Poland, Israel, Yugoslavia, Brazil, Germany, Japan	578	typhimurium 51 enteritidis 6 montevideo 6 newport 6 tennessee 5 thompson 5 infantis 4 heidelberg 2 kentucky 2	oranienburg 2 bareilly 1 give 1 aberdeen binza bovis morbificans bredeney cerro chester	chicago cholerae-suis derby hessarek irumu jerusalem manhattan meleagridis	michigan mikawasima muenchen onderstepoort potsdam saphra senftenberg worthington

[a] Extracted from Public Health Laboratory Service weekly summaries.

[b] Figures indicate percentage of strains. No figure indicates percentage of < 1.

it was rarely, if ever, isolated from native domestic animals before its establishment in native fowl in 1942. In retrospect the source of human infection with *S. thompson* in 1923–1939 would appear to be Chinese egg products.

   c. Sweetmeats.   In England and Wales for many years milk had been regarded as the most common vehicle of paratyphoid fever. In the early 1950's, however, the vehicle of infection most commonly associated with paratyphoid fever was noted to be bakers' confectionery (Hobbs and Smith, 1954). Chinese egg products were identified as the vehicle carrying contamination into bakeries (Newell, 1955*a, b;* Newell *et al.,* 1955).

   Desiccated coconut also carries contamination into bakeries. Table XVI shows the serotypes isolated from desiccated coconut in England and Wales, 1959–1962. Four serotypes predominate: *S. waycross* (16% of strains), *S. ferlac* (14%), *S. paratyphi* B (12%), and *S. bareilly* (9%). Desiccated coconut is used in cakes as a substitute for dried fruit, in confectionery as a decoration, and in sweets both as filling and as decoration. Thomson (1953) showed that in cream cakes the cake, moistened with water from the cream filling, formed an excellent medium for the multiplication of organisms. Cream cakes sprinkled with contaminated desiccated coconut thus form an excellent medium for the transfer of paratyphoid infection. *Salmonella typhi* has been isolated from Papuan desiccated coconut (Wilson and MacKenzie, 1955), and *S. paratyphi* B (Anderson, 1960) and many other serotypes from Ceylon desiccated coconut (Kovacs, 1959; Galbraith *et al.,* 1960).

   Gelatin is used in confectionery as a substitute for egg albumen, and as a filling for meat and pork pies after the casing and meat contents have been cooked at sterilizing temperatures. *Salmonella meleagridis* was isolated frequently from gelatin in England and Wales in 1960–1962 (Table XVII). The process of adding warm gelatin solution by hand to pies after cooling carries a risk of infection from human excretors.

   d. Milk.   Milk and milk products formed the vehicle of infection in about 5% of outbreaks (Table XIV). Knox *et al.* (1963) reported an explosive outbreak of *S. heidelberg* infection following the consumption of raw milk from a cow with a symptomless salmonella mastitis, and discussed recorded milk-borne outbreaks in the United Kingdom in 1942–1963.

   Raw milk formed the vehicle in 31 outbreaks, pasteurized milk in 2, and raw and pasteurized milk in 1. Pasteurized milk was contaminated

TABLE XVI

SALMONELLA SEROTYPES ISOLATED FROM DESICCATED COCONUT, 1959–1962[a]

Strains	Serotypes[b]					
1618	waycross 15	litchfield 2	aqua	ilala	minnesota	shangani
	ferlac 14	nchanga 1	bootle	infantis	mount pleasant	simsbury
	paratyphi B 12	thompson 1	braenderup	javiana	muenchen	stanley
	bareilly 9	muenster 1	bredeney	lanka	onderstepoort	takoradi
	hvittingfoss 7	solna 1	bukavu	lansing	oranienburg	taksony
	senftenberg 6	cubana 1	butantan	lexington	oslo	tennessee
	perth 5	chittagong 1	cerro	lomita	poona	treforest
	kotte 4	weltevreden 1	charity	maron	richmond	vancouver
	angoda 3	abaetetuba	chester	matopeni	san diego	virchow
	typhimurium 3	abony	chingola	menston	san juan	welikada
	rubislaw 3	anatum	give	meleagridis	schwarzengrund	westhampton
	newport 2	angola	heidelberg			

[a] Extracted from Public Health Laboratory Service weekly summaries.
[b] Figures indicate percentage of strains. No figure indicates percentage of < 1.

TABLE XVII

Salmonella Serotypes Isolated from Human Foods, 1960–1962[a]

Product	Description	Serotypes			
Milk	Fresh and dried	bredeney 1 dublin 4	heidelberg 10	senftenberg 2	typhimurium 1
Confectionery	Flour	senftenberg 1			
	Cornflour	newport 1			
	Soya flour	tennessee 1	typhimurium 1		
	Gelatin	meleagridis 177			
	Cake mix	montevideo 2			
	Sweets, biscuits, etc.	enteritidis 1 ferlac 1 hvitingfoss 1	senftenberg 1 litchfield 1 meleagridis 27 paratyphi B 52	senftenberg 1 stanley 1	typhimurium 2 waycross 2
Fish	Shellfish	bareilly 1	derby 2	typhimurium 3	
	Shrimps, frozen	typhimurium 1			

[a] Extracted from Public Health Laboratory Service weekly summaries.

after treatment—in one case by raw milk in the bottling machine, in another through capping bottles of pasteurized milk by hand with cardboard caps contaminated by mouse feces. In the third outbreak, contamination of pasteurized milk occurred at the center where the milk was consumed. *Salmonella typhimurium* was isolated from patients in 18 outbreaks, *S. dublin* in 9, *S. enteritidis* and *S. newport* each from 2 outbreaks; and *S. heidelberg, S. oranienburg,* and *S. thompson* each from 1 outbreak. Of the raw milk outbreaks, the source of contamination was considered to be sick animals in 18 outbreaks, human or animal symptomless excretors in 2, and sick or symptomless human excretors in 2. The source of contamination was not determined in the remaining outbreaks.

e. Fish. Marine and fresh water vertebrates, crustaceans, and shellfish living in waters not polluted by the discharge of crude sewage or effluent are normally free from salmonellae. Oysters and mussels, which feed by filtering organic matter from 20 to 40 liters of sea water hourly from beds in sewage-polluted waters form a well-recognized source of infection with *S. typhi* (Gay, 1918).

F. Salmonellae in Human Foods

1. TYPES AND STRAINS—ENGLAND AND WALES

Table XVIII presents serotypes and number of strains of each isolated from meat and meat products in England and Wales, 1960–1962. Most of the isolations from poultry were made at processing plants. During the processing of large numbers of fowl from many sources, opportunities for the transfer of contamination occur during the cooling of eviscerated carcasses in chilling tanks (Galton *et al.,* 1955*b;* Dixon and Pooley, 1961, 1962). The evisceration of infected fowl carcasses in butchers' shops has led to outbreaks of human infection of the contamination of cooked meats intended for consumption without further cooking (Galbraith *et al.,* 1962*a*), through the use of a common preparation room and equipment. The isolations from beef and veal show the dissemination of contamination produced by the handling of meat, as the greatest number of serotypes were isolated from boneless imported meat. During slaughter, contamination of carcasses with intestinal contents is inevitable, and transfer of surface contamination between carcasses results from contamination of working surfaces and equipment. In the boning of carcasses, surface contamination is transferred to all cut surfaces. Offals, pie meat, and sausages are heavily contaminated in the same way. In sausage manufacture, opportunities for multi-

TABLE XVIII

SALMONELLA SEROTYPES ISOLATED FROM HUMAN FOOD—ENGLAND AND WALES, 1960–1962[a]

Product	Source	Serotypes			
Poultry	Duck	amager 4 anatum 1	derby 1 infantis 1	oranienburg 2	typhimurium 116
	Chicken	anatum 2 cubana 1 enteritidis 1	kentucky 3 manhattan 1	menston 3 thompson 16	typhimurium 61 worthington 1
	Turkey	saint-paul 4	typhimurium 15		
Beef	Bechuanaland[b]	abortus-bovis 1 agbeni 1 amsterdam 1 anatum 4 bareilly 1 bovis morbificans 1 braenderup 3 cerro 2	chester 6 dublin 1 germiston 1 gilbert 7 hamburg 1 irumu 2 kaapstad 6	loma linda 4 meleagridis 1 montevideo 1 nachshomin 1 newport 23 onderstepoort 10 oranienburg 1	portbeck 1 potsdam 1 saint-paul 2 suberu 2 tshiongwe 17 typhimurium 24 westpark 15
	Australia	typhimurium 3			
	Argentina	dublin 1	newport 2		
	England	dublin 1			
Veal	New Zealand[b]	bovis morbificans 3	heidelberg 2 chester 1 dublin 2	saint-paul 1	typhimurium 151
	Australia	anatum 1			
	Holland	bareilly 2			
	Unspecified	paratyphi B 1			

Food	Source				
Mutton	New Zealand	bovis morbificans 1	typhimurium 1		
	Australia	bovis morbificans 2	typhimurium 1	newport 1	
	Argentina	bovis morbificans 1	minnesota 2		
Offal	Unspecified	bareilly 1	give 4	menston 2	san diego 2
		bredeney 1	heidelberg 1	oranienburg 1	weltevreden 1
		chingola 1	livingstone 1	typhimurium 15	worthington 1
		dublin 15			
Pie meat					
Sausages	United Kingdom	adelaide 2	butantan 6	lomita 1	reading 5
		agama 18	derby 5	mbandaka 1	schwarzengrund 7
		anatum 12	dublin 2	meleagridis 8	senftenberg 1
		bonariensis 1	give 22	menston 1	stanley 1
		bovis morbificans 1	heidelberg 25	montevideo 1	typhimurium 55
		brancaster 1	kiambu 3	newington 4	tennessee 1
		brandenburg 12	lexington 1	oranienburg 1	taksony 5
		bredeney 3	livingstone 15	paratyphi B 4	worthington 39

[a] Abstracted from Public Health Laboratory Service weekly summaries.
[b] Boneless.

plication of organisms occur, owing to the rise in temperature resulting from the grinding of whole meats to sausage meats. In factory practice this can be minimized by the addition of solid carbon dioxide "ice" to the chopping bowls. In England and Wales, however, some of the sausage eaten is produced by individual butchers in small quantities.

Table XVII presents serotypes isolated from foods other than meat products. The serotypes isolated from sweets, biscuits, etc., contain types present in contaminated raw materials, gelatin, egg products, and desiccated coconut.

2. Food Involved in Human Infection — England and Wales

Table XIX presents the vehicles of infection in general and family outbreaks from meat products in England and Wales, 1954–1963. Pork

TABLE XIX

Outbreaks and Family Outbreaks Associated with Processed and Made-up Meats, 1954–1963[a]

Description	Pork, ham, bacon	Poultry	Beef	Veal	Mutton	"Meat"	Total
Canned			1				1
Stew, mince						1	1
Sandwiches	3	1				1	5 (4)[b]
Pressed			6	1			7 (6)
Reheated	1	7		1			9 (8)
Cold	16	7	4			2	29 (24)
Sausages	3					21	24 (20)
Processed[c]	6					16	22 (19)
Pies	14		1			6	21 (18)
	43 (36)	15 (13)	12 (10)	2 (1)		47 (39)	119

[a] Abstracted from "Food Poisoning in England and Wales, 1954–1963" (Report, 1950, 1951, 1954, 1955, 1956, 1957, 1958, 1959, 1960, 1961, 1962, 1963, 1964).

[b] Percentages in parentheses.

[c] Stuffed meat, brawn, potted meat, meat roll, meat in gelatin, etc.

products were responsible for 36% of outbreaks, poultry for 13%, and beef and veal 11%. Mutton was not incriminated as a vehicle. This is in keeping with the known absence of salmonella infection with serotypes other than host-specific types in sheep. Cold meats, sausages, processed meats, and meat pies were responsible for more than 80% of outbreaks.

3. Human Salmonellosis

Tables XX and XXI present the serotypes isolated from human incidents (outbreaks, family outbreaks, and sporadic cases) in England and Wales in the five years 1960–1964.

During this period 168 serotypes were identified, of which 96 were isolated in two or more years, 72 in only one year in the period. Thirty-eight serotypes were isolated consistently in every year.

A few serotypes, however, were responsible for most of the incidents: *Salmonella typhimurium* was isolated from 10,896 out of 16,171 incidents (67%); *S. brandenburg, S. bredeney, S. enteritidis, S. heidelberg, S. menston, S. newport, S. saint-paul, S. stanley,* and *S. thompson* from 3154 incidents (20%).

Taylor (1965c) discussed salmonellosis in the United Kingdom in the period 1958 to 1962, and considered that the major source of the most common serotypes isolated from man was home-produced domestic animals and poultry.

In Table XX the reduction in human infections with *S. typhimurium, S. menston,* and *S. thompson* may almost certainly be attributed to the introduction of pasteurization for home-produced and imported whole egg mixtures. This effect is shown most strikingly in the fall in human infection with *S. thompson,* which since 1923 has consistently been one of the ten most common serotypes isolated from human infection in England and Wales.

III. THE ARIZONA GROUP

The Arizona group of organisms is closely allied both in biochemical reactions and in antigenic structure to the salmonellae and is defined in the report of the Enterobacteriaceae Subcommittee (1958). Kauffmann (1966) divided the salmonellae into subgenus I or typical *Salmonella,* subgenus II or atypical *Salmonella,* subgenus III or *Arizona (Salmonella arizonae),* and subgenus IV, atypical *Salmonellae.*

The Arizona group has been studied very fully in the United States by Edwards. In a review by Edwards *et al.* (1959), an account was given of strains isolated from man and the type of symptoms caused. Some outbreaks of food poisoning were discussed, and clearly the evidence points to some of these as being caused by Arizona. Earlier outbreaks implicating these organisms must be viewed with some caution, as the importance of *Cl. welchii* was recognized more recently. Verder (1961) gives an account of an outbreak due to Arizona in which 51 nurses be-

TABLE XX

SALMONELLA SEROTYPES ISOLATED FROM HUMAN INFECTIONS IN MORE THAN ONE YEAR—ENGLAND AND WALES, 1960–1964[a]

Serotype	1960	1961	1962	1963	1964
typhimurium	2943	2544	1864	1820	1725
adelaide	2	3	1	2	18
agama	2	9	2	5	3
anatum	17	31	19	17	58
bareilly	18	14	8	7	5
blockley	9	7	4	12	19
bovis morbificans	30	16	11	22	18
brandenburg	39	53	7	90	312
bredeney	3	14	103	33	40
chester	4	6	3	1	26
cholerae-suis	10	4	4	19	10
derby	13	8	11	11	14
dublin	27	24	10	7	23
enteritidis	145	90	93	64	122
give	9	10	8	10	50
havana	1	1	1	1	3
heidelberg	118	289	133	150	109
infantis	28	2	12	19	16
kiambu	8	1	3	14	8
livingstone	4	2	5	1	15
london	1	4	1	3	3
manhattan	2	44	7	7	3
meleagridis	19	56	9	4	2
menston	25	45	33	38	3
montevideo	12	19	11	12	16
muenchen	7	3	2	17	18

newington	1	6	1	5	3
newport	54	61	64	62	28
oranienburg	8	1	5	4	39
panama	9	7	8	12	39
reading	16	8	31	8	27
saint-paul	42	69	32	20	38
schwarzengrund	8	3	2	6	6
senftenberg	11	8	4	25	19
stanley	68	42	50	65	19
tennessee	9	1	10	28	37
thompson	122	54	39	41	20
virchow	7	5	2	25	25
glostrup	—	2	4	15	8
irumu	2	1	2	—	1
johannesburg	1	—	4	2	2
minnesota	2	1	—	3	1
poona	7	—	3	2	—
worthington	3	4	3	4	9
abony	—	5	2	—	2
bahati	—	1	—	1	2
bispebjerg	2	—	1	—	1
bradford	1	—	1	—	2
braenderup	—	2	1	—	5
butantan	5	—	—	1	1
cubana	—	3	—	3	5
durham	—	—	4	1	1
haifa	—	—	1	2	5
kottbus	6	—	2	—	1
lexington	7	6	1	—	—
manchester	5	2	1	—	—

TABLE XX (continued)

SALMONELLA SEROTYPES ISOLATED FROM HUMAN INFECTIONS
IN MORE THAN ONE YEAR—ENGLAND AND WALES, 1960–1964[a]

Serotype	1960	1961	1962	1963	1964
mikawasima	2	2	—	—	1
nagoya	—	3	5	3	—
onderstepoort	1	—	1	4	—
potsdam	4	—	3	—	1
pullorum	1	—	1	—	1
richmond	1	2	—	—	2
san diego	—	2	—	3	3
stanleyville	2	2	—	—	6
taksony	1	1	—	3	—
vejle	—	5	1	—	1
wangata	—	6	1	2	—
waycross	1	2	1	—	—
weltevreden	2	2	2	—	—
westerstede	—	—	2	4	2
aberdeen	—	3	—	—	2
albany	—	—	—	1	1
bonariensis	—	—	1	—	1
coeln	1	6	—	—	—
drypool	—	—	1	2	—
ealing	3	1	—	—	—
eastbourne	—	—	1	17	—
emek	—	—	—	4	3
essen	1	—	—	3	—
hartford	—	—	—	1	1

hvittingfoss	1	—	1	—	—
ibadan	1	—	—	—	4
indiana	—	—	—·—	1	2
kaapstad	—	1	—	4	—
kentucky	—	6	—	—	2
larochelle	—	—	—	2	2
liverpool	—	—	1	—	1
mendoza	—	—	1	—	1
mission	—	—	1	1	—
norwich	—	1	—	—	—
onireke	1	1	—	—	1
orion	1	—	2	—	—
rubislaw	—	—	1	1	—
ruiru	—	—	—	2	—
singapore	—	—	—	3	7
takoradi	1	3	—	—	—

ª Reports on Food Poisoning, England and Wales, 1960–1964.

TABLE XXI

SALMONELLA SEROTYPES ISOLATED FROM HUMAN INFECTIONS
IN ONLY ONE YEAR—ENGLAND AND WALES, 1960–1964[a]

abadina (2)	california (1)	godesberg (1)	kunduchi (1)	ndola (1)	simsbury (1)
agodi (1)	canastel (1)	hadar (4)	lethe (1)	nima (1)	sofia (1)
amager (1)	chailey (3)	hindmarsh (1)	limete (1)	new brunswick (1)	stourbridge (3)
ank (1)	chicago (2)	huvudista (3)	lindenburg (1)	nyborg (1)	teshie (1)
arachavaleta (1)	christiansborg (1)	idikan (1)	lisboa (2)	ohio (2)	treforest (1)
bergen (1)	coleypark (1)	jaffna (1)	litchfield (2)	oslo (6)	tuebingen (2)
berta (1)	colindale (1)	javiana (1)	mapo (3)	othmarschen (1)	uppsala (1)
binza (2)	concord (1)	jedburg (1)	mons (1)	perth (1)	victoria (1)
bleadon (1)	durban (1)	kennedy (1)	monschaui (1)	pomona (2)	vitkin (1)
blomfontein (1)	gaminara (2)	liel (1)	muenster (2)	saarbrucken (1)	vom (2)
bonn (1)	gatow (2)	kinshasa (1)	nashua (1)	salford (1)	wagenia (1)
brancaster (1)	georgia (1)	kotte (2)	nchanga (1)	seegefeld (2)	wandsworth (1)

[a] Number of incidents in parentheses.

came acutely ill within 12 to 24 hours of eating custard containing the same organism. Serum samples were taken from some of those affected during the acute stage of the illness, then after lapses of 6 weeks, 14 weeks, and 26 weeks. All patients from whose feces Arizona was isolated showed a rise in titer to the infecting organism, followed by a fall, which is strong evidence of the etiology in this outbreak.

Many attempts have been made to isolate Arizona from sporadic cases and outbreaks of food poisoning in the United Kingdom, but so far none has been isolated. One must conclude that Arizona food poisoning exists, that it is probably more prevalent in the United States than in other parts of the world, but that nowhere is it as common as salmonellae food poisoning.

The Arizona group has been responsible for a number of outbreaks in poultry, particularly turkeys and hens, in the United States where the death rate may be high in some incidents. This group has also been frequently isolated from reptiles, which are regarded as one of the reservoirs of infection.

IV. THE CONTROL OF SALMONELLOSIS

Poultry, cattle, and pigs form the permanent reservoirs of which human food-borne salmonellosis is derived. In general, host-specific serotypes of fowl and animals play little part in human infection. In the United Kingdom, however, host-associated serotypes of fowl and cattle are responsible for the majority of human infections. In pigs, no host association with particular serotypes has yet been demonstrated. Animal foods have been shown to play no part in the maintenance of fowl and cattle salmonellosis due to host-specific and host-associated types.

The elimination of epidemic typhoid fever from indigenous sources in England and Wales was accomplished simply by the protection of potable waters against pollution with human excrement during collection and distribution. This elimination, it should be noted, took place before the introduction of routine sterilization of potable waters prior to distribution to the consumer. At no time has any attempt been made to eliminate the reservoir of human chronic carriers of the typhoid fever. Indeed, no certain methods exist for the cure of chronic carriers. For this reason, and for other reasons given earlier, we believe that the most important method for the control of the spread of infection among animals lies in simple hygienic measures designed to reduce the herding together of susceptible animals and their contact with the excreta of infected animals. These measures apply equally to markets, transport, dealers' premises, and rearing farms. Sterilization of animal food can be confidently expected to produce no reduction in the incidence of animal salmonellosis due to host-associated serotypes, the major cause of human infection, or host-specific types.

Control of the transfer of contamination from the intestine of symptomless excretor animals to clean carcasses during slaughter may again be limited by simple hygienic measures at the abattoir designed to limit contact with excreta. These include limitation of the time spent in lairages at the abattoir before slaughter, the daily cleaning of lairages, the provision of clean water for animals awaiting slaughter, the general introduction of the line system of dressing for carcasses, the sterilization of implements, and the abolition of wiping cloths.

In the United Kingdom liquid egg served as the vehicle for some human salmonella infections from fowl; routine pasteurization of all liquid egg before distribution has now been introduced. Imported dried egg products still remain heavily contaminated with salmonellae. The general introduction of pasteurization to egg products before drying would eliminate this source of human infection. In general there is no single measure which, applied to one link in the chain of infection, will eliminate human food-borne salmonella infection. The greatest reduction in human infection is expected to follow the application of proven simple hygienic principles at all stages from the birth of the food animal to the final appearance of the cooked meal on the plate of the human consumer.

REFERENCES

Albertson, V. E. (1957). *Monog. Ser. Wld. Hlth. Org.* **33**, 263.
Allen, G. R., Combs, G. F., and Petersen, P. (1961). *In* "Future Developments in the Production and Utilization of Fish Meal," Vol. II. Food and Agriculture Organization, Rome.
Anderson, E. S. (1960). *Monthly Bull. Min. Health Lab. Serv.* **19**, 172.
Anderson, E. S. (1962). "Food Poisoning." Royal Society of Health, London.
Anderson, E. S., Galbraith, N. S., and Taylor, C. E. D. (1961). *Lancet* **i**, 854.
Bartel, H. (1938). *Tierarztl. Rsch.* **44**, 669.
Bernstein, A. (1952). *Monthly Bull. Min. Health Lab. Serv.* **11**, 64.
Blaxland, J. D., Sojka, W. J., and Smither, A. M. (1958). *Vet. Record* **70**, 374.
Buxton, A. (1957). "Salmonellosis in Animals." Farnham Royal: Commonwealth Agricultural Bureaux.
Callow, B. R. (1959). *J. Hyg.* **57**, 346.
Camps, F. E. (1947). *Monthly Bull. Min. Health Lab. Serv.* **6**, 89.
Carpenter, K. P. (1963). *Intern. Bull. Bacteriol. Nomen. Tax.* **13**, 139.
Clarenburg, A., and Vink, H. H. (1949). *Rept. 14th Intern. Vet. Congr., London* **2**, 262.
Contango. (1962). "People and Profits." *Sunday Times,* 21/10/62.

Cook, G. T. (1963). Personal communication.

Couper, W. R., Newell, K. W., and Payne, D. J. H. (1956). *Lancet* i, 1057.

Craigie, J., and Yen, C. H. (1938a). *Can. Public Health J.* 29, 448.

Craigie, J., and Yen, C. H. (1938b). *Can. Public Health J.* 29, 484.

Datta, N., Pridie, R. B., and Anderson, E. S. (1960). *J. Hyg.* 58, 229.

Dixon, J. M. S. (1965). *Brit. Med. J.* ii, 1343.

Dixon, J. M. S., and Pooley, F. E. (1961). *Monthly Bull. Min. Health Lab. Serv.* 20, 30.

Dixon, J. M. S., and Pooley, F. E. (1962). *Monthly Bull. Min. Health Lab. Serv.* 21, 138.

Edwards, P. R. (1958). *Ann. N.Y. Acad. Sci.* 70, 598.

Edwards, P. R., and Ewing, W. H. (1962). "The Identification of Enterobacteriaceae," 2nd ed. Burgess, Minneapolis.

Edwards, P. R., Bruner, D. W., and Moran, A. B. (1948). *Kentucky Agr. Expt. Sta. Bull.* 525.

Edwards, P. R., Fife, M. A., and Ramsey, C. H. (1959). *Bacteriol. Rev.* 23, 155.

Enterobacteriaceae Subcommittee Report (1958). *Intern. Bull. Bacteriol. Nomen. Tax.* 8, 25.

Felix, A. (1955). *Bull. World Health Organ.* 13, 109.

Felix, A., and Callow, B. R. (1943). *Brit. Med. J.* ii, 127.

Felix, A., and Callow, B. R. (1951). *Lancet* ii, 10.

Field, H. I. (1948). *Vet. J.* 104, 251, 294, 323.

Fry, R. M. (1951). Personal communication.

Fussell, G. E. (1962). *Nature* 195, 750.

Galbraith, N. S. (1961). *Vet. Record* 73, 1296.

Galbraith, N. S., Hobbs, B. C., Smith, M. E., and Tomlinson, A. J. H. (1960). *Monthly Bull. Min. Health Lab. Serv.* 19, 99.

Galbraith, N. S., Archer, J. F., and Tee, G. H. (1961). *J. Hyg.* 59, 133.

Galbraith, N. S., Manson, K. N., Maton, G. E., and Stone, D. W. (1962a). *Monthly Bull. Min. Health Lab. Serv.* 21, 209.

Galbraith, N. S., Taylor, C. E. D., Cavanagh, P., Hagan, J. G., and Patton, J. L. (1962b). *Lancet* i, 372.

Galto, P. (1939). *Rev. Med. Vet. Parasito, Caracas* 1, 34.

Galton, M. M., Smith, W. V., McElrath, H. B., and Hardy, A. B. (1954). *J. Infect. Diseases* 95, 236.

Galton, M. M., Harless, M., and Hardy, A. V. (1955a). *J. Am. Vet. Med. Assoc.* 126, 57.

Galton, M. M., Mackell, D. C., Lewis, A. L., Haire, W. C., and Hardy, A. V. (1955b). *Am. J. Vet. Res.* 16, 132.

Gärtner, A. H. (1888). *KorrespBl Ärztl. Ver. Thüringen* 17, 573.

Gay, F. P. (1918). "Typhoid Fever." Macmillan, New York.

George, T. C. R., Harvey, R. W. S., and Thomson, S. (1953). *J. Hyg.* 51, 532.

Gibson, E. A. (1958). Ph.D. Thesis. University of London.

Gibson, E. A. (1961). *Vet. Record* 73, 1284.

Gibson, E. A. (1965). *J. Dairy Res.* 32, 97.

Gitter, M. (1959). *Vet. Record* 71, 234.

Gordon, R. F., and Buxton, A. (1946). *Vet. J.* 102, 187.

Gray, D. F., Lewis, P. F., and Gorrie, J. R. (1958). *Australian Vet. J.* 34, 345.

Greenberg, B. (1964). *Am. J. Hyg.* 80, 149.

Grunsell, C. S., and Osborne, A. D. (1948). *Vet. Record* 60, 85.

Harmsen, H. (1953). *Stadtehygiene* 4, 48.

Harmsen, H. (1954). *Stadtehygiene* 5, 54.

Harvey, R. W. S.., and Price, T. H. (1962). *Monthly Bull. Min. Health Lab. Serv.* **21**, 54.
Hauge, S., and Borre, K. (1958). *Nord. Veterinarmed.* **10**, 255.
Henning, M. W. (1939). *Onderstepoort J. Vet. Res.* **13**, 79.
Hinshaw, W. R., Upp, C. W., and Moore, J. M. (1926). *J. Am. Vet. Med. Assoc.* **68**, 631.
Hinshaw, W. R., McNeil, E., and Taylor, T. J. (1944). *Am. J. Hyg.* **40**, 264.
Hirsch, W., and Sapiro-Hirsch, R. (1958). *J. Med. Assoc. Israel* **3**, 59.
Hobbs, Betty C. (1965). *Monthly Bull. Min. Health Lab. Serv.* **24**, 123.
Hobbs, Betty C., and Smith, Muriel E. (1954). *J. Hyg.* **52**, 230.
Hobbs, Betty C., and Wilson, J. G. (1959). *Monthly Bull. Min. Health Lab. Serv.* **18**, 198.
Hormaeche, E., and Salsamendi, R. (1936). *Arch. Urug. Med.* **9**, 665.
Hormaeche, E., and Salsamendi, R. (1939). *Arch. Urug. Med.* **14**, 375.
Intelligence Bulletin (1964*a*). **16**, No. 12, p. 22. Commonwealth Economic Committee, London.
Intelligence Bulletin (1964*b*). **16**, No. 12, p. 25. Commonwealth Economic Committee, London.
Ingelligence Bulletin (1965*a*). **17**, No. 9, p. 45. Commonwealth Economic Committee, London.
Intelligence Bulletin (1965*b*). **18**, No. 2, p. 46. Commonwealth Economic Committee, London.
Kauffmann, F. (1966). "The Bacteriology of Enterobacteriaceae" Munksgaard, Copenhagen.
Knox, W. A., Galbraith, N. S., Lewis, M. J., Hickie, G. C., and Johnston, H. H. (1963). *J. Hyg.* **61**, 175.
Kovacs, N. (1959). *Med. J. Australia* **I**, 557.
Kwantes, W. (1952). *Monthly Bull. Min. Health Lab. Serv.* **11**, 239.
Lennox, M., Harvey, R. W. S., and Thomson, S. (1954). *J. Hyg.* **52**, 311.
Levine, N. D., Peterson, E. H., and Graham, R. (1945). *Am. J. Vet. Res.* **6**, 242.
Little, L. A. (1964). Personal communication.
Ludlam, G. B. (1954). *Monthly Bull. Min. Health Lab. Serv.* **13**, 196.
Lundbeck, H., Plazikowski, V., and Silverstolpe, L. (1955). *J. Appl. Bacteriol.* **18**, 535.
Lutje, F. (1938). *Deut. Tieraerztl. Wochschr.* **46**, 798.
McCoy, J. H. (1961–1963). Unpublished data.
McCoy, J. H. (1962). *Intern. J. Air Water Pollution* **7**, 597.
McCoy, J. H. (1963). *J. Roy. Soc. Health* **83**, 154.
McCoy, J. H. (1965). *Sanitarian* **74**, 79.
McCullough, N. B., and Eisele, C. W. (1951*a*). *J. Infect. Diseases* **88**, 278.
McCullough, N. B., and Eisele, C. W. (1951*b*). *J. Infect. Diseases* **89**, 209.
McCullough, N. B., and Eisele, C. W. (1951*c*). *J. Infect. Diseases* **89**, 259.
McDonagh, V. P., and Smith, J. G. (1958). *J. Hyg.* **56**, 271.
McElrath, H. B., Galton, M. M., and Hardy, A. V. (1952). *J. Infect. Diseases* **91**, 12.
Meseck, G. (1961). *In* "Future Developments in the Production and Utilization of Fish Meal," Vol. II. Food and Agriculture Organization, Rome.
Moore, B. (1957). *J. Hyg.* **55**, 414.
Mullaney, P. E. (1949). *Irish Vet. J.* **3**, 386.
Muller, J. (1954). *Medlemebl. Danske Dyrlaegeforen.* **37**, 73.
Muller, J. (1956). *Medlemebl. Danske Dyrlaegeforen.* **39**, 489.
Murdock, C. R. (1954). *Monthly Bull. Min. Health Lab. Serv.* **13**, 43.
Murdock, C. R., and Gordon, W. A. M. (1953). *Monthly Bull. Min. Health Lab. Serv.* **12**, 72.

Newell, K. W. (1955a). *J. Appl. Bacteriol.* **18**, 462.

Newell, K. W. (1955b). *Monthly Bull. Min. Health Lab. Serv.* **14**, 146.

Newell, K. W., Hobbs, B. C., and Wallace, E. J. G. (1955). *Brit. Med. J.* **ii**, 1296.

Newell, K. W., McClarin, R., Murdock, C. R., MacDonald, W. N., and Hutchinson, H. L. (1959). *J. Hyg.* **57**, 92.

de Nobele, J. (1898). *Ann. Soc. Med. Gand.* **77**, 281.

Penha, A. M., and D'Apice, M. (1946). *Arch. Inst. Biol. S. Paulo* **17**, 239.

Petersen, P. (1961). *In* "Future Developments in the Production and Utilization of Fish Meal," Vol. II. Food and Agriculture Organization, Rome.

Pritulin, P. I. (1959). *Veterinariya* **36**, No. 9, 26.

Quiroga, S. S., and Monteverde, J. J. (1945). *Rev. Med. Vet. B. Aires* **27**, 226.

Report (1934). *J. Hyg.* **34**, 333.

Report (1947). *Spec. Rept. Ser. Med. Res. Council London* **260**.

Report (1950). *Monthly Bull. Min. Health Lab. Serv.* **9**, 254.

Report (1951). *Monthly Bull. Min. Health Lab. Serv.* **10**, 228.

Report (1954a). *Monthly Bull. Min. Health Lab. Serv.* **13**, 38.

Report (1954b). *Monthly Bull. Min. Health Lab. Serv.* **13**, 12.

Report (1955). *Monthly Bull. Min. Health Lab. Serv.* **14**, 34, 203.

Report (1956). *Monthly Bull. Min. Health Lab. Serv.* **15**, 263.

Report (1957). *Monthly Bull. Min. Health Lab. Serv.* **16**, 233.

Report (1958a). *Monthly Bull. Min. Health Lab. Serv.* **17**, 133.

Report (1958b). *Monthly Bull. Min. Health Lab. Serv.* **17**, 252.

Report (1959). *J. Hyg.* **57**, 435.

Report (1959a). *Monthly Bull. Min. Health Lab. Serv.* **18**, 26.

Report (1959b). "Report of the Committee of Inquiry on Anthrax." Her Majesty's Stationery Office, London.

Report (1959c). *Monthly Bull. Min. Health Lab. Serv.* **18**, 169.

Report (1960). *Monthly Bull. Min. Health Lab. Serv.* **19**, 224.

Report (1961a). *Monthly Bull. Min. Health Lab. Serv.* **20**, 73.

Report (1961b). *Monthly Bull. Min. Health Lab. Serv.* **20**, 160.

Report (1962). *Monthly Bull. Min. Health Lab. Serv.* **21**, 180.

Report (1963). *Monthly Bull. Min. Health Lab. Serv.* **22**, 200.

Report (1964a). "The Aberdeen Typhoid Outbreak, 1964." Her Majesty's Stationery Office, London.

Report (1964b). *Vet Record* **76**, 1139.

Report (1964c). *J. Hyg.* **62**, 283.

Report (1964d). *Monthly Bull. Min. Health Lab. Serv.* **23**, 189.

Report (1965). *J. Hyg.* **63**, 223.

Rhode, R., and Bischoff, J. (1956). *Zentr. Bakteriol. Abt. 1* (Ref.) **159**, 145.

Rittger, L. F., and Stonebrun, F. H. (1909). *Bull. Storrs Agr. Expt. Sta.* **60**, 29.

Salisbury, R. M. (1958). *New Zealand Vet. J.* **6**, 76.

Simmons, G. C., and Sutherland, A. K. (1950). *Australian Vet. J.* **26**, 57.

Smith, H. Williams (1960). *J. Hyg.* **58**, 381.

Smith, H. Williams (1961). *J. Appl. Bacteriol.* **24**, 235.

Smith, H. Williams, and Crabb, W. E. (1961). *J. Pathol. Bacteriol.* **82**, 53.

Sojka, W. J., and Gitter, M. (1961). *Vet. Rev. Annot.* **7**, 11.

Solowey, M., McFarlane, V. H., Spaulding, E. H., and Chemerda, C. (1947). *Am. J. Public Health* **37**, 971.

Strozzi, P. (1934). *Clin. Vet. Milano* **57**, 337.

Swahn, C., and Rutquist, L. (1957). *Medlemsbl. Sverig. Vet. Forb.* **9,** 377.

Taylor, J. (1960). *Bull. World Health Organ.* **23,** 763.

Taylor, J. (1965a). *Proc. Roy. Soc. Med.* **58,** 167.

Taylor, J. (1965b). *Practitioner* **195,** 12.

Taylor, J. (1965c). *Proc. Natl. Conf. Salmonellosis, Mar. 11–13, 1964.* U.S. Department of Health, Education, and Welfare, pp. 18.

Thomas, W. E., Stephens, T. H., King, G. J. G., and Thomson, S. (1948). *Lancet* **ii,** 270.

Thomson, S. (1953). *Monthly Bull. Min. Health Lab. Serv.* **12,** 187.

Thomson, S. (1954). *J. Hyg.* **52,** 67.

Thomson, S. (1955). *J. Hyg.* **53,** 217.

Varela, G., and Zozaya, J. (1941). *Rev. Inst. Salutr. Enfermedad. Trop.* **2,** No. 3–4, pp. 318 (abstracted in *Bull. Hyg.* [1942], **17,** 721).

Verder, E. (1961). *J. Food Sci.* **26,** 618.

Walker, J. H. C. (1957). *Lancet* **ii,** 283.

Weaver, W. (1952). The Treatment of Animal Wastes. The Institute of Public Cleansing, London.

Wilson, M. M., and MacKenzie, E. F. (1955). *J. Appl. Bacteriol.* **18,** 510.

Withers, F. W. (1952). *Brit. Vet. J.* **108,** 315, 382.

Withers, F. W. (1953). *Brit. Vet. J.* **109,** 65, 122.

Wormald, P. J., and Greasby, A. S. (1965). *Monthly Bull. Min. Health Lab. Serv.* **24,** 58.

Zakula, R. I. (1956). *Acta Vet. Belgrade* **6,** 90.

CHAPTER II | VIRAL INFECTIONS

Dean O. Cliver

I. INTRODUCTION

The present chapter is evidently the first on viruses to appear in a book on food-borne diseases. It is intended to serve three purposes: to summarize what is already known of virus transmission in foods, to

73

collect applicable information from related work, and to indicate what has been learned concerning detection methods for viruses in foods.

At least five reviews concerned with viruses in foods have appeared to date. Brown (1949) has reviewed transmission of viruses in milk and water and concluded that only infectious hepatitis is conveyed to a significant extent in these media. Berg (1964) suggests that a variety of viruses, some of which are of nonhuman origin, may occur in foods and that the potential of these agents for producing infections in man is in need of additional study. Lemon (1964) has suggested that the agents of avian leukosis and bovine lymphosarcoma may be present in eggs and milk, respectively, and that these may be a threat to the farmer and, to a lesser extent, to the consumer. Cliver (1966, 1967c) has reviewed the epidemiology of infectious hepatitis and of viruses in general, as transmitted by foods. Enough food-associated outbreaks of some virus diseases are now on record to suggest that analogous transmission of other viruses in foods occurs.

Although virus transmission in foods may well be nothing new, it is possible that recent developments in food technology have served to accentuate the problem. Certainly great progress has been made in recent years in the area of food sanitation, but Goresline (1963) has described how innovations in food processing methods may cause established sanitation procedures to be inadequate. The increased use of refrigeration and freezing to extend the commercial distribution of uncooked foods may preserve viruses while suppressing bacterial growth. Centralization of food production, together with more efficient and widespread distribution, may serve to transport viruses from an area where they are endemic and relatively innocuous to populations where they are exotic and therefore highly pathogenic. Finally, the trend to production of "convenience foods," which leave little preparation and cooking to be done by the consumer, provides additional opportunities for virus to contaminate foods and not be inactivated prior to ingestion.

II. VIRUSES

The animal viruses have a number of unique properties which warrant their consideration apart from other infectious and toxigenic agents which may occur in foods.

A. Fundamental Properties

The true viruses of animals range in size from approximately 12 mμ to the limit of microscopic visibility, or approximately 200 to 300 mμ in diameter, and are simply constructed, consisting of two or at most three classes of chemical components. The nucleic acid moiety of the virus particle contains the genetic information which is essential to infectivity. A particular virus will be found to contain a single molecule either of deoxyribonucleic acid (DNA) or of ribonucleic acid (RNA), but no true virus has been found to contain both kinds of nucleic acid. The proteins of the known viruses form a coat, around the nucleic acid, which is evidently made up of identical subunits. Some viruses have, in addition to the components just described, an envelope which surrounds the protein coat of the virus and contains an essential structural lipid without which the infectivity of the particle is lost. The true viruses of animals are replicated only within living animal cells.

B. Corollary Properties

Viruses often show a marked selectivity in the species of hosts and the tissues within those species which they are capable of infecting. The functional site of this specificity in the enteroviruses is thought to be the coat protein, since it has been shown that infectious ribonucleic acid prepared from human enteroviruses which normally infect only primate cells is fully infectious for cultured cells of a variety of other warm-blooded species *in vitro* (Holland *et al.,* 1959*a,b*). The mechanism of specificity in the viruses which have lipid envelopes is less well known. A single virus type may infect the majority of humans with no symptoms at all, but other human infections with the same agent may have very dire clinical consequences. The tissues and organs affected may differ markedly from one infection to the next with a single strain of virus (Rose, 1964). In contrast to the array of biochemical tests which are available for the identification of bacteria, final distinctions among virus types rest upon serology, which is relatively expensive and time-consuming.

C. Groups of Animal Viruses

Individual types of viruses which have properties in common have been placed, where possible, within arbitrary "groups." Not all viruses have been assigned to groups, nor will all established groups be listed in this section.

1. ENTEROVIRUSES

The enteroviruses are presently considered to be a subgroup of the picornaviruses, all of which have a diameter of approximately 28 mμ, cubic symmetry, an RNA core, and are ether-resistant (indicating the lack of essential structural lipid). Within the group of picornaviruses, the enteroviruses are distinguished from the rhinoviruses by the ability of the former to withstand exposure to pH 3. The rhinoviruses will not be discussed further. In addition to the properties cited above, cationic stabilization, or ability of these agents to retain infectivity after incubation for 60 minutes at 50°C in molar magnesium chloride solution, is considered to be a common attribute (Committee on Enteroviruses, 1962). Enteroviruses of humans are presently divided into four sets. The polioviruses have been defined upon the basis of their ability to produce clinical poliomyelitis, although members of each of the remaining sets have at times been found to cause central nervous system infections with residual paralysis. The Coxsackie A and B sets are distinguished by their pathogenicity for suckling mice, and from each other by the distinctive symptoms they induce in these animals. The ECHO viruses are distinguished from the other three sets on a deductive basis.

2. REOVIRUSES

The reovirus group was defined by Sabin (1959), its members having previously been included among the enteroviruses. It is made up of ether-resistant RNA viruses whose diameter is approximately 72 mμ.

3. ARBOVIRUSES

The definitive properties of the arthropod-borne, or arbovirus, group are essential structural lipid, as demonstrated by high-ether and deoxycholate lability, and an RNA core. Most of these agents have diameters in the range of 20 to 30 mμ, but some have been reported to be as large as 60 to 180 mμ (Hammon and Work, 1964).

4. MYXOVIRUSES

The myxoviruses have the definitive common properties of a helical ribonucleoprotein core surrounded by an envelope containing essential structural lipid (Hirst, 1965). The influenza viruses are apparently the smallest of the group, having an average diameter of approximately 100 mμ. Subtypes within the influenza A type have been identified in humans; and additional type A viruses infect swine, horses, and poultry. Other members of the myxovirus group are somewhat larger and include the agents of mumps, Newcastle disease of poultry, measles, and a heterogeneous group of parainfluenza viruses.

5. ADENOVIRUSES

The properties of the adenovirus group have been reviewed by Ginsberg (1962) and are said to include a DNA core, a diameter of 60 to 80 mμ, icosahedral shape, and ether resistance.

6. PAPOVAVIRUSES

These agents form a group of ether-resistant DNA viruses whose diameter is approximately 50 mμ. Infections with agents of this group are associated with neoplastic growth in a number of host species (Melnick, 1965).

D. The Nonviruses: Chlamydia and Rickettsiae

The obvious properties which distinguish the chlamydia and rickettsiae from the true viruses are their relatively large size, the fact that they contain both DNA and RNA, and the presence of structures resembling the cell walls of bacteria. In addition, members of each of these groups multiply by binary fission and appear to be capable of carrying out a limited range of biochemical processes in the absence of living host cells. The extracellular form of the chlamydia has a diameter of approximately 0.3 μ. The group includes agents of ornithosis of birds and pneumonia of man, and Koprowski (1958) cites chlamydial agents of diarrhea in cattle and of enzootic abortion in sheep among counterparts of human virus diseases occurring in animals. The rickettsiae are short rods measuring about 0.3 $\times$ 1 μ. *Coxiella burnetii* is the rickettsia of Q fever of man and domestic animals.

E. Epidemiology of Viruses

1. MODES OF TRANSMISSION

Viruses are thought typically to be transmitted by direct contact. However, these agents may be transmitted indirectly if they are sufficiently stable when outside of host tissues. An important limitation upon the potential for indirect transmission of an agent relates to the manner in which it is shed by infected individuals. No virus is shed in pure form, but always in suspension in some product of the host's body such as sputum or feces. Arboviruses and most of the rickettsiae may not, in the strict sense, be shed at all. Members of both of these groups of agents have been shown to be strongly dependent upon the action of arthropod vectors for natural transmission; but they are very hazardous to handle in the laboratory, in the absence of the natural arthropod vectors (Hammon and

Work, 1964; Fox, 1964). It might therefore be surmised that the essential function of the arthropod vector is not the inoculation of new hosts but the provision of a mode of egress for the agent from its previous host. The portal of entry of a virus is determined by its stability, as well as by its primary tissue tropism. A virus whose tissue specificity renders it quite capable of infecting the tissues of the intestines is limited in its ability to do so if it is incapable of withstanding the rigors of passage down the digestive tract. Viral agents capable of infecting by the oral route would appear to be of greatest concern in the context of transmission in foods, but only under experimental conditions is it possible to introduce a virus into the digestive tract without some contamination of the upper respiratory tract. Further, contaminated foods may be handled in such a way as to cause the generation of infectious aerosols with consequent infection by the respiratory route. Until more extensive evidence has been accumulated, it would appear that the significance in foods of any viral or rickettsial agent known to be infectious for man should not be minimized.

2. HERD IMMUNITY PHENOMENA

The immune status of the population as a whole, or herd immunity, influences the ability of a virus to establish itself in a community. A highly contagious viral agent may persist within a circumscribed host population only for relatively brief periods of time because it virtually exhausts the population of susceptible individuals. When an agent is endemic or enzootic in a defined host community, it may approach ideal parasitism in that it infects a large proportion of the population without causing high mortality or even recognizable clinical symptoms. As has been shown in studies of isolated populations, outbreaks of clinical disease may be induced even by viruses of limited pathogenic potential when introduced into an inexperienced host community.

3. THE FOOD VEHICLE

Food has already proved an extremely efficient vehicle in the transmission of infectious hepatitis virus, and there is reason to believe that it may serve in the same manner with other agents as well. Foods may provide an unusually hospitable environment for virus both outside the body and while traversing the gastrointestinal tract, and thus enhance the potential of labile agents for indirect transmission. Even agents incapable of withstanding cooking may be of significance if by transmission in foods they reach persons preparing foods or if cooked foods are contaminated while being served.

III. VIRUSES ASSOCIATED WITH FOODS

A. Agents Primarily Infectious for Humans

1. INFECTIOUS HEPATITIS VIRUS

Infectious hepatitis predominates among reported outbreaks of food-associated viral diseases. It is clinically distinctive to a degree which usually permits differential diagnosis without isolation of the agent. Cliver (1966) has reviewed 22 food-associated outbreaks of infectious hepatitis, in 7 of which the vehicle of transmission was evidently raw shellfish. A more recent review (Cliver, 1967c) brings the total of recorded outbreaks of food-borne hepatitis to at least 36. Vehicles implicated have included shellfish (10 outbreaks), milk and dairy products (4 outbreaks), and a great variety of other foods. There is no indication that the hepatitis virus can withstand thorough cooking: cooked products could usually be shown to have been handled by an infected individual just prior to serving. Food handlers implicated as sources of food contamination have frequently been shown to have worked while ill and, in some instances, jaundiced. The source of virus in the shellfish-associated outbreaks has been shown or assumed to be polluted water.

Two factors make unlikely the isolation of infectious hepatitis virus from foods. The first is the restricted host range of the agent. At present, no laboratory host is available for this virus. Work in tissue cultures appears promising, but no system yet developed has yielded readily reproducible results when tried in other laboratories (Kissling, 1967). The second factor is the extremely long incubation period of the disease, which averages 30 days with a range of 15 to 50 days. This makes it difficult to obtain accurate food histories from persons involved in outbreaks, and in many cases food samples will not be available by the time an outbreak is detected.

2. POLIOMYELITIS VIRUS

Food-associated outbreaks of poliomyelitis have also been reported. The declining incidence of this agent in foods and in the population as a whole makes it seem at this time that the principle significance of these outbreaks is what they illustrate of how other enteroviruses may be spread in foods.

Transmission of poliovirus in raw milk was suggested by Jubb (1915) in a report of a possible common source outbreak involving 4 cases in children. Dingman (1916) described an outbreak of 6 cases within a 5-day period among children at three vacation boarding houses in Spring Valley, New York. All of the affected children drank raw milk from an unsanitary dairy farm at which a case of poliomyelitis had occurred in a child 14 days previously. Children who drank the same milk after boiling or milk from other sources showed no symptoms. Knapp *et al.* (1926) reported 8 cases in an 11-day period in children in Cortland, New York. All were thought to have consumed raw milk from one dealer, and a 16-year-old milker had been stricken with poliomyelitis 7 days previously on one of the farms from which this dealer obtained his milk. Sixty-two cases with onsets in a 16-day period in Broadstairs, England, were discussed by Aycock (1927) Most of the patients were students at local boarding schools and had consumed raw milk from a single dealer. Goldstein *et al.* (1946) have reported a rather explosive outbreak of poliomyelitis at a west coast naval training station. There were 18 confirmed cases and possibly 100 milder ones which had onsets within an 8-day period. Tentatively incriminated was unpasteurized milk supposedly contaminated by flies from human or animal feces. Mathews (1949) has reported an outbreak of 11 cases of clinical poliomyelitis among 1400 persons at a naval station in Portland, Oregon. The presumed vehicle of transmission was pasteurized milk, but in this case the evidence available was a good deal less definitive. Hargreaves (1949) has described 2 possibly food-associated outbreaks which occurred in Cornwall, England. The first included 4 cases of poliomyelitis and 27 cases of "24-hour flu," principally in children, which were compatible with a common source of raw milk from one dealer in the village of North Tamerton. The other comprised 23 cases with onsets in regular cycles over a 5-month period, primarily among workers and customers of a bakery in St. Austell. Cream-filled pastries contaminated in preparation were considered to be among the possible vehicles. A poliomyelitis outbreak of 22 cases associated with pasteurized milk bottled under unsanitary conditions was reported by Lipari (1951), and one in which the presumed vehicle was lemonade has been reported by Piszczek *et al.* (1941).

The frequent association of poliomyelitis with milk is rather hard to explain. Koprowski (1958) has reported the isolation of poliomyelitis virus from a calf, but no similar reports have appeared. Mitchell *et al.* (1958) attempted without success to infect the mammary glands of cows with poliovirus. Bellelli (1963) has reported an unsuccessful attempt to isolate poliovirus or other human enteroviruses from 100 market samples of butter. The possibility that enteroviruses might be spread in foods in a manner analogous to infectious hepatitis virus has been discussed by Cliver (1966). The similarities in epidemiology of the other enteroviruses to the polioviruses have been pointed out by Melnick *et al.* (1964). Among the food-associated outbreaks of disease tabulated by Dauer (1961) for the years 1952 through 1960, nearly 45% were classified as gastroenteritis of undetermined etiology. Cockburn (1960) has reported that 38% of "food poisoning" outbreaks occurring in England and Wales during the years 1949 through 1958 were of unrecorded etiology. Gastrointestinal symptoms resulting from enterovirus infections of human volunteers as described by Buckland et al. (1959) might under other circumstances have been diagnosed as food poisoning.

B. Agents Potentially Infectious for Humans

1. REOVIRUSES

These agents have an extremely wide host range and may, when isolated from children, be associated with mild upper respiratory infections, diarrhea, steatorrheic enteritis, or no symptoms at all (Macrae, 1962). All three recognized types of reoviruses have been isolated from cattle (Abinanti, 1961). The occurrence of hemagglutination-inhibiting antibodies in the sera of swine is thought to indicate that this species may also be infected. Human strains of all three types of reoviruses have been used to infect calves (Abinanti, 1961), and a strain of reovirus type 1 of bovine origin has been used to infect human volunteers (Kasel et al., 1963).

2. PARAINFLUENZA VIRUSES

Parainfluenza virus type 1 (Sendai) infects swine, and parainfluenza type 3 (SF-4) infects cattle (Abinanti, 1964). The parainfluenza viruses of humans most frequently infect children and are generally associated with mild upper respiratory symptoms and occasional fever, or no symptoms, in first infections (Hilleman, 1962). Chanock and Johnson (1964) have discussed reported human infections with Sendai virus and state that the authenticity of these reports has been questioned. A survey of sera from cattle in nine states by Abinanti (1961) showed that 75% had antibody to parainfluenza type 3. Abinanti (1961) has also discussed the antigenic differences between the bovine (SF-4) and human (HA-1) strains of parainfluenza virus type 3. Although these differences seem to be relatively constant, it has not been demonstrated directly that the SF-4 strain cannot infect humans. When the bovine (SF-4) strain of parainfluenza 3 virus was introduced into a mammary gland of a goat by Cliver and Bohl (unpublished), the agent propagated and titers greater than 10^7 infectious units per milliliter of milk were obtained on the second through the fifth days postinoculation. The possibility that this virus might have localized in the mammary gland following normal respiratory infection was not tested.

3. CHLAMYDIA OF ORNITHOSIS

Delaplane (1958) has described outbreaks of ornithosis in poultry plant workers dressing turkeys and in laboratory workers. Meyer and Eddie (1964) state that most human infections occur by the aerosol route or by contact with infected birds or human beings. They further suggest that other chlamydia infect cattle, sheep, and goats and may cause unrecognized infections in man. Although turkeys are most frequently involved in outbreaks in the United States, cases of ornithosis in humans in Europe have resulted from contact with ducks, geese, chickens, pigeons, and pheasants as well. Most human cases occur in poultry plant and farm workers, but the agent has been known to infect cooks (Meyer, 1964).

4. RICKETTSIA OF Q FEVER

Q fever has a wide and increasing distribution in cattle in the United States (Luoto, 1960) and also infects sheep and goats (Welsh et al., 1958). The usual route of infection for man is respiratory. *Coxiella burnetii* is more resistant to desiccation, heat, and chemical disinfectants than other rickettsiae, and less dependent on insect vectors for its transmission (Stoenner, 1964). The titer of the agent shed in milk usually does not exceed 10^3 infectious units per milliliter and declines in succeeding lactations. The stability of *C. burnetii* in milk has been studied by Enright *et al.* (1957), and the results of these studies were responsible for an increase in the temperature recommended for low-temperature, long-time pasteurization. A study of a penitentiary group using raw milk from a Q fever-infected herd by Benson *et al.* (1963) indicated that 35% of the persons sampled had complement-fixing antibody to the agent and at least 10% showed a diagnostic rise in titer indicative of infection.

5. VIRUSES OF THE TICK-BORNE ENCEPHALITIS COMPLEX

An outbreak of approximately 600 human cases of encephalitis occured in Czechoslovakia in 1951, apparently as a result of a defect in apparatus used to pasteurize infectious goats' and cows' milk; and smaller outbreaks have been observed among families drinking raw milk from infected goats and cattle (H. Libíková, personal communication). Dairy animals in natural foci of these viruses evidently become infected by the bites of carrier ticks (Libíková et al., 1963). Powassan, a virus of the tick-borne encephalitis group, is known to occur in North America; but its incidence in dairy animals apparently has not been studied.

6. OTHER VIRUSES

Shahan and Traum (1958) state that vesicular stomatitis virus, which occurs sporadically in parts of the western hemisphere in cattle and horses, can cause an influenza-like syndrome in man, and that Rift Valley fever virus, which is infectious for cattle and sheep, can cause an influenza-like disease in persons handling the diseased animals or carcasses. The reservoir of the virus of lymphocytic choriomeningitis is thought to be house mice, but naturally infected swine have been reported, and transmission to man by contaminated food has been suggested (Gordon, 1965). Encephalomyocarditis virus infects humans, in whom it causes a disease of the central nervous system, as well as swine (Buescher, 1964). Machupo virus, the agent of Bolivian hemorrhagic fever, causes chronic infections in the rodent *Calomys callosus* and is thought to infect humans as a result of contamination of food, water, or air by virus in rodent urine (Johnson, 1965).

C. Agents Not Primarily Infectious for Humans

1. ANIMAL VIRUSES PRINCIPALLY OF ECONOMIC SIGNIFICANCE

The agents discussed in this section are responsible primarily for animal health problems, some of which are of such significance as to have caused the erection of important barriers to commerce.

a. Foot and Mouth Disease Virus. Infections of cattle, swine, sheep, and goats with the virus of foot and mouth disease may be responsible for relatively high mortality in young animals, as well as losses in milk production and weight and, at times, abortion (Merchant and Packer, 1961). The virus may be spread to swine in infectious milk and uncooked meat scraps and has been shown to be transmitted internationally in meat from infected animals. Survival of the virus in meat products occurs principally in lymphoid tissues. Heidelbaugh and Graves (1968) have evaluated heating, freezing and thawing, salt curing, and proteolytic enzyme treatments as possible means of accelerating the inactivation of this virus in lymph node tissue. Heating to 155°F was found to destroy the virus, but none of the other techniques appeared efficient enough to have commercial value.

b. Rinderpest Virus. The rinderpest virus is a large myxovirus related antigenically to the agents of human measles and canine distemper (Plowright, 1962). Infections occur in sheep, goats, and swine but are most frequent in cattle, in which the disease is severe and frequently fatal. The disease is usually transmitted in animals by relatively direct contact but may be transmitted in infected meat products as well (Merchant and Packer, 1961).

c. Viruses of Hog Cholera and African Swine Fever. The virus of hog cholera causes an acute febrile disease of swine. It is readily spread among swine by direct contact but may be introduced, by feeding infectious pork scraps in uncooked garbage, into areas from which it has been eliminated (Merchant and Packer, 1961). African swine fever virus causes disease in swine which is quite similar to hog cholera. Its transmission in Spain has been attributed to the practice of feeding uncooked garbage (Kamphans, 1963).

d. Picornaviruses of Cattle and Swine. Reports of isolations of bovine enteroviruses prior to 1962 have been reviewed by McFerran (1962). Each of the subgroups of picornaviruses which infect man has its counterpart among the agents which infect cattle or swine. Abinanti (1964) has described rhinoviruses of cattle. Teschen disease of swine in Europe is clinically similar to human poliomyelitis (Mayr, 1962), and an immunologically distinct agent of polioencephalomyelitis has been isolated from swine in the United States (Kasza, 1965). A bovine enterovirus strain which caused death with Coxsackie A-type lesions in suckling mice has been isolated by Kunin and Minuse (1958). On the other hand, bovine enteroviruses isolated by Moll and Finlayson (1957) showed no pathogenesis in suckling or weanling mice and can therefore be considered counterparts of the ECHO viruses.

Definitive studies concerning the relationship of these animal enteroviruses to human enteroviruses are still wanting. Bovine enteroviruses isolated by Moll and Davis (1959) were found to be infectious for human cells, as were three of nine strains isolated by Moscovici *et al.* (1961). One of two strains isolated by Cliver and Bohl (1962) was shown to form plaques on human kidney cells. However, five strains of porcine enteroviruses studied by Bohl *et al.* (1960) did not infect human amnion or HeLa cells. No direct serologic cross reactions among human, bovine, and porcine enteroviruses have yet been reported.

e. Adenoviruses. Adenoviruses infect cattle, swine (Kasza, 1966), and chickens (Sharpless, 1962). Although generally regarded as respiratory viruses, they are most frequently associated with the gastrointestinal tract in each host species (Rowe and Hartley, 1962), and human adenoviruses have been associated with gastrointestinal symptoms in man (Duncan and Hutchinson, 1961). Avian adenoviruses have no antigens in common with human adenoviruses (Sharpless, 1962). Bovine adenoviruses share the group complement-fixing antigen with human adenoviruses but do not infect human cells *in vitro* (Klein *et al.,* 1959, 1960).

Two types of bovine adenoviruses studied by Klein (1962) are neutralized by many individual adult human sera but not by infant sera nor by standard typing sera. Human adenoviruses have been used to infect pigs under experimental conditions (Jennings and Betts, 1962). Although there is no evidence that adenoviruses are capable of crossing host species, the possibility has not been excluded (Rose, 1964).

f. Newcastle Disease Virus. The myxovirus of Newcastle disease in chickens and other poultry is efficiently transmitted directly among chickens by the air-borne route but is also found in eggs laid early in infection, in viscera (Merchant and Packer, 1961), and in muscle tissue (Hanson and Brandly, 1958). It occasionally infects humans, most frequently eviscerating plant and laboratory workers, in whom it has caused conjunctivitis and occasionally systemic infections. Human infections as a result of eating or handling dressed poultry have not been reported (Hanson and Brandly, 1958).

2. ONCOGENIC VIRUSES OF UNCERTAIN HOST RANGE

Oncogenic viruses of animals may be of significance to human health, as suggested by Lemon (1964). Huebner (1963) states that cancer in animals is an infectious disease and that it would be "unbiological" to assume that the same is not true for man.

a. Polyoma virus. The polyoma agent, a papovavirus, infects mice and may be shed in mouse urine in titers as high as 10^5 infectious units per milliliter. Forty-nine of 110 specimens of mouse-infested grain in four states yielded polyoma virus. Humans and cattle exposed to such mice and grain have not shown rises in serum antibody titers, but laboratory animals in which tumors are formed by this agent frequently show little antibody response (Huebner, 1963).

b. Avian Leukosis Virus. Beard (1963) states that there has evidently been an enormous increase in the incidence of lymphomatosis in chickens. Vertical transmission, apparently through the female line in eggs, is common in avian leukosis (Huebner, 1963).

c. Bovine Lymphosarcoma Virus. Bovine lymphosarcoma occurs predominantly in cattle 5 years or more of age (Reisinger, 1963). A virus which may be the etiologic agent of this disease has been isolated from the milk of a leukemic cow (McKercher *et al.,* 1963), and Dutcher *et al.* (1964) have detected virus-like particles in milk from a herd of cattle with a high incidence of leukemia by means of electron microscopy.

IV. SOURCES OF VIRUSES IN FOODS

Knowledge of the possible sources of virus in foods is clearly necessary if virus contamination of foods is to be prevented. Such knowledge may also prove useful in other ways, for the mode of contamination will influence the quantity and location of virus occurring in food. The quantity and location of the contaminant will determine the methods necessary to detect it and will govern the probability that the virus will escape inactivation until it reaches the consumer.

A. Primary Contamination

Primary contamination will, in the present usage, be defined as contamination of a foodstuff prior to the time of harvest or slaughter.

1. INFECTED ANIMALS

Although many of the agents capable of infecting animals are thought to have limited infectivity for man, it is in this category of contamination that the highest concentrations of infectious agents may be expected to occur. Further, although solid foods which become contaminated with viruses during handling will probably be found to harbor the contaminating agents only upon their surfaces, food products of infected animal origin are likely to be contaminated within the deep tissues. Dimopoullos (1960) has reported that the survival of foot and mouth disease virus in infected tissues differs with the rate of acid production in various organs during the process of rigor mortis and that the agent may be protected when harbored within lymphoid tissues. Newcastle disease virus in muscle tissue of fowl may survive for 6 months under trade chilling conditions (Hanson and Brandly, 1958). Enright *et al.* (1957) have ascertained that a concentration of *Coxiella burnetii* in milk and cream of naturally infected cattle does not exceed 10^3 infectious doses per 2 ml. Myxoviruses in the milk of ruminants under experimental conditions as studied by Mitchell *et al.* (1953*a,b,* 1954, 1956) were found to persist for 10 to 14 days and to reach titers as high as 10^6 to 10^8 infectious units per milliliter. A peak of 10^6 infectious units per milliliter has been observed in the milk of goats infected with tick-borne encephalitis virus under experimental conditions (H. Libíková, personal communication).

2. FECAL POLLUTION

Although it is conceivable that plants used as foods might become contaminated in regions where the use of night soils as fertilizers is prevalent, the contamination of shellfish harvested or held in waters polluted with sewage is of greatest interest in this connection. Statistical data presented by the Communicable Disease Center (1965) seem to indicate that the transmission of infectious hepatitis through shellfish in this fashion is by no means unusual. Metcalf and Stiles (1965) have reported isolations of enteroviruses from oysters taken as far as 4 miles from the nearest outlet of raw sewage into estuary waters.

B. Contamination in Handling by Humans

It must be assumed that the most significant sources of viruses infectious for humans, as they are likely to occur in foods, is the human food handler. Food handlers serving as sources of virus in foods may include symptomatic, asymptomatic, and incubating cases of virus disease; convalescents; and vaccinees. Virus being shed by almost any route might conceivably contaminate foods, but analogy to other known enteric pathogens suggests that virus shed in feces will most certainly find its way into foods at times. Adenoviruses have been reported in feces at concentrations as high as 10^6 infectious units per gram (Duncan and Hutchinson, 1961). Plotkin *et al.* (1962) state that the median duration of fecal shedding of poliovirus is 5 weeks for cases, contacts, and nonimmunes fed vaccine, while about 10% of such individuals shed virus for 8 weeks. Experimental studies have indicated that in individual cases shedding of enteroviruses in feces may persist for longer periods of time. It would appear that the infected food handler whose symptoms are not sufficiently severe to keep him from working would have considerable opportunity to contaminate foods with viruses such as these.

C. Other Sources of Contamination

1. ANIMAL VECTORS

The contamination of grains with polyoma virus and of other foods with the viruses of lymphocytic choriomeningitis and Bolivian hemorrhagic fever by infected mice has been suggested. More significant, however, is the role of insects in the carriage of a number of viruses. Polio and Coxsackie viruses were isolated by Ward (1952) from flies in Egypt, and Syverton *et al.* (1952) have isolated polioviruses from cockroaches. ECHO viruses have also been recovered from flies (Wenner, 1962*b*). Gudnadóttir (1961) reports that the polioviruses may multiply in flies or at least survive for considerable periods of time. These results have been confirmed and extended by Davé and Wallis (1965*a*). Davé and Wallis (1965*b*) have also shown that poliovirus can persist for long periods in cockroaches fed the agent, and Tarshis (1962) has cited evidence to suggest that infectious hepatitis virus may be carried by cockroaches.

2. CONTAMINATED WATER USED IN FOOD PROCESSING

There is no food product whose microbiological quality transcends that of the water used in producing it. The transmission of infectious hepatitis in water has been reviewed by Mosley (1959), and the potential for transmission of enteroviruses in water has been discussed by Berg (1966). Sweet-water cooling systems used in dairy plants and other food-producing facilities are of especial interest in this regard.

V. STABILITY OF VIRUSES IN FOODS

The stability of a contaminating virus, the concentration of the virus, the composition of the food, and the conditions under which it is handled all serve to determine whom the virus may infect. A labile virus of animals may infect slaughterhouse workers but no one else, while a food contaminated shortly before consumption with an equally labile virus may cause infections only in consumers. At the other extreme, the hog cholera virus can apparently remain active through the entire cycle of food processing and distribution, and beyond the consumer in garbage to infect other swine. Although the intrinsic stability of the virus is certainly important, the environmental factors to be discussed below probably play an even larger part in determining the targets of food-borne virus.

A. Stabilization by Foods

Certain components of foods may serve to protect virus from inactivation. Kaplan and Melnick (1952, 1954) have shown that polioviruses and Coxsackie viruses withstand greater periods of heating when in dairy products than in water and that cream has a greater protective effect than fluid milk. Tick-borne encephalitis virus stored at 4°C showed at least a 99.9% loss of titer during 2 weeks in saline, but essentially no decline during 2 weeks in milk or 2 months in butter at that temperature (Grešíková-Kohútová, 1959a). On the other hand, Lynt (1966) found that three types of enteroviruses were no better stabilized in a selection of foods other than dairy products than in buffer solution at room temperature, 10°C, and −20°C. This may be taken to suggest that, in addition to other protective components of dairy products, milk fat or lipid may have a protective influence. Work (1964) states that arboviruses are stabilized in the presence of protein; and Hahon and Kozikowski (1961) have shown that variola virus was stabilized when heated at 50°C in the presence of 10% skim milk. Both proteins and salts in foods may act as buffers, and salts appear to have a direct action upon the stability of viruses apart from their buffering activity. Speir (1961) demonstrated that inactivation of poliovirus at 56°C was much slower in the presence of sodium ions at concentrations up to 2 M than in physiological salt solutions. The sodium ion at 2 M concentration has also been shown to have a protective effect upon vaccinia, *Herpes simplex,* and adenoviruses (Wallis *et al.,* 1962*a*). Molar concentrations of divalent cations, such as magnesium and calcium, have been shown by Wallis and Melnick (1962*b*) to protect most enteroviruses for 60 minutes at 50°C; and reoviruses have also been shown to be protected (Melnick, 1962*b*). Molar divalent cations also stabilize polioviruses at room temperature (Melnick, 1962*a*). Melnick (1962*b*) states that adenoviruses, papovaviruses, herpesviruses, myxoviruses, arboviruses, and poxviruses are more labile at 50°C in molar magnesium chloride than in water. However, Barnes and Gordon (1965) have demonstrated that molar magnesium sulfate at 50°C protected an arbovirus from thermal inactivation. Grausgruber (1963) showed the practical significance of these findings in the context of foods when he demonstrated that pickling salts used in the preparation of Krakow sausage were responsible for the difficulty experienced in inactivating the swine poliovirus which occurred in central nervous system tissue used in sausage production.

B. Physical Inactivation

1. THERMAL EFFECTS

A number of animal viruses have been heated experimentally, with varying results. Among the avian adenoviruses studied by Clemmer (1964), one strain retained infectivity after 40 minutes at 60°C and another for at least 30 minutes at 70°C. Half-lives determined by Hahon and Kozikowski (1961) for variola virus in saline at pH 4.5 ranged from 12.1 minutes at 40°C to 2 minutes at 55°C. Hanson and Brandly (1958) state that various strains of Newcastle disease virus survive for 30 to 180 minutes at 56°C, while a bovine leukemia agent isolated from milk by McKercher *et al.* (1963) was inactivated in 20 minutes at 63°C in milk. Inactivation of 99.99% of tick-borne encephalitis virus in milk was obtained in 30 minutes at 62°C (Grešíková-Kohútová, 1959*b*) and in less than 10 seconds at 72°C (Grešíková *et al.,* 1961). Youngner (1957) studied inactivation of polioviruses in clarified tissue culture fluid at pH 8.3 to 8.4 and 50°C. Approximately 90% inactivation was observed after

1 hour with common laboratory strains of types 1 and 2, and 99.9% inactivation with type 3. Hancock *et al.* (1959) showed that two of four strains of porcine enteroviruses had retained infectivity after 30 minutes at 65°C. Starting with approximately 10^7 infectious units per milliliter in serum-free tissue culture medium, Medearis *et al.* (1960) detected residual infectivity in suspensions of all three types of polioviruses after 60 minutes at 65°C. Swine poliovirus in salted pieces of infected brain and spinal cord tissue was still detectable after 10 minutes at 75°C (Grausgruber, 1963).

Barnes and Gordon (1964) have shown that heating causes the protein coat of an arbovirus (Sindbis) to open and to expose the infectious RNA to the action of ribonuclease. Degradation of the protein coat of poliovirus as a result of heating, with consequent change in complement-fixing antigenicity, has been demonstrated by Hummeler and Hamparian (1958), Hummeler and Tumilowicz (1960), and Roizman *et al.* (1959). Infectious nucleic acid from poliovirus, heated in phosphate saline by Gordon *et al.* (1963), showed 90% loss of infectivity at 65°C, 80°C, and 100°C in 100, 35, and 5 minutes, respectively. Similar experiments by Norman and Veomett (1960) gave 90% inactivation in 23 minutes at 60°C and in about 7 minutes at 70°C. At temperatures below 40°C, the rates of inactivation of picornaviruses seem to parallel those of their nucleic acids, suggesting that the protein is not the prime site of thermal inactivation in this temperature range (Dimmock, 1967). Environmental conditions which tend to protect the protein coat may result in an apparent increase in the stability of the virus during heating, but may have quite a different effect at ambient and lower temperatures.

As with many other reactions, the inactivation of viruses is appreciably slowed at temperatures below ambient. Rhim *et al.* (1961) found that reovirus type 1 in tissue culture maintenance medium was 90% inactivated in 14 days at 4°C and more than 99% inactivated after the same period at 24°C. The time for 90% inactivation of adenoviruses in culture fluid has been found to be approximately 30 days at 4°C and 15 days at 20°C (Dossena, 1961). Meyer and Eddie (1964) state that specimens of chlamydia may be held for several days in the refrigerator without appreciable loss in potency of the agent.

2. DRYING EFFECTS

The enteroviruses are directly degraded by desiccation and cannot usually be freeze-dried successfully (Plotkin *et al.*, 1962; Plager, 1962). On the other hand, the preservation of poxviruses such as variola and vaccinia by air-drying has long been practiced. Hanson and Brandly

(1958) state that drying Newcastle disease virus in tissues or secretions may prolong survival of the virus by months or years. Dried foot and mouth disease virus shows greater resistance to heat (Dimopoullos, 1960), to the sterilizing action of ethylene oxide (Fellowes, 1960), and to cobalt-60 gamma rays (Baldelli *et al.*, 1963) than the same virus in the wet form. In general, it would appear that viruses that efficiently withstand desiccation are protected in the dried state from the actions of a number of other inactivating agents.

3. IRRADIATION EFFECTS

Viruses have been shown to be susceptible to inactivation by both ionizing and nonionizing radiations. However, considerable differences in the mode of action of low-energy photons in the ultraviolet range and of high-energy photons in the gamma-ray range have been demonstrated. Low-energy radiations may cause demonstrable degradation of the virus coat. Intact poliovirus is about twice as sensitive to ultraviolet inactivation as its ribonucleic acid (Norman, 1960), but the target volume for radiation inactivation at ionizing energies corresponds to the nucleic acid volume of the virus (Lauffer, 1960).

A limited number of studies has been reported on the inactivation of viruses by various dosages of ionizing radiation. Trump and Wright (1957) irradiated fifteen viruses at dry ice temperatures with high-energy electrons. Poliovirus, influenza virus, and Japanese B encephalitis virus were found to be completely inactivated at 1.7 megarads. All the other viruses, including those of rabies, herpes, vaccinia, three other arboviruses, and mumps virus were shown to be or would have been completely inactivated at 3.3 megarads. Unfortunately, the concentrations of virus initially present in these suspensions were not specified. Jordan and and Kempe (1957) performed cobalt-60 gamma-ray irradiations of poliovirus, St. Louis encephalitis virus, western equine encephalomyelitis virus, and vaccinia virus, all of which had been prepared in mouse brains. The dosages required to bring about at least 99.999% inactivation were found to be 4.0, 4.0, 3.5, and 2.5 megarads, respectively, for these viruses. If the brain suspensions were partially purified by centrifuge clarification and Seitz filtration, 1 megarad less of irradiation was required to produce the same effect. Baldelli *et al.* (1963) found that wet suspensions of foot and mouth disease virus containing more than 10^6 infectious units per milliliter were inactivated to below detectable levels by 3 megarads of cobalt-60 gamma rays but that the virus was somewhat more stable in the dry state. On the other hand, Lauffer (1960) reports that tobacco mosaic virus is inactivated at the same rate whether wet, dry, or frozen. Kaplan and Moses (1964) have published formulas with which to predict the quantity of radiation which should be required to inactivate any given quantity of a virus. Predictions based on these formulas, for the viruses which have been discussed above, have been slightly higher than the dosages which were required in practice. Studies on irradiation of viruses whose results are intended to be directly applicable to the question of irradiation of foods are presently being performed by the U.S. Public Health Service under the direction of Dr. Robert Sullivan.

On the basis of presently available information, Cliver (1965b) has concluded that levels of irradiation intended to inactivate spores of *Clostridium botulinum* in the range of 3.0 to 4.5 megarads will probably prove adequate to inactivate concentrations of virus likely to occur in foods. However, pasteurizing doses of ionizing radiation in the range of 100,000 to 500,000 rads are likely to permit a considerable proportion of a contaminating virus to elude inactivation. Attention should be given to the question of whether the viruses found to have withstood such irradiation are genetically typical of the input population. The possibilities of induced mutations affecting host range, virulence, and serologic specificity have been suggested. Baldelli *et al.* (1964) have used cobalt-60 gamma-ray dosages of 500,000 rads in an apparently successful attempt to attenuate foot and mouth disease virus. There is no evidence to indicate that modification in the direction of heightened virulence could not also take place.

C. Chemical Inactivation

The prime site of chemical action seems likely to be the virus surface, either through denaturation of the coat protein or by removal of essential structural lipid where present. The actions of acids and bases upon viruses have been most extensively studied in the cases of foot and mouth disease and polioviruses. The inactivation of foot and mouth disease virus in infected tissues has been shown to differ with the extent of acid production in various organs during the process of rigor mortis (Dimopoullos, 1960), but acid inactivation of foot and mouth disease virus has been shown to differ at the same pH, depending upon the types of anions present (Fellowes, 1960). Plotkin *et al.* (1962) report that poliovirus remains infectious after 3 hours at 37°C in hydrochloric acid at pH 2.0. Wallis and Melnick (1962c) reported that the inactivation of poliovirus at 50°C and various pH's was materially influenced by the type of acid employed to achieve any given pH level.

Since oxidation has been thought to play a part in the inactivation of viruses, it might be supposed that oxidizing agents would enhance the inactivation of viruses and that reducing agents or antioxidants might cause the virus to be stabilized. The kinetics of inactivation of poliovirus by chlorine have been studied by Weidenkopf (1958). Pohjanpelto (1962) has suggested that the stabilizing effect of L-cystine on poliovirus may be an antioxidant action. Organic solvents such as ether and chloroform are routinely used to test viruses for the presence of essential structural lipids. However, other organic solvents, such as acetone, are effective in inactivating viruses lacking essential structural lipid, presumably by degrading the protein coat of the virus. Detergents and emulsifying agents, such as sodium deoxycholate, may also serve to remove essential structural lipid when present.

D. Biological Inactivation

1. ANTIBODY

The phenomenon of neutralization, in which virus combines with antibody with resultant loss of infectivity, has long been known. The neutralization of enteroviruses by antibody apparently does not result in any degradation of the virus, and it seems possible that quantities of virus neutralized by antibody are shed during the late period of infection with enteroviruses. The method of fluorocarbon treatment used by Howe (1962) to demonstrate the presence of poliovirus in convalescent chimpanzee feces is the same as that used by Ketler et al. (1961) to reactivate poliovirus neutralized by antibody. Lipton and Steigman (1963) have reported the presence of antibody against polioviruses in the feces of immune individuals. Neutralized poliovirus may adsorb to susceptible host cells and reactivate spontaneously (Mandel, 1958), or it may be reactivated by a pH of 2.5 (Mandel, 1960, 1961), such as might be found in the stomach. Since it seems possible that neutralized enteroviruses may be fully infectious when ingested with food, means of reactivating neutralized virus in food extracts are under study.

2. ENZYMES

The action of digestive enzymes upon viruses has not been extensively studied, but it appears that trypsin and chymotrypsin have relatively little effect upon the enteroviruses. Gomatos and Tamm (1962) reported greater than 99% inactivation of reovirus type 3 by treatment with trypsin. However, more recent reports indicate that trypsin and chymotrypsin treatments activate reoviruses (Newlin and McKee, 1965; Spendlove et al., 1965). Information regarding the interactions of other enzymes and viruses has not been found.

3. Microbial Action

Direct evidence of the action of bacteria and fungi upon viruses is limited, but studies by Kelly *et al.* (1961) and Squeri *et al.* (1963) suggest that certain bacterial species are capable of inactivating enteroviruses, and perhaps other viruses as well. However, Lynt's (1966) experiments in which a variety of perishable foods were held for 7 days at room temperature gave no indication that the inactivation of inoculated enteroviruses was accelerated by the action of the natural foodspoilage organisms.

VI. DETECTION AND IDENTIFICATION OF VIRUSES IN FOODS

The isolation and identification of viruses in foods may be regarded as a specialized area of diagnostic virology, and as such its problems are basically similar to those of diagnostic virology in general. However, the level of virus in a food is likely to be low initially and to go lower still during the time consumed by incubation of human infections and deductive implication of virus etiology. For these reasons, methods more sensitive than those currently available are being sought for the detection of viruses in foods.

A. Collection of Samples

The sampling of foods for virus isolation may be undertaken from two general approaches: market sampling to determine the prevalence of viruses in foods, and the retrospective study of samples associated with outbreaks of food-borne disease. In the case of market sampling, evidence presently available indicates that the foods most likely to contain viruses are those which have been handled by humans and which, since the last previous handling, have not been heated to an extent greater than that involved in low-temperature long-time pasteurization of milk. In investigation of food-associated outbreaks, the criteria for selection of food samples are the same as those which would be used in the case of food-associated outbreaks of bacterial etiology. Clinical samples obtained from patients involved in such outbreaks may also prove of value. Lennette (1964) has tabulated virus-induced clinical syndromes and indicated the appropriate clinical samples to be taken from patients and the modes of testing for various viral and rickettsial agents. Kalter (1963) suggests that, in storing samples for virus isolations, the best temperature to employ is usually the lowest available. This is a particularly critical factor if samples are to be held for periods longer than 24 hours. In addition to clinical samples taken for possible virus isolations, sera obtained from patients during the acute and convalescent phases of the disease are likely to prove of value.

1. FLUIDS

In the absence of toxic components, many fluids can be inoculated directly into suitable laboratory host systems, sometimes without dilution. Because of the capability of many viruses to adsorb to surfaces, fluid samples should be collected whenever possible in silicone-treated glassware, and any sediments which are present should also be sampled.

2. SOLIDS

Solid foods must be put into fluid suspension prior to testing in laboratory hosts by preparation either of a slurry or an extract. Bellelli (1963) has reported a method for extracting enteroviruses from experimentally contaminated butter with a loss of as little as 0.1 to 0.2 $\log_{10}$ of the input virus concentration. Extraction methods of a comparable degree of efficiency which are applicable to other foods need to be developed. If food particles are to be removed from a slurry, care must be taken to free adsorbed virus prior to the elimination of the solids.

3. SURFACES

Virus contamination of foods and of food containers, particularly that which takes place in handling, may often be limited to the surface. At present the swab method appears to be the only one available which has broad applicability. Enteroviruses in particular have shown a tendency to adsorb to cotton, and Wenner (1964) has suggested that cotton swabs should be freed of enteroviruses by extraction with saline buffered at pH 8. However, freeing adsorbed virus from the surface being sampled is quite another matter, and one to which further study should be devoted.

B. Processing of Samples

1. CLARIFICATION

This step in the processing of samples is intended to remove gross food particles and microbial contaminants. If the food particles present are not objectionable and the food sample or extract is not toxic in the host system to be employed, this step may be bypassed by the use of available methods for suppressing microbial contaminants in the sample. Treatment of samples with ether or chloroform has been employed for this purpose with sewage samples, but this is applicable only when viruses being sought lack essential structural lipid. Bourgeois and Branche (1963) have described a combination of antibiotics which was found to be effective against test bacteria and fecal contaminants, nontoxic to three types of tissue culture, and noninhibitory to a number of virus types. Lynt (1966) has tested another antibiotic mixture and shown it to be effective in suppressing even heavy growths of food spoilage organisms in food slurries from which enteroviruses were to be isolated.

a. Centrifugation. Wenner (1964) has suggested the gross clarification of clinical samples for virus isolation by centrifugation at 2000 rpm for 20 minutes and states that virtually all contaminating bacteria can be removed by centrifugal forces of $14,000 \times g$. However, if colloidal protein or emulsified lipid is objectionable, higher gravitational forces may be necessary to eliminate these components and may result in the sedimentation of a certain portion of the virus as well. Addition of Freon TF (DuPont) and bentonite to a solid food during homogenization with an aqueous diluent has been found to reduce significantly the centrifugal force required to sediment food solids (Herrmann and Cliver, unpublished data). The aqueous diluent must be buffered well to the alkaline side, or the virus sediments with the bentonite.

b. Membrane Filtration. Filter membranes may prove an aid in the isolation of viruses from foods. Atoynaton and Hsiung (1964) have used membrane filters of 220-mμ porosity or Seitz filters in the clarification of virus suspensions, with losses of virus titer from approximately none to more than 95%. Cliver (1965a) has studied methods whereby such losses of virus in filtration may be minimized. The techniques described have since been modified slightly in the interest of simplicity, and at present filter membranes are simply being soaked for 10 minutes either in undiluted agamma chicken serum or in a 2% solution of gelatin prior to use in filtration. Care must also be taken that virus is not retained mechanically by the filter. Selection of a pore size large enough to permit the virus to pass is obviously required, but filtrations should not be continued beyond the point at which flow rate is retarded because of plugging by food particles. When flow rate has been retarded, the effective pore size of the membrane is consequently reduced; and virus particles may be retained at the surface of the membrane. Problems of membrane plugging and reduced flow rate can be minimized if membrane filtration is used in conjunction with preliminary centrifuge clarification of sample suspensions.

2. CONCENTRATION OR CHANGE OF DILUENT

Studies are presently in progress to develop methods whereby virus in dilute suspensions of foods may be concentrated prior to testing, or the diluent in which a toxic primary food extract is contained may be changed. The procedures employed to achieve these two ends are basically the same and will be discussed together here. Each of these techniques is subject to mechanical interference by small food particles, so clarification of food extracts to be subjected to such procedures is much more demanding than if the same extracts were to be tested directly in the same host system.

a. Ultracentrifugation. The technique of preparative ultracentrifugation has been said by Clarke and Kabler (1964) to be the method of choice among those currently available for concentrating enteroviruses from sewage. Baron (1957) recommended adding 3% calf serum or 0.06% gelatin to a suspension to be centrifuged to improve the efficiency with which poliovirus formed a pellet at the bottom of the tube. Cliver and Yeatman (1965) have performed quantitative studies of the ultracentrifuge concentration of enteroviruses, determining both the input concentration of virus and the quantity recovered, and employing only membrane-filtered suspensions of enteroviruses so as to minimize the possibility that input virus included undetected aggregates. The technique employed a "trap" of 0.1 ml of 2% gelatin solution applied at the point at which the pellet forms in tubes of the number 30 and number 50 rotors of the Spinco model L preparative ultracentrifuge. Enteroviruses were chosen as model agents in these studies because they are among the smallest known to infect man, and it was assumed that conditions adequate to sediment enteroviruses would also serve to sediment all the larger agents. The technique has been applied successfully to reoviruses by Gibbs and Cliver (1965). Peizer *et al.* (1961) also noted, when using the ultracentrifuge in concentrating enteroviruses from stool suspensions, that cytotoxic components of the original stool suspension were thereby eliminated.

b. Dialysis with Hydrophilic Compounds. Soller (1961) described a technique for the concentration of bacteriophage suspension in dialysis tubing surrounded with polyethylene glycol. Gibbs and Cliver (1965) applied a modification of this technique to the concentration of reovirus suspensions. In this modification a dialysis tube containing 100 ml of virus suspension was immersed in a solution of 100 gm of polyethylene glycol (molecular weight 20,000) in 100 ml of water and concentrated for 2 to 3 hours at room temperature. The method has since been adapted to the concentration of enteroviruses for detection (Cliver, 1967a).

c. Membrane Filtration. Since the results of Atoynatan and Hsiung (1964) and of Cliver (1965a) indicated that membrane filters had a considerable ability to retain viruses, means were sought whereby filter membranes could be used in the concentration of viruses for detection purposes. Enteroviruses were again used as model agents in these studies. They could be retained mechanically on membranes whose pores were smaller than the virus, but flow rates were far too slow. An alternative method, in which filter membranes of relatively large pore size serve as a matrix to which enteroviruses are adsorbed and consequently recovered in a small volume of eluent, has since been developed (Cliver, 1967b). The application of this method to the detection of viruses in food extracts has been complicated by the tendency of soluble protein in the virus suspension to interfere with adsorption of the virus to the filter matrix.

d. *Adsorption to Erythrocytes.* The ability of certain viruses to adsorb to erythrocytes can be exploited to concentrate the virus from dilute suspension. Johnson and Lang (1962) adsorbed hemagglutinating variants of Coxsackie A-21 virus onto human type O red blood cells and sedimented the red cells with the centrifuge; they achieved a concentration factor of approximately 1000. Since Eggers *et al.* (1962) had indicated that agglutination of bovine red blood cells was a general characteristic of strains of reovirus type 3, the reovirus type 3–bovine red blood cell system was applied by Gibbs and Cliver (1965) as a concentration method in the detection of this virus. The great liability of this technique is the high degree of specificity which viruses show in the spectrum of red blood cells to which they will adsorb.

e. *Chromatography.* Conventional techniques of column chromatography may be useful in changing the diluent of a virus suspension, although no single method is applicable to all groups of viruses. It appears in general that concentration factors achievable by column chromatography are not comparable to those of the methods described above. Additional problems include the difficulty of sterilizing chromatography columns and the fact that flow rates may not prove significantly better than those obtained with small porosity membrane filters.

f. *Differential Solubility Methods.* Methods for concentration of viruses have been devised which take advantage of the peculiarities in solubility of viruses in fluid suspension. Steinman and Murtaugh (1959) reported that a simian adenovirus could be precipitated at a pH 3.4 with a concentration factor of 1000. A modification of this concentration method has been applied successfully by Brandt *et al.* (1963) to six types of human adenoviruses. Philipson *et al.* (1960) have described a technique employing aqueous polymer phase systems which was applied successfully to enteroviruses and adenoviruses and is said to be applicable to myxoviruses and bacteriophages as well. Concentration factors achieved ranged from 10- to 1000-fold. Kitano *et al.* (1961) concentrated ECHO virus types 5 and 7 from hypertonic phosphate buffer by treatment with a mixture of 2-ethoxyethanol and 2-butoxyethanol. The concentration factor achieved was approximately 8, with over 90% efficiency. Schwerdt and Schaffer (1956) achieved a 50- to 60-fold concentration of poliovirus with 80 to 100% efficiency by precipitation with cold methanol at pH 4.

C. Testing Samples for Virus

1. AVAILABLE SUBSTRATES

Virus can be said to have been detected only when its infectivity has been demonstrated. Since virus is replicated only by living cells, this requires a suitable living host system.

a. Whole Organisms. The method employed by Enright *et al.* (1957) for the detection of *Coxiella burnetii* in milk involves inoculation of guinea pigs and later testing the sera of these animals for complement-fixing antibody. Guinea pigs, as well as adult mice, are recommended by Smadel and Jackson (1964) for the isolation of other rickettsiae from clinical specimens. Meyer and Eddie (1964) recommend testing for the agent of ornithosis in both mice and embryonated eggs. A number of other agents detectable in embryonated eggs have been tabulated by Lennette (1964), who also gives a detailed discussion of virologic methods employing embryonated eggs. Among the enteroviruses, it has been common to test for the Coxsackie viruses in suckling mice. However, Kalter (1963) states that the majority of Coxsackie viruses producing apparent disease show a predilection for tissue culture over suckling mice. It can be seen that whole host organisms are the best or the only living systems applicable to the detection of certain viral and rickettsial agents.

b. Tissue Culture. The principle advantages of tissue cultures as host systems for the detection of viruses are relative sensitivity, reproducibility, and economy. Work (1964) suggests the isolation of arboviruses in suckling mice or in embryonated eggs but states that the laboratory not equipped with such substrates may use hamster kidney tissue culture instead. In general, the host range of virus in tissue culture tends to mirror the species specificity, though not the tissue tropism, of the same virus in the whole host organism. Primary monkey kidney cells, usually from the rhesus monkey (*Macaca mulatta*), and various established cell lines of human origin are most frequently used. Primary cells cultivated from human amnion and human kidney have also been recommended for the detection of viruses of human origin, although these may not be available to many laboratories. Primary monkey kidney cell cultures are said to be equal or superior to embryonated eggs in the detection of influenza viruses (Davenport and Minuse, 1964) and are recommended for the detection of parainfluenza viruses (Chanock and Johnson, 1964).

2. CHOICE OF SUBSTRATES

The discussion in the preceding section was based upon known properties of known viruses; but where test substrates are to be selected for the detection of viruses in foods, the virus being sought is unknown. In investigations of outbreaks, clinical symptoms in infected humans may provide a guide to the types of tests to be performed. Even if a certain amount of such information is available, however, it would seem advisable to select a test substrate whose spectrum of susceptibility to viruses is as broad as possible.

3. TYPES AND METHODS OF TISSUE CULTURE

a. Primary versus Established Cell Types. Primary cultures are prepared by enzyme dispersion of tissues, with subsequent growth of the cells in the final vessels in which they are to be used for virus propagation. They cannot be propagated in suspension, and any particular batch of such cultures has a finite life in the laboratory. Therefore, a laboratory basing its operation on the routine use of primary cell cultures must ensure that it has a reliable source of suitable tissues. Primary cell cultures are quite hardy and are easily maintained for considerable periods of time after they have grown to confluent monolayers. Cells grown in primary culture may be harvested and stored to make secondary cultures.

Wenner (1964) states that primary kidney cultures of rhesus and cynomolgus (*Macaca cynomolgus* or *Macaca irus*) monkeys are the cultures of choice for the isolation of most enteroviruses; and they have been recommended by Macrae (1962) for the detection of reoviruses and by Hsiung and Henderson (1964) and Canchola *et al.* (1964) for myxoviruses. Primary human amnion cell cultures are relatively inexpensive, but their routine use requires the ready availability of fresh amnions free of soap and disinfectants. Lehmann-Grube (1961) demonstrated that virtually every enterovirus type which would propagate with cytopathic effect in primary monkey kidney cells or in HeLa established human cells would also propagate in primary human amnion cells. Nardi (1964) has found that primary human amnion cultures served well for the isolation of adenoviruses and parainfluenza virus types 2 and 3, but were not susceptible to influenza viruses or to parainfluenza type 1 virus. Hsiung (1959a) and Hsiung et al. (1959) report that primary cultures from kidneys of human infants 2 years of age or less do not harbor adventitious viruses and are approximately as susceptible as primary monkey kidney to human enteroviruses. Adult kidneys showed a much more limited range of virus susceptibilities. Vargosko et al. (1964) recommended the use of primary human embryonic kidney cells in the isolation of all human adenovirus types.

An established cell culture line is usually one which has been transplanted in tissue culture ten or more times since the cells were first prepared from the donor tissue. The cells may have been obtained from malignant or normal tissues; but by the time that they have become established in tissue culture, their chromosome complement is usually not typical of the donor species. They grow quite rapidly in culture and may be propagated in fluid suspension, so that they may prove less expensive than primary cultures if they are to be used in quantity. They also tend to be highly reproducible and may be cloned from a single cell periodically in order to enhance the homogeneity of the population. Problems have been encountered with mycoplasma which seems to have a predilection for established cell types and may interfere with virus propagation. Attempts have been made to establish cell lines from primary cell types which have proved useful in virus isolation. These have met with limited success because the virus susceptibility spectrum of the established cell line has often been found to differ from that of the comparable primary cell type.

Euploid cell lines have been prepared from human embryonic tissues. These mimic, to some extent, the properties of primary human cell cultures in that they have a limited potential for growth and can be maintained in culture only through a limited number of passages. They are presently in use for the isolation of rhinoviruses and the quantitation of chlamydia and may have some potential as a substrate for the isolation of candidate infectious hepatitis viruses.

b. Microscopic versus Macroscopic Methods. The present discussion is intended only to supplement such extensive presentations as that of Schmidt (1964). Microscopic methods consist basically in the inoculation of monolayer cultures, most frequently in tubes, with the suspected virus suspension, followed by microscopic observation for cytopathic effects. The principle advantages of this technique are the broad spectrum of virus susceptibility which may be attained with one basic procedure and the possibility that distinctive cytopathic effects may provide early evidence of the type of virus which is present. In the case of minimal inocula or of slow-growing viruses, the maintenance medium can be changed as the cells require it until such time as the virus effect is manifest. "Blind" passages, or serial subcultures of wild-type viruses which initially multiply in cell cultures without visible effects, are easily performed if fluid medium has been used.

The cytopathic effects of some viruses can best be distinguished if the infected cells are stained prior to microscopic observation. This may necessitate growing the cells on coverslips to permit manipulation and mounting. Standard methods for staining tissue sections have been adapted to this purpose. Acridine orange fluorescent dye may also serve to bring out highly distinctive patterns of cytopathic effects such as those of reovirus; this was used for detection by Gibbs and Cliver (1965). Coverslip cultures of primary monkey kidney cells were harvested on the fourth day after inoculation. The coverslips were passed through the following solutions for the indicated periods of time: cold 95% ethanol (4 minutes); $M/15$ phosphate buffer, pH 6 (30 seconds); 0.01% acridine orange in $M/15$ phosphate buffer, pH 6 (3 minutes); $M/10$ $CaCl_2$ (30 seconds); $M/15$ phosphate buffer, pH 6 (rinse five times). Excess buffer used in mounting was taken up with bibulous paper, and the coverslip was ringed with clear fingernail polish to retard drying during examination with ultraviolet light. Antiviral antibody labeled with fluorescein is also employed in specific staining of virus-infected cells. Alternatively, a great deal of additional detail and most kinds of viral cytopathic effects can be demonstrated by means of phase-contrast microscopy.

The essence of macroscopic methods in animal virology is the plaque technique (Dulbecco, 1952). Virus is usually added in a small volume of inoculum to a monolayer of cells from which medium has been discarded. After an appropriate period of incubation to allow the virus to adsorb to the cells, a semisolid overlay medium is added to confine the action of the virus to local areas. These areas of degeneration, which gradually become macroscopically visible, are called plaques. Viable cells under the overlay medium are usually distinguished by their ability to take up and hold the vital dye, neutral red; and Hsiung (1959b) has suggested that this is a more sensitive measure of cell viability than is direct microscopic observation for cytopathic effects. Gabrielson and Hsiung (1964) report the detection by the plaque technique of enteroviruses in clinical specimens too dilute to induce the formation of cytopathic effects under fluid medium.

The conditions of adsorption are critical to the sensitivity of the plaque technique. Studies have indicated that the rate of adsorption of entero-viruses can be increased by minimizing the volume of inoculum (McLaren *et al.,* 1959), incubating at 36°C, and overlaying without washing the cells (Youngner, 1956). Our own tests show that sensitivity is enhanced by use of larger volumes of inoculum (0.02 ml/cm² of cell surface) and adsorption for 2 hours at room temperature. If the cell sheet is not washed prior to overlaying, unadsorbed virus caught in the medium may later reach the cells and form plaques with as much as 38% efficiency (Mc-Laren *et al.,* 1959). The adsorption period is eliminated in the agar cell suspension plaque technique of Cooper (1955, 1961).

Plaque formation by some viruses has been found to be inhibited by the agar, serum, or neutral red in the overlay medium. It has been suggested that agar inhibition be avoided by substituting fibrin, starch gel, or methyl cellulose as gelling agents or by changing agar preparations (Wallis *et al.,* 1962*b*). Alternatively, DEAE dextran (Liebhaber and Takemoto, 1961) or protamine sulfate at 400 μg/ml (Tytell *et al.,* 1962; R. L. Heberling, personal communication) added to the overlay may minimize the effect of agar inhibitors. Inhibition by serum may be due to antibody or to other factors. Chicken serum, agamma calf serum, and skim milk (Rhim and Melnick, 1961) are generally noninhibitory. Neutral red inhibition of plaque formation by enteroviruses and arboviruses (Darnell *et al.,* 1958), reoviruses (Rhim and Melnick, 1961), and adeno-viruses (Tytell *et al.,* 1962) may be avoided by adding the dye after the plaques have formed. However, Mosley and Enders (1961) have shown that the addition of neutral red or of second overlays precludes obtaining pure subcultures of poliovirus from individual plaques. Heberling and Cheever (1961) have reported that the addition of L-cysteine to overlays for some enteroviruses has made possible plaque formation in the presence of neutral red. Magnesium chloride at 25 m*M* concentration (Wallis and Melnick, 1962*a*) has also been found to enhance the sensi-tivity of the plaque technique for prototype ECHO-6 virus in rhesus kidney cultures (Cliver, unpublished).

D.   Identification of Isolated Viruses

1.   PRELIMINARY GROUPING

A general scheme by which most unknown viruses may be assigned to groups has been described by Hsiung (1964). It is based upon the type of nucleic acid of the virus, its approximate diameter as determined by membrane filtration at 50-mμ and 100-mμ porosity, and the presence of essential structural lipid as determined by the ether sensitivity test. The detailed method for the membrane filtration tests has been described by Atoynatan and Hsiung (1964). The type of nucleic acid contained in the virus may be determined by tests for inhibition by 5-fluorodeoxyuridine, which is a specific inhibitor of the synthesis of DNA (Salzman, 1960). The presumptive identification obtained by the above method may be verified by staining infected tissue culture cells to determine the type of cytopathic effect present (Hsiung and Henderson, 1964).

Because there are more than sixty serological types of human entero-viruses, a number of methods have been developed for subdividing the group prior to serological testing (Wenner, 1964). Subgroupings based upon laboratory host range (Hsiung and Henderson, 1964; Marchetti and Gelfand, 1963; Barron and Karzon, 1959), plaque type (Hsiung and Melnick, 1957; Hsiung and Henderson, 1964), hemagglutinating ability, inactivation by sulfhydryl reagents, and inhibition by other chemicals have been proposed. With the possible exception of inactivation by sulfhydryl reagents such as parachloromercuribenzoate (Philipson and Choppin, 1960; Choppin and Philipson, 1961), partially oxidized 2,3-dimercaptopropanol (Philipson and Choppin, 1962), and iodoacetamide (Allison et al., 1962), none of these tests is entirely reliable. Observed host ranges may vary among strains of a single virus type or among strains of a single cell type. Variations in plaque types may be due to inhibitors in the overlay medium (Barron and Karzon, 1962) which may be elim-inated. Hemagglutination by an enterovirus type may differ among strains (Bussell et al., 1960; Lahelle, 1958) and may be altered by tissue culture passage (Johnson et al., 1961; Johnson and Lang, 1962). Inhibition by 2-(α-hydroxybenzyl)benzimidazole (Eggers and Tamm, 1961b) and guanidine (Tamm and Eggers, 1962) may allow resistant strains to de-velop (Eggers and Tamm, 1961a) or may be lost through mutation (Eggers and Tamm, 1965). Aside from the use of plaque types to indicate mixed isolates (Hsiung and Melnick, 1958), the need for these methods could be obviated by the development of simpler serological techniques.

2. SEROTYPING

Once an unknown virus has been assigned to a group or subgroup, it may by typed by means of standard antisera. A virus isolated in an out-break may be tested with patient sera to verify that it has actually infected the persons involved. In the case of viruses isolated from market samples, serologic testing with human gamma globulin may be used to provide presumptive evidence that the unknown agent is of human origin. Methods which are applicable virtually to the entire range of viruses and rickettsiae are the neutralization (Nt) and complement fixation (CF) tests.

The Nt reaction may be defined as one in which molecules of antibody combining with the virus particle result in loss of infectivity as determined in suitable host systems. The reaction is generally quite type-specific, with few exceptions (Krech, 1957; Wenner, 1964), and has been recommended in the typing of arboviruses (Work, 1964), adenoviruses (Rose, 1964), and enteroviruses (Wenner, 1962a). The use of mixtures of antisera to several types of enteroviruses in an intersecting scheme has been reported by Lim and Benyesh-Melnick (1960) and modified by Schmidt *et al.* (1961). Neutralization methods involving the diffusion of antibody, enteroviruses, or both through agar media have been devised by Burt and Cooper (1961), DeSomer and Prinzie (1957), Gravelle and Chin (1962), Woods *et al.* (1962), and Kalter (1963). Diffusion Nt tests are simple to perform, but final results are obtained rather slowly. A technique for enterovirus typing which employs mixtures of antisera in an agar-diffusion Nt test has been devised by Cliver and Engeseth (unpublished).

The CF test is identical in principle with the CF test used in general serology and is relatively rapid. It has been recommended by Smadel and Jackson (1964) for the identification of *Coxiella burnetii,* and both direct and indirect CF tests are available for the identification of chlamydia (Meyer and Eddie, 1964). The CF reaction of adenoviruses is generally group-specific (Rowe and Hartley, 1962), although under certain conditions type-specific CF reactions may also be employed (Katz *et al.,* 1957; Binn *et al.,* 1958; Ginsberg, 1962). The type specificity of the CF test for enteroviruses has been reported by Halonen *et al.* (1959), and methods for producing suitable CF antigens for typing enteroviruses have been described by Schmidt (1964). The CF reaction is group-specific among the reoviruses (Macrae, 1962). Because CF antibody may have declined within 2 years after an enterovirus infection, the CF test has been recommended for demonstrating a diagnostic rise in antibody titer in patient sera (Plotkin *et al.,* 1962). However, heterotypic CF antibody responses to enterovirus infections have been reported by Lennette *et al.* (1961), Schmidt *et al.* (1962), and Bussell *et al.* (1962b).

Another common technique, applicable only to hemagglutinating agents, is that of hemagglutination inhibition (HI), which has been recommended by Davenport and Minuse (1964) for typing influenza viruses. Rosen (1964) has described a method for preparing HI typing antisera for reoviruses. Vargosko *et al.* (1964) and Rosen *et al.* (1961) suggest that in their tests the HI technique differentiates more adenovirus serotypes than the neutralization method. The HI test has been found applicable by Bussell *et al.* (1962a) to typing enteroviruses which hemagglutinate. The HI test is also quite rapid. A modification of the HI test in which antisera are employed to inhibit hemadsorption by virus-infected cells has been recommended for the identification of myxoviruses by Canchola *et al.* (1964).

A serologic technique of more recent derivation is the fluorescent antibody (FA) test, in which antibody labeled with fluorescein or some other fluorescent dye is used to demonstrate the presence of infecting virus in cells. The method has been employed for diagnostic purposes with a variety of viruses because it can be used with infected cells in tissue culture, blood cells, or cells obtained from exudates of virus-induced lesions, and it has been suggested as a possible approach to the diagnosis of infectious hepatitis by the reaction of labeled convalescent human sera with "buffy-coat" cells from patients (Coons, 1964). The FA technique has been recommended for the identification of poliovirus isolates by Hatch *et al.* (1961); and by mixing labeled antisera Shaw *et al.* (1961) have devised a method for typing almost all of the enteroviruses with a limited number of tests. An indirect FA test has been applied to the typing of enteroviruses by Riggs and Brown (1961).

VII. THE FUTURE OF FOOD VIROLOGY

It is appropriate that the discussion of a field with as limited a past as that of food virology conclude with a look to the future. As the needs cited below are met, it would appear that the information regarding viruses in foods which will result may prove of considerable significance to human health and well-being.

A. Reporting

In reviewing food-borne diseases in the United States and in England and Wales, Dauer (1961) and Cockburn (1960) both have concluded that outbreaks of food-borne disease were not nearly as extensively reported as they should be. This observation relates to food-borne diseases in general and may be assumed to be even more significant in the case of food-borne viruses. Until recently, virus etiology and food transmission have been regarded as mutually exclusive. Now that there is a sufficient body of literature to refute this, surveillance of food-borne virus diseases needs to be extended to clinical entities less spectacular than infectious hepatitis and poliomyelitis.

B. Testing Facilities

Relatively few laboratories are equipped to isolate and identify viruses, and those which are available are already overworked. This problem could be eased somewhat by simplifying procedures and by giving priority to specimens from common source (for example, food-associated) outbreaks.

C. Research

The unique aspects of viruses as they may occur in foods are in need of further investigation. Improved detection methods, further evaluation of the influence of foods in protecting viruses from inactivation, and the possibility that serologically neutralized virus which contaminates foods may be reactivated in the gastrointestinal tract are areas of particular significance. The species specificity of animal viruses has perhaps been overemphasized (Klein, 1960), and viruses of animal origin which occur in foods may be of considerable significance. The overemphasis on species specificity appears to have led to an unfortunate dichotomy between "human" and veterinary virologists, and it is hoped that additional research on food-borne viruses will tend to counteract this to some extent.

VIII. SUMMARY

Available information regarding the occurrence of viruses and rickettsiae in foods and of resultant human infections and the influence of foods on the survival of virus contaminants has been reviewed. Methods applicable to the isolation and identification of viruses in foods have been summarized. Relatively little direct knowledge of viruses in foods exists as yet, but a great deal of potentially useful information is available. Although much of what has been said on the subject in the present discussion is still speculative, the dissemination of viruses as such in foods is no longer debatable.

REFERENCES

Abinanti, F. R. (1961). *Public Health Rept. (U.S.)* **76,** 897.
Abinanti, F. R. (1964). *In* "Occupational Diseases Acquired from Animals," pp. 53–71. The University of Michigan School of Public Health, Ann Arbor.
Allison, A. C., Buckland, F. E., and Andrewes, C. H. (1962). *Virology* **17,** 171.
Atoynatan, T., and Hsiung, G. D. (1964). *Proc. Soc. Exptl. Biol. Med.* **116,** 852.

Aycock, W. L. (1927). *Am. J. Hyg.* **7**, 791.

Baldelli, B., Begliomini, A., Frescura, T., and Massa, D. (1963). Atomic Energy Research Establishment, Harwell, AERE Trans 1000.

Baldelli, B., Begliomini, A., Frescura, T., and Massa, D. (1964). *Bull. Off. Int. Epiz.* **64**, 717.

Barnes, R., and Gordon, I. (1964). *Bacteriol. Proc.* p. 143.

Barnes, R., and Gordon, I. (1965). *Bacteriol. Proc.* p. 109.

Baron, S. (1957). *Proc. Soc. Exptl. Biol. Med.* **95**, 760.

Barron, A. L., and Karzon, D. T. (1959). *Proc. Soc. Exptl. Biol. Med.* **100**, 316.

Barron, A. L., and Karzon, D. T. (1962). *Bacteriol. Proc.* p. 134.

Beard, J. W. (1963). *Ann. N.Y. Acad. Sci.* **108**, 1057.

Bellelli, E. (1963). *Ann. Sclavo* **5**, 739.

Benson, W. W., Brock, D. W., and Mather, J. (1963). *Public Health Rept. (U.S.)* **78**, 707.

Berg. G. (1964). *Health Lab. Sci.* **1**, 51.

Berg, G. (1966). *Health Lab. Sci.* **3**, 86.

Binn, L. N., Hilleman, M. R., Rodriguez, J. E., and Glabere, R. R. (1958). *J. Immunol.* **80**, 501.

Bohl, E. H., Singh, K. V., Hancock, B. B., and Kasza, L. (1960). *Am. J. Vet. Res.* **21**, 99.

Bourgeois, L. D., and Branche, W. C., Jr. (1963). *Bacteriol. Proc.* p. 143.

Brandt, C. D., Neal, A. L., Owens, R. E., and Jensen, K. E. (1963). *Proc. Soc. Exptl. Biol. Med.* **113**, 281.

Brown, G. C. (1949). *Am. J. Public Health* **39**, 764.

Buckland, F. E., Bynoe, M. L., Philipson, L., and Tyrrell, D. A. J. (1959). *J. Hyg.* **57**, 274.

Buescher, E. L. (1964). *In* "Diagnostic Procedures for Viral and Rickettsial Diseases" (E. H. Lennette and N. J. Schmidt, eds.), pp. 719–722. American Public Health Association, New York.

Burt, A. M., and Cooper, P. D. (1961). *J. Immunol.* **86**, 646.

Bussell, R. H., Karzon, D. T., Barron, A. L., and Hall, F. T. (1960). *Bacteriol. Proc.* p. 105.

Bussell, R. H., Karzon, D. T., and Hall, F. T. (1962*a*). *J. Immunol.* **88**, 38.

Bussell, R. H., Karzon, D. T., Barron, A. L., and Hall, F. T. (1962*b*). *J. Immunol.* **88**, 47.

Canchola, J. G., Chanock, R. M., Jeffries, B. C., Christmas, E. E., Kim, H. W., Vargosko, A. J., and Parrott, R. H. (1964). *Bacteriol. Proc.* p. 132.

Chanock, R. M., and Johnson, K. M. (1964). *In* "Diagnostic Procedures for Viral and Rickettsial Diseases" (E. H. Lennette and N. J. Schmidt, eds.), pp. 470–486. American Public Health Association, New York.

Choppin, P. W., and Philipson, L. (1961). *Bacteriol. Proc.* p. 165.

Clarke, N. A., and Kabler, P. W. (1964). *Health Lab. Sci.* **1**, 44.

Clemmer, D. I. (1964). *J. Infect. Diseases* **114**, 386.

Cliver, D. O. (1965*a*). *Appl. Microbiol.* **13**, 417.

Cliver, D. O. (1965*b*). *In* "Radiation Preservation of Foods," pp. 269–274. Publication 1273. National Academy of Sciences–National Research Council, Washington, D.C.

Cliver, D. O. (1966). *Public Health Rept. (U.S.)* **81**, 159.

Cliver, D. O. (1967*a*). *In* "Transmission of Viruses by the Water Route" (G. Berg, ed.), pp. 109–120. Interscience, New York.

Cliver, D. O. (1967*b*). *In* "Transmission of Viruses by the Water Route" (G. Berg, ed.), pp. 139–141. Interscience, New York.

Cliver, D. O. (1967*c*). *Health Lab. Sci.* **4**, 213.

Cliver, D. O., and Bohl, E. H. (1962). *J. Dairy Sci.* **45**, 921.

Cliver, D. O., and Yeatman, J. (1965). *Appl. Microbiol.* **13**, 387.

Cockburn, W. C. (1960). *Roy. Soc. Health J.* **80**, 249.
Committee on Enteroviruses (1962). *Virology* **16**, 501.
Communicable Disease Center (1965). U.S. Department of Health, Education, and Welfare, Public Health Service, Hepatitis Surveillance Report No. 23, June 30.
Coons, A. H. (1964). *Bacteriol. Rev.* **28**, 397.
Cooper, P. D. (1955). *Virology* **1**, 397.
Cooper, P. D. (1961). *Virology* **13**, 153.
Darnell, J. E., Jr., Lockart, R. Z., Jr., and Sawyer, T. K. (1958). *Virology* **6**, 567.
Dauer, C. C. (1961). *Public Health Rept. (U.S.)* **76**, 915.
Davé, K. H., and Wallis, R. C. (1965a). *Proc. Soc. Exptl. Biol. Med.* **119**, 121.
Davé, K. H., and Wallis, R. C. (1965b). *Proc. Soc. Exptl. Biol. Med.* **119**, 124.
Davenport, F. M., and Minuse, E. (1964). *In* "Diagnostic Procedures for Viral and Rickettsial Diseases" (E. H. Lennette and N. J. Schmidt, eds.), pp. 455–469. American Public Health Association, New York.
Delaplane, J. P. (1958). *Ann. N.Y. Acad. Sci.* **70**, 495.
DeSomer, P., and Prinzie, A. (1957). *Virology* **4**, 387.
Dimmock, N. J. (1967). *Virology* **31**, 338.
Dimopoullos, G. T. (1960). *Ann. N.Y. Acad. Sci.* **83**, 706.
Dingman, J. C. (1916). *N.Y. State J. Med.* **16**, 589.
Dossena, G. (1961). *Ann. Sclavo* **3**, 178.
Dulbecco, R. (1952). *Proc. Natl. Acad. Sci. (U.S.)* **38**, 747.
Duncan, I. B. R., and Hutchinson, J. G. P. (1961). *Lancet* p. 530, March 11.
Dutcher, R. M., Larkin, E. P., and Marshak, R. R. (1964). *J. Natl. Cancer Inst.* **33**, 1055.
Eggers, H. J., and Tamm, I. (1961a). *J. Exptl. Med.* **113**, 657.
Eggers, H. J., and Tamm, I. (1961b). *Virology* **13**, 545.
Eggers, H. J., and Tamm, I. (1965). *Science* **148**, 97.
Eggers, H. J., Gomatos, P. J., and Tamm, I. (1962). *Proc. Soc. Exptl. Biol. Med.* **110**, 879.
Enright, J. B., Sadler, W. W., and Thomas, R. C. (1957). *Public Health Monograph* **47**,
Fellowes, O. N. (1960). *Ann. N.Y. Acad. Sci.* **83**, 595.
Fox, J. P. (1964). *In* "Occupational Diseases Acquired from Animals," pp. 98–109. The University of Michigan School of Public Health, Ann Arbor.
Gabrielson, M. O., and Hsiung, G. D. (1964). *Bacteriol. Proc.* p. 132.
Gibbs, T., and Cliver, D. O. (1965). *Health Lab. Sci.* **2**, 81.
Ginsberg, H. S. (1962). *Virology* **18**, 312.
Goldstein, D. M., Hammon, W. McD., and Viets, H. R. (1946). *J. Am. Med. Assoc.* **131**, 569.
Gomatos, P. J., and Tamm, I. (1962). *Bacteriol. Proc.* p. 145.
Gordon, J. E., ed. (1965). "Control of Communicable Diseases in Man." American Public Health Association, New York.
Gordon, M. P., Huff, J. W., and Holland, J. J. (1963). *Virology* **19**, 416.
Goresline, H. E. (1963). *Public Health Rept. (U.S.)* **78**, 737.
Grausgruber, W. (1963). *Wiener Tierärztl. Monatsschr.* **50**, 678.
Gravelle, C. R., and Chin, T. D. Y. (1962). *Bacteriol. Proc.* p. 151.
Grešíková, M., Havránek, I., and Görner, F. (1961). *Acta Virol.* **5**, 31.
Grešíková-Kohútová, M. (1959a). *Česk. Epidemiol. Mikrobiol. Immunol.* **8**, 26.
Grešíková-Kohútová, M. (1959b). *Acta Virol.* **3**, 215.
Gudnadóttir, M. G. (1961). *J. Exptl. Med.* **113**, 159.
Hahon, N., and Kozikowski, E. (1961). *J. Bacteriol.* **81**, 609.
Halonen, P., Rosen, L., and Huebner, R. J. (1959). *Proc. Soc. Exptl. Biol. Med.* **101**, 236.

Hammon, W. McD., and Work, T. H. (1964). In "Diagnostic Procedures for Viral and Rickettsial Diseases" (E. H. Lennette and N. J. Schmidt, eds.), pp. 268–311. American Public Health Association, New York.

Hancock, B. B., Bohl, E. H., and Birkeland, J. M. (1959). Am. J. Vet. Res. 20, 127.

Hanson, R. P., and Brandly, C. A. (1958). Ann. N.Y. Acad. Sci. 70, 585.

Hargreaves, E. R. (1949). Lancet p. 969.

Hatch, M. H., Kalter, S. S., and Ajello, G. W. (1961). Proc. Soc. Exptl. Biol. Med. 107, 1.

Heberling, R. L., and Cheever, F. S. (1961). Bacteriol. Proc. p. 156.

Heidelbaugh, N. D., and Graves, J. H. (1968). Food Technol. 22, 120.

Hilleman, M. R. (1962). Ann. N.Y. Acad. Sci. 101, 564.

Hirst, G. K. (1965). In "Viral and Rickettsial Infections of Man" (F. L. Horsfall, Jr., and I. Tamm, eds.), pp. 685–688. Lippincott, Philadelphia.

Holland, J. J., McLaren, L. C., and Syverton, J. T. (1959a). J. Exptl. Med. 110, 65.

Holland, J. J., McLaren, L. C., and Syverton, J. T. (1959b). Proc. Soc. Exptl. Biol. Med. 100, 843.

Howe, H. A. (1962). Proc. Soc. Exptl. Biol. Med. 110, 110.

Hsiung, G. D. (1959a). Proc. Soc. Exptl. Biol. Med. 102, 612.

Hsiung, G. D. (1959b). Virology 9, 717.

Hsiung, G. D. (1964). Bacteriol. Proc. p. 131.

Hsiung, G. D., and Henderson, J. R. (1964). "Diagnostic Virology." Yale University Press, New Haven.

Hsiung, G. D., and Melnick, J. L. (1957). J. Immunol. 78, 128.

Hsiung, G. D., and Melnick, J. L. (1958). Ann. N.Y. Acad. Sci. 70, 342.

Hsiung, G. D., Black, F. L., and Paul, J. R. (1959). Bacteriol. Proc. p. 71.

Huebner, R. J. (1963). Ann. N.Y. Acad. Sci. 108, 1129.

Hummeler, K., and Hamparian, V. V. (1958). J. Immunol. 81, 499.

Hummeler, K., and Tumilowicz, J. (1960). J. Immunol. 84, 630.

Jennings, A. R., and Betts, A. O. (1962). Ann. N.Y. Acad. Sci. 101, 485.

Johnson, K. M. (1965). Am. J. Trop. Med. Hyg. 14, 816.

Johnson, K. M., and Lang, D. J. (1962). Proc. Soc. Exptl. Biol. Med. 110, 653.

Johnson, K. M., Bloom, H. H., Rosen, L., Mufson, M. A., and Chanock, R. M. (1961). Virology 13, 373.

Jordan, R. T., and Kempe, L. L. (1957). In "Hepatitis Frontiers" (F. W. Hartman, ed.), pp. 343–354. Little, Brown, Boston.

Jubb, G. (1915). Lancet p. 67.

Kalter, S. S. (1963). "Procedures for Routine Laboratory Diagnosis of Virus and Rickettsial Diseases." Burgess, Minneapolis.

Kamphans, S. (1963). Tierärztl. Umschau 18, 538.

Kaplan, A. S., and Melnick, J. L. (1952). Am. J. Public Health 42, 525.

Kaplan, A. S., and Melnick, J. L. (1954). Am. J. Public Health 44, 1174.

Kaplan, H. S., and Moses, L. E. (1964). Science 145, 21.

Kasel, J. A., Rosen, L., and Evans, H. E. (1963). Proc. Soc. Exptl. Biol. Med. 112, 979.

Kasza, L. (1965). Am. J. Vet. Res. 26, 131.

Kasza, L. (1966). Am. J. Vet. Res. 27, 751.

Katz, S., Jordan, W. S., Jr., Badger, G. F., and Dingle, J. H. (1957). J. Immunol. 78, 118.

Kelly, S., Sanderson, W. W., and Neidl, C. (1961). J. Water Pollution Control Federation 33, 1056.

Ketler, A., Hinuma, Y., and Hummeler, K. (1961). J. Immunol. 86, 22.

Kissling, R. E. (1967). In "Transmission of Viruses by the Water Route" (G. Berg, ed.), pp. 337–346. Interscience, New York.

Kitano, T., Haruna, I., and Watanabe, I. (1961). *Virology* **15**, 503.
Klein, M. (1960). *J. Am. Vet. Med. Assoc.* **137**, 670.
Klein, M. (1962). *Ann. N.Y. Acad. Sci.* **101**, 493.
Klein, M., Early, E., and Zellat, J. (1959). *Proc. Soc. Exptl. Biol. Med.* **102**, 1.
Klein, M., Zellat, J., and Michaelson, T. C. (1960). *Proc. Soc. Exptl. Biol. Med.* **105**, 340.
Knapp, A. C., Godfrey, E. J., Jr., and Aycock, W. L. (1926). *J. Am. Med. Assoc.* **87**, 635.
Koprowski, H. (1958). *Ann. N.Y. Acad. Sci.* **70**, 369.
Krech, U. (1957). *Virology* **4**, 185.
Kunin, C. M., and Minuse, E. (1958). *J. Immunol.* **80**, 1.
Lahelle, O. (1958). *Virology* **5**, 110.
Lauffer, M. A. (1960). *Ann. N.Y. Acad. Sci.* **83**, 727.
Lehmann-Grube, F. (1961). *Arch. Virusforsch.* **11**, 276.
Lemon, H. E. (1964). *Bacteriol. Rev.* **28**, 490.
Lennette, E. H. (1964). *In* "Diagnostic Procedures for Viral and Rickettsial Diseases" (E. H. Lennette and N. J. Schmidt, eds.), pp. 1–66. American Public Health Association, New York.
Lennette, E. H., Schmidt, N. J., and Magoffin, R. L. (1961). *J. Immunol.* **86**, 552.
Libíková, H., Grešíková, M., Řeháček, J., Ernek, E., and Nosek, J. (1963). *Bratislav. Lekárske Listy* **43**, 40.
Liebhaber, H., and Takemoto, K. K. (1961). *Virology* **14**, 502.
Lim, K. A., and Benyesh-Melnick, M. (1960). *J. Immunol.* **84**, 309.
Lipari, M. (1951). *N.Y. State J. Med.* **51**, 362.
Lipton, M. M., and Steigman, A. J. (1963). *J. Infect. Diseases* **112**, 57.
Luoto, L. (1960). *Public Health Rept. (U.S.)* **75**, 135.
Lynt, R. K., Jr. (1966). *Appl. Microbiol.* **14**, 218.
Macrae, A. D. (1962). *Ann. N.Y. Acad. Sci.* **101**, 455.
Mandel, B. (1958). *Virology* **6**, 424.
Mandel, B. (1960). *Ann. N.Y. Acad. Sci.* **83**, 515.
Mandel, B. (1961). *Virology* **14**, 316.
Marchetti, G. E., and Gelfand, H. M. (1963). *Public Health Rept. (U.S.)* **78**, 813.
Mathews, F. P. (1949). *Am. J. Hyg.* **49**, 1.
Mayr, A. (1962). *Ann. N.Y. Acad. Sci.* **101**, 423.
McFerran, J. B. (1962). *Ann. N.Y. Acad. Sci.* **101**, 436.
McKercher, D. G., Wada, E. M., Straub, O. C., and Theilen, G. H. (1963). *Ann. N.Y. Acad. Sci.* **108**, 1163.
McLaren, L. C., Holland, J. J., and Syverton, J. T. (1959). *J. Exptl. Med.* **109**, 475.
Medearis, D. N., Jr., Arnold, J. H., and Enders, J. F. (1960). *Proc. Soc. Exptl. Biol. Med.* **104**, 419.
Melnick, J. L. (1962a). *Am. J. Public Health* **52**, 472.
Melnick, J. L. (1962b). *Ann. N.Y. Acad. Sci.* **101**, 331.
Melnick, J. L. (1965). *In* "Viral and Rickettsial Infections of Man." (F. L. Horsfall, Jr., and I. Tamm, eds.), pp. 841–859. Lippincott, Philadelphia.
Melnick, J. L., Wenner, H. A., and Rosen, L. (1964). *In* "Diagnostic Procedures for Viral and Rickettsial Diseases" (E. H. Lennette and N. J. Schmidt, eds.), pp. 194–242. American Public Health Association, New York.
Merchant, I. A., and Packer, R. A. (1961). "Veterinary Bacteriology and Virology." Iowa State University Press, Ames.
Metcalf, T. G., and Stiles, W. C. (1965). *J. Infect. Diseases* **115**, 68.
Meyer, K. F. (1964). *In* "Occupational Diseases Acquired from Animals," pp. 4–35. University of Michigan School of Public Health, Ann Arbor.

Meyer, K. F., and Eddie, B. (1964). *In* "Diagnostic Procedures for Viral and Rickettsial Diseases" (E. H. Lennette and N. J. Schmidt, eds.), pp. 603–639. American Public Health Association, New York.

Mitchell, C. A., Walker, R. V. L., and Bannister, G. L. (1953a). *Can. J. Comp. Med.* **17**, 97.

Mitchell, C. A., Walker, R. V. L., and Bannister, G. L. (1953b). *Can. J. Comp. Med.* **17**, 218.

Mitchell, C. A., Walker, R. V. L., and Bannister, G. L. (1954). *Can. J. Comp. Med.* **18**, 426.

Mitchell, C. A., Walker, R. V. L., and Bannister, G. L. (1956). *Can. J. Microbiol.* **2**, 322.

Mitchell, C. A., Nordland, O., and Walker, R. V. L. (1958). *Can. J. Comp. Med.* **22**, 154.

Moll, T., and Davis, A. D. (1959). *Am. J. Vet. Res.* **20**, 27.

Moll, T., and Finlayson, A. V. (1957). *Science* **126**, 401.

Moscovici, C., LaPlaca, M., Maisel, J., and Kempe, H. (1961). *Am. J. Vet. Res.* **22**, 852.

Mosley, J. W. (1959). *New Engl. J. Med.* **261**. 703; 748.

Mosley, J. W., and Enders, J. F. (1961). *Proc. Soc. Exptl. Biol. Med.* **108**, 406.

Nardi, G. (1964). *Ann. Sclavo* **6**, 379.

Newlin, S. C., and McKee, A. P. (1965). *Bacteriol. Proc.* p. 101.

Norman, A. (1960). *Virology* **10**, 384.

Norman, A., and Veomett, R. C. (1960). *Virology* **12**, 136.

Peizer, L. R., Mandel, B., and Weissman, D. (1961). *Proc. Soc. Exptl. Biol. Med.* **106**, 772.

Philipson, L., and Choppin, P. W. (1960). *J. Exptl. Med.* **112**, 455.

Philipson, L., and Choppin, P. W. (1962). *Virology* **16**, 405.

Philipson, L., Albertsson, P. A., and Frick, G. (1960). *Virology* **11**, 553.

Piszczek, E. A., Shaughnessy, H. J., Zichis, J., and Levinson, S. O. (1941). *J. Am. Med. Assoc.* **117**, 1962.

Plager, H. (1962). *Ann. N.Y. Acad Sci.* **101**, 390

Plotkin, S. A., Carp, R. I., and Graham, A. F. (1962). *Ann. N.Y. Acad Sci.* **101**, 357.

Plowright, W. (1962). *Ann. N.Y. Acad. Sci.* **101**, 548.

Pohjanpelto, P. (1962). *Virology* **16**, 92.

Reisinger, R. C. (1963). *Ann. N.Y. Acad. Sci.* **108**, 855.

Rhim, J. S., and Melnick, J. L. (1961). *Virology* **15**, 80.

Rhim, J. S., Smith, K. O., and Melnick, J. L. (1961). *Virology* **15**, 428.

Riggs, J. L., and Brown, G. C. (1961). *Bacteriol. Proc.* p. 145.

Roizman, B., Mayer, M. M., and Roane, P. R., Jr. (1959). *J. Immunol.* **82**, 19.

Rose, H. M. (1964). *In* "Diagnostic Procedures for Viral and Rickettsial Diseases" (E. H. Lennette and N. J. Schmidt, eds.), pp. 434–454, American Public Health Association, New York.

Rosen, L. (1964). *In* "Diagnostic Procedures for Viral and Rickettsial Diseases" (E. H. Lennette and N. J. Schmidt, eds.), pp. 259–267. American Public Health Association, New York.

Rosen, L., Baron, S., and Bell, J. A. (1961). *Proc. Soc. Exptl. Biol. Med.* **107**, 434.

Rowe, W. P., and Hartley, J. W. (1962). *Ann. N.Y. Acad. Sci.* **101**, 466.

Sabin, A. B. (1959). *Science* **130**, 1387.

Salzman, N. P. (1960). *Virology* **10**, 150.

Schmidt, N. J. (1964). *In* "Diagnostic Procedures for Viral and Rickettsial Diseases" (E. H. Lennette and N. J. Schmidt, eds.), pp. 78–176. American Public Health Association, New York.

Schmidt, N. J., Guenther, R. W., and Lennette, E. H. (1961). *J. Immunol.* **87**, 623.

Schmidt, N. J., Dennis, J., and Lennette, E. H. (1962). *Proc. Soc. Exptl. Biol. Med.* **109,** 364.

Schwerdt, C. W., and Schaffer, F. L. (1956). *Virology* **2,** 665.

Shahan, M. S., and Traum, J. (1958). *Ann. N.Y. Acad. Sci.* **70,** 614.

Sharpless, G. R. (1962). *Ann. N.Y. Acad. Sci.* **101,** 515.

Shaw, E. D., Newton, A., Powell, A. W., and Friday, C. J. (1961). *Virology* **15,** 208.

Smadel, J. E., and Jackson, E. B. (1964). *In* "Diagnostic Procedures for Viral and Rickettsial Diseases" (E. H. Lennette and N. J. Schmidt, eds.), pp. 743–772. American Public Health Association, New York.

Soller, A. (1961). *Virology* **13,** 267.

Speir, R. W. (1961). *Virology* **14,** 382.

Spendlove, R. S., Lennette, E. H., Knight, C., and Chin, J. (1965). *Bacteriol. Proc.* p. 110.

Squeri, L., Iolt, A., and Garrani, A. (1963). *Ann. Sclavo* **5,** 729.

Steinman, H. G., and Murtaugh, P. A. (1959). *Virology* **7,** 291.

Stoenner, H. G. (1964). *In* "Occupational Diseases Acquired from Animals," pp. 36–52. University of Michigan School of Public Health, Ann Arbor.

Syverton, J. T., Fischer, R. G., Smith, S. A., Dow, R. P., and Schoof, H. F. (1952). *Federation Proc.* **11,** 483.

Tamm, I., and Eggers, H. J. (1962). *Virology* **18,** 439.

Tarshis, I. B. (1962). *Am. J. Trop. Med. Hyg.* **11,** 705.

Trump, J. G., and Wright, K. A. (1957). *In* "Hepatitis Frontiers" (F. W. Hartman, ed.), pp. 333–341. Little, Brown, Boston.

Tytell, A. A., Torop, H. A., and McCarthy, F. J. (1962). *Proc. Soc. Exptl. Biol. Med.* **109,** 916.

Vargosko, A. J., Jeffries, B. C., Chanock, R. M., Kim, H. W., and Parrott, R. H. (1964). *Bacteriol. Proc.* p. 131.

Wallis, C., and Melnick, J. L. (1962*a*). *Virology* **16,** 122.

Wallis, C., and Melnick, J. L. (1962*b*). *Virology* **16,** 504.

Wallis, C., and Melnick, J. L. (1962*c*). *Proc. Soc. Exptl. Biol. Med.* **111,** 305.

Wallis, C., Yang, C.-S., and Melnick, J. L. (1962*a*). *J. Immunol.* **89,** 41.

Wallis, C., Melnick, J. L., and Bianchi, M. (1962*b*). *Texas Rept. Biol. Med.* **20,** 693.

Ward, R. (1952). *Federation Proc.* **11,** 486.

Weidenkopf, S. J. (1958). *Virology* **5,** 56.

Welsh, H. H., Lennette, E. H., Abinanti, F. R., and Winn, J. F. (1958). *Ann. N.Y. Acad. Sci.* **70,** 528.

Wenner, H. A. (1962*a*). *Ann. N.Y. Acad. Sci.* **101,** 343.

Wenner, H. A. (1962*b*). *Ann. N.Y. Acad. Sci.* **101,** 398.

Wenner, H. A. (1964). *In* "Diagnostic Procedures for Viral and Rickettsial Diseases." (E. H. Lennette and N. J. Schmidt, eds.), pp. 243–258. American Public Health Association, New York.

Woods, W. A., Weiss, R. A., and Robbins, F. C. (1962). *Proc. Soc. Exptl. Biol. Med.* **111,** 401.

Work, T. H. (1964). *In* "Diagnostic Procedures for Viral and Rickettsial Diseases" (E. H. Lennette and N. J. Schmidt, eds.), pp. 312–355. American Public Health Association, New York.

Youngner, J. S. (1956). *J. Immunol.* **76,** 288.

Youngner, J. S. (1957). *J. Immunol.* **78,** 282.

CHAPTER III | **HALOPHILIC**

VIBRIO INFECTIONS

Riichi Sakazaki

I. HISTORY

In the summer season, food poisoning with symptoms of acute gastro-enteritis frequently occurs. Sea fish and fish products contribute to these outbreaks in Japan, where it is a custom for the people to eat raw fish. The causative agent of the food poisoning has been unknown for a long time, but a group of facultatively halophilic organisms has been recently designated as the most important culprit. Now it is said that over 70% of the cases of bacterial gastroenteritis in Japan may be caused by these organisms.

The enteropathogenic, halophilic organisms were first isolated from autopsy materials, which were collected from an outbreak of food poisoning by Fujino and his co-workers (1951). They considered that the organisms were members of the genus *Pasteurella* and suggested the species

115

name *Pasteurella parahemolytica,* but they were unaware of the halophilism of the organisms. Takikawa and Fujisawa (1956) found an halophilic bacterium in an outbreak of gastroenteritis. Later, Takikawa (1958) encountered several cases similar to that described above, and he compared his isolates with the strain isolated by Fujino *et al.* He believed that *P. parahemolytica* was in the same category as the organisms of his collection and proposed the name *Pseudomonas enteritis* for these halophilic organisms, although the specific epithet "parahemolytica" *(P. parahemolytica,* Fujino *et al.,* 1951) had priority to the epithet "enteritis" *(P. enteritis,* Takikawa, 1958), which was a grammatical error and should be replaced by "enteritidis." On the other hand, Miyamoto and his co-workers (1961*a*) proposed a new genus, *Oceanomonas,* for the halophilic organisms. Recently, however, Sakazaki and his co-workers (1963) carried out morphological, physiological, and biochemical studies of a total of 1702 cultures of the organisms. They classified these organisms into the genus *Vibrio* and suggested a species name *Vibrio parahaemolyticus* for the organisms. The nomenclature has been widely accepted among many Japanese bacteriologists.

While disputing the taxonomic rank of the halophilic organisms, many Japanese investigators have studied these organisms, which have been isolated repeatedly from other outbreaks of gastroenteritis, and they have clarified their pathogenicity to human beings and their ecology. At present, there is no doubt that members of the halophilic organisms produce gastroenteritis in human beings.

II. BACTERIOLOGY

A. Morphology

Vibrio parahaemolyticus is a gram-negative, straight, and occasionally slightly curved rod which varies from 1 to 3 μ in length and from 0.4 to 0.6 μ in width. The vibrio exhibits pleomorphism, and undulating filament and spheroplast forms which show single polar flagellation are seen.

B. Physiological Characteristics

The vibrios are facultatively anaerobic. They can be cultured readily on or in ordinary media if 1 to 3% sodium chloride is added, but they grow very poorly or do not grow at all on or in media containing no salt. When the vibrios are inoculated into 1% peptone water (pH 7.0), they grow if 0.5 to 9.0% salt is added; the most abundant growth is obtained from media containing 2 to 4% salt. Figure 1 shows the response curves of growth of the vibrios in the presence of various concentrations of sodium chloride in comparison to those of *Vibrio cholerae, Escherichia*

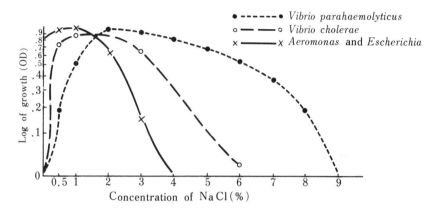

FIG. 1. Response curves of *Vibrio parahaemolyticus* to various concentrations of sodium chloride.

coli, and *Aeromonas hydrophila.* However, the optimum concentration of salt necessary to induce growth of the vibrios may be influenced by the kinds of media employed and the temperature of incubation. For example, the vibrios grow well on blood agar or brain heart infusion agar plates without additional salt.

The vibrios can be cultured readily in a pH range from 5.0 to 9.6, but they grow best at pH 7.5 to 8.5. They grow at temperatures between 15° and 40°C, but maximum growth is obtained at 37.5°C.

Colonies of the vibrios freshly isolated on ordinary agar containing 3% sodium chloride are moist, smooth, circular, and opaque in appearance, and they attain a size of 2 to 3 mm after 24 hours. Following several subcultures, however, various colonies which are translucent, mucoid, or wrinkled dissociate from the original colonies mentioned above. On blood agar plate, a hemolytic zone is produced around the colonies. In a broth medium, the vibrios produce clouding and, sometimes, a pellicle on the surface of the broth.

The vibrios grow relatively well on MacConkey and Aronson agars containing 3% sodium chloride, but they fail to grow on SS agar.

With few exceptions, the vibrios are sensitive to streptomycin, tetracycline, chloramphenicol, and novobiocin, and are resistant to polymyxin B and colistin. They are also sensitive to the vibriostatic agent 0-129 (2,4-diamino-6,7-diisopropylpteridine), which was employed for the identification of the genus *Vibrio* by Shewan *et al.* (1954).

C. Biochemical Characteristics

The biochemical characteristics of *V. parahaemolyticus* are summarized in Table I. All the vibrios reduce nitrate to nitrite, fail to produce hydrogen sulfide in the butt of Kligler's iron agar, liquefy gelatin, produce indole, and give a positive cytochrome oxidase test of Gaby and Hadley (1957) and negative Voges-Proskauer and phenylalanine deaminase (PPA) tests. Generally, citrate is utilized as a sole carbon source, and urea and malonate are not decomposed.

The vibrios ferment glucose, maltose, trehalose, and mannitol without forming gas in the modified Hugh-Leifson medium (MOF medium according to Leifson, 1963), and they fail to acidify lactose, rhamnose, sucrose, xylose, adonitol, dulcitol, inositol, and salicin. Starch is not hydrolyzed, and cellobiose is not fermented within 24 hours of incubation. Fermentation of arabinose is different with biotypes. Very rarely does a strain acidify sucrose.

A majority of the vibrios give a positive reaction in the cholera red test. No luminescence is observed.

D. Serological Characteristics

Three antigenic components can be recognized in organisms of *V. parahaemolyticus*. The somatic (O) antigen is thermostable, and it is not destroyed by 50% alcohol and *n*-HCl treatment at 37°C for 24 hours. The K antigen is an envelope or capsular antigen which is thermolabile. The vibrios, freshly isolated, possess well-developed K antigen, and such organisms are not agglutinated in the living state with the homologous O antiserum.

Antigenic studies of *V. parahaemolyticus* have been carried out by Sakazaki (1965a), and an antigenic scheme that includes ten O groups and thirty-two K antigen types has been established (Table II). This scheme will contribute much in epidemiological studies.

It was demonstrated by Sakazaki (1965a) and by Oomori and Iwao (1963) that the K antigen appearing within the vibrios is of the same category as the B antigen of *Escherichia coli* in its physical properties. However, it is more thermostable than the B antigen. In some cases, when cultures are boiled at 100°C for 2 hours, they lose their agglutinability to the homologous K antiserum, but they still retain their O-

TABLE I
BIOCHEMICAL CHARACTERISTICS OF
Vibrio parahaemolyticus

Test (substrate)	Reaction[a]
Indole	+
Methyl red	+
Voges-Proskauer	−
Ammonium citrate	+
Hydrogen sulfide (Kligler agar)	−
Gelatin liquefaction	+
Nitrate reduction	+
Cytochrome oxidase test	+
Phenylalanine deamination	−
Urease	−
Organic acid medium	
Citrate	−
D-Tartrate	+
Mucate	+
Malonate	−
Catalase	+
Casein hydrolysis	+
Alginate utilization	−
Tyrosine dissolution	+
Tributyline utilization	+
Luminescence	−
Cholera red reaction	d
Hugh-Leifson medium	F
Fermentation tests	
Glucose	+
Arabinose	d
Cellobiose	−, (+)
Lactose	−
Maltose	+
Rhamnose	−
Sucrose	−
Trehalose	+
Xylose	−
Adonitol	−
Dulcitol	−
Inositol	−
Mannitol	+
Sorbitol	−
Salicin	−
Gas from glucose	−

[a] + = positive; (+) = late positive; − = negative;
d = different with subspecies; F = fermentation.

TABLE II
ANTIGENIC SCHEMA OF *Vibrio parahaemolyticus*[a]

O group	K antigen	O group	K antigen
1	1	4	8
	25		9
	26		10
	32		11
2	2		12
	3		13
	27	5	15
	28		17
3	4	6	18
	5		
	6	7	19
	7		20
	29	8	21
	30		22
	31	9	23
		10	24

[a] K antigens 14 and 16 were discarded from the schema, because the test strains of these K antigens were lost.

inagglutinability. In addition, each K antigen is specific in the individual O group; for example, K antigens 2 and 3 are demonstrated only in O group 2. It is believed, therefore, that the determination of O groups is impracticable for use in routine serological typing of the vibrios.

Flagellar (H) antigen determination of the vibrios is very difficult because the organism flagellates monotrichously. The antigenic study of the flagellum of the vibrio has not been carried out as yet.

No significant antigenic relationship between *V. parahaemolyticus* and *V. cholerae,* or Gardner and Venkatraman's vibrio, has been observed as far as the agglutination test is concerned.

E. Pathogenicity

Since 1960, many Japanese investigators have studied the vibrios. The organisms have been detected in a large number of outbreaks of food poisoning and in sporadic cases where the vibrios appeared in the stools in pure cultures and disappeared quickly as symptoms subsided. No other significant causative agents were detectable from these cases. The vibrios are not found in the feces of healthy persons. Judging from the results of feeding experiments with human volunteers performed by

Takikawa (1958) and Aiiso and Fujiwara (1963), there is no doubt that *V. parahaemolyticus* produces gastroenteritis.

Zen-Yoji (1964) and Sakazaki and his co-workers (1963) demonstrated that, when the broth cultures of the vibrios were inoculated into the ligated rabbit gut, by the method of De and Chatterje (1953), severe enteritis was produced in the loop of the gut. When 3-week-old mice of the CFW strain received intraperitoneally 0.5 ml of an overnight broth culture of the vibrios, almost all of the animals died within 24 hours, but a majority of the mice inoculated with 0.5 ml of a 10-fold dilution of the broth cultures survived.

Under natural conditions, *V. parahaemolyticus* may be pathogenic only to man. Aiiso (1961) stated that *V. parahaemolyticus* could produce experimental gastroenteritis in dogs and monkeys, but other workers (Zen-Yoji, 1964) reported that the vibrios produced no illness in any animals.

It has been uncertain whether all members of the species are entero-pathogenic to human beings. Sakazaki (1965a) observed that most of all K antigen types were found not only in sea water and sea fish but also in patients, and that K antigen types frequently isolated from sea water and sea fish were also found predominantly in cultures from patients. From these results, he considered that all members of the *V. para-haemolyticus* may be able to produce human gastroenteritis.

Fujino *et al.* (1951), who first found *V. parahaemolyticus* in food poisoning, described a hemolytic activity of the vibrios but did not note the pathogenic significance of the hemolytic activity. Recently, Kato *et al.* (1965) found that the vibrio cultures isolated from human patients affected with gastroenteritis were hemolytic and those from sea water and sea fish were nonhemolytic when the vibrios were placed on brain heart infusion agar containing human blood. Although all cultures of *V. parahaemolyticus* lysed red cells on blood agar, Sakazaki, Kato, and Obara (1966, unpublished data) found that the hemolytic activity of most of the vibrios originating from sea fish was inhibited if 0.2% glucose was added to blood agar after the basal medium was sterilized. Using blood agar plates which contained 0.5% Thiotone (BBL), 1% NaCl, 0.2% $Na_2HPO_4 \cdot H_2O$, 1.5% agar, 0.2% glucose, and 1% washed human red cells, Sakazaki *et al.* (1968) demonstrated that 96% of 2720 cultures which were isolated from diarrheal stools were hemolytic, whereas only 1% of 650 cultures from sea fish showed such activity. The same results were obtained by Kato *et al.* (1966) and Asakawa *et al.* (1966). No re-lationships were found between biochemical and serological properties and the hemolytic activity of the vibrios.

Feeding experiments with the hemolytic and nonhemolytic vibrios were carried out on human adult volunteers by Sakazaki *et al.* (1968). When a volunteer received 10 million organisms of the hemolytic vibrios isolated from diarrheal stools, he revealed clinical signs of abdominal ache and diarrhea after 5 hours of incubation. However, 14 volunteers who received nonhemolytic vibrios originating from sea fish revealed no signs of illness, although over a thousand million organisms were consumed.

The hemolytic activity of the vibrios is demonstrable in supernatant from a centrifuged broth culture of the organisms. The hemolytic substance in the supernatant was studied by Kato *et al.* (1966, 1967) cooperating with the author. A hemolytic substance toxic for mice and Hela cells was precipitated in a sugar-free and nucleic acid-free protein fraction by 30 to 60% saturation of ammonium sulfate at pH 4.5 and zone chromatography on a Sephadex G-200. The hemolytic activity of the fraction remained after heating at 100°C for 30 minutes at pH 6.0, but was destroyed by trypsin digestion. The fraction revealed no activity of lethicinases A and B or gelatinase.

It is clear, from the results described above, that only the hemolytic culture is enteropathogenic. However, the role of the enteritis-producing power of the toxic substance obtained from the hemolytic vibrios is uncertain. Kato *et al.* (1967), in cooperation with the author, demonstrated with intraduodenal dog inoculation tests that the enteritis-producing power of the toxic substance was much weaker than that of live organisms. They also found that a variant of the vibrios which lacked the K antigen could not produce enteritis in inoculated dogs, although the vibrios were strongly hemolytic. From results of epidemiological and clinical investigations of vibrio food poisoning, there is no doubt that gastroenteritis caused by the vibrios is of the infectious and not the toxic type. It is probable, therefore, that the enteritis-producing power of the hemolytic vibrios may be in the living bacterial cell itself. The antigen and toxic substance may have a supplemental effect in producing enteritis.

According to recent studies on the hemolytic activity of the vibrios mentioned above, a contradiction has arisen about the epidemiology of vibrio food poisoning. As stated previously, it is evident from epidemiological investigations carried out by many workers that sea fish and their products are the main causative foodstuffs of the food poisoning. Nevertheless, almost all cultures of the vibrios originating from sea fish are apathogenic for man. On the other hand, it has been demonstrated by many workers that the vibrios do not infect from person to person. There will be many future problems to study on the pathogenicity of the vibrios.

F. Distribution

Miyamoto *et al.* (1960, 1961*b,c*), Sekine *et al.* (1962), Horie *et al.* (1963), Wagatsuma (1961), and Noguchi (1964) demonstrated that the vibrios are widely distributed in the coastal sea of Japan. Tubosaki and Sakazaki (1962, unpublished data), Takahira (1965), and Yasunaga (1964) found the vibrios in the coastal sea waters of the United States, the Philippines, Taiwan, Hong Kong, and Singapore. It may well be that *V. parahaemolyticus* is an inhabitant of the coastal sea.

G. Vibrios Similar to *V. parahaemolyticus*

Different halophilic vibrios are found in sea water and sea fish. Two kinds of organisms, *V. alginolyticus* (Sakazaki, 1965*a*) and *V. anguillarum*-like organism, possess biochemical and ecological properties similar to those of *V. parahaemolyticus.*

Vibrio alginolyticus is more frequently isolated from coastal sea water and sea fish than *V. parahaemolyticus,* and sometimes it is found in the feces of human patients affected with gastroenteritis. The vibrios grow well in peptone water containing 10% sodium chloride, ferment sucrose, and give a positive reaction in the Voges-Proskauer test, but they behave similarly to *V. paraheaemolyticus* in most other biochemical tests. In addition, the growth of *V. alginolyticus* is characterized by swarming on the surface of ordinary solid media containing sodium chloride of suitable concentration.

Vibrio anguillarum-like organism is a halophilic sea inhabitant which may be pathogenic to sea fish (Akazawa *et al.,* 1963). The vibrios are often mistaken for *V. parahaemolyticus* in routine isolation procedure because a majority of the cultures of the vibrios fail to ferment sucrose and to produce acetoin. However, they do not grow in peptone water containing 7% sodium chloride, whereas *V. parahaemolyticus* grows well in the peptone water. The vibrios are found only in sea water and in sea fish shortly after capture, and they are never found in the feces of patients suffering from gastroenteritis.

The important differential points of these three vibrios are listed in Table III.

III. EPIDEMIOLOGY

A. Symptoms

Food poisoning due to *V. parahaemolyticus* is of the infectious type, and gastroenteritis develops in most cases of food poisoning. Symptoms

TABLE III

DIFFERENTIATION AMONG *Vibrio parahaemolyticus, Vibrio alginolyticus,*
AND A *Vibrio* SPP. RESEMBLING *Vibrio anguillarum*[a]

Characteristics	V. parahaemolyticus	V. alginolysticus	Vibrio spp. resembling V. anguillarum
Growth in peptone water containing 7% NaCl	+	+	−
Growth in peptone water containing 10% NaCl	−	+	−
Voges-Proskauer reaction	−	+	−
Sucrose fermentation	−	+	− or +
Arabinose fermentation	+ or −	− or +	−
Cellobiose fermentation within 24 hours	−	−	+ or −

[a] + = positive; − = negative; + or − = usually positive, occasionally negative; − or + = usually negative, occasionally positive.

usually appear 12 hours after the infected food has been eaten, although the interval may be as short as 2 hours or as long as 48 hours.

The outstanding symptoms are abdominal pain and diarrhea, which are usually associated with nausea and vomiting. Mild fever, chills, and headache are also seen in most of the cases. The symptoms of vibrio gastroenteritis are very similar to those of *Salmonella* infection, but abdominal pain is generally more than that in *Salmonella* gastroenteritis. The pain may be felt in the stomach rather than in the abdomen.

A second clinical type of vibrio infection is a dysentery-like disease. Patients with this type of infection have fever, excretion of stools with mucus and blood, and symptoms closely simulating those of bacillary dysentery.

Recovery is usually complete within 2 to 5 days. The fatality rate is very low, and the majority of deaths from this type of food poisoning occurs in old, debilitated adults.

B. Incidence

It is difficult to assess the real incidence of food poisoning due to *V. parahaemolyticus.* Food poisoning was not made reportable until 1951, although a lot of food poisoning implicating sea fish and their products broke out during the summer season. Food poisoning due to *V. parahaemolyticus* was first reported by Fujino *et al.* in 1951. The outbreak involved 272 persons, including 20 fatal cases. Boiled and semi-dried young sardine was believed to be the causative foodstuff. However, no attention was paid to their paper for several years until the second outbreak of food poisoning, involving 120 persons consisting of 73 in-patients and 47 hospital employees of the Yokohama National Hospital, was described by Takikawa and Fujisawa (1956). The causative foodstuff was salted cucumber, which is called "asazuke" in Japan. Later, Takikawa (1958) encountered several endemic and sporadic cases of similar food poisoning. In 1960, an explosive epidemic of food poisoning due to horse mackerel, affecting several thousands of persons, broke out along the Pacific coast of Japan. It was confirmed by many workers that the causative agent of the epidemic was *V. parahaemolyticus*. Since then, the isolation of vibrios from cases of gastroenteritis has been performed in almost all bacteriological laboratories in Japan as routine procedure. Consequently, it has been clarified that vibrio gastroenteritis is one of the diagnoses encountered in bacterial food poisoning. According to statistics prepared by the Ministry of Welfare of Japan, over 70% of food poisoning cases in 1963 were caused by the vibrios (Fig. 2). In-

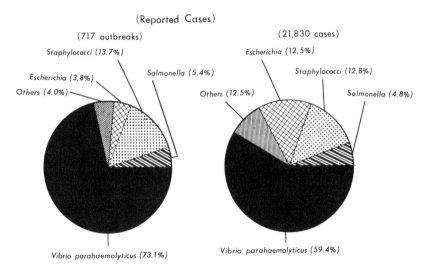

FIG. 2. Incidence of bacterial food poisoning in Japan (1963).

cidentally, most of the larger outbreaks were recorded by the Ministry of Welfare, but the smaller outbreaks and sporadic cases were probably overlooked. There is a general impression, therefore, that the real incidence of food poisoning is much higher than the number reported in the statistics.

The incidence of vibrio food poisoning is strictly associated with the atmospheric temperature. Food poisoning due to organisms other than the vibrios—*Salmonella, Escherichia,* etc.—also vary in their seasonal incidence, but food poisoning due to the vibrios is strictly confined to the warmer months of the year—May to October, in Japan. Vibrio food poisoning is never found during the cold season. It was demonstrated by Miyamoto *et al.* (1960, 1961*b,c*), and Noguchi (1964) that the number of vibrios along the coastal sea water decreases during the winter season.

The causative foodstuff in vibrio food poisoning is associated directly or indirectly with sea fish or sea water; raw sea fish meat is the most important foodstuff. Sometimes salted vegetables are the source of food poisoning. In this case, the vegetables become contaminated with the vibrios from a chopping board or a kitchen knife with which the sea fish has been prepared for cooking.

The incidence of vibrio food poisoning may be restricted to Japan as far as present knowledge is concerned. However, food poisoning due to the vibrios may exist in other countries, especially in tropical and subtropical zones. Sakazaki and Gomez (unpublished data in 1965) have found several cultures of vibrios among the so-called NAG (nonagglutinable) vibrios isolated from cholera cases in the Philippines.

IV. ISOLATION AND IDENTIFICATION

The diagnosis of gastroenteritis due to *V. parahaemolyticus* rests on the isolation of the vibrios from the feces and vomit of patients and from the responsible food.

A. Isolation

The isolation of vibrios from specimens of patients is not difficult. Several selective plating media have been developed for the isolation of the vibrios. Among them, BTB–salt–Teepol agar of Akiyama *et al.* (1963), modified by Sakazaki (1965*c*), and TCBS (thiosulfate–citrate–bile salt–sucrose) agar of Kobayashi *et al.* (1963) may give excellent results.

The BTB–salt–Teepol agar modified contains (per 100 ml): 1 gm of peptone, 0.3 gm of beef extract, 4 gm of sodium chloride, 2 gm of sucrose, 1.5 gm of agar, 0.2 ml of Teepol, which is a commercial anionic detergent (Shell), or 0.02 gm of sodium lauryl sulfate, 0.004 gm of bromthymol blue, and 0.004 gm of thymol blue (pH 9.0). The TCBS agar medium is higher in selectivity than the BTB–salt–Teepol agar, and it is satisfactory for the isolation not only of *V. parahaemolyticus* but of *V. cholerae* when used in the following composition (per 100 ml): 0.5 gm of yeast extract, 1 gm of peptone, 1 gm of sodium chloride, 1 gm of sodium citrate, 1 gm of sodium thiosulfate, 2 gm of sucrose, 0.5 gm of sodium cholate, 0.5 gm of ox gall, 1.5 gm of agar, 0.004 gm of thymol blue, and 0.04 gm of bromthymol blue (pH 8.6).

The growth of gram-positive organisms, a majority of the coliform bacteria, and many strains of *Proteus* is inhibited on these plating media. Isolated colonies of *V. parahaemolyticus,* after 18 to 24 hours of incubation, are round and 2 to 3 mm in diameter with a green or blue center which is stained by the alkaline bromthymol blue indicator. Colonies of *V. alginolyticus,* which are yellow-colored due to fermentation of sucrose, are usually larger than those of *V. parahaemolyticus.* Few strains of coliform bacteria, *Proteus* and enterococci, may grow on these agar media, but they form very small, translucent colonies which are also distinguishable from those of *V. parahaemolyticus.*

For the isolation of the vibrios from fecal specimens of convalescent patients, sea water, and sea fish, it is necessary to use the enrichment procedure. The salt–colistin broth of Sakazaki (1965c), which contains 0.3% yeast extract, 1% peptone, 2% sodium chloride, and 500 units of colistin metansulfonate per milliliter and the glucose–salt–Teepol broth of Akiyama *et al.* (1963), which contains 1% peptone, 0.3% beef extract, 2% sodium chloride, 0.4% Teepol, 0.5% glucose, and 0.0002% methyl violet (pH 9.2), are recommended for the enrichment of the vibrios.

B. Identification

Colonies suspected of being *V. parahaemolyticus* are usually picked into TSI medium containing 2 or 4% sodium chloride, in which the vibrios form acid in the butt, an alkaline slant, and no hydrogen sulfide. For additional biochemical tests, the cytochrome oxidase test of Gaby and Hadley (1957), the indole test, and the Voges-Proskauer test using Barritt's method, are recommended. In these tests, 2 or 4% sodium chloride should be added to the media.

If cultures to be examined have been isolated from fecal specimens of patients affected with diarrhea, the vibrios can be accurately identified with the use of only the tests mentioned above. If cultures are isolated from sea fish, sea water, and foodstuff, however, additional biochemical and physiological tests are necessary to identify *V. parahaemolyticus,* because many similar vibrios may exist in these materials. Other tests include the examination of halophilism and salt tolerance, by using peptone water in which 0%, 3%, 7%, and 10% sodium chloride are added, the nitrate reduction test, the gelatin liquefaction test, the motility test, and the fermentation test using the MOF medium of Leifson (1963). Slide agglutination tests with thirty-two K antisera are performed to determine the serotype of the vibrio which is valuable for etiological study.

For details of the isolation and identification of the vibrios, the reader should refer to the procedure of Sakazaki (1965b).

REFERENCES

Aiiso, K. (1961). *Shokuhin Eisei Kenkyu* **11,** 65 (text in Japanese).
Aiiso, K., and Fujiwara, K. (1963). *Ann. Rept. Inst. Food Microbiol. Chiba Univ.* **15,** 34.
Akazawa, I., Ichiura, Y., Ibuki, H., and Miura, U. (1963). *J. Japan. Assoc. Infect. Diseases* **37,** 21 (text in Japanese).
Akiyama, S., Takizawa, K., Ichinoe, H., Enomoto, S., Kobayashi, T., and Sakazaki, R. (1963). *Japan. J. Bacteriol.* **18,** 255 (text in Japanese).
Asakawa, Y., Akabane, M., and Noguchi, M. (1966). The 23rd Meeting of the Japanese Society of Public Health, Chiba.
De, S. N., and Chatterje, D. N. (1953). *J. Pathol. Bacteriol.* **66,** 559.
Fujino, T., Okuno, Y., Nakada, D., Aoyama, A., Fukai, K., Murai, K., and Ueho, T. (1951). *J. Japan. Assoc. Infect. Diseases* **25,** 11 (text in Japanese).
Gaby, W. Z., and Hadley, C. (1957). *J. Bacteriol.* **74,** 356.
Horie, S., Saheki, K., Kojima, T., and Sekine, Y. (1963). *Bull. Japan. Soc. Sci. Fisheries* **29,** 37.
Kato, T., Obara, H., Ichinoche, H., Akiyama, S., Takizawa, K., and Matsushima, Y. (1965). *Shokuhin Eisei Kenkyu* **15,** 81 (text in Japanese).
Kato, T., Obara, H., Ichinoche, H., Yamai, S., Miyamoto, Y., and Sakazaki, R. (1966). The 23rd Meeting of the Japanese Society of Public Health, Chiba.
Kato, T., Obara, H., Yamai, S., Hobo, K., Sakazaki, R., and Tamura, K. (1967). The 25th Meeting of the Japanese Society of Public Health, Sendai.
Kobayashi, S., Enomoto, S., Sakazaki, R., and Kuwahara, S. (1963). *Japan. J. Bacteriol.* **10–11,** 387 (text in Japanese).
Leifson, E. (1963). *J. Bacteriol.* **85,** 1183.
Miyamoto, Y., Nakamura, K., Takizawa, K., and Kodama, T. (1960). *Japan. J. Public Health* **7,** 587 (text in Japanese).
Miyamoto, Y., Nakamura, K., and Takizawa, K. (1961a). *Japan. J. Microbiol.* **5,** 477.
Miyamoto, Y., Nakamura, K., Takizawa, K., and Kodama, T. (1961b). *Japan. J. Public Health* **8,** 673 (text in Japanese).
Miyamoto, Y., Nakamura, K., Takizawa, K., and Kodama, T. (1961c). *Japan. J. Public Health* **8,** 703 (text in Japanese).
Noguchi, M. (1964). *In* "Vibrio parahaemolyticus" (T. Fujino and H. Fukumi, eds.), pp. 289–311. Isseido, Tokyo.
Oomori, G., and Iwao, M. (1963). *Japan. J. Bacteriol.* **18,** 379 (text in Japanese).
Sakazaki, R. (1965a). *Report of the SEATO Cholera Research Symposium,* Honolulu.

Sakazaki, R. (1965b). "Vibrio parahaemolyticus. Isolation and Identification." Nihon Eiyo-Kagaku, Tokyo.

Sakazaki, R., Iwanami, S., and Fukumi, H. (1963). Japan. J. Med. Sci. Biol. 16, 161.

Sakazaki, R., Tamura, K., Kato, T., and Obara, H. (1968). In press.

Sekine, Y., Horie, S., and Saheki, K. (1962). Bull. Japan. Soc. Sci. Fisheries 28, 920.

Shewan. J. M., Hodgikiss, W., and Liston, J. (1954). Nature 173, 208.

Takahira, Y. (1965). Personal communication.

Takikawa, I. (1958). Yokohama Med. Bull. 2, 313.

Takikawa, I., and Fujisawa, T. (1956). Shokuhin Eisei Kenkyu 6, 15 (text in Japanese).

Wagatsuma, S. (1962). Modern Media 8, 236 (text in Japanese).

Yasunaga, S. (1964). Endemic Disease Bull. Nagasaki Univ. 6, 201 (text in Japanese).

Zen-Yoji, H. (1964). In "Vibrio parahaemolyticus" (T. Fujino and H. Fukumi, eds.), pp. 181–193. Isseido, Tokyo (text in Japanese).

CHAPTER IV | **CLOSTRIDIUM PERFRINGENS**

AND BACILLUS CEREUS INFECTIONS

Betty C. Hobbs

I. INTRODUCTION AND HISTORY

A. *Clostridium perfringens*

Pathological conditions in man and animals caused by *Clostridium perfringens* have been described since 1892. The diseases in man vary from gas gangrene due primarily to α-toxin, to enteritis of varying severity including necrosis and mild diarrhea. The diseases in animals are the dysentery of lambs and the enterotoxemia of sheep and calves, but it is not proposed to describe these diseases in this chapter.

The association of *C. perfringens* with outbreaks of mild but chronic diarrhea was suggested by Klein (1895) and by Andrews (1899). They said that abdominal pain was common but vomiting rare, and these symptoms are similar to those of *C. perfringens* food poisoning today. Dunham

(1897) reported five cases of infection with *"Bacillus aerogenes capsula-tus (welch)"* and gave a time and temperature of 1 minute at 98°C for the heat resistance of the spores. Wild (1898) said that spores were more heat-resistant in stools than in cultures. Von Hibler (1906) and Rodella (1910) stated that the spores of *C. perfringens* would survive 1 hour at 100°C.

Simonds (1915) made pertinent statements about the sporulation of *C. perfringens*. In feces it was related to intestinal disturbances; it was influenced by the reaction of the meat medium in the presence or absence of fermentable carbohydrates; it occurred promptly in pure cultures in sterilized alkaline and neutral but not acid suspensions of feces; and temperature was important.

Later there were investigations on toxin production in relation to classi-fication (Oakley, 1943, 1949). Oakley and Warrack (1951, 1953) give most of the toxicological facts.

The first warning of food poisoning came from Knox and Macdonald (1943), who described outbreaks in which children were ill after school meals; gravy made the previous day was heavily contaminated with anaerobic sporing bacilli including *C. perfringens*. McClung (1945) was more precise; he described four outbreaks of food poisoning after the con-sumption of chickens steamed the previous day. The predominant symp-toms were abdominal cramp, nausea, and diarrhea, and *C. perfringens* was isolated from the cooked chickens.

In 1946 Cravitz and Gillmore tested strains of *C. perfringens,* sent by McClung and by the American Type Culture Collection (including types A, B, C, and D), by oral administration of filtrates and living broth cultures to man and animals. Some of McClung's strains were likely to have been heat-resistant. This is the earliest record of human volunteer experiments with cultures and filtrates of *C. perfringens*. Sterile culture filtrates heated for 1 hour at 95°C produced symptoms of nausea and emesis, and occasionally cramps and diarrhea, in human subjects from 45 to 80 minutes after administration. Young cats given heated and un-heated sterile filtrates developed similar symptoms in ½ to 4 hours. It was assumed from the results that enterotoxin was acting on peripheral structures of the viscera, the impulses passing through the vagus nerve to the vomiting center; but the symptoms which followed the intake of filtrates were unlike those produced in outbreaks. However, when sterile filtrates were given to dogs, diarrhea sometimes developed, but the re-sults were variable. When rabbits were given living cultures, the majority developed diarrhea, but no symptoms followed the ingestion of sterile filtrates. Living cultures from McClung given in milk to three human volunteers induced in one "bloating and cramps in 4 hours and flatulent diarrhoea several hours later"; two were unaffected. These symptoms were similar to those commonly obtained in food poisoning. The animal tests were carried out not only with McClung's strains but with others, including types A, B, C, and D. They gave similar results, indicating that all types could cause symptoms of food poisoning. But the symptoms arising from the filtrates seemed to be unrelated to those of the typical food poisoning, and they were not confirmed. Investigations by later workers indicated that most volunteers given living cultures actively growing in meat broth media developed the typical food poisoning symptoms of abdominal pain and diarrhea, sometimes accompanied by nausea but rarely by vomiting (Østerling, 1952; Hobbs et al., 1953; Dische and Elek, 1957).

Dack et al. (1954) fed four strains of C. perfringens (two from outbreaks of food poisoning in England) to 32 people with negative results. Looking back on Dack's experiments and the method of administration, one may assume that the dose, the medium in which the organism was suspended and fed, and the variation between strains, aggravated perhaps by laboratory storage, all contributed to the negative results. Whatever the reasons for the results, the recognition of C. perfringens as a food poisoning entity was retarded in the United States.

In 1948 there appeared to be a big outbreak in Hamburg, Germany, at least 400 cases, of an intestinal disease described as enteritis necroticans and reported in a number of papers (Ernst, 1948; Schutz, 1948; Marcuse and König, 1950).

The clinical and bacteriological findings were clarified in a series of papers by Zeissler and Rassfeld-Sternberg (1949), Oakley (1949), and Hain (1949a) concerned with a small outbreak of hemorrhagic enteritis after a family had eaten tinned rabbit meat or fish; this outbreak was apparently part of the much larger episode. The symptoms were predominantly abdominal pain and diarrhea, and the strain of C. perfringens isolated from the food and patients survived 100°C for 1 to 4 hours and produced much β-toxin. At this time Oakley gave the strain the new type letter F, although it is now regarded as a subgroup (heat-resistant) of type C (Brooks et al., 1957; Warrack, 1963; and Sterne and Warrack, 1964).

Soon after this incident 19 of 108 samples of feces from normal people in Hamburg were found to contain heat-resistant C. perfringens (Hain, 1949b), although it was not clear at the time whether these strains were the same as those causing the current outbreak of enteritis necroticans or whether they were the particular type A organism later found to be causing outbreaks of gastroenteritis in the United Kingdom.

Østerling (1952) claimed that *C. perfringens* was the predominant organism in food products suspected to be the vehicle of infection in 15 of 33 outbreaks of food poisoning. He gave the period of incubation as 10 to 12 hours, the symptoms as diarrhea and abdominal pain, rarely vomiting, and the duration as generally not more than 24 hours.

Hobbs and her colleagues (1953) described epidemiological and laboratory studies of outbreaks of food poisoning due to particular strains of *C. perfringens,* type A, which were similar with regard to heat resistance but dissimilar in their toxicology to those described by the German workers. Many outbreaks were investigated, and in general they conformed to the pattern given by previous workers. After an incubation period of 8 to 24 hours there was abdominal pain and diarrhea, sometimes nausea but rarely vomiting, but no evidence of infection such as pyrexia and headache.

The food vehicle was fairly consistent — a cold or warmed-up meat dish cooked at a temperature not greater than 100°C for 2 to 3 hours on the previous day or even a few hours before required and allowed to cool slowly in the kitchen or larder. The meat when eaten was almost always described as normal in appearance, taste, and smell; occasionally there were rumors of gas bubbles in stews and pies.

The great majority of outbreaks reported subsequently and up to the present time (Table I) have been due to heat-resistant strains of *C. perfringens,* but McKillop (1959) reported outbreaks of food poisoning from precooked chickens in which the predominant organisms were β-hemolytic heat-labile strains of *C. perfringens* with a minority of α-hemolytic heat-resistant strains. Boiled chickens in liquor were decanted while hot into deep metal receptacles so that spores and bacilli from the air and in the surface dust from the container would have ideal conditions for multiplication in the warm mass of chicken and liquor.

Hall *et al.* (1963) claimed that strains of *C. perfringens* causing food poisoning in the United States did not always possess the characteristics of those hitherto associated with food poisoning in the United Kingdom. Taylor and Coetzee (1966) reported an outbreak due to a nonhemolytic strain which was destroyed by heating for 10 to 15 minutes at 100°C. Sutton and Hobbs (1968) isolated a variety of strains from food poisoning outbreaks including nonhemolytic heat-resistant (surviving at least 1 hour at 100°C), nonhemolytic heat-sensitive, and β-hemolytic heat-sensitive strains not surviving 15 minutes at 100°C).

The figures relating to heat resistance quoted above refer to the heating of feces in cooked meat in broth media in universal containers. They must not be confused with the results from experiments in which suspensions of spores are heated in sealed ampoules.

B. *Clostridium bifermentans*

Clostridium bifermentans was described as the agent of a food poisoning episode by Duncan (1944). The history was similar to that of *C. perfringens* food poisoning, and the food vehicle was a meat and potato pie cooked ahead of requirement.

Clostridium bifermentans is not an uncommon organism, and it may be questioned why it is not implicated in food poisoning more frequently, considering that it sporulates far more readily than *C. perfringens*. The degree of heat resistance of *C. bifermentans* spores does not appear to be quoted in the literature. It is possible that they may be relatively sensitive to heat, and this could explain why the organism is so rarely described as a cause of food poisoning. In a similar way the more heat-sensitive strains of *C. perfringens* have been less frequently found to cause outbreaks in the United Kingdom, although hitherto in many laboratories the techniques used for examining feces from food poisoning outbreaks may have excluded the isolation of heat-sensitive strains of *C. perfringens*.

Gibbs (1964) found that spores of *C. bifermentans* required activation by heat shock, 85°C for 10 minutes, before rapid and complete germination took place and that, although strictly anaerobic conditions were required for the vegetative growth, these conditions were not necessary for the germination of the spores. The same author (Gibbs, 1967) found that heat-activated spores of *C. bifermentans* had an obligate requirement for sodium ions for germination.

Apart from *C. botulinum,* which is the subject of a separate chapter, other anaerobes of the genus *Clostridium* have not been established as agents of food poisoning (see also Section II, K in Chapter 6, Infections Due to Miscellaneous Microorganisms).

C. *Bacillus cereus* and *Bacillus mesentericus*

Bacillus cereus as a food poisoning agent has been the subject of papers by Hauge (1950, 1955). A national food habit, that of packing precooked cornflour sauce to eat with fruit at picnics, was responsible for outbreaks occurring in Norway; Christiansen *et al.* (1951) described an outbreak due to cornflour pudding.

TABLE 1

Examples of Outbreaks of *Clostridium perfringens* Food Poisoning, 1951–1965

Author	Year	Place	Number at risk	Number affected	Food (serological type)	Stools No + ve (serological type)	Comment (symptoms and incubation period)
Johns (United Kingdom)	1951	Schools	1537	352	Galantine (type 6)	Not found	Diarrhea and abdominal pain; 9 to 10 hours
Collee (United Kingdom)	1954	Army camp	138	48	Reheated mutton stew	35/35 0/17 (not at risk)	Diarrhea and abdominal pain; 6 to 12 hours
Beck *et al.* (United Kingdom)	1954	Hospital	360	44	Precooked warmed-up pork (type 1)	18/40 (type 1 and untypable)	Diarrhea and abdominal pain; 8 to 16 hours
Stocks (United Kingdom)	1955	Outing	104	33	Precooked cold turkey	8/15 (type 7)	Diarrhea and abdominal pain, few vomited; 10 to 18 hours
Stocks (United Kingdom)	1955	Nurses' home	45	24	Precooked cold brisket of beef (type 4)	8/12 (type 4)	Diarrhea and abdominal pain, few vomited; 9 to 18 hours
Collee (United Kingdom)	1955	Army camp	300	25	Cold boiled mutton (type 6)	11/18 (type 6)	Diarrhea and abdominal pain; 6 to 12 hours

Smith and Wallace (United Kingdom)	1956	Old people's home	55	21	Precooked cold rolled brisket	17/21 (type 4)	Diarrhea and abdominal pain, few vomited; 10 to 20 hours
Linzenmeier (Germany)	1956	Canteen	390	300	Precooked pork (type 3)	7/7 (type 3)	Diarrhea and abdominal pain; 8 to 10 hours
Dickie and Smith (Scotland)	1957	Old people's home	24	21	Precooked boiled shin of beef	9/10 (type 6)	Diarrhea and abdominal pain; 9 to 24 hours
Dickie and Smith (Scotland)	1957	Old people's home	24	6	Precooked boiled rolled meat	Not examined	Diarrhea and abdominal pain; 9 to 24 hours
Main (New Zealand)	1959	Outing	14	11	Cold chicken	3/4 (not typed)	Diarrhea and abdominal pain; up to 12 hours
Wade (United Kingdom)	1959	Coach party	41	38	Cold chicken	5/? (not typed)	Diarrhea and abdominal pain; 6 to 12 hours
		Wedding party	115	88	Precooked cold chicken	9/20 (not typed)	
Yamagata et al. (Japan)	1959	Wedding party	46	11	Precooked boiled fish (type 6)	Not examined	Diarrhea and abdominal pain; 11 to 14 hours

TABLE I (*Continued*)

EXAMPLES OF OUTBREAKS OF *Clostridium perfringens* FOOD POISONING, 1951–1965

Author	Year	Place	Number at risk	Number affected	Food (serological type)	Stools No + ve (serological type)	Comment (symptoms and incubation period)
Edinburgh Public Health Department (Scotland)	1960	Wedding party	?	30	Reheated steak in pie	8/? (not typed)	Diarrhea and abdominal pain, few vomited; 11 to 18 hours
		School	?	91	Precooked rolled mutton	6/? (not typed)	Diarrhea and abdominal pain; 8 to 18 hours
Hayashi *et al.* (Japan)	1961*a*	Unknown	?	309	Fried fish paste "agekamaboko"	9/21 (type 2)	Diarrhea, colic, fever, few vomited; 1 fatal case; incubation unknown
Hayashi *et al.* (Japan)	1961*b*	Family	?	4	Fried fish paste "agekamaboko"	?/? (type 6)	Diarrhea, colic, few vomited; incubation unknown
Dam-Mikkelson *et al.* (Denmark)	1962	Canteen Canteen Home (3 outbreaks)			Chicken in asparagus, meat broth, and mutton in parsley sauce		

Yamagata and Kawaguchi (Japan)	1962	13 outbreaks (1957 to 1961) summarized					
Parry (United Kingdom)	1963	Geriatric hospital Welfare home	202	41	Sliced and minced ham	24/26 (type 3/4)	Diarrhea, abdominal pain, few vomited; 12 to 24 hours
Robertson (United Kingdom)	1963	School	259	91	Precooked rolled boiled beef	32/49 (not typed)	Diarrhea and abdominal pain
Ager and Ploeger (United States)	1964	College	?	300	Lamb stew	Not examined	Diarrhea, abdominal pain, some vomiting; 12 to 24 hours
Howitt (Scotland)	1964	Bus party	34	29	Roast beef and cold ham	3/9 (not typed)	Diarrhea and abdominal pain; 8 to 12 hours
McCroan and Mixson (United States)	1965	High school	447	256	Cold roast beef	Not examined	Diarrhea, abdominal cramp, and nausea; 14½ hours
Herron et al. (United States)	1965	Two public schools	301	181	Turkey à la king	Not examined	Diarrhea, abdominal pain, nausea, few vomited; 3½ to 13½ hours

It is not uncommon to find spores of aerobic organisms, particularly *B. cereus,* in cornflour and other dried foods, and it is easy to imagine the excellent growth conditions in the rehydrated and prepared food. Again, heat resistance (Hauge, 1955) must be a prerequisite of survival and growth after cooking. Likewise the actual food ingredients and the pH must be conducive to the growth of the organism.

Bacillus mesentericus (synonym for *Bacillus subtilis*), an organism responsible for ropy bread, has been described as the agent of food poisoning in those eating bread heavily contaminated with the organism. Nikodemusz (1964) has studied 51 outbreaks in Hungary, involving 1400 people and thought to be due to aerobic sporing bacilli (see also Section II, I, in Chapter 6, Infections Due to Miscellaneous Micro-organisms).

II. THE ORGANISMS

A. *Clostridium perfringens*

1. IDENTIFICATION AND MORPHOLOGY

The general characteristics of *C. perfringens* are well known, but methods of isolation and identification are variable.

a. Morphology. The square-ended gram-positive bacilli give a typical morphological appearance in direct smears of meat suspected to have caused food poisoning. In foods the bacilli may be shorter and fatter than those seen in smears from colonies on blood agar plates. Spores are rarely if ever seen in smears from foods, although structures resembling spores are occasionally seen in smears from colonies.

b. Classification. The classification of *C. perfringens* according to soluble antigens is important in relation to the distinctive pathogenicity of the different types in man and animals. The routine typing methods are described by Oakley and Warrack (1953); more detailed descriptions of the tests used are given in earlier papers (Oakley *et al.,* 1946, 1948; Oakley and Warrack, 1951).

There are six types, A to E, arising from differences in toxicological behavior and concerned with the soluble antigens detected in culture filtrates. Twelve such antigens have been found, and each has been given a Greek letter. The type letters A to E refer to the predominant antigen, so-called toxin, produced by the organism.

The type F originally associated with the Rassfeld-Sternberg strain causing enteritis necroticans in Germany (Oakley, 1949) has been abandoned and replaced by the designation "special type C" in view of the similarity between the types C and F. The Rassfeld-Sternberg strain may also be regarded as special on the grounds that it causes enteritis necroticans. Strains designated type C have been associated with gastroenteritis called "Pig-bel" following pig feasts in New Guinea (Murrell *et al.*, 1966). The strains survived 90° to 100°C for 1 minute only, and thus were more sensitive to heat than the German strains; they also differed antigenically. The classical type C strains cause an enteritis, "struck," of sheep.

In a similar way the type A strains are subdivided. The classical strains producing abundant α-toxin as well as θ- and κ-toxins, cause gas gangrene; on the whole they are heat-sensitive and less frequently implicated in food poisoning. The majority of strains which cause food poisoning produce small amounts of α-toxin, though occasionally they appear not to form α-toxin under laboratory conditions. The θ-toxin is rarely found, and the production of κ-toxin is variable. These strains are usually heat-resistant, surviving more than 30 minutes of boiling.

Type B strains cause dysentery in lambs, type D enterotoxemic and pulpy kidney diseases of sheep, and type E an enterotoxemia of calves.

Table II, reproduced from the paper by Oakley and Warrack (1953), describes the various soluble antigens and gives the properties and distribution of the antigens among the designated types.

Table III relates the pattern of toxin production to the type and clinical effect.

It is worthwhile considering colonial appearance, hemolysis, and heat resistance of type A strains implicated in food poisoning in more detail, and also to consider the characteristics of this organism in relation to its role as the causative agent of food poisoning of a particular pattern.

c. Colonial Appearance and Hemolysis of Type A Strains. The appearance of strains on blood agar varies according to the toxicological type and within each type. On 5% horse blood agar (nutrient), layered on a plain 1.5% agar base, the classical type A strains are round and somewhat opaque with a clear zone of β-hemolysis. The colonies from the heat-resistant type A strains more frequently found to cause food poisoning in the United Kingdom may have a similar morphology, or they may be irregular with a vine leaf appearance; the colonies are usually smooth, and rarely rough; they are sometimes umbonate. On first isolation many are nonhemolytic, or they may have a faint and hazy zone of hemolysis. After standing on the bench the hemolysis of the colonies from different strains varies from nonhemolytic to clear β-hemolysis.

TABLE II

THE SOLUBLE ANTIGENS SO FAR DETECTED IN *Clostridium welchii* CULTURE FILTRATES[a]

Designation	Activity	Occurrence in filtrates of *C. welchii* types[b]					
		A	B	C	D	E	F
α	Lethal, necrotizing, hemolytic, lecithinase C	+++	+	+	+	+	+
β	Lethal, necrotizing	−	+++	+++	−	−	+
γ	Lethal	−	+	+	−	−	+
δ	Hemolytic, lethal	−	+	++	−	−	−
ϵ	Lethal, necrotizing (activated by trypsin)	−	++	−	+++	−	−
η	Lethal (validity doubtful)	(+)	−	−	−	−	−
θ	Hemolytic (oxygen-labile, ?lethal)	++−	+	+	+	+	−
ι	Necrotizing, lethal (activated by trypsin)	−	−	−	−	++	−
κ	Collagenase (lethal, necrotizing gelatinase)	++	−	+	+−	+	−
λ	Proteinase, disintegrates azocoll and hide powder but not collagen; gelatinase	−	+++	−	+−	+	−
μ	Hyaluronidase	+−	+	−	+−	−	−
ν	Deoxyribonuclease	+	+	+	+	+	+

[a]Reproduced from Oakley and Warrack. (1953).
[b]++− or +− = present in some strains.
(+) = limited to very few strains.

Nevertheless, in general, the classical type A strains are β-hemolytic, and the strains predominantly causing food poisoning are non- to α-hemolytic; the appearance of the colonies on blood agar may be a useful guide to significance in relation to food poisoning.

d. Heat Resistance of Type A Strains. It is pertinent to consider heat resistance in relation to hemolysis. Strains able to survive heating to 100°C for more than 1 hour are almost always non- to α-hemolytic. Heat-sensitive strains, on the other hand, may be either β-hemolytic or non- to α-hemolytic. It should be noted, therefore, that almost all β-hemolytic strains are heat-sensitive but that non- to α-hemolytic strains may be either heat-resistant or heat-sensitive.

The heat resistance of *C. perfringens* spores is important both for purposes of classification and in the sequence of events which leads to food poisoning. Historical accounts confirm the present-day facts about the variation between strains in their survival times at 100°C or below. Dunham (1897) reported survival for 1 minute at 98°C. Barnes *et al.* (1963) observed growth after 5 hours at 100°C, and they estimated that the count was reduced sevenfold for each hour of steaming. The same authors found that only 3 to 20% of spores of *C. perfringens* (F2095/50 NCTC 8797) germinated without heat shock; however, they tested one strain only, and later work suggests that the effect of heat shock varies among strains. Roberts (1968) found that without heat-treatment the germination of spores was 30 to 50% for classical type A strains and 0.13 to 3.6% for "food poisoning" strains. Roberts also investigated the decimal reduction time (*D* value), which is the time in minutes to effect a 90% reduction of organisms, for various strains of *C. perfringens* suspended in distilled water and sealed in ampoules. He found that non-hemolytic food poisoning strains gave a value for D_{100} of 6 to 17 minutes. Comparable figures for a β-hemolytic strain were less than 1 for D_{100}, and 3 to 5 minutes for D_{90}; there were variations between strains.

Collee *et al.* (1961) found that heat-resistance tests with spores produced from cultures of stock strains revealed a protective effect of cooked meat which was most pronounced with the more heat-resistant strains; this has been confirmed by Sutton (1966). M. E. Smith (personal communication) and Sutton (personal communication) both comment on the reduced heat resistance of spores from cultures grown in media designed to increase sporulation—for example, Ellner's medium (Ellner, 1956).

The findings of Collee *et al.* (1961) indicated that the heat-stable property of spores from typical food poisoning strains was genetically determined and was not primarily dependent upon the presence of large numbers of spores.

e. Nagler Reaction. The production of the enzyme lecithinase C by all types of *C. perfringens* is used to distinguish *C. perfringens* from other species of clostridia with the exception of *C. bifermentans*, which also produces lecithinase. A combination of lecithin from egg yolk and lecithinase produces a complex material manifested by an opaque halo precipitated around colonies growing on media containing either egg emulsion or plasma; the precipitation is inhibited by antitoxin to the α-antigen.

Clostridium bifermentans may be differentiated from *C. perfringens* by its proteolytic behavior, its ready sporulation, and the absence of lactose fermentation.

f. Fermentation Reactions. One percent peptone water sugars incubated anaerobically are of value in differentiating members of the clostri-

TABLE III

Characters of the Antigens and Their Distribution in the Clostridium perfringens Types[a,b]

Type	Occurrence	α	β	γ	δ	ε	θ	ι	κ	λ	μ	ν
A												
Classical, β-hemolytic, heat-sensitive	Gas gangrene (man) Intestinal commensal (man and animals) Patients and food in outbreaks of food poisoning	+++	−	−	−	−	++−	−	++−	−	+−	++−
"Food Poisoning" Non- to α-hemolytic, heat-resistant	Patients and food in outbreaks of food poisoning	+	−	−	−	−	(+)	−	+	−	+−	++−
B												
Classical	Lamb dysentery Foals	+	+++	++−	+	++−	+	−	−	+	+++	++−
Iranian	Hemorrhagic enteritis of goats and sheep, Iran	+	+++	?	−	+	+	−	+	−	−	+
C												
Classical	"Struck" in sheep	+	+++	++−	+++	−	+++	−	+++	−	−	+
Colorado	Neonatal hemorrhagic enteritis in calves and lambs	+	+++	?	−	−	+	−	+	−	−	+

Piglets	Enteritis in piglets	+	+++	?	−	−	+	−	−	+	(+)	++
Special, heat-resistant and heat-sensitive		+	+	+	−	−	−	−	−	−	−	+
D	Enterotoxemia in adult sheep, lambs, goats, and bovines	+	−	−	+++	++	++	++	++	+	+	++
	Intestine of man											
E	Sheep, cattle (? pathogenic)	+	−	−	−	+−	++−	++−	++	(+)	+−

[a] Modified from Brooks *et al.* (1957).

[b] − Not produced. +, ++, +++, produced by all strains. ++−, +−, produced by most or some strains. (+) produced by very few strains.

dia, and the characteristic pattern of carbohydrate fermentation by *C. perfringens* is glucose, maltose, lactose, and sucrose, with variable reactions in salicin and glycerol. Acid is formed in litmus milk, but the stormy clot phenomenon varies between strains.

g. The *nitrate reduction test* may be carried out in potassium nitrate broth.

h. *Nonmotility* may be demonstrated in a special motility medium containing 0.1% of peptone or gelysate.

i. *Serology.* Since *C. perfringens* was established as a food poisoning agent, strains isolated in large numbers from foodstuffs suspected to be the vehicle in authentic outbreaks have been used to immunize rabbits given killed suspensions intravenously (Hobbs *et al.,* 1953). The technique of immunization was originally described by Henderson (1940).

The antisera were required to agglutinate not only the homologous food strains but also strains isolated from a large majority of stool samples obtained from as many patients as possible. Thus 17 serological types have been established among heat-resistant strains known to cause food poisoning. A high proportion of heat-resistant strains isolated from food poisoning outbreaks may be characterized in this way. Since 1956 most strains selected for serology have been those directly isolated from foods and in the same outbreak grown from feces after exposure to 100°C for 1 hour. More recent work indicates that, although the majority of strains from outbreaks will be represented in this way, some strains causing food poisoning are β-hemolytic and heat-sensitive and do not belong to these serological types. It has therefore been necessary to prepare antisera against the "classical" strains. To date twenty such antisera have been prepared, but it is most probable that there will be more to follow (Sutton, personal communication).

Although there are some cross reactions between "classical" and heat-resistant strains, Henderson (1940) found a wide diversity in the specificity of somatic antigens of the classical type A strains which appeared to be strain-specific.

j. *Hemagglutinins.* A study of the nature and properties of the hemagglutinin of *C. perfringens* was carried out by Collee (1961) using several different strains representing types A to E. Collee characterized the factor as possessing affinity for a wide range of red cells and being active over wide pH (5.5 to 9.4) and temperature (4° to 37°C) ranges. It was thermolabile in 30 to 45 minutes at 55°C, and it was most rapidly inactivated at an acid pH. Normal human red cells were most strongly agglutinated at pH 6.8 and 20°C, and they were exceptionally sensitive after the removal of their myxovirus receptors with the *Vibrio cholerae* receptor-destroying enzyme. The hemagglutinin could cause clumping of red cells when liberated particles became absorbed onto the surface of the red cells.

Hemagglutination was inhibited by a heat-stable factor (C) in cooked meat broth, and the factor A of blood group-A isoagglutination, blood group-A substance, present in commercial peptone. *Clostridium perfringens* cultures could destroy inhibitors C and A by different mechanisms. Hemagglutinin was not produced by any of more than 100 freshly isolated strains of *C. perfringens*. Good yields were obtained from 48-hour blood agar cultures of old stock strains which had been subcultured for months; best yields of diffusible hemagglutinin occurred in sporing cultures in Ellner's medium (Ellner, 1956). Collee suggests that the hemagglutinin may be a degraded form of one or more enzymes normally active on cell-surface polysaccharides.

2. Growth Characteristics

The temperature of growth and survival of *C. perfringens* is important in relation to growth in food. Collee *et al.* (1961) demonstrated optimum growth at temperatures between 43° and 47°C; there was some growth at 50°C but none at 55°C. White and Hobbs (1963) could not demonstrate growth at temperatures lower than 6.5°C up to 7 days. Barnes *et al.* (1963) showed that growth was restricted at temperatures below 15° to 20°C but improved at 25°C and 37°C. They could not recover the organism after freezing at −5°C and −20°C for 6 months; their experiments were carried out in meat blocks or in Robertson's cooked meat medium; spores and vegetative cells were used in most experiments, and the number of strains tested was limited.

Using five type A strains of *C. perfringens,* Strong and Canada (1964) tested the survival after freezing (−17.7°C) of vegetative cells which had grown in gravy, and spores produced in Ellner's medium dried in soil and added to the gravy. If we assume that predominantly vegetative cells were present after growth, 4.29% and 3.69% of viable cells survived for 90 to 180 days, respectively. The survival of spores was 37.9% after 90 days and 10.9% after 180 days.

The effects of pH and salt on the growth of *C. perfringens* are important in relation both to spoilage and to food poisoning. Barnes *et al.* (1963) showed that at 20°C growth of one strain of *C. perfringens* was erratic at pH 5.7 to 5.8 but regular and good at pH 7.2; thus meat of high pH was more likely to encourage growth of *C. perfringens* at ambient temperatures. Hobbs and Reeves (unpublished) found that, with a spore inoculum of 1300 per milliliter, *C. perfringens* grew in media containing all concentrations of sodium chloride tested (up to 5%).

Gough and Alford (1965) investigated the effect of different combinations of curing salts, including NaCl, $NaNO_3$, and $NaNO_2$, on growth, survival, and heat resistance of several strains of *C. perfringens*. They used reproducible inocula of aged spores, which showed much greater resistance to heat after storage in meat brine mixtures than did spores from young cultures.

They found that 6% (w/v) NaCl, 10,000 ppm of $NaNO_3$, or 400 ppm of $NaNO_2$ was required to inhibit the growth of *C. perfringens* significantly. When used in conjunction with heat, the concentrations of curing salts required to affect cell survival were still well above those usually employed. Various strains of *C. perfringens* consistently survived normal curing and smoking processes. The authors pointed out that not only would the organisms survive but they might actually grow if the suspending medium and temperature were suitable.

Spencer (1964) concluded from his experiments that the risk of *C. perfringens* food poisoning from cooked cured meats could be reduced by ensuring that (1) products were stored only at temperatures of 20°C or below, and for no longer than 7 days; (2) the content of *C. perfringens* spores was less than 1 per gram; (3) the pH was 6.1 or below; and (4) the meat contained 25 ppm of nitrite (28 ppm in the brine phase) with 5.3% sodium chloride (as a brine strength).

The examination of a number of commercial samples of ham and tongue indicated that conditions 3 and 4 were possible but not always met.

3. HABITAT

Sporing organisms must survive longer and have a more ubiquitous distribution in nature than those that rely on vegetative cells for their existence.

a. Human Beings. The investigation of normal stools has indicated that type A *C. perfringens* may be isolated from 100% of feces after repeated examination and by use of enrichment techniques, but the majority of strains are β-hemolytic, and 6 to 8% of the strains are non-hemolytic and heat-resistant (Taylor and Gordon, 1940; Collee *et al.,* 1961). Reports from other workers have given figures of 2.2% to 30% for the isolation of heat-resistant strains from samples of normal feces (Hobbs *et al.,* 1953; Dische and Elek, 1957; Leeming *et al.,* 1961). The larger percentages were associated with persons in institutions. Turner and Wong (1961) demonstrated heat-resistant spores in 63% of Chinese hospital patients in Hong Kong; the same workers using the same technique found heat-resistant spores in 9% of a hospital population of Leeds.

Sutton (1966) investigated the distribution of heat-resistant strains of *C. perfringens* in a rural area of Australia. The carrier rate for the general population was low (1.5 to 6.0%), but persons associated with communal feeding and poor hygiene were shown to have higher rates of carriage (15.1 to 25%). The excretion of any one serological type of *C. perfringens* appeared to be transient.

Immediately following an outbreak, 90% (Hobbs *et al.*, 1953) or 97% (Dische and Elek, 1957) of affected persons have been found positive for heat-resistant strains, and 50% were found still to be positive 2 weeks later (Hobbs *et al.*, 1953). Sutton (1966) carried out viable counts of *C. perfringens* on feces of persons in the general population and those associated with an outbreak of *C. perfringens* food poisoning. Analysis of these quantitative studies indicated that in healthy people heat-resistant, nonhemolytic *C. perfringens* strains were present in smaller numbers than the classical type A hemolytic strains.

After one outbreak investigated in this way there was a significant increase in the viable count of heat-resistant organisms which could be readily isolated by direct culture; quantitative studies of subsequent outbreaks have indicated that in some instances there was an increase in viable colony count of both the causative strain and other strains of *C. perfringens* present in the intestine before the outbreak occurred.

Subsequent work confirms the fact that strains of *C. perfringens* assumed to be responsible for the outbreaks are present in vastly increased numbers in the feces of patients (Sutton and Hobbs, 1968). The significance in feces of strains of varying heat resistance and hemolysis in relation to their role as causal agents of food poisoning should be investigated, particularly when food poisoning due to heat-sensitive *C. perfringens* is suspected. The presence of large numbers of β-hemolytic heat-sensitive *C. perfringens* in a series of feces is suggestive and would warrant the preparation of an antiserum in order to establish a serological relationship between strains.

Klotz (1965) and Hall and Hauser (1966) have shown that normal persons may carry more than one strain (differences in hemolysis, heat resistance, serology, and fluorescent antibody) of *C. perfringens* at the same time; thus it is safer for the preparation of antisera to use a strain isolated, preferably in large numbers, from the incriminated meat dish rather than one selected at random from the feces. It seems that strains normally present in the feces prior to ingestion of food contaminated with *C. perfringens* may also show some multiplication and thus be present in numbers approaching those of the causative agent.

Furthermore, whereas the heat-resistant, nonhemolytic, so-called food-poisoning strains seem to form a fairly circumscribed serological group, the serology of the heat-sensitive β-hemolytic strains appears to be diffuse and to include a large number of specific serotypes. Hall and Hauser (1966) could type only one-third of their isolates of β-hemolytic and nonhemolytic strains of *C. perfringens* using a battery of 67 antisera prepared from various strains including heat-resistant nonhemolytic and also β-hemolytic organisms. Klotz (1965) while investigating fluorescent antibody techniques for the detection of 150 strains of *C. perfringens,* found that 34 of 79 cultures could be typed by using five pools of fluorescent antibody reagents, and Hobbs serotypes were observed 21 times. With one exception, antisera and conjugates were specific for *C. perfringens* when tested against cultures of 10 other species of clostridia and 4 species of the genus *Bacillus.*

b. Animals. Heat-resistant strains of *C. perfringens* have been found in 18.4% of 76 samples from normal pig feces, 14.6% of 41 samples of feces from rats, and 1.7% of 113 samples of normal cattle feces (Hobbs *et al.,* 1953). Smith and Crabb (1961) found *C. perfringens,* often in large numbers, in the alimentary tract of most farm and domestic animals, including dogs, cats, pigs, and cows, and also in 80% of the human stools examined. Narayan (1966) isolated 25 strains of *C. perfringens* from numerous samples of muscle, gland and offal tissue taken immediately after slaughter from 125 clinically healthy cattle. Only three strains formed spores which withstood boiling for 1 hour; two were isolated from mesenteric lymph nodes and liver of the same animal, and the third from the mesenteric lymph nodes of another animal. *Clostridium perfringens* type C was also found. It is interesting to note that Narayan reports the isolation of various species of clostridia from 85% of 103 cattle prepared for slaughter according to the usual practice in Hungary, whereas in a control group of 26 cattle rested for 24 to 48 hours and treated with due preslaughter care and management, clostridia could be demonstrated, localized and limited to the mesenteric glands, in 7.7% of cattle only. The majority of identified strains belonged to *C. perfringens, C. bifermentans,* and *C. lentoputrescens.* It was emphasized that a period of 24 to 48 hours rest before slaughter was indispensable to reduce the invasion and spread of clostridia in muscles and internal organs. The mesenteric lymph node samples were found to be contaminated in the maximum number of instances, owing, it was thought, to the glands' acting as a reservoir for clostridia migrating from the intestine. The same species of clostridia could be isolated from more than one organ of the same animal, and often more than one species could be demonstrated in one organ. These findings suggested endogenous invasion of the carcass. Wijewanta (1964) found that two heat-resistant strains isolated from soil (as well as one known to cause food poisoning in man) were of very low pathogenicity for day-old chicks; month-old chickens that received the same cultures continued to excrete them for 72 hours.

c. *Foods.* The contamination with heat-resistant strains of *C. perfringens* of samples of raw meat including pork, beef, veal, mutton, and lamb has been shown to vary from 1.5 to 42.7% (Hobbs *et al.,* 1953; Sylvester and Green, 1961). McKillop (1959), investigating raw products of meat, fish, and poultry, found hemolytic strains of *C. perfringens* in 72% and nonhemolytic strains in 18% of samples; 2.2% of the strains were heat-resistant. In 173 cooked foods of the same kind the equivalent figures were 19.6%, 10.4%, and 0.6%.

Hobbs and Wilson (1959) found heat-resistant strains of *C. perfringens* in 4.3% of samples of chilled carcass meat and 10.3% of samples of frozen boneless meat imported into the United Kingdom.

Strong *et al.* (1963) isolated *C. perfringens* as black colonies in sodium sulfite medium from 6.1% of 510 samples of various foods, including 16% of 122 samples of raw meats, poultry, and fish.

d. *Flies.* Hobbs and her colleagues (1953) isolated heat-resistant *C. perfringens* from every batch of blowflies examined. The flies were collected from different sources including a hospital, a butcher's shop, a fried-fish shop, a slaughterhouse, and a refuse sorting depot. The majority were *Lucilia* (greenbottles), but some *Calliphora* and *Phormia* (bluebottles) were also included. The results suggest that the blowfly plays an important role in the spread of *C. perfringens* during breeding and feeding on meat and other animal products.

e. *Soil and Dust.* Taylor and Gordon (1940) found *C. perfringens* in 190 of 196 soil samples. McKillop (1959) isolated hemolytic and nonhemolytic *C. perfringens* from nearly 90% of samples of kitchen dust.

4. ISOLATION

General principles for the isolation of anaerobes include fairly strict anaerobiosis, enriched media, and a means of demonstrating characters of identification such as hemolysis, lecithinase production, sulfite production, and heat resistance. It is important to remember that heat-resistant spores of *C. perfringens* may be demonstrated readily in feces and raw meat but rarely if ever in cooked meat vehicles of food poisoning. Thus the examination of food and feces from outbreaks may require different methods – direct examination and enrichment culture for unheated cooked food, and direct and enrichment techniques for unheated and heated feces. It is essential to investigate the state of heat resistance of strains from the patient's stool rather than of strains from food because, whereas spores are found in the feces, they are not readily formed in cooked food where there are young cultures from freshly germinated spores. Also, to cover all degrees of heat resistance of the strains of *C. perfringens* which may be present in the feces, different times of heating of fecal samples may be required.

   a. Food. Direct smears of cooked foods, usually meats, involved in outbreaks of *C. perfringens* food poisoning may show square-ended gram-positive bacilli almost exclusively, unless there has been a long storage period between cooking and eating or between the outbreak and sampling. Direct cultures of the food sample on horse blood agar incubated anaerobically at 37°C show *C. perfringens* of the non- to α-hemolytic or β-hemolytic types, whereas comparative horse blood agar plates incubated aerobically from the same sample may show a few colonies only of saprophytic organisms.

   A profuse growth of any organism on the anaerobic plate requires further investigation. It is valuable to know the approximate number of *C. perfringens* in the sample. Counts may be simply carried out by the Miles and Misra (1938) technique of inoculating drops of a known volume of food dilution on the surface of duplicate blood agar plates incubated anaerobically and aerobically for comparison. Neomycin, 3 drops of a 1% solution, spread over the surface of the blood agar before inoculation helps to reduce the growth of contaminants. Iron sulfite agar with polymyxin (Mossel *et al.,* 1956), and sulfadiazine (Angelotti *et al.,* 1962) in pour plates, plastic pouches (Waart and Smit, 1967) or deep agar tubes which may be flattened (Miller *et al.,* 1939) or round with a black rod inserted into the center of the agar for ease of counting colonies (Ingram and Barnes, 1956), are used for general clostridial counts of black colonies or for pure culture studies (Collee *et al.,* 1961). This method of counting may be valuable also for factory surveys of clostridial invasion. But for the investigation of food poisoning it is useful to differentiate strains of *C. perfringens* by colonial appearance and hemolysis.

   Liquid enrichment media such as Robertson's cooked meat, liver broth, and trypticase glucose broth (Schmidt *et al.,* 1962), with or without the addition of neomycin, may be used for the isolation of small numbers of *C. perfringens* undetectable by direct culture; but the results of enrichment cultures from foods suspected to have caused food poisoning should be assessed carefully in relation to the type of food eaten, the incubation period, the symptoms, and, in particular, the results from stool samples. It is thought that *C. perfringens* must be present in food in very large numbers (millions per gram) to cause food poisoning. Nevertheless, the meat or other food samples received in the laboratory may not have come from the particular container, among many, responsible for the food poisoning; also the dish from which the sample was taken may have been heated after the relevant portions were eaten. A large block of meat or poultry may be contaminated in one part and not in another, so that the distribution of contamination may vary from part to part and from slice to slice. Furthermore, there may be a rapid drop in the count of vegetative cells due to adverse conditions such as low pH or freezing. Thus it may be difficult or impossible to obtain the agent from the foodstuff, and the examination of feces from patients is essential.

b.   Feces.   Feces contain numerous organisms in large numbers so that the isolation of anaerobic or even aerobic agents of food poisoning directly on plain blood agar is difficult or impossible unless the agent is predominant.

It is suggested that a small portion of feces be transferred to a bijou bottle and emulsified with a swab in broth or other diluent. Direct cultures should be made on blood agar spread with three drops of a 1% solution of neomycin; control plates without neomycin should be used for other organisms. Likewise, plates may be made for aerobic as well as for anaerobic incubation. Enrichment cultures should be made by heating a portion of feces in cooked meat medium at 100°C for 60 minutes followed by overnight incubation at 37°C before plating onto neomycin blood agar. For fresh samples of feces — that is, collected within 3 days following the food poisoning — direct cultures on blood agar and enrichment cultures after steaming for 1 hour ought to be adequate to isolate any variety of *C. perfringens*. But after 3 days the number of causative organisms may be reduced, and a second enrichment technique after heating for 10 minutes at 80°C should be carried out in addition to direct culture to isolate heat-sensitive strains. When antibiotics or other drugs have been given, the same procedure may be necessary.

The causal organism should be present in a high proportion of those at risk and in a low proportion only of stools from healthy control people; the numbers are important also. The same serological type of *C. perfringens* isolated from a large majority of feces, even when it has been impossible to isolate the organism from food, may be regarded as evidence of *C. perfringens* food poisoning. Although the carriage rate of *C. per-*

fringens among those living in institutions may be high, it is likely that many different types will be found, and it would be essential to find the same type prevalent throughout all fecal samples.

B. *Clostridium bifermentans*

Clostridium bifermentans is a large gram-positive, anaerobic rod, motile by peritrichate flagella, and readily forming central or subterminal oval spores; it probably grows best at about 37°C. Some strains are toxigenic, and the organism produces lecithinase C resembling the α-toxin of *C. perfringens,* and antigenically related to it (Miles and Misra, 1947, 1950; Lewis and Macfarlane, 1953). It is readily differentiated from *C. perfringens* by its proteolytic behavior in cooked meat and ready spore formation.

The organism grows well anaerobically on ordinary media, and fermentation and other biochemical reactions are used for identification. A characteristic feature is the late deposit of small rounded masses of white crystals in cooked meat medium which is digested and blackened.

C. *Bacillus cereus* and *Bacillus mesentericus*

Bacillus cereus is a large gram-positive, rod-shaped, aerobic, spore-bearing organism capable of growing also under anaerobic conditions. It is widely distributed, being found in air, soil, water, milk, dust, and other situations. The organism grows at temperatures between 10° and 45°C with an optimal temperature of 35°C. The spores germinate equatorially and are fairly resistant to heat. It has been generally held not to be pathogenic to man, but some strains may produce localized ulceration or generalized infection in mice and guinea pigs, and more recently Hauge (1950; 1955) has described a number of outbreaks of food poisoning in which *B. cereus* was incriminated as the cause. A short account of his findings is given on p. 166.

The following method has been described for enumerating the vegetative cells and spores of *B. cereus* in foods (International Association of Microbiological Societies [in press]):

Ten percent of food is macerated in peptone–saline and diluted further, 1–10 and 1–100, in peptone–saline. The macerated food and the two dilutions are examined by the spread drop method using the following medium and incubation:

D-Mannitol	10 gm
Egg yolk (10% suspension Oxoid)	100 ml
Peptone	10 gm

Beef extract	1 gm
Agar	15 gm
Phenol red	0.025 gm
Water	1000 ml

Incubate at 30°C for 24 to 48 hours.

When foods are known to be heavily contaminated, the addition of 5% sodium chloride will help to suppress other organisms. The salt does not affect the appearance of the colonies of B. cereus, but it reduces their number.

Appearance of colonies: The colonies can be easily recognized and a quantitative estimate of numbers obtained. They are irregular, dry, and gray on a red background surrounded by a halo of cloudiness.

Bacillus mesentericus belongs to the subtilis group of organisms, and is a small gram-positive, strictly aerobic spore-bearing organism. Like B. cereus it is widely distributed in such natural habitats as hay, dust, milk, soil, and water. It grows at temperatures between 12°C and 50° or 55°C, with an optimum temperature of 37°C. The spores show polar germination and are fairly resistant to boiling. Most strains are non-pathogenic, but some cause eye infections in man and may occasionally invade the bloodstream in cachetic diseases.

The organism grows freely on ordinary media such as nutrient or blood agar. It grows with ammonia as the sole source of nitrogen and without added growth factors.

Fermentation and other biochemical reactions are used for identification. There is little exact knowledge available on the antigenic structure.

III. THE OUTBREAKS

A.   Statistics

Food poisoning in England and Wales is reviewed yearly by the Epidemiological Research Laboratory of the Public Health Laboratory Service (Vernon, 1966, 1967). It is a notifiable disease, and the annual figures hitherto published in the Monthly Bulletin of the Ministry of Health and Public Health Laboratory Service are based on reports made by pathologists in public health and hospital laboratories to the Public Health Laboratory Service and on returns submitted by medical officers of health to the Ministry of Health.

Table IV, reproduced from the paper by Vernon (1967), gives data for food poisoning incidents of all types in 1966; general outbreaks (mem-

TABLE IV

Food Poisoning of All Types, 1966

General Outbreaks, Family Outbreaks and Sporadic Cases According to Causal Agent[a,b]

Presumed causal agents	General outbreaks		Family outbreaks		Sporadic cases		All incidents		All cases[c]	
	No.	%	No.	%	No.	%	No.	%	No.	%
S. typhimurium	51	36	217	56	1139	55	1407	54	2346	36
Other salmonellas	34	24	145	37	910	44	1089	42	1868	29
Staphylococci	8	6	19	5	27	1	54	2	262	4
C. welchii	47	33	6	2	10	–	63	2	1947	30
Other organisms	1	1	1	–	–	–	2	–	10	–
Chemical	–	–	–	–	1	–	1	–	1	–
All agents	141	100	388	100	2087	100	2616	100	6434	100
Cause not discovered	40		175		913		1128		2350	
Totals	181		563		3000		3744		8784	

[a]Reproduced from Vernon (1967).

[b]– = less than 0.5%.

[c]Includes 316 symptomless excreters of *Salmonella typhimurium* and 299 symptomless excreters of other salmonellas.

bers of more than one family), family outbreaks (two or more persons in one family), and sporadic cases (individuals not associated with other cases) are related to the causal agents. The figures indicate that *C. perfringens* food poisoning occurs predominantly in general outbreaks and that the estimated number of persons per outbreak including family outbreaks is 37, whereas the average number of persons affected in salmonella outbreaks is 5 and there is a preponderance of sporadic cases compared with the number for *C. perfringens*. In England and Wales, *C. perfringens* appears to be responsible for a third of the general outbreaks of food poisoning and more than a third of the total number of cases of gastroenteritis. Table V, reproduced also from the paper by Vernon (1967), gives incidents in which the causal agent was discovered and shows the occurrence of food poisoning and the fatalities in England and Wales from 1957 to 1966.

Table VI has been adapted from Vernon (1967) to show that the food vehicle is found in a higher proportion of outbreaks due to *C. perfringens* than of outbreaks due to the other well-known agents of food poisoning, staphylococci and salmonellae. The factors leading to *C. perfringens* food poisoning are clear cut and well understood. The vehicle is nearly always a meat dish prepared in such a way that anaerobic conditions are provided for the germination of spores which have survived cooking; the period of time allowed during cooling and storage encourages multiplication of the organisms from germinated spores or from vegetative cells reaching the meat after cooking. In contrast, the sources of salmonellae in relation to animals and foods is not yet fully appreciated, and failure to correlate sporadic cases as part of one outbreak from a single vehicle or from multiple vehicles coming from a single origin limits the proportion of incidents traced to the source. Thus the vehicle and paths of spread may continue to present hazards.

Table VII, from the same paper, gives an analysis of outbreaks associated with meat products and demonstrates the association between precooked meats and *C. perfringens* food poisoning, in contrast to staphylococcal food poisoning in which the vehicle is mostly cooked and cured meats eaten cold. Such cooked meats are contaminated by hands or equipment, and poor storage conditions for some hours allow growth and toxin production. Staphylococci will tolerate concentrations of salt inhibitory to many other organisms including *C. perfringens*.

There are no United Kingdom statistics for food poisoning due to *C. bifermentans* and to the *Bacillus* group of organisms, although there may be incidents due to these agents among those placed in the category "undiscovered outbreaks."

TABLE V

FOOD POISONING OF ALL TYPES, 1957–1966

INCIDENTS ACCORDING TO YEAR AND CAUSAL AGENT[a]

Presumed causal agents	1957	1958	1959	1960	1961	1962	1963	1964	1965	1966
S. typhimurium	2,931	3,329	3,198	2,907	2,503	1,864	1,820	1,725	1,721	1,407
Other salmonellas	1,287	1,512	1,840	1,047	1,268	982	1,149	1,368	1,224	1,089
Staphylococci	128	125	118	98	91	143	74	107	74	54
C. welchii	93	88	110	97	83	84	82	85	64	63
Other agents	2,632	2,246	2,580	2,279	1,442	1,448	1,340	1,087	1,008	1,131
All causes										
Incidents	7,071	7,300	7,846	6,428	5,387	4,521	4,465	4,372	4,091	3,744
Cases[b]	15,100	14,900	16,600	13,700	12,750	9,696	13,104	9,975	11,317	8,784
Deaths	36	30	27	25	22	23	27	19	19	26

[a] Reproduced from Vernon (1967).

[b] For the years 1957–1960 the number of cases is approximate.

TABLE VI

Food Poisoning of All Types, 1966
Vehicles of Infection and Causal Agents in General and Family Outbreaks[a]

Vehicle of infection	Presumed causal agents						
	S. typhimurium	Other salmonellas	Staphylococci	C. welchii	Other organisms	Not discovered	All agents
Meat:							
Fresh	–	–	–	3	–	1	4
Canned	–	–	7	1	–	–	8
Processed and made-up[b]	9	4[d]	5	37	1	9	65 } 85
"Meat"[c]	2	2	–	1	–	–	5
Gravy, soup, meat sauce	–	–	–	2	–	1	3
Milk:							
Unpasteurized	2	2	–	–	–	–	4
Fish:							
Canned	–	–	2	–	–	–	2 } 4
Processed and made-up[b]	–	–	2	–	–	–	2
Sweets	3	1[d]	1	–	–	–	5
All foods	16	9	17	44	1	11	98
General and family outbreaks	268	179	27	53	2	215	
Food vehicle discovered	6%	5%	63%	83%			

[a] Adapted from Vernon (1967).
[b] Other than canned.
[c] Meat not further described.
[d] In one outbreak both cooked meats and sweets were incriminated.

TABLE VII

Food Poisoning of All Types, 1966

General and Family Outbreaks Associated with Processed or Made-up Meats[a]

Vehicle of infection	Presumed causal agents					All agents
	Salmonellas	Staphylococci	C. welchii	Other organisms	Not discovered	
Reheated meat:						
Poultry	6	—	6	—	1	13
Stew, including mince	—	—	4	—	2	6
Beef	—	—	5	—	—	5
Pork	—	—	3	—	—	3
Liver	—	—	2	—	1	2
Mutton	—	—	—	—	1	1
"Meat"[b]	—	—	1	—	1	2
						(32)
Cold meat						
Processed meat:[c]						
Ham, boiled bacon	1	1	1	—	1	3
Tongue	1	—	1	—	1	3
Fresh meat[d]						
Beef	—	1	3	—	1	5
Poultry	2	—	1	—	1	4
Pork	—	—	1	—	—	1
"Meat"[d]	—	1	—	—	—	1
						(17)

Meat pies					
Manufactured					
Pasties	–	2	–	–	2
Other	1	–	1	–	2 } 10
Made at home or canteen:					
Steak[e]	–	3	–	–	3
Shepherd's	–	3	–	–	3 }
Sausage	1	1	–	–	2
Various cooked meats	4	–	–	–	4
All meat dishes	13	37	1	9	65

[a] Reproduced from Vernon (1967).
[b] Meat not further described.
[c] Semipreserved meat.
[d] Uncured meat.
[e] Steak pies consist essentially of meat in a dish with pastry on top. Shepherd's pies are similar but with potato in place of pastry.

B. Clostridium perfringens

1. FOOD AND PREPARATION

Various forms of meat with histories of storage at a warm temperature
for at least 2 hours after cooking are common factors in almost all outbreaks of clostridial food poisoning. The presence of precooked cold or
warmed-up meat dishes, particularly in large bulk, in a meal eaten at the
relevant time before the onset of food poisoning symptoms should arouse
suspicion that *C. perfringens* may be the agent. Usually, raw meat or
poultry, already contaminated with *C. perfringens* spores from a variety
of sources, is cooked at a temperature which allows the survival of the
spores which are heat-shocked into germination as soon as the cooling
mass reaches a suitable temperature; multiplication of the young vegetative cells will continue at rates dependent on the storage times and
temperatures and, obviously, on the situation of the organisms within
the joint or carcass. Growth will be encouraged when the organism is in
parts of the meat where the Eh is low after removal of oxygen by cooking, and also by long slow cooling periods.

In a liquid mass of meat there will be no difficulty for the organism to
find anaerobic conditions. Both home and canteen-cooked fresh meats
such as joints and poultry (turkeys particularly), as well as minced or
cut-up meats for stew and pies, have been implicated whether eaten cold
or warmed-up. Reheating may increase the multiplication if not carried
out at temperatures inhibitory to the growth or even survival of the
organisms. Rolled joints are particularly susceptible because the outside surface contamination is rolled into the middle of the mass of meat
where heat penetration and heat loss are poor and anaerobic conditions
are good. Salted meats are not usually involved, but boiled brisket of
beef, rolled and salted, has frequently been a vehicle of *C. perfringens*
food poisoning in canteen outbreaks; presumably the salt is diluted out
either by soaking overnight before cooking or by dilution in the actual
water used for boiling.

Thus in any method of cooking where the temperature is not greater
than 100°C, spores will survive even the roasting of rolled joints and of
large bulks of poultry—for example, turkeys. Subsequent long slow cooling and ambient storage with or without reheating are the common faults
which give rise to this type of food poisoning. Examples of outbreaks
and the foods incriminated are given in Table I.

A recent outbreak due to roast turkey illustrates the faults that will
automatically give rise to food poisoning if the raw material is con-

taminated with *C. perfringens*. Two 21-pound frozen turkeys were thawed in their plastic bags, stuffed, and roasted for 3 ½ hours only at about 350°F. The cooked turkeys were left for 2 hours on top of the cooker to cool and were stored overnight in the oven. The following day slices were cut and the portions warmed in the oven before serving with gravy to persons attending a celebration supper. There was one fatality among about 50 cases out of 150 persons at risk.

2. Clinical Symptoms, Incubation Period, and Postmortem Findings

Clostridium perfringens food poisoning is characterized by severe diarrhea and abdominal pain, usually without vomiting, beginning 8 to 24 hours after a meal. Pyrexia, shivering, headache, and other signs of infection are seldom described.

The illness is of short duration, one day or less, but in elderly debilitated patients and even in younger debilitated persons fatalities may occur. The results of postmortem examination of two of three fatal geriatric cases in a hospital outbreak are described by Parry (1963). There were changes in the large and small bowel. Distention, possibly due to antemortem gas formation, congestion, and infection were found.

3. Human Volunteer Experiments

It is popularly believed that the symptoms are caused by a toxin or toxoid produced by the organism in the food before ingestion, but there is no evidence from volunteer experiments with filtrates that such a preformed toxin system exists. Neither is there evidence of heat-resistant endotoxins from dead cells causing symptoms of food poisoning.

Most successful volunteer experiments have followed the ingestion of live organisms growing in a meat medium (Hobbs *et al.*, 1953; Dische and Elek, 1957). Hauschild and Thatcher (1967) carried out an experiment in which 6 volunteers were given centrifuged cells (resuspended in cooked milk) of a culture of *C. perfringens* isolated from roast beef suspected to have caused food poisoning. By 4 to 9 hours after cell ingestion, 5 of the 6 persons developed diarrhea which continued for periods of 4 to 21 hours. Furthermore, 7 of 7 lambs produced loose stools or developed severe diarrhea after receiving vegetative cells of the same strain of *C. perfringens*. The organism produced only traces of α-toxin and no β-toxin; it was partially hemolytic on sheep-blood agar and nonhemolytic on horse-blood agar. Heat-resistant tests on feces from the volunteer cases showed survival of 0.2% after 4 minutes but no survival after 10 minutes of heating at 100°C. In Hall and colleagues (1963) and Ellner's (1956) spore media, only 0.1 to 0.2% spores survived 2 minutes at 100°C.

4. MECHANISM

Prior to Hauschild and Thatcher's positive experiment with volunteers drinking *C. perfringens* cells suspended in cooked milk, volunteer experiments had succeeded only when the live organisms sporing actively in a meat medium were ingested. Nevertheless, it would be interesting to know what the Canadian volunteers ate for lunch, as the paper states that the dose of cells in milk was taken immediately prior to lunch. If meat were included in the menu, the cells of *C. perfringens* would have accompanied the meat through the intestinal tract, and this may not be very different from swallowing a cooked meat culture of the organism.

Nygren (1962), in an interesting and detailed report entitled "Phospholipase C-Producing Bacteria and Food Poisoning, an Experimental Study of *C. perfringens* and *B. cereus,*" suggests that lecithin in food may be hydrolyzed by the enzyme phospholipase C, produced by both *C. perfringens* and *B. cereus,* to form phosphorylcholine. It is claimed that this substance will initiate diarrhea in 8 to 10 hours. A similar effect on the intestine exposed to synthetic phosphorylcholine both *in vivo* and *in vitro* was demonstrated for various animals; no human experiments were carried out. Nelson *et al.* (1966) fed phosphorylcholine to human volunteers, but there were no symptoms.

The long incubation period and lack of symptoms indicating infection suggests a mechanical effect on the lower intestine. Gas production alone might cause pain and discomfort; but the unusual products of a predominating microbe feeding on partially digested meat or other foodstuffs might well disturb metabolism. On the whole, the mechanism is considered to be a live reaction of the actively multiplying organism on a particular substrate; it is probably enzymic in action, and dependent on initiation by a large dose of organisms. It has not so far been associated with any soluble or filtrable substance nor with the internal products of dead cells. Thus, any organisms producing phospholipase C or other relevant enzymes would be potentially able to cause food poisoning provided they were consumed in large numbers with the right substrate. It is assumed that food poisoning due to *C. bifermentans* and *B. cereus* could occur by the same mechanism.

Hauschild *et al.* (1967) in a second paper again reported diarrhea in lambs after introduction of *C. perfringens* into the digestive tract of the animals. Heat-resistant and heat-sensitive strains were used for these experiments, with similar results. They noted that the heat-resistant strains produced traces of α-toxin only and that it could not be the main food poisoning factor. Furthermore, the incidence of diarrhea after challenge with cells of *C. perfringens* was the same in immunized and non-immunized lambs, and thus they assumed that the factors involved in the food poisoning process were nonantigenic—although it is pointed out that any antibodies formed do not come in contact with the corresponding antigens in intestinal infections.

C. *Clostridium bifermentans*

Duncan (1944), in the introduction to a short paper, describes an outbreak of food poisoning of special interest because it added to the list of outbreaks resulting "not from the presence of a pathogenic bacterium or of a specific toxin, but from the ingestion of food in which abundant growth of a non-pathogenic organism has occurred—in this instance *C. bifermentans.*"

All members of a camp, 70 to 80 people, were taken ill with diarrhea 6 to 7 hours after eating meat and potato pie; vomiting was not a feature of the outbreak. The condition of the pie was poor; it was in a state of fermentation and smelled and tasted unpleasant, so that small portions only were eaten. A sample liquid gave a rich and almost pure culture of *C. bifermentans* and no other significant organisms. A few weeks before there had been a similar incident associated with the same kind of pie.

The organisms were thought to have come from the potatoes which were washed and peeled by the same machine, the eyes removed, sliced by hand, partly cooked in a steamer for 15 minutes and added to cooked meat and other ingredients in a pie dish which was then covered with pastry and put in the oven for 15 minutes. After cooking the warm pie was placed in an insulated container where it remained warm for 6 to 7 hours before eating.

Clostridium bifermentans was isolated from the raw sliced potatoes as well as from the cooked pie. The author suggested that spores of *C. bifermentans* would survive the two-stage cooking, the pie crust would encourage anaerobic conditions beneath, and the period of warm storage would encourage multiplication. The cause of the diarrhea was thought to be the large dose of living *C. bifermentans* or a small amount of toxin. There is no mention of the possibility that sporing organisms were present in the meat before cooking, nor of how long the meat was cooked in the first stage.

D. *Bacillus cereus* and *Bacillus mesentericus*

As there have been no authentic or published accounts of food poisoning due to *B. cereus* or other organisms of the *Bacillus* group in the United Kingdom, reference must be made again to the description given by Hauge (1955), who investigated outbreaks of food poisoning in three hospitals and an old people's home. Approximately 600 persons were affected in four outbreaks quoted; it is interesting to note that all the outbreaks occurred after a Sunday meal when preparation had occurred the day before. A description of one of these outbreaks is given. The menu for the Sunday dinner included a meat dish with vegetables, and chocolate pudding with vanilla sauce for dessert. Both the chocolate pudding and the sauce had been prepared the previous day and stored in a large container at room temperature. Of 99 persons, 80 were affected; of the 19 who escaped, 11 had not eaten the dessert, and 8 had a small quantity only. The average incubation period was 10 hours for the patients and 12 ½ hours for 20 members of the staff who ate the meal 2 ½ hours later than the patients; the staff were more severely affected and their symptoms lasted longer. The symptoms included abdominal pain and profuse watery diarrhea, rectal tenesmus, and moderate nausea seldom giving rise to vomiting; fever was uncommon, and the duration of the illness was about 12 hours.

Vanilla sauce prepared from similar constituents the day before it was served was also the cause of the other outbreaks. In all four instances, on examination the sauce was found to contain 25 to 100×10^6 *B. cereus* per milliliter, although it was not changed much in odor, taste, or consistency. There were numerous spores of *B. cereus*—up to 10^4 per gram—in the cornstarch used to make the sauce. Hauge thought that the organisms might have multiplied selectively in the starch manufactured over a period of 4 days.

Direct microscopic examination of the food, vanilla sauce, showed large numbers of large gram-positive rods about 0.9 μ in diameter. Aerobic and anaerobic blood agar plates gave pure cultures of an aerobic gram-positive sporing rod which produced hemolysis and lecithinase by the method of McGaughey and Chu (1948). Seitz filtrates from cultures of *B. cereus* did not produce food poisoning symptoms, nor were filtrates very toxic for mice and guinea pigs. However, 200 ml of a vanilla sauce inoculated with 10^4 *B. cereus* per milliliter and incubated for 24 hours at room temperature, when the count was 92×10^6 per milliliter, was consumed by the author. Symptoms started after 13 hours and were similar to those already described. Again, few colonies of *B. cereus* developed from fecal samples taken during the illness, but the number of entero-bacteriaceae was also low.

Hauge points out that, in the case of confusion with *C. perfringens* food poisoning, anaerobic incubation of blood agar from food should be followed by aerobic incubation. He says that, unlike *C. perfringens* food poisoning, *B. cereus* is not excreted in large numbers in the feces after an outbreak.

Another outbreak reported by Christiansen *et al.* (1951) due to corn-flour pudding involved 121 of 154 persons, mostly children. It appears that national food habits may encourage food poisoning from certain organisms growing in particular food substrates.

Nikodemusz (1964) has published work on food poisoning ascribed to aerobic sporing bacilli; from 1958 to 1964 he studied 51 outbreaks involving 1400 people. Another paper (Nikodemusz and Gonda, 1966) describes the action on cats of long-term administration of food products and drinking water contaminated with *B. cereus* (10^5 to 10^8 per gram). The result was continuous diarrhea, occurring quickly after feeding; there were 10^5 to 10^7 *B. cereus* per gram in the feces. Vomiting occurred 4 to 5 hours after feeding, and there appeared to be abdominal pain and tenderness; there was also loss of body weight. After 2 to 3 weeks, three of six cats died, and the remaining three were killed after 30 days. The postmortem findings showed abnormal conditions in the ileum and colon, with much gas and hyperemia. The livers of four cats were examined histologically, and the wall of the hepatic vein showed fat deposits; the kidneys of three cats were parenchymatous.

Rodents were thought to be relatively insensitive to the enteropathogenic effect of the aerobic sporing bacilli, as even after 50 days there was no recognizable damage.

The relationship of *Bacillus mesentericus* to the condition known as ropy bread is described in books on food technology. It is discussed by Tanner (1944), who cites a considerable number of related publications as far back as 1885. W. Hirsch (personal communication) says that food poisoning from ropy bread seems to have become very rare in Europe and in the United States, and that this is so in Israel, too. The last sample of ropy bread seen by him was in September 1955. Details of food poisoning from ropy bread were not found.

IV. PREVENTION

As it would be almost impossible to reduce the incidence of *C. perfringens* or *B. cereus* in nature, it must be accepted that both organisms will continue to be present in food. Neither organism is dangerous when ingested in small numbers; thus the whole field of prevention must be concerned not only with destruction but also with the control of germination of spores and subsequent multiplication of vegetative cells in heat-treated foods. So far as *C. perfringens* is concerned, the greater danger is slow cooling and storage of cooked but nonsterile meat and poultry masses.

Freshly cooked meat eaten hot immediately after cooking is safe. Of the various methods of cooking, steaming under pressure, thorough roasting, frying, and grilling are most likely to destroy cells and spores alike. All forms of cooking at temperatures not exceeding 100°C will allow survival of the spores of *C. perfringens,* and the more resistant types will give most trouble.

As with *C. perfringens,* prevention of food poisoning due to aerobic sporing bacilli is concerned principally with confining the organism to small numbers by preparing the food a short time before serving or, if it has to be stored, by rapid and adequate cooling to a temperature sufficiently low to prevent growth of the organism.

The prevention of this type of poisoning from food prepared on a large scale by institutional cookery is difficult, and control measures are based on the assumption that in most incidents *C. perfringens* spores are already present on or within the meat at the time of purchase. Thus, codes of practice must prevent multiplication of the organisms in post-cooking periods, as well as reduce the likelihood of survival by correct cooking procedures.

Ideally, meat and poultry dishes should be freshly cooked and eaten hot. If this is impossible, the following procedures are recommended for cooking meat, and for keeping the periods between cooking and refrigeration, and removal from the refrigerator and serving, minimal:

1. If the portions of meat are large, cut them into smaller pieces (3 to 6 pounds). Ensure that frozen products are completely thawed.

2. Cook the meat well, making sure that the oven or steamer temperatures are correct. Slow, overnight cooking is unsatisfactory.

3. Separate the meat from the broth immediately after cooking.

4. Place the meat portions on shallow metal trays, or on a trolley, and cover them loosely with a piece of waxed paper or foil. Do not leave them in hot, deep pans where cooling will be slow.

5. If a large cold room is available, the cooked meat can be placed there immediately. Otherwise, transfer to a refrigerator within 1 hour of cooking, and leave there until shortly before serving.

6. If it is necessary to slice the meat before serving, this should be done immediately after removal from the refrigerator.

7. This meat may now be served cold—for example, with salads—or hot with gravy.

8. If the meat is to be served hot, it should be brought to the required serving temperature as quickly as possible, either in a steamer or by placing it in a container with gravy that is maintained at almost boiling temperature. Alternatively, the cold slices may be covered with boiling hot gravy immediately before serving.

9. The habit of pouring warm gravy on the sliced meat and then putting both into a "warming cabinet," with the temperature below 140°F, to await serving with vegetables, must be condemned.

The following practices should be avoided:

1. Keeping the meat with broth immediately after cooking.
2. Long delay between cooking and refrigeration.
3. Placing the meat in "warm" ovens, particularly with gravy, for periods longer than 1 hour before serving.

It is good practice to remember that cooked meat should be kept either cold—below 41°F (5°C)—or above 140°F (60°C) at all times.

Recommendations for the preparation, cooking, and storage of frozen turkeys are given below:

1. Keep in the bag until ready for thawing, then remove the bag and defrost thoroughly, either at ambient temperature or by means of hot water allowed to run on the outside and through the inside of the bird. It should be noted that a turkey which has been thawed slowly may still be at a temperature of 30°F when outwardly it is apparently thawed.

2. Do not mop the inside of the bird.

3. Do not stuff the inside of the bird. Cook the stuffing separately.

4. Cook well; for example, a 20-pound turkey should be cooked for 6 to 7 hours at approximately 350°F.

5. *Catering:* After cooking, put birds immediately into the cold room and keep cold until required. *Housewife:* After cooking, cool rapidly in cold, draughty place. Keep cold until required.

6. If large numbers of slices are required, slice as quickly as possible, and if needed hot, heat rapidly and thoroughly throughout the mass of slices—for example, by boiling in gravy—or keep the slices cold and add boiling hot gravy at the time of serving.

ADDENDUM
QUESTIONNAIRE FOR *Clostridium perfringens*
FOOD POISONING OUTBREAKS

Place: ———————— *Examining laboratory:* ————————
Date of outbreak: ————————
Date of collection of food/feces samples ————————————

Meat: Description of meat, e.g., mince, rolled roast, poultry, etc.
_____

Size of joint(s), poultry ————————————————————
Was the meat received frozen ————————————————
If so, was it thawed *before* cooking ————————————
How thawed ————————————————————————
Was stuffing used ————————————————————————
Cooking: Method of cooking ————————————————————
Time of cooking ———————— Temperature of cooking ————
Were several pieces cooked in one container ————————
When served, did the inside of the meat appear well cooked
_____

(brown/red/pink color) ————————————————————
Storage: Was the meat or poultry removed from cooking vessel when it
was taken from the stove ————————————————
Where cooked ————————————————————————
Was it refrigerated ————————————————————
Time lapse between cooking and refrigeration ————————
If not refrigerated, what was the storage temperature ————
How long stored ————————————————————————
Was it covered — e.g., cloth, paper, foil, lid on vessel during
storage ————————————————————————————
Serving: Time delay between removal from refrigerator and serving
_____

Was it rewarmed ———— How ———— Was gravy used ————
If so, was it prepared from commercial gravy powder ————
Food: Samples obtained ————————————————————————
Results of examination to be added later ————————————
(Count, organisms, serotypes)
Patients: No. at risk ———— No. ill ———— No. of feces examined ————
No. positive for *C. perfringens* (and serotypes) ————————
Incubation period ————————————————————————
Symptoms ———————————— Diarrhea
                                Vomiting
                                Fever
Duration of illness ————————————————————————

REFERENCES

Ager, E., and Ploeger, E. O. (1964). *Morbidity Mortality* **13**, 234.

Andrews, F. W. (1899). *Lancet* **i**, 8

Angelotti, R., Hall, H. E., Foter, M. J., and Lewis, K. H. (1962). *Appl. Microbiol.* **10**, 193.

Barnes, E. M., Despaul, J. E., and Ingram, M. (1963). *J. Appl. Bacteriol.* **26**, 415.

Beck, A., Foxell, A. W. H., and Turner, W. C. (1954). *Brit. Med. J.* **i**, 686.

Brooks, M. E., Sterne, M., and Warrack, G. H. (1957). *J. Pathol. Bacteriol.* **74**, 185.

Christiansen, O., Koch, S. O., and Madelung, P. (1951). *Nord. Veterinarmed.* **3**, 194.

Collee, J. G. (1954). *J. Roy. Army Med. Corps* **100**, 296.

Collee, J. G. (1955). *J. Roy. Army Med. Corps* **101**, 46.

Collee, J. G. (1961). *J. Pathol. Bacteriol.* **81**, 297.

Collee, J. G., Knowlden, J. A., and Hobbs, B. C. (1961). *J. Appl. Bacteriol.* **24**, 326.

Cravitz, L., and Gillmore, J. D. (1946). Project I-756, Report No. 2. Bethesda, Naval Medical Research Institute.

Dack, G. M., Sugiyama, H., Owens, F. J., and Kirsner, J. B. (1954). *J. Infect. Diseases* **94**, 34.

Dam-Mikkelsen, H., Petersen, P. J., and Skovgaard, N. (1962). *Nord. Veterinarmed.* **14**, 200.

Dickie, G. C., and Smith, J. (1957). *Health Bull., Edinburgh* **15**, 81.

Dische, F. E., and Elek, S. T. (1957). *Lancet* **ii**, 71.

Duncan, J. T. (1944). *Monthly Bull. Min. Health* **3**, 61.

Dunham, E. K. (1897). *Bull. Johns Hopkins Hosp.* **8**, 68.

Edinburgh Public Health Department (1960). *Annual Report of the Public Health Department for the Year 1959.* Hugh Paton & Sons Ltd., Edinburgh, 129 pp.

Ellner, P. D. (1956). *J. Bacteriol.* **71**, 495.

Ernst, O. (1948). *Deut. Gesundheitsw.* **3**, 262.

Gibbs, P. A. (1964). *J. Gen. Microbiol.* **37**, 41.

Gibbs, P. A. (1967). *J. Gen. Microbiol.* **46**, 285.

Gough, B. J., and Alford, J. A. (1965). *J. Food Sci.* **30**, 1025.

Hain, E. (1949a). *Brit. Med. J.* **i**, 271.

Hain, E. (1949b). *Brit. Med. J.* **i**, 271.

Hall, H. E., and Hauser, G. H. (1966). *Appl. Microbiol.* **14**, 928.

Hall, H. E., Angelotti, R., Lewis, K. H., and Foter, M. J. (1963). *J. Bacteriol.* **85**, 1094.

Hauge, S. (1950). *Nord. Hyg. Tidskr.* **31**, 189.

Hauge, S. (1955). *J. Appl. Bacteriol.* **18**, 591.

Hauschild, A. H. W., and Thatcher, F. S. (1967). *J. Food Sci.* **32**, 467.

Hauschild, A. H. W., Niilo, L., and Dorward, W. J. (1967). *Bacteriol. Proc.* p. 6.

Hayashi, K., Kugita, Y., Tawara, M., and Yamagata, H. (1961a). *Endemic Diseases Bull. Nagasaki Univ.* **3**, 1.

Hayashi, K., Kugita, Y., Tawara, M., and Yamagata, H. (1961b). *Endemic Diseases Bull. Nagasaki Univ.* **3**, 87.

Henderson, D. W. (1940). *J. Hyg.* **40**, 501.

Herron, J. T., Bunch, W. L., Lawson, M. C., and Cairns, J. M. (1965). *Morbidity Mortality* **14**, 385.

Hobbs, B. C., and Wilson, J. G. (1959). *Monthly Bull. Min. Health* **18**, 198.

Hobbs, B. C., Smith, M. E., Oakley, C. L., Warrack, G. H., and Cruickshank, J. C. (1953). *J. Hyg.* **51**, 75.

Howitt, L. F. (1964). *Health Bull., Edinburgh* **22**, 42.

Ingram, M., and Barnes, E. M. (1956). *Lab. Pract.* **5,**145.

International Association Microbiological Societies. "Microorganisms in Foods: Their Significance and Methods of Enumeration." Univ. of Toronto Press. (in Press).

Johns, A. W. (1951). *Med. Offr.* **86,**85.

Klein, E. (1895). *Zentr. Bakteriol. Parasitenk., (I Abt. Orig.)* **18,**737.

Klotz, A. W. (1965). *Public Health Rept. (U.S.)* **80,**305.

Knox, R., and Macdonald, E. K. (1943). *Med. Offr.* **69,**21.

Leeming, R. L., Pryce, J. D., and Meynell, M. J. (1961). *Brit. Med. J.* **i,**50.

Lewis, G. M., and Macfarlane, M. G. (1953). *Biochem. J.* **54,**138.

Linzenmeier, G. (1956). *Offic. GesundhDienst.* **17,**708.

McClung, L. S. (1945). *J. Bacteriol.* **50,**229.

McCroan, J. E., and Mixson, B. W. (1965). *Morbidity Mortality* **14,**187.

McGaughey, C. A., and Chu, H. P. (1948). *J. Gen. Microbiol.* **2,**334.

McKillop, E. J. (1959). *J. Hyg.* **57,**31.

Main, B. W. (1959). *J. New Zealand Assoc. Bacteriol.* **14,**50.

Marcuse, K., and Konig, I. (1950). *Zentr. Bakteriol. Parasitenk. (Abt. I Orig.)* **156,**107.

Miles, A. A., and Misra, S. S. (1938). *J. Hyg.* **38,**732.

Miles, E. M., and Misra, S. S. (1947). *J. Gen. Microbiol.* **1,**385.

Miles, E. M., and Misra, S. S. (1950). *J. Gen. Microbiol.* **4,**22.

Miller, N. J., Garrett, O. W., and Prickett, P. S. (1939). *Food Res.* **4,**447.

Mossel, D. A. A., de Bruin, A. S., van Diepen, H. M. J., Vendrig, C. M. A., and Zoutewelle, G. (1956). *J. Appl. Bacteriol.* **19,**142.

Murrell, T. G. C., Egerton, J. R., Rampling, A., Samels, J., and Walker, P. D. (1966). *J. Hyg.* **64,**375.

Narayan, K. G. (1966). *Acta Vet. Acad. Sci. Hung.* **16,**65.

Nelson, K. E., Ager, E. A., Marks, J. R., and Emanuel, I. (1966). *Am. J. Epidemiol.* **83,** 86.

Nikodemusz, I. (1964). "Ar aerob sporas babstericuuock etelmorgesest doiders kepessegeoch hiserlates rizogalata es ligveves ertekeles." Budapest.

Nikodemusz, I., and Gonda, G. (1966). *Zentr. Bakteriol. Parasitenk. (Abt. I Orig.)* **199,**64.

Nygren, B. (1962). *Acta Pathol. Microbiol. Scand. Suppl.* **160,**88 pp.

Oakley, C. L. (1943). *Bull. Hyg.* **18,**781.

Oakley, C. L. (1949). *Brit. Med. J.* **i,**269.

Oakley, C. L., and Warrack, G. H. (1951). *J. Pathol. Bacteriol.* **63,**45.

Oakley, C. L., and Warrack, G. H. (1953). *J. Hyg.* **51,**102.

Oakley, C. L., Warrack, G. H., and Heyningen, W. E. van (1946). *J. Pathol. Bacteriol.* **58,**229.

Oakley, C. L., Warrack, G. H., and Warren, M. E. (1948). *J. Pathol. Bacteriol.* **60,**495.

Østerling, S. (1952). *Nord. Hyg. Tidskr.* **33,**173.

Parry, W. H. (1963). *Brit. Med. J.* **ii,**1616.

Roberts, T. (1968). *J. Appl. Bacteriol.* **31,**133.

Robertson, J. S. (1963). *Monthly Bull. Min. Health* **22,**144.

Rodella, A. (1910). *Wien Klin. Wochschr.* **23,**1383.

Schmidt, C. F., Nank, W. K., and Lechowich, R. V. (1962). *J. Food Sci.* **27,**77.

Schutz, F. (1948). *Deut. Med. Wochschr.* **73,**176.

Simonds, J. P. (1915). *Monographs Rockefeller Inst.* No. 5.

Smith, H. W., and Crabb, W. E. (1961). *J. Pathol. Bacteriol.* **82,**53.

Smith, J., and Wallace, J. M. (1956). *Health Bull., Edinburgh* **14,**31.

Spencer, R. (1964). *Second Interim Technical Progress Report on the Vacuum-packaging of Cooked Cured Meats. Technical Circular No. 271.* British Food Manufacturing Industries Research Association, Leatherhead, England.

Sterne, M., and Warrack, G. H. (1964). *J. Pathol. Bacteriol.* **88,** 279.

Stocks, A. V. (1955). *Med. Offr.* **93,** 191.

Strong, D. H., and Canada, J. C. (1964). *J. Food Sci.* **29,** 479.

Strong, D. H., Canada, J. C., and Griffiths, B. B. (1963). *Appl. Microbiol.* **11,** 42.

Sutton, R. G. A., and Hobbs, B. C. (1968). *J. Hyg.* **66,** 135.

Sutton, R. G. A. (1966). *J. Hyg.* **64,** 65.

Sylvester, P. K., and Green, J. (1961). *Med. Offr.* **105,** 231.

Tanner, F. W. (1944). *In* "The Microbiology of Foods," Garrard Press, Champaign, 707 pp.

Taylor, A. W., and Gordon, W. S. (1940). *J. Pathol. Bacteriol.* **50,** 271.

Taylor, C. E. D., and Coetzee, E. F. C. (1966). *Monthly Bull. Min. Health* **25,** 142.

Turner, G. C., and Wong, M. M. (1961). *J. Pathol. Bacteriol.* **82,** 529.

Vernon, E. (1966). *Monthly Bull. Min. Health* **25,** 194.

Vernon, E. (1967). *Monthly Bull. Min. Health* **26,** 235.

Von Hibler, E. (1906). *Zentr. Parasitenk. Bakteriol. (I Abt. Orig.)* **37,** 545.

Waart, J. de., and Smit, F. (1967). *Lab. Pract.* **16,** 1098.

Wade, C. H. T. (1959). *Med. Offr.* **102,** 334.

Warrack, G. H. (1963). *Bull. Off. Int. Epizoot.* **59,** 1393.

White, A., and Hobbs, B. C. (1963). *Roy. Soc. Health J.* **83,** 111.

Wijewanta, E. A. (1964). *J. Pathol. Bacteriol.* **88,** 339.

Wild, O. (1898). *Zentr. Parasitenk., Bakteriol. (I Abt. Orig.)* **23,** 913.

Yamagata, H., Iwafune, Y., and Shimamura, Y. (1959). *Japan. J. Microbiol.* **3,** 365.

Zeissler, J., and Rassfeld-Sternberg, L. (1949). *Brit. Med. J.* **i,** 267.

CHAPTER V | **PARASITIC INFECTIONS**

George R. Healy and Neva N. Gleason

I. INTRODUCTION

Many types of food ordinarily ingested by humans serve as the mechanism by which a great variety of helminth and some protozoan parasites are acquired. Such foods include those of animal origin: pork, beef, lamb, freshwater and saltwater fish, plus various molluscs such as crabs and crayfish. In some cases more exotic varieties are involved: snakes, frogs, and freshwater and terrestrial snails and slugs.

Various plant foods also function as the source whereby parasitic organisms are acquired. The parasite may be contained in or on the plant food as an obligate part of the parasite life cycle, or the infective stages may be present as a result of contamination. Included in the former category are such foods as watercress, water caltrops, and bamboo shoots, containing the encysted metacercariae of flukes. In the latter category are a great variety of leafy vegetables such as lettuce, celery, or parsley which become contaminated with infective stages of parasites liberated in human excrement used to fertilize crops.

The limited geographic distribution of some of the food-borne helminths is due to ethnic customs among the people living in the areas where such parasites are found. For example, susceptible snails (first-intermediate hosts) and crab or crayfish (second-intermediate hosts) are essential for the life cycle of *Paragonimus westermani,* and are found in the United States. However, the habit of eating crabs and crayfish raw or partially cooked is not found in this country, whereas in certain parts of the Far East, because of such a practice, paragonimiasis is an endemic disease.

Whatever the requirements for the life cycle of a particular parasite, it is essential that there be continuous reinfection of both the food and man for the parasite to perpetuate itself. Inasmuch as some of the food-borne helminth parasites have limited environmental tolerances, restricted host specificity, and complicated life cycles, it is amazing that they are able to persist, to say nothing of causing disease and even death.

Almost without exception, the success of these food-borne organisms is due to the practice of eating infected foods raw or insufficiently cooked. Reservoir vertebrate hosts which harbor the stages found in man serve also to perpetuate the parasite in nature. It can be said, however, that the major cycle which perpetuates most of the human food-borne parasites is a man–food–man one.

II. NEMATODES

A. *Trichinella spiralis* (Scheme 1)

The causative organism of trichinosis is by far the best known of the food-borne parasites. *Trichinella spiralis* has a worldwide distribution in animals and occurs not only in pigs, rats, and other mammals of man's domestic environment but in more than forty species of wild animals (Kozar, 1962).

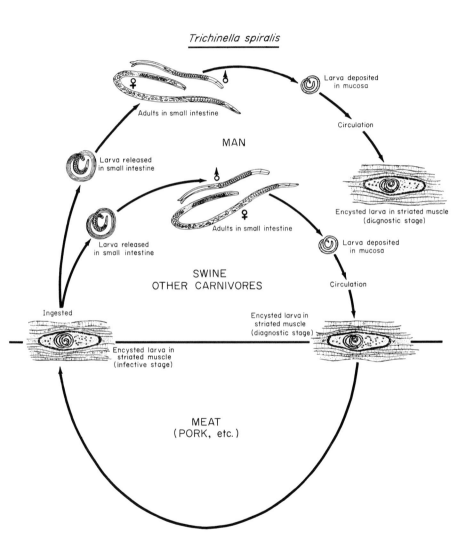

SCHEME 1. Life cycle of *Trichinella spiralis*.

The life cycle of the parasite is simple and direct. Encysted larvae contained in striate muscle are released during digestion in the host stomach. The male and female larvae mature in the stomach in five days, embed themselves in the intestinal mucosa, and copulate. The female then begins the deposition of larvae into the lymphatics. Larvae are carried to all parts of the body, and on penetration of striate muscle they become encysted. They remain viable and infective for many months. The cycle is completed when the flesh of such a host is eaten raw by a predator or scavenger in nature or when it is improperly cooked and eaten by man.

In the majority of human cases of trichinosis, especially those which are serious enough to come to medical attention, pigs have been the major source of infection. Although large-scale epidemics of trichinosis like those in Germany in the 1800's no longer occur (Gould, 1945), there are still small foci of human infections in the United States wherein raw or insufficiently cooked pork is the suspected source. Of increasing importance are reports of disease occurring as a result of the ingestion of meat other than pork. Among game animals, the bear has assumed some epidemiological importance. Maynard and Pauls (1962) recorded an outbreak among Alaskan natives who had consumed black or brown bear meat. Roselle et al. (1965) also presented interesting facts concerning a small cluster of six cases involving bear meat which was sold commercially in Vermont.

The prevalence of trichinosis in humans (mostly subclinical) is often quoted as 16%, based largely on the studies of Wright et al. (1943). This figure for the United States is believed to be high, since studies carried out by Zimmerman (unpublished) on human diaphragms indicate a 5% level of infection in individuals 40 years of age and younger. In the past four years (1962–1965) 799 cases of human trichinosis were reported in the Morbidity and Mortality Weekly Report of the U.S. Public Health Service (1966). In contrast, during the period 1944–1948 there were 1999 cases reported. Since pigs are the primary source of human trichinosis, the decline in human infections can be explained in part by the decline of the infection in pigs, as shown by studies of Zimmerman and his co-workers. Recently Zimmerman and Brandly (1965) reported the latest in a series (1961–1965) of studies on the occurrence of trichinosis in swine. Results of the examination of diaphragms disclosed that only 11 (0.12%) of 9495 butchered swine and 15 (0.22%) of 6881 farm-raised breeder swine were infected. Garbage-fed hogs showed a much higher infection rate, with 131 (2.6%) of the diaphragms of 5041 animals positive for *T. spiralis.*

Although data on the occurrence of trichinosis in swine on a national basis were not being tabulated years ago, Zimmerman and Brandly (1965) noted that in Iowa in 1944–1946 the incidence of trichinosis in samples of bulk pork sausage was 12.4%, while in the period 1953–1960 only a 1.0% incidence was recorded. The reduced prevalence of trichinosis in swine with a concomitant reduction in human infections has come about for a number of reasons. In the early 1950's a nationwide outbreak of vesicular exanthema, a pathogenic virus disease in pigs, resulted in the passage of legislation prohibiting the feeding of uncooked garbage to swine. Added to the above has been the resultant decline of trichinosis in pork due to improvements in swine management, a decline in home processing of pork, and the widespread use of home freezers.

The principal methods for killing trichinellae in muscle are heating, freezing, and curing. All pork products which are to be consumed uncooked are subject to regulations of the U.S. Department of Agriculture (1960). In heating, all parts of the pork muscle tissue must be subjected to a temperature not lower than 137°F (58.3°C). In freezing, for pieces not exceeding 6 inches in thickness or arranged on separate racks with the layers not exceeding 6 inches in depth, the following temperatures and times are required: 5°F (−15°C) for 20 days; −10°F (−24°C) for 10 days; or −20°F (−29°C) for 6 days. For pork products in pieces, in layers, or within containers, the thickness of which exceeds 6 inches but not 27 inches, times for lethal effects of freezing are longer: 5°F (−15°C) for 30 days; −10°F (−24°C) for 20 days; or −20°F (−29°C) for 12 days. The above times and temperatures are indicated for use after the meat has been subjected to a preparatory cooling at a temperature not above 40°F (4.4°C). The several methods for curing sausages and other pork products to render trichinellae nonviable can be found in the Federal Regulations as set forth by the U.S. Department of Agriculture (1960).

Since individual pig carcasses at slaughterhouses are not examined for trichinosis, there is still no guarantee that the meat is trichinella-free, and the consumer must freeze or cook it at the recommended temperatures to be assured or noninfective pork.

The clinical course of trichinosis in humans is divided into three stages. The first stage is an intestinal one which occurs during the first four to five days when the larvae are ingested, molt, become sexually mature, and embed themselves in the mucosa. During this period, symptoms of gastroenteritis may occur. The second stage, lasting from a few days to weeks, occurs when larvae, deposited by the female, begin the penetration of muscle tissue. This period is characterized by muscle pain, fever, periorbital edema, generalized toxemia, and transient myocarditis. The severity of the reaction to the presence of the larvae depends somewhat on the number ingested or the size of the parasitizing dose. In the third stage of the infection, categorized as the stage of convalescence, the larvae become encapsulated in the muscles and the symptoms abate. It should be noted that the majority of cases of trichinosis are subclinical in nature, with only transient symptoms giving evidence of infection.

1. DETECTION OF TRICHINOSIS IN ANIMALS

As noted above, U.S. Federal meat inspection procedures do not guarantee trichinella-free pork, nor is the meat examined microscopically except in specific circumstances. In some countries a selected portion of each pig carcass, usually the diaphragm, is compressed between glass plates (a trichinoscope), and the thinly squashed meat is examined microscopically for cysts. In survey procedures such as those of Zimmerman and his colleagues, weighed samples of diaphragm (45 gm) are subjected to pepsin–hydrochloric acid digestion, and the liberated larvae are counted after Baiermann sedimentation (Zimmerman *et al.,* 1961). Neither of these procedures is practical on a routine basis in large slaughter operations, particularly in this country where several million hogs are slaughtered annually. The prospect of using a serologic method to detect infected hogs before slaughter has been under consideration for several years. Although not yet employed on a routine basis, the recent work of some investigators indicates that serology for detecting infections in pigs is possible. Suessenguth *et al.* (1965) reported on the usefulness of the Suessenguth-Kline (SK) slide flocculation test in experimentally infected pigs and were able to detect antibody in 51 of 53 of them. An inverse relation was noted between the size of the infecting dose and the time of serologic response. Reactivity was noted in 44 swine up to the time of slaughter. In recent comparable work, Scholtens *et al.* (1966) evaluated the fluorescent antibody, charcoal agglutination, Suessenguth-Kline (SK), latex agglutination, and bentonite flocculation tests in experimentally infected hogs. From their studies, the authors concluded that the charcoal agglutination and fluorescent antibody tests gave the most accurate indication of infection. Problems associated with use of serologic tests for the detection of infections in hogs are largely associated with sensitivity in detecting low-magnitude infections resulting from the ingestion of as few as ten larvae.

2. DETECTION OF INFECTIONS IN HUMANS

The ideal diagnosis of suspected trichinosis in humans combines a clinical, epidemiologic, and laboratory approach. Symptoms of the infection are legion, and Gould (1945) noted 50 diseases which were mistakenly diagnosed in persons ill with trichinosis. In a patient with a history of consuming poorly cooked pork, symptoms are suggestive of trichinosis. Laboratory diagnosis can be divided into direct and indirect evidence. The demonstration of larvae in a biopsy of gastrocnemius or deltoid muscle is *prima facie* evidence. The biopsied tissue is either compressed between glass slides and the transluscent tissue searched microscopically for larvae, or the biopsied muscle is fixed, sectioned, and stained according to routine procedures.

Indirect evidence of infection can be obtained from peripheral blood smears, serum enzyme determinations, and serology. The presence of eosinophilia in peripheral blood is suggestive or corroborative in suspected trichinosis, although even in some severe infections an increase in eosinophils is not always present. Elevations in glutamic, oxalacetic, and pyruvic transaminases are present in the second through the fourth week of illness (Maynard and Kagan, 1963). A variety of serologic tests for detecting antibody to *T. spiralis* have been developed and are currently employed in diagnosis (Kagan, 1960, 1965). These include the complement fixation test, the bentonite flocculation test, the latex agglutination test, the Suessenguth-Kline flocculation test, the fluorescent antibody test, and the charcoal–cholesterol–lecithin test.

B. *Gnathostoma spinigerum*

Although there are about seven species of this genus, *G. spinigerum* is believed to be the only one of medical importance (Miyazaki, 1960). The worms are widely distributed in parts of Asia (China, Japan, Thailand, Malaya, Indonesia, Philippines, Burma, India, Indochina, and Palestine) and are acquired by man through the ingestion of raw, fermented, or partially cooked freshwater fish containing third-stage larvae of the parasite. The life cycle of this worm involves at least two hosts. The adult *G. spinigerum,* which is 2 to 3 cm long, is found primarily in the stomach of dogs and cats, although worms have also been reported from tigers and leopards. The adult parasites live in gastric tumors formed in response to their presence, and eggs released by the female are passed in the feces. After an embryonation period lasting about seven days, an active free-swimming larva hatches from the egg and is ingested by a copepod, several species of which serve as first intermediate hosts. The infected copepods are in turn eaten by any of several species of fish, amphibians, reptiles, birds or mammals which serve as second intermediate hosts. The third-stage larvae, which encyst in the muscles of the second intermediate host, can be passed from one such intermediate host to another in predator–prey relationships without any change except for slight increase in length and color when in warm-blooded vertebrates. The parasite resides in the stomach and becomes active only in natural definitive hosts, such as cats and dogs. When man inadvertently ingests third-stage larvae in uncooked fish such as "Sashimi" fillets, the larvae migrate to various parts of the body rather than becoming encysted in gastric tumors in the stomach.

Immature adult worms have been recovered from the skin of infected persons, either emerging naturally to the outside in their wanderings or by excision from cutaneous nodules or abscesses. They also have been coughed up or have emerged from natural orifices such as the vagina. They have been found in the abdominal cavity and in the vitreous chamber of the eye (Bovornkitti and Tandhanand, 1959).

Symptoms exhibited by infected persons may include epigastric pain, nausea, and vomiting in early stages of the infection as the larvae, liberated from the infected fish, make their way from the stomach to the peritoneal cavity and begin migrating throughout the body. Disturbances of liver function are common as the larvae migrate through that organ. Symptoms elicited are believed to be due both to mechanical damage to tissues and to toxic substances secreted by the worms. In their later wanderings in human tissues, the larvae eventually move to subcutaneous tissues and evoke a variety of symptoms such as migrating, intermittent edema which is erythematous and pruriginous. Fever usually accompanies these subcutaneous migrations, and it may be slight or, in rare cases, quite high.

Diagnosis of gnathostomiasis in humans is based on recovery of spontaneously emerging worms or symptoms associated with a recent or past history of eating raw fish, chicken, or snake from endemic areas. Clinically, fever and eosinophilia are present. In Japan, a skin test employing antigen prepared from third-stage larvae has been used (Miyazaki, 1960).

Treatment may be only symptomatic in the early stages of gnathostomiasis, while in later stages of peripheral migration removal of the worms by excision from the skin is a simple procedure. Many cases are self-limiting, since larvae occasionally emerge spontaneously from various sites as previously mentioned. However, infections have been known to persist for as long as ten years.

Human gnathostomiasis is not found in the United States because of the absence of the custom in this country of eating raw, partially cooked, or fermented fish. A closely related gnathostome, *G. procyonis,* occurs as a natural parasite of raccoons in the United States. First intermediate hosts are copepods. In Louisiana, Ash (1962) found that snakes, turtles, alligators, and fish serve as second intermediate hosts for this parasite.

C. *Angiostrongylus cantonensis*

During the past several years this nematode parasite, which lives in the pulmonary artery of rats, has been implicated as one possible etiologic agent of a syndrome known as eosinophilic meningoencephalitis which occurs in the Pacific and parts of the Far East. As a food-borne parasite it appears to be associated with the custom of eating uncooked molluscs.

The worm has been found as a naturally occurring parasite of rats in Hawaii, Australia, Thailand, Malaya, Canton China, Taiwan, the Philippines, and a number of other Pacific islands—Carolinas, Cook, Guam, Loyalty, Mariana, Marshall, New Caledonia, New Hebrides, and Tahiti (Alicata, 1965; Nishimura and Yogore, 1965).

The life cycle of the parasite, as described by Mackerras and Sandars (1955), is as follows: adult worms live in the pulmonary artery, but in heavy infections, they may also be found in the right ventricle of the heart. Eggs are deposited into the bloodstream where they lodge in the small vessels of the lungs, forming emboli. As larvae hatch from the eggs, they break through the vessel wall and travel up the trachea and are swallowed. They are then passed out of the body with the feces. Under natural conditions various species of terrestrial snails and slugs serve as intermediate hosts. They become infected either by ingestion of the rat feces or through active penetration of the molluscan cuticle by the larvae.

The larvae molt twice within the mollusc and increase in size. In about two weeks they have molted and grown to an infective stage. Rats become parasitized by eating infected molluscs. In the rat, the third-stage larvae pass from the stomach to the intestine and thence to the bloodstream by penetration of the intestinal wall. The larvae eventually come to the brain, for which they have a strong affinity. Here two additional molts occur, and the larvae then migrate to the surface of the brain and the subarachnoid space where they remain for about two weeks. Thereafter they penetrate the wall of the cranial venules and travel via the right ventricle of the heart to the pulmonary artery where they lodge and the female begins egg production.

Man acquires *Angiostrongylus* by eating raw, infected molluscs. A wide variety of terrestrial snails, slugs, land crabs, and planarians as well as amphibians, freshwater snails, and prawns have been found naturally infected throughout endemic areas (Alicata, 1965). The custom of eating molluscs raw, or ingesting them inadvertently as contaminants of vegetables (Alicata and Brown, 1962), is the main method of man's acquisition of the parasite. In Tahiti, for example, ground freshwater prawn stomachs and surrounding tissue are added to lime juice which is then mixed with grated coconut to prepare a local delicacy called "taioro."

Evidence to date indicates that the worms in humans may not proceed in development beyond the brain-inhabiting juvenile stage. However, in two recent reports from Thailand, a sexually mature male *Angiostrongylus* was recovered from the eye in one case and a sexually mature female from the eye of another case (Prommindaroj *et al.*, 1962; Ketsuwan and Pradatsundarasar, 1966).

The first description of a case of eosinophilic meningoencephalitis with concomitant *Angiostrongylus* present was reported by Nomura and Lin (Beaver and Rosen, 1964). Larvae and immature adults were recovered from the spinal fluid of a patient during his hospitalization and at autopsy. Subsequent human infections, with the demonstration of worms in the brain, were reported by Rosen *et al.* (1962) and by Jindrak and Alicata (1965). In addition to these cases, there have been several in which typical signs and symptoms were present and the clinical, laboratory, and epidemiological evidence strongly suggested infection with *Angiostrongylus* as the cause of the encephalitis (Horio and Alicata, 1961; Linaweaver, 1966).

Infective larvae have been recovered from molluscs by teasing apart the bodies of the snails, slugs, etc., or by artificial digestion of the tissues. Such procedures are of limited use as a measure of control or examination prior to ingesting the food. The small size of the larvae (about 550 μ) makes it extremely difficult to be sure that molluscs from endemic areas are not infected. Thorough cooking is the only practical method of control if the diet includes such potential intermediate hosts.

The diagnosis of *Angiostrongylus* infections in humans is based on clinical and epidemiological evidence, with the laboratory offering ancillary information. No specific immunodiagnostic test is as yet available for detecting infections caused by larvae or immature adults in the brain. Neither is there a specific chemotherapeutic compound available for treatment, although Cuckler *et al.* (1965) have shown that thiabendazole was effective in experimentally infected rats.

D. *Anisakis* sp.

During the past five years there has been increasing evidence of the possible relationship of a food-borne parasite to a syndrome of undescribed etiology which has been recognized for some 25 years. In 1960 Van Thiel *et al.* in Holland recorded the presence of small larval nematodes in tissues of intestine removed at surgery from six out of ten patients with episodes of violent abdominal colic and fever. The worms were tentatively identified as larval *Eustoma* but later designated as species of *Anisakis* (Van Thiel, 1962). Members of this group of ascarid-like worms mature in fish-eating mammals. Ashby *et al.* (1964) in England reviewed the reported cases of eosinophilic granuloma of the gastrointestinal tract and added two cases of their own. Although larvae were not seen in the surgically removed sections of intestine in Ashby's cases, the syndrome was compatible with that reported by Van Thiel *et al.* In 1965, Asami *et al.* and Yokogawa and Yoshimura (1965) in Japan added two more cases of stomach granuloma in which *Anisakis*-like larvae were discovered in the histological sections of resected stomachs.

Although more than one species of parasite may be involved, the essential mechanism for acquiring the parasite is believed to be the same for all. The round worms live as adults in the intestines of fish-eating marine mammals and birds and possibly in species of predator fish. The worms are ascarid-like forms with a two-host life cycle, the adult in the mammal and infective larval stages in fish such as herring or cod. In fish, the larvae lie coiled in various parts of the abdominal cavity. When fish are eaten raw by natural predators, the ingested larvae mature in the intestine of the mammal, bird, or predator fish. When man consumes such infected fish raw, partially cooked, pickled, or smoked, the larvae apparently do not mature but penetrate the stomach or intestine. The number of persons consuming raw, pickled herring far exceeds the number of reported cases of eosinophilic granuloma or severe abdominal colic which would cause persons to seek medical attention. The present belief is that the clinical symptoms exhibited by the patients are the result of local tissue reaction, a hypersensitivity to the burrowing action of the worms in areas of the stomach or intestine which have been sensitized by the penetration of previously ingested larvae.

The practice of consuming raw, pickled, or smoked herring is known to be widespread. Reports originating from Holland, Japan, and England may be followed by others from various parts of the world as knowledge is gained of the etiology of this peculiar syndrome.

Van Thiel *et al.* (1960) were partially able to explain the rare occurrence of this disease in the people of Holland despite their long-standing habit of eating raw herring. Prior to 1955, herring were gutted (and cleaned) at sea shortly after being caught. Since then, some fishing boats have harvested their catch and kept them on ice until returning to the harbor factories where the fish are cleaned. It is believed that the cold temperatures of the ice-packing stimulate the *Anisakis* larvae to leave the abdominal cavity of the fish and migrate into the muscles of the body wall. In such a situation, the larvae are not detected and consequently are not removed when the fish are gutted. Since fish are kept cold from the time of catch until they are consumed, the viability of the nematodes is assured, and man inadvertently becomes an unsuitable substitute for the normal definitive host such as the sperm whale or dolphin. The eventual fate of larvae which are ingested by man and penetrate the stomach without eliciting symptoms is not known.

Treatment of the infection has included surgical removal of portions of the stomach or intestine, or the use of corticosteroids for symptomatic relief. Some patients have given a history of transient bouts of severe colic, and the prevalence of this peculiar type of food-borne parasite may be higher than is indicated at present.

Diagnosis, to date, has been confined to recognition of the larvae and the severe tissue reaction noted in portions of resected stomachs and intestines. Clinically a tentative diagnosis of anisakiasis can be made in the presence of symptoms coupled with a history of consuming raw, pickled, or smoked herring.

III. CESTODES

Three species of this group of flatworms are common adult tapeworms in man and no strangers to the American scene. All three are found in the United States, although they are not as prevalent today as they were years ago. Infections with these worms result from eating raw or insufficiently cooked beef, pork, and fish.

A. *Taenia saginata* (Scheme 2)

The beef tapeworm of man has a wide geographic distribution and is absent only in areas of the world where beef is not eaten. The adult worm, which may be several feet in length, is composed of hundreds of proglottids or "tapes" and lives in the small intestine of man, the only definitive host. The worm is attached to the wall of the intestine by the unarmed scolex or head from which many proglottids are produced. When the terminal proglottids become gravid, they break off and are passed in the feces. Occasionally proglottids may rupture in the intestine and release the eggs, which are passed out with the feces.

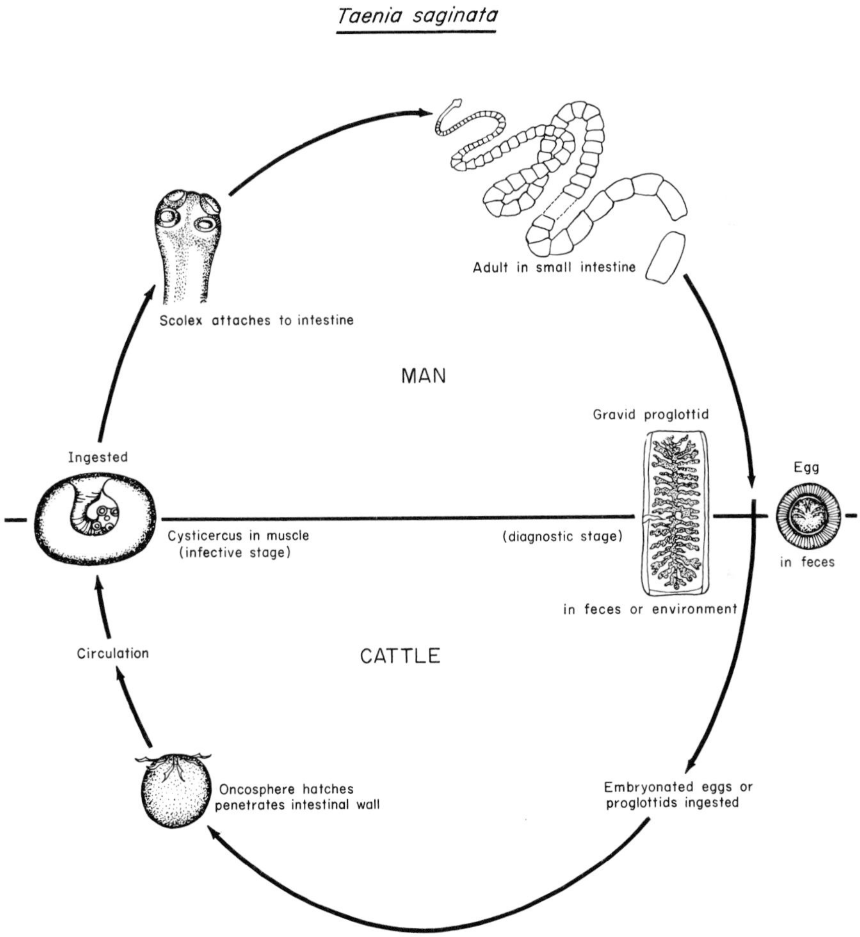

SCHEME 2. Life cycle of *Taenia saginata*.

Proglottids or eggs in human excrement are eaten by cattle grazing on polluted ground. In the intestine of the cattle, onchospheres emerge from the eggs, penetrate the intestinal wall, and are carried to all parts of the body via the lymphatics and venous system. The onchospheres metamorphose in striate and cardiac muscle and eventually become the fluid-filled "bladderworms" or cysticerci. In reality, they are miniature tapeworms with the scolex invaginated into the fluid-filled bladder. Work by McIntosh and Miller (1960) indicates that these cysticerci must be at least 10 to 12 weeks old before they have matured sufficiently to be infective for man. Although the size may vary, the bladderworms are, on the average, 4.5 mm long by 3.7 mm wide. Human autochthonous *T. saginata* infections are rare in the United States, but bovine cysticercosis is reported occasionally from isolated farms (McIntosh and Miller, 1960). Infection in cattle has been traced to pollution of the soil, especially in vegetable-raising areas where infected workers defecate on the ground rather than use latrines. Another source of infection in cattle has been from effluents of overburdened sewage disposal systems which drain into pastures where stock grazes. Miller (1956) in the United States and Silverman and Griffith (1955) in England have called attention to the spread of bovine cysticercosis from contaminated human feces to which cattle have access through sewage. The latter two authors state that the increased use of detergents inhibits the sedimentation of sewage by overworked disposal plants and allows *T. saginata* eggs to be discharged with the effluent. Birds and filth flies which can pass the eggs through their bodies are also believed responsible for the spread of the infection to cattle (Round, 1961). Eggs of *T. saginata* have been shown to survive in stored hay for periods of up to three weeks, depending on temperature and humidity (Lucker and Douvres, 1960).

Human infections with *T. saginata* may be asymptomatic, with moderate hunger pains and loss of weight the only disorders noted. On the other hand, symptoms of ulcers, nausea, vomiting, intestinal colic, diarrhea, and systemic toxicity, apparently due to the absorption of products of the worms' metabolism, may be present. Of no small importance is the psychological impact upon the infected person who finds that he is passing proglottids in his stools!

Although a *Taenia* egg is easily identified, one cannot distinguish between the eggs of *T. solium, T. saginata,* or other *Taenia* species. This is of great importance in the diagnosis and treatment, as will be shown below. Specific diagnosis depends on recovery of a gravid proglottid and identification of the species based on the number of lateral branches of the uterus. The proglottid may be cleared in lactophenol, xylene, or beechwood creosote, and the side branches of the uterus counted. The uterine branches in *T. saginata* number between 16 and 22, with an average of about 18.

Taenia saginata infections are usually treated with quinacrine hydrochloride. The patient is kept on a liquid diet overnight and given a saline purge two hours after receiving the drug. Many patients are nauseated by the quinacrine, and sodium bicarbonate or chlorpromazine (Brown, 1960) can be given to reduce side effects. The authors have knowledge of three cases of *T. saginata* infection successfully treated by the technique of Rosen and Kiefer (1958). This employs glycerin, magnesium sulfate, and physiological saline. Vomiting is not a problem with this treatment.

Infections with *T. saginata* can be prevented by thorough cooking of all beef or by the use of freezing temperatures as noted for trichinosis. Cattle should be kept from grazing areas where human feces or the effluent from sewage disposal plants is allowed to run off or accumulate on grazing pastures. Federal meat inspection in slaughterhouses prescribes specific steps which are taken to insure bladderworm-free carcasses, but it is possible for very light infections to escape detection. Unfortunately Federal regulations do not apply to cattle slaughtered for local consumption or cattle not involved in interstate commerce. Human *T. saginata* infections in the United States are few in number at the present time. However, infections with the beef tapeworm are often acquired by Americans visiting countries such as Africa, where both human and bovine infections are more prevalent.

B. *Taenia solium* (Scheme 3, p. 190)

The pork tapeworm is also worldwide in distribution, and human infections are found with varying frequency wherever pork is eaten. The adult worm superficially resembles *T. saginata,* but the scolex is armed with a double row of hooklets which help anchor the worm to the intestinal wall. The life cycle is similar to that of *T. saginata* except that pigs are the intermediate hosts and harbor the bladderworms. The term "measly pork" refers to pork infected with such bladderworms. Swine infected with cysticerci constitute an important economic and public health problem in Central America and Panama (Acha and Aguilar, 1964).

Eggs of *T. solium* cannot be distinguished from those of *T. saginata.*

Taenia solium

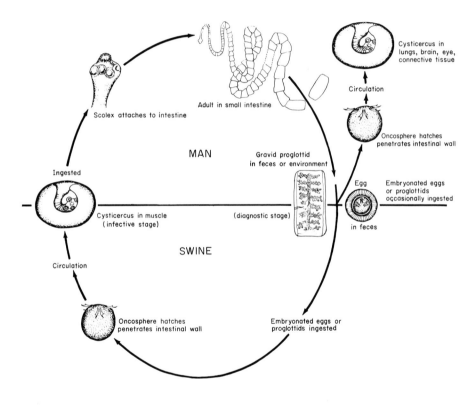

SCHEME 3. Life cycle of *Taenia solium*.

Specific identification depends on enumeration of the lateral uterine branches in a gravid proglottid, recovered from a stool specimen, which has been fixed and cleared for microscopic examination. The uterine tree of *T. solium* has from 8 to 12 lateral branches with an average of 9. Symptoms in pork tapeworm infections may be similar to those of beef tapeworm. However, the inherent danger in *T. solium* infections is that man can harbor the adult worm and also serve as an intermediate host; that is, if eggs are accidentally ingested, the bladderworm may become lodged in muscles or other vital organs, resulting in human cysticercosis. This "internal autoinfection" is believed to occur most commonly by the ingestion of eggs regurgitated from the intestine to the stomach. For this reason, specific identification of a *Taenia* worm is important, and great care should be taken to avoid nausea and vomiting when therapy is given for *T. solium* infections.

When *T. solium* eggs are ingested from polluted soil or when gravid proglottids in the intestine of an infected person rupture, the liberated eggs are swept back to the stomach. The eggs hatch in the intestine, and the liberated onchospheres penetrate the intestinal wall and are carried to all parts of the body via the venous system. Many make their way to skeletal muscles and mature to bladderworms as in pork. Others are distributed to various organs. Numerous cases of human cysticercosis of the eye, heart, liver, lungs, and brain have been described. Cysticercosis of the brain is by far the most serious of the larval infections. Cohen (1962), in reviewing the literature, indicated that 26 cases of cerebral cysticercosis have been diagnosed in the United States since 1915. The multiple symptoms and pathological processes involved are illustrated in the case reports of Cohen (1962), White *et al.* (1957), Dent (1957), and Haining and Haining (1960). Cerebral cysticercosis is important in the differential diagnosis of various neurologic and mental disorders in several countries. For example, Lombardo and Mateos (1961) reported a cysticercosis prevalence of 2.6% in 2202 consecutive autopsies at the Pathology Unit of the School of Medicine in Mexico City.

A diagnosis of cerebral cysticercosis is difficult to make, since the infection may resemble brain tumor, pseudomotor (meningeal hydrops), Alzheimer, or Picks disease, and syphilitic meningitis (White *et al.,* 1957). Concurrent cysticercosis in muscles, verified by X-ray, may be helpful but the worms may not always be demonstrated by this technique. A complement fixation test (Nieto, 1956) and a hemagglutination test (Biagi *et al.,* 1961) are useful in cases of suspected cerebral involvement, but they are not always positive in etiologically diagnosed cases.

C. *Diphyllobothrium latum* (Scheme 4)

The broad fish tapeworm is ubiquitous but spotty in its distribution. It has been reported from parts of Europe, Asia, Australia, and North and South America. Autochthonous infections were reported in the Great Lakes region of the United States for many years, and a small focus existed some years ago in Florida (Summers and Weinstein, 1943). Infection in man is acquired by the ingestion of raw or insufficiently cooked freshwater fish. Species found to be naturally infected in the United States include walleyed pike, barred pike, and American burbot.

The adult cestode lives in the small intestine of man and other reservoir hosts such as dogs, cats, leopards, foxes, mink, bears, and domestic pigs. Worms may attain lengths of 3 to 10 feet.

Eggs formed within individual proglottids are released through a uterine pore and pass out of the body with feces. When deposited in water, one to two weeks are required for the development of the onchospheres or ciliated embryo stages. The onchosphere escapes through the operculum of the egg as a free-swimming organism. The life cycle continues when the embryos are eaten by any of several species of copepods of the genera *Cyclops* and *Diaptomus*. Each larval organism transforms into a procercoid in the body cavity of the copepod. When infected copepods are eaten by fish, the larvae develop into plerocercoids or sparganum forms. When smaller fish are eaten by larger carnivorous species, the plerocercoid makes its way into the muscles where it remains as an unencysted, glistening white worm. This larval form may pass through a number of fish until consumed by a member of the pike family which is, in turn, eaten by man or other suitable vertebrate host. The plerocercoid develops into the hermaphroditic adult tapeworm in the definitive mammalian host, and the life cycle is completed.

Diphyllobothrium latum infections in man may be asymptomatic or produce nausea, vomiting, weakness, dizziness, diarrhea, or constipation alternating with diarrhea. A severe form of broad fish tapeworm infection occurs in a small percentage of those infected wherein a "tapeworm anemia" hematogenically identified with genuine pernicious anemia is demonstrated (Saarni *et al.*, 1963). Workers in Finland have reported that, even in many carriers who do not show overt symptoms, 50% have pathologically low serum vitamin B_{12} concentrations (Nyberg *et al.*, 1961). In his review of *D. latum* and anemia, von Bonsdorff (1956) indicated that the worm competes with the host for the supply of vitamin B_{12}. No severe tapeworm anemias have been recorded among the autochthonous North American cases.

Diphyllobothrium latum

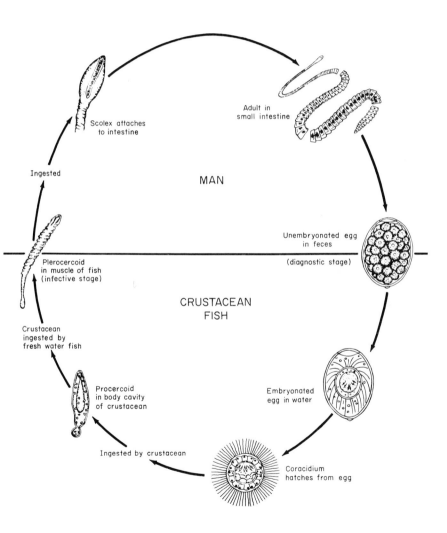

SCHEME 4. Life cycle of *Diphyllobothrium latum*.

Diagnosis of *D. latum* infections in man depends on the demonstration of characteristic eggs in the feces, or occasionally the identification of proglottids which may be dislodged from the strobila of the worm and passed spontaneously in the feces.

Control of *D. latum* infections in man can be accomplished by the thorough cooking of fish, particularly of pike and pickerel, which are caught in endemic areas. Although, as mentioned above, several species of animals have been found to be infected as reservoir hosts, it is the consensus that human *D. latum* infection is largely a man-to-man process whereby promiscuous defecation into ponds and lakes, containing copepods and fishes, keeps the life cycle operating. The pollution of such bodies of water with human sewage emanating from faulty or over-burdened disposal plants may be another means whereby eggs of *D. latum* reach the proper intermediate hosts.

The treatment of *D. latum* infections involves the use of atebrine (quinacrine hydrochloride) as noted above for other tapeworm infections.

D. Sparganosis

This disease is not a classical food-borne entity but one which, so far as is known, is acquired inadvertently from drinking water containing procercoid-infested copepods or ingesting raw meat from amphibia, reptiles, or mammals containing plerocercoids. The causative organism or *Sparganum* is the same form as the plerocercoid of *D. latum* found in fishes. Other species of pseudophyllidean tapeworms occur in a variety of animals, and apparently when man ingests either infected copepods containing procercoids, or fish, frogs, water snakes, or other animals containing the plerocercoid forms, these larvae are incapable of maturing to the adult tapeworm stage in man. For the most part, they migrate in the skin or in the subcutaneous tissues of the body, although some larvae have been recovered from deeper muscles (Swartzwelder *et al.,* 1964).

The wandering *Sparganum* may cause considerable irritation and swelling. Several of the cases reported in recent years were brought to medical attention because of the presence of a migratory swelling which caused sufficient annoyance for the patient to see a physician. On the basis of case histories, it is apparent that the larvae may be present for several years before they elicit enough discomfort or curiosity on the part of the infected individual to seek medical attention.

Mueller *et al.* (1963*a*), in a recent review of the infections reported in the United States, is of the opinion that the cases of sparganosis acquired in the United States most probably arose from individuals drinking water containing procercoid-laden copepods. An adult tapeworm, *Spirometra mansonoides,* parasitizing cats in this country, is believed to be the adult form responsible for sparganosis. Although 38 years elapsed between the first two reported cases of sparganosis in the United States (Moore, 1915; Stiles, 1908) and the third one (Read, 1952), there has been an increasing number since 1952. To date, a total of 43 cases has been reported in the United States, 17 tabulated by the review of Mueller in 1963*b*, 23 reported by Swartzwelder *et al.* in 1964, and one case each reported by Markell and Haber (1964), Short and Lewis (1964), and Corkum (1966). With the exception of three infections, all cases were confined to the Southeastern United States.

The diagnosis of sparganosis is generally made when the white, glistening, ribbon-like larva is removed from the human tissues or its presence is noted in histological sections of surgically excised abscesses or nodules. There is no treatment except surgical removal of the worm.

E. Laboratory Methods

1. DETECTION OF CESTODES IN FOOD

Although cattle and pig carcasses are examined in accordance with U.S. Department of Agriculture meat inspection procedures (1960), light infections of bovine or porcine cysticercosis may escape detection. This fact was emphasized in the second Joint FAO/WHO Expert Committee on Meat Hygiene (World Health Organization, 1962) which states the following: "Normal meat inspection procedures (Masseter and cardiac muscle cuts) even when performed with meticulous care cannot give complete assurance of safe meat when carcasses are only lightly infested with cysticerci. Even in highly enzootic areas, all feasible inspection procedures used to date – i.e., cuts in additional different portions of the carcass (diaphragm, shoulder muscle, etc.) – are fallible in that they may miss moderately and, more rarely, heavily infected carcasses." "Differences in the site of predilection of cysts in carcasses have been reported by various workers." Thus, even though pork and beef may have passed Federal meat inspection and be so stamped, there is no assurance, as in trichinosis, that the meat is parasite-free. Individual examination of pork and/or beef products is not feasible, since cysticerci are no greater than 10 mm in length and 5 mm in width and thus may be difficult to see.

Examination of pike, pickerel, or other *D. latum* intermediate host fish species may disclose the presence of the plerocercoid or sparganum. However, even in the case of thin fish fillets, the plerocercoid forms, up to 6 mm in length, may easily escape notice. The organisms are whitish in color and easily overlooked in the muscle. It is also possible that the sparganum may be inside the thin strip of fillet and thus escape detection. Infected fish or fillets which are thoroughly cooked or frozen offer no health hazard, although their obvious presence makes the food undesirable for consumption.

2. Detection of Cestode Infections in Humans

With the exception of sparganosis, which is generally diagnosed at surgery, the cestode infections discussed above can be detected by examination of stool specimens. The simplest procedure is to preserve a portion of the feces in 10% formalin, generally in the ratio of one part stool to five parts formalin. Direct wet mount preparations and formalin-ether concentrations (Ritchie, 1948) may then be examined for the presence of eggs. *Taenia* eggs are dark brown, spherical to subspherical, measuring from 30 to 43 μ by 29 to 38 μ. When they are found in fecal specimens, it is important to obtain a proglottid to determine whether it is *T. solium* or *T. saginata*. As has been pointed out, the species cannot be identified from the eggs. Proglottids are often passed in normal stools, or they can be obtained after a mild cathartic.

Eggs of *D. latum* are inconspiculously operculated, oval to ellipsoidal, and often with a small knob on the posterior end. They are thin-shelled and measure from 55 to 76 μ long by 37 to 56 μ wide. Eggs are immature when found in the feces. The appearance of food-borne cestode and trematode eggs can be found in The Color Atlas of Intestinal Parasites by Spencer and Monroe (1961) and the Microscopic Diagnosis of the Parasites of Man by Burrows (1965).

IV. TREMATODES

Members of this group of flatworms are noteworthy in that they all have complicated life cycles which involve one or two intermediate hosts. Domestic or wild animals perpetuate the parasites in nature and, in many cases, are the principal definitive hosts. Man's penchant for eating raw or insufficiently cooked freshwater fish, crabs, crayfish, and various raw vegetables (water chestnuts, caltrops, bamboo roots) provides adequate opportunity for infection, regardless of the source of the parasites. Promiscuous or, in some cases, deliberate defecation into ponds, lakes, etc., as fertilizer, serves to furnish a constant supply of parasite eggs which, in the presence of suitable intermediate hosts, continue the life cycle.

A. *Clonorchis sinensis, Opisthorchis* sp., *Metagonimus yokogawai,* and *Heterophyes heterophyes* (Scheme 5, p. 198)

These are the better known and most prevalent members of a large group of trematodes which show a common epidemiological character-istic; that is, they are acquired through eating raw or insufficiently cooked fish. All four species are found in the Far East. *Heterophyes heterophyes* is also known from the Nile delta and Turkey, and *M. yokogawai* has been found in the Balkan States, Siberia, and Spain.

Clonorchis sinensis is a small (10 to 20 mm long) lanceolate fluke which lives in the bile duct. Eggs, voided with the feces, are eaten by certain freshwater snails. A reproductive cycle takes place in the mollusc, with the eventual liberation of cercariae. The cercariae penetrate under the scales of certain fish and encyst in the flesh. Man acquires the worms when he eats improperly cooked, salted, dried, or raw fish. Carp and some 80 other species of fish have been found to be naturally infected in endemic areas of China, Japan, and Korea (Yoshimura, 1965).

The symptoms evoked and the pathology produced by *C. sinensis* are referable to the hypertrophy of the biliary epithelium and the consequent inflammatory and fibrotic changes in the bile duct, gallbladder and liver. A chronic type of clonorchiasis is found in man living in endemic areas. This parasite is not found in nature in the United States for a number of reasons: (1) the lack of suitable snail intermediate hosts; (2) the absence of promiscuous defecation in ponds; and (3) the absence of the custom of ingesting raw fish as a normal part of the diet. However, chronic and asymptomatic infections with *Clonorchis* have been diagnosed in the United States among Orientals who came to this country as immigrants, students, or visitors. As pointed out by Strauss (1962), infections have been found in both Caucasians and Orientals in San Francisco, but these were acquired in endemic areas. They, therefore, constitute a medical rather than a public health problem. Infected persons may harbor the worms for many years.

A closely related liver fluke, *Opisthorchis viverrini,* has been found to be widespread in northern Thailand (Sadun, 1955). At least nine species of freshwater fish which constitute a part of the normal human diet have been found to harbor the metacercariae. Conservative estimates place the number of people infected at about three and one-half million in Thailand with several additional million in China, Laos, and Vietnam (Wykoff and Winn, 1965).

Metagonimus yokogawai and *Heterophyes heterophyes* are very small (1 to 2.5 mm in length) flukes whose life cycle is similar to that of *C.*

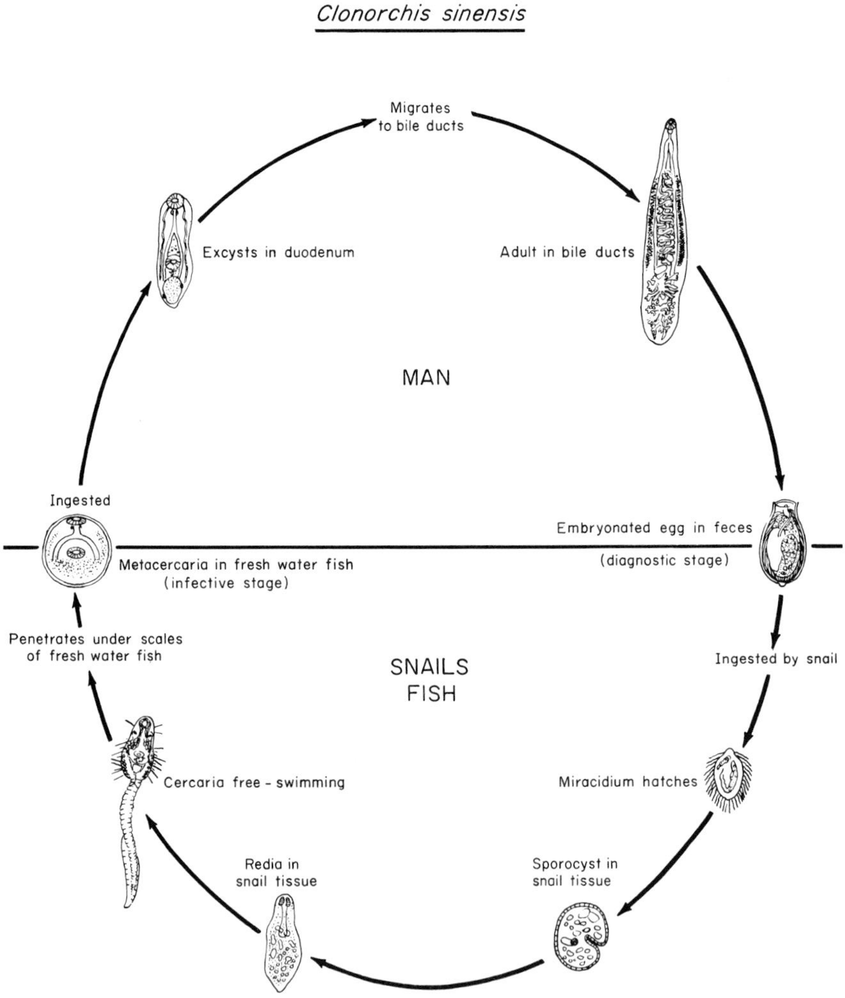

Clonorchis sinensis

Migrates
to bile ducts

Excysts in duodenum

Adult in bile ducts

MAN

Ingested

Embryonated egg in feces
(diagnostic stage)

Metacercaria in fresh water fish
(infective stage)

Penetrates under scales
of fresh water fish

SNAILS
FISH

Ingested by snail

Cercaria free - swimming

Miracidium hatches

Redia in
snail tissue

Sporocyst in
snail tissue

SCHEME 5. Life cycle of *Clonorchis sinensis*.

Paragonimus westermani

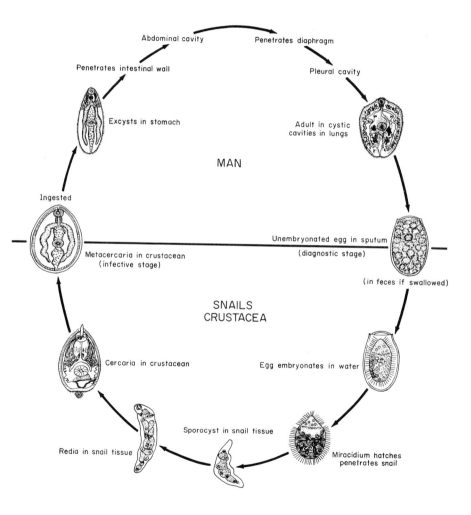

SCHEME 6. Life cycle of *Paragonimus westermani*.

sinensis. Man becomes infected by eating any of several genera of fish which form a part of the normal diet in the Far East. These two worms and other closely related species differ from *Clonorchis* in that they live in the intestine. They burrow into the mucosa of the small intestine where their presence produces irritation, inflammation, excessive mucus formation, and symptoms of diarrhea with resulting abdominal pain. The severity of symptoms is often related to the number of worms present. The small size of the eggs and the deeply burrowed habitat of the worms often cause the eggs to be liberated into the lymphatics or capillaries whence they are carried to many parts of the body. They may be deposited in vital organs such as the myocardium or brain with the resultant effect of granulomatous lesions (Africa *et al.,* 1935).

B. *Paragonimus westermani* (Scheme 6, p. 199)

The genus *Paragonimus* contains a number of species of which *P. westermani,* the oriental lung fluke, is the most important human species. Infection is acquired by eating raw or inadequately cooked crabs or crayfish which constitute a part of the diet in the Far East. Comprehensive reviews of *Paragonimus* and paragonimiasis were published recently by Yokogawa *et al.* (1960) and Yokogawa (1965). The adult flukes, which are 8 to 12 mm in length, live encapsulated in cysts in the lungs of the definitive host. Not only man, cats, dogs, and pigs, among domestic animals, but also tigers, lions, leopards, panthers, and other wild carnivores harbor this parasite. Under normal conditions, the fibrous capsule around the parasite in the lung has a small opening into the respiratory tract so that eggs laid by the worms are passed up the bronchioles, coughed up and discharged by expectoration, or swallowed and voided in the feces. The process of egg release seems to be aided by the paroxysms of coughing which occur in patients with paragonimiasis. The eggs passed into water require several days to embryonate. The emerging miracidium penetrates various species of freshwater snails where reproduction takes place and from which cercariae emerge to penetrate the soft parts of the exoskeleton of crabs and crayfish. There they encyst in various parts of the body of the crustacean.

Man and other carnivores acquire *P. westermani* infections by eating the infected crustaceans, several species of which serve as intermediate hosts in endemic areas. Thus the cycle may continue indefinitely. During the period when liberated metacercariae are released in the intestine and begin their migration to the lungs there may be no outward signs of infection. As the worms develop in the lungs, a chronic pulmonary disease is seen. In endemic areas this is known as "lung fluke disease" or "endemic hemoptysis" (Yokogawa et al., 1960). The symptoms evoked and the pathology produced are similar to those caused by other helminth parasites. As stated by Yokogawa et al., "the extent of the injuries to the host depends on the number of worms present." In very light infections most of the damage is centered in the lungs, while in heavier infections other organs may be involved. In cases involving large numbers of worms, severe disease and even death can occur.

One of the complicating factors in lung fluke infections is the tendency of the worms to migrate to ectopic sites such as the skin, liver, scrotum, or, more important, the brain (Klemme, 1966). The latter type of infection may result in mental retardation, meningitis syndrome, and various motor disturbances. Yun (1960) gives an interesting report of paragonimiasis in eleven Korean children who became infected after ingesting the juice extracted from mashed, strained raw crayfish. The potion was administered by a local herb doctor as a remedy for measles! Raw crab juice is also given for whooping cough and diarrhea in some parts of the world.

Therapy for paragonimus infections is adequately reviewed by Yokogawa (1965). A number of drugs have been tried, but, as yet, no single preparation or combination has gained general acceptance. Those which have been used with some success include emetine hydrochloride, chloroquine, and various sulfonamides, used in conjunction with emetine. As reported by Buck et al. (1958), success with a particular drug or drugs may depend on the length of time the infection has persisted. They reported good response to chloroquine in patients infected for two years or less, but considerably poorer response in patients with infections lasting for five years or more.

Therapy with emetine and sulfonamides requires long-term drug administration (10 to 14 days), and a recurrence rate of 70 to 80% has been reported. Effective therapy with chloroquine phosphate is attained only after prolonged drug administration of three to seven months, and side effects are common. In recent years, bithionol has shown great promise as an effective therapeutic agent for pulmonary paragonimiasis (Yokogawa et al., 1963). Long-term follow-up studies by these workers have indicated that this may be the drug of choice when ten treatments are given orally at the rate of 30 to 40 mg/kg daily on alternate days. The above-mentioned drugs have been tried in the treatment of cerebral paragonimiasis with limited success, and the present feeling is that surgical removal is the method of choice.

Paragonimus westermani is not endemic in the United States, although a close relative, *P. kellicotti,* has been found in several species of animals (opossum, wildcat, muskrat, and mink). At one time *P. kellicotti* was of some importance to mink ranchers in the northern United States. Numerous deaths occurred among mink which were fed crayfish infected with metacercariae of *Paragonimus* (Ameel, 1934), and recent reports indicate that wild mink can still be found parasitized (Beckett and Gallicchio, 1966). Bisgard and Lewis (1964) recently found *P. kellicotti* in the United States in dogs and cats.

C. *Fasciola hepatica* (Scheme 7)

This fluke has a cosmopolitan distribution in sheep- and cattle-raising areas. Although it was described initially from sheep, whence it gets its common name "sheep liver fluke," it is also found in cattle and has been reported in goats, hogs, and deer. The parasite is present in both cattle and sheep in the United States and is of some economic importance, since livers infected with the flukes are condemned for human consumption.

Although human infections have not been reported from the United States, cases have been recorded from Central and South America, Europe, the Middle East, and parts of Africa and Asia.

The adult *F. hepatica* lives in the bile duct where it produces inflammatory and fibrotic changes in the biliary epithelium. Eggs released by the worms are passed out in the feces, and, after several days of embryonation, miracidia hatching from the eggs penetrate certain species of *Lymnaeid* snails. The reproductive cycle of the intramolluscan stages also occupies several days. Cercariae which emerge from the snail quickly encyst on plants or inanimate objects after a short free-swimming period. The mammalian hosts acquire the fluke when plants such as grass and various weeds are ingested raw. Man usually acquires the infection from watercress, used in the diet either as a tasty staple or as an esthetic garnish.

Facey and Marsden (1960) recorded six cases of human fascioliasis in the town of Ringwood, England. Epidemiologically, this localized outbreak seemed to be associated with three important factors: (1) the high prevalence of *F. hepatica* in cattle (so high, in fact, that sheep cannot be raised there); (2) the fact that there are many watercress beds in the area; and (3) an unusually wet summer in 1958 which produced flooding and contamination of the watercress beds with snails and consequent encysted fluke metacercariae. All six patients gave a history of eating watercress. As pointed out by the authors, such outbreaks are possible when environmental conditions are right for dissemination of infected snails

Fasciola hepatica

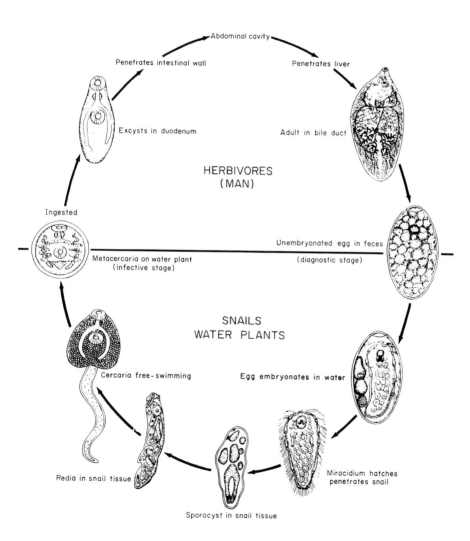

Abdominal cavity

Penetrates intestinal wall

Penetrates liver

Excysts in duodenum

Adult in bile duct

HERBIVORES
(MAN)

Ingested

Unembryonated egg in feces
(diagnostic stage)

Metacercaria on water plant
(infective stage)

SNAILS
WATER PLANTS

Cercaria free-swimming

Egg embryonates in water

Redia in snail tissue

Miracidium hatches
penetrates snail

Sporocyst in snail tissue

SCHEME 7. Life cycle of *Fasciola hepatica*.

or metacercariae — wet, warm weather and flooding of otherwise protected watercress beds. Although *F. hepatica* has been known for years as a parasite of sheep and cattle in many parts of the world, autochthonous human cases occasionally occur. Oterdoom in 1961 reported the first such instance from the Netherlands.

Both emetine hydrochloride and chloroquine (chloroquine sulfate and hydroxychloroquine sulfate) have been employed in recent years with varying success in the treatment of human fascioliasis. Some workers report symptomatic relief after chloroquine therapy, although fluke eggs persisted in the stools. Other investigators report some success with emetine therapy followed by a course of chloroquine.

Another species, *F. gigantica,* has been reported in man in Hawaii (Stemmermann, 1953). This fluke differs morphologically from *F. hepatica.* It is longer, has a different arrangement and size of certain internal organs, and produces larger eggs. It is a parasite of cattle in Hawaii and was originally found in areas of the world where Zebu, Brahmin, and similar types of cattle originated.

D. *Fasciolopsis buski* (Scheme 8)

This intestinal fluke has a life cycle similar to that of *F. hepatica*. Man becomes infected by eating plants which harbor encysted metacercariae. These include water bamboo, water chestnuts, and lotus plant roots. Infections often occur when the outer skin of water chestnuts, containing the encysted metacercariae, is bitten into and peeled off with the teeth in order to obtain the more succulent food within the husk. Human infections and those in the principle reservoir hosts (pigs and dogs) are confined to China, Thailand, Indonesia, Taiwan, Vietnam, and India. The cultivation of edible plants in ponds and streams, together with continual defecation into such bodies of water, ensures a constant source of infection. Eggs liberated in the feces of man or other hosts hatch, and the life cycle is completed in the presence of the snail intermediate host and the cultivated caltrops. The studies of Sadun and Maiphoom (1953) detailed many facets of the epidemiology of this parasite in Thailand.

The adult flukes of *F. buski* are more robust than those of *F. hepatica* and measure from 25 to 75 mm in length. They live in the duodenum and jejunum attached by the ventral and oral suckers. The worms can cause ulceration of the mucosa at these points of attachment, and large numbers of them may cause interference with absorption and passage of food in the intestine. Symptoms in heavy infections include diarrhea, abdominal pain, nausea, vomiting, and edema of the face and abdomen, presumably due to toxic products liberated by the worms.

Fasciolopsis buski

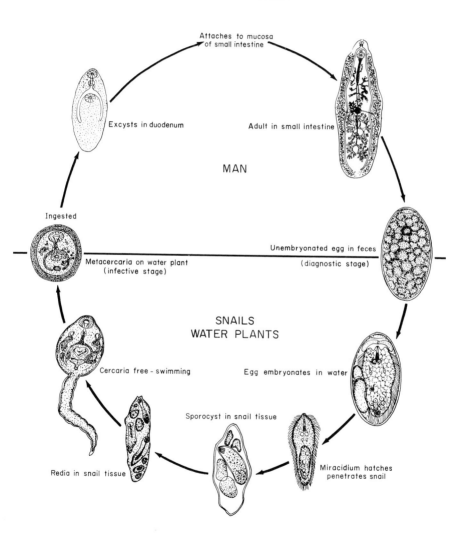

Attaches to mucosa of small intestine

Excysts in duodenum

Adult in small intestine

MAN

Ingested

Metacercaria on water plant (infective stage)

Unembryonated egg in feces (diagnostic stage)

SNAILS
WATER PLANTS

Cercaria free-swimming

Egg embryonates in water

Sporocyst in snail tissue

Redia in snail tissue

Miracidium hatches penetrates snail

SCHEME 8. Life cycle of *Fasciolopsis buski*.

Infections with *Fasciolopsis* are perhaps the best examples of human parasites perpetuated by local custom. Were it not for the practice of defecating into the ponds where the caltrops are grown and the custom of removing the husks or skin with the lips and teeth, the prevalence and clinical importance of this parasite might be appreciably reduced.

E. Laboratory Methods

1. DETECTION OF TREMATODES IN FOOD

Attempts to detect the infective stages of trematodes in food are impractical as a means of preventing infection. The small size of the metacercariae and their distribution in the skin and flesh of fish, crabs, and crayfish would necessitate literally tearing the animals apart, piece by piece, in order to detect the infection. Aside from studies designed to show the various species of fish and crustaceans serving as intermediate hosts, there is no advantage in such attempts at detection, and they would be economically unfeasible.

Fasciolopsis metacercariae may be found on water caltrops, chestnuts, bamboo, etc., but the process of detection is again purely academic. The outer skin or husks of plants in endemic areas should be removed with a knife or the entire husk dropped in boiling water to kill the cysts. The metacercariae of *Fasciola* may be found on grasses, plants, pieces of wood, and other objects in an endemic area. Encysted metacercariae of *Fasciola* are also difficult to see with the unaided eye. A practical means of treating watercress prior to eating is to dip it in boiling water. This is particularly important in areas where such food is grown and harvested in the presence of cattle or sheep.

2. DETECTION OF TREMATODE INFECTIONS IN HUMANS

Since eggs of all the flukes mentioned above may be liberated in the feces, the infections can be diagnosed by microscopic examination of stool specimens utilizing the techniques mentioned in the section on cestodes. Eggs of the flukes are also pictured in photomicrographs in the book by Spencer and Monroe (1961). Eggs of *Clonorchis* are ovoid. yellowish-brown in color, 27 to 35 μ in length by 11 to 20 μ wide. The anterior end is somewhat narrower than the posterior and has a small operculum with slight shoulders. At the posterior end there is often a small knob which is comma-shaped. The eggs, when passed in feces, contain a fully developed miracidium.

Metagonimus eggs are ovoid, without shoulders at the opercular end. They are yellowish-brown, 25 to 30 μ long by 15 to 20 μ wide, and contain a mature miracidium. The eggs of *Heterophyes* are ovoid, thick-shelled and light brown in color, 25 to 30 μ long by 15 μ wide. The egg contains a fully developed miracidium when passed in feces. *Paragonimus* eggs are large, 65 to 115 μ long by 40 to 65 μ wide. They are yellow-brown to dark brown in color and, when found in sputum or feces, are immature, containing an ovum surrounded by yolk cells. In patients with clinically suggestive paragonimiasis, the sputum should be examined, particularly if it is bloody or rusty in color, since it often contains many eggs.

Fasciola hepatica eggs are thin-shelled with an indistinct operculum. They are light brown in color and measure 120 to 182 μ long by 63 to 102 μ wide. *Fasciola gigantica* eggs are slightly larger than *F. hepatica*, being 156 to 197 μ long by 78 to 95 μ wide. Eggs of *Fasciolopsis* are large, thin-shelled, ovoidal, and yellowish-brown in color. They are slightly smaller than *F. hepatica*, 130 to 150 μ long by 78 to 95 μ wide, with an indistinct operculum. The eggs, when passed in feces, are immature, containing an ovum surrounded by yolk cells.

A problem, occurring more often in *Fasciola* and *Paragonimus* than in the other fluke infections, is that infected patients come to medical attention with symptoms at a time when eggs are not demonstrable in the feces. This is due to infection with immature flukes which are not yet producing eggs or adults located in ectopic sites where eggs cannot reach the bowels to be passed out in the feces. In such situations the diagnosis must be based on clinical symptoms, history, and/or use of available immunodiagnostic tests.

The intradermal and complement fixation tests have been employed for paragonimiasis with some success. However, as is true for most helminth immunodiagnostic tests, more-sensitive and specific antigens are needed. Good results have been reported by Sadun *et al.* (1959), and the subject has been reviewed recently by Yokogawa (1965). An all-encompassing diagnostic procedure for suspected paragonimiasis would include stool examination, sputum examination (X-ray for lung cases), and intradermal and complement fixation tests, all pursuant to a clinical history of the patient.

Many cases of fascioliasis, including those in which adult flukes have been removed from ectopic sites at surgery, are characterized by the failure to find fluke eggs despite typical symptoms. For these cases the complement fixation and intradermal tests are useful, their efficacy depending on the familiarity and experience of qualified workers in endemic areas.

In like manner, intradermal and complement fixation tests have been shown to be of use in *Clonorchis* infections. More sensitive and specific antigens are also needed for these tests. The work of Sawada *et al* (1965), indicating that a polyglucose is one of the active substances in the antigen for complement fixation tests, augurs well for the development of better immunodiagnostic tests.

V. *TOXOPLASMA GONDII*

The causative organism of toxoplasmosis is a small crescent or banana-shaped protozoan 4 to 6 μ long by 2 to 3 μ wide. Cystic stages of the parasite, containing from a few to hundreds of the organisms, are found in a variety of host tissues and cells. Recent knowledge of various aspects of toxoplasmosis has been reviewed by Jacobs (1963).

It is pertinent to consider *Toxoplasma* among the food-borne diseases, since there is good evidence that improperly cooked meats, such as pork and lamb, may harbor cysts in various tissues and thus be a source of human infection (Weinman and Chandler, 1956; Rawal, 1959; Jacobs *et al.*, 1960). However, as pointed out by Jacobs (1963) and others, the prevalence of toxoplasmosis among vegetarians and certain ethnic groups who do not consume pork because of religious beliefs precludes meat as the predominant factor in the transmission of the infection. Both respiratory and oral transmission may occur, since cysts have been found in human tonsils and saliva (Siim, 1961). The exact methods by which the parasite is transmitted are not known, and current knowledge indicates that food-borne transmission may be a minor one.

The presence of the parasite in man and animals is determined parasitologically by isolation of the organisms from tissues or immunodiagnostically by various serologic and skin tests. Although there is widespread and, in some areas, high prevalance of infection with *Toxoplasma,* overt disease is relatively rare. The most severe form of the disease in man is the acute neonatal type acquired by congenital transmission. A number of organs and tissues such as the eyes, brain, liver, heart, and spleen may be involved. Jaundice, hydrocephalus, and chorioretinitis are but a few of the symptoms seen (Feldman, 1958). The spectrum of symptoms in the fetus or newborn depends somewhat on the time that the infection is acquired from the mother.

Acute and chronic forms of toxoplasmosis occur in children and adults. It may be fulminating and fatal, involving the lymph nodes, heart, spleen, brain, liver, etc., or of a more chronic form with lymphadenitis, chorioretinitis, or posterior uveitis as the predominant clinical feature. Toxoplasmosis in children and adults has been reviewed by Frenkel and Jacobs (1958), Remington *et al.* (1960), Desmonts (1960), Siim (1960), and Perkins (1961). The parasite has been isolated from human heart (Cathie, 1955), from uteri (Remington *et al.*, 1960) and from one of fifteen brains of persons who died from causes unrelated to toxoplasmosis (Walls *et al.*, 1963).

As an addendum to the role of food in the transmission of toxoplasmosis, its presence in wild and domestic mammals and birds has been demonstrated by serologic surveys and isolation of the parasite (Habegger, 1953; Jacobs, 1963). Such epidemiologic surveys have been concerned with eliciting evidence of infection as well as documenting the role of the parasite in producing disease in these animals (Hartley and Kater, 1963; McAllister, 1964; and Walton and Walls, 1964). The chance of acquiring toxoplasmosis from infected meat can be minimized by thorough cooking, as recommended for trichinosis. *Toxoplasma* cysts in meat are killed by thorough cooking, and studies by Jacobs *et al.* (1960) have shown that the cysts are also destroyed by freezing and thawing.

The treatment of toxoplasma infections involves the use of pyrimethamine and sulfadiazine. The drugs used together exert a synergistic action which permits use of much smaller concentrations of each than are required when the drugs are administered singly. Recent papers of Frenkel and Jacobs (1958), Eyles (1960), and Jacobs (1963) present current information on the therapy of toxoplasmosis.

A. Laboratory Methods

1. DETECTION OF TOXOPLASMA IN FOODS

At the present time, any attempt to isolate *Toxoplasma* from foodstuffs is an involved process, not feasible for routine use. Although there are minor differences in techniques, the method given by Jacobs *et al.* (1963) for the isolation of *Toxoplasma* from sheep and cattle serves as a model. Briefly, tissue from diaphragm, muscle, brain, uterus, etc., is ground in a suitable meat grinder or tissue grinder (for small samples) and digested in artificial gastric juice (pepsin, HCl) for 1 hour at 37°C with agitation or stirring. The digested tissue is strained through gauze, centrifuged, washed with saline, and resuspended in saline. Approximately 1 ml of the solution is inoculated into each of ten or more mice. The mice are kept under observation for one month. During this time, smear preparations are made from lungs, brain, liver, spleen, and peritoneal exudate of any weakened or dead mice. The slides are stained with Giemsa and examined for *Toxoplasma* organisms. Sera collected from mice surviving for one month are tested by the Sabin-Feldman dye test for *Toxoplasma* antibodies. Mice having a positive dye test titer are sacrificed, and smears of tissues are stained and searched for proliferative *Toxoplasma* organisms. In addition to the above procedures, selected tissues from the brain may be inoculated into clean mice after digestion, washing, etc., and the smear technique performed on any serologically positive mice of the second passage.

2. ISOLATION AND DIAGNOSIS OF *Toxoplasma* IN HUMANS

Attempts to isolate organisms from humans are limited by the amount of tissue available for digestion and inoculation. The involved process of isolation has hindered the routine use of such techniques in human infections, and cases of toxoplasmosis are diagnosed by immunodiagnostic techniques. In suspected cases of disease, enlarged lymph nodes may be biopsied, and mouse inoculation, serodiagnosis, and identification may be attempted.

The diagnosis of toxoplasmosis in humans is accomplished with the aid of skin tests and serology. The skin test indicates previous or present infection and may be of some help in cases of suspected (present) disease. However, in acute, recently acquired infections, the skin test may be negative. The Sabin-Feldman dye test, the complement fixation test, and the indirect hemagglutination test are utilized in the routine diagnosis of toxoplasmosis. Other tests such as the fluoresence inhibition and indirect test, the acrylic flocculation test, and the direct agglutination test have also been employed with success by various authors. The techniques for performing these tests, their relative merits in diagnosis, their use in acute and chronic cases or for survey purposes, and the significance of titers obtained are too numerous to include in this section. Details for performing the various serologic techniques can be found in the published works of Sabin and Feldman (1948), Jacobs and Lunde (1957), Frenkel and Jacobs (1958), Siim and Lind (1960), Piekarski *et al.* (1961), Lewis and Kessel (1961), Goldman *et al.* (1962), Kelen *et al.* (1962), Fulton (1965), and Fulton and Fulton (1965).

VI. PROTOZOA AND HELMINTHS DISSEMINATED IN FECALLY CONTAMINATED FOOD OR WATER

While the organisms mentioned above are most widely known as the major food-borne parasites, they are not the most prevalent nor the most important from the public health or medical viewpoint. Although coprophagous tendencies may be found among the mentally retarded and pica sometimes occurs in children, man does not ordinarily ingest substances except for purposes of satisfying hunger or thirst. In spite of this, there are several protozoan and helminth parasites which are acquired by the ingestion of infective forms passed in feces.

The dissemination of such fecally transmitted parasites may be uncomplicated and direct as when raw human excrement or "night soil" is used as fertilizer for crops. Under such conditions a constant reinfection of the populace takes place and the parasite burden becomes a matter of course. Persons ingesting raw vegetables or drinking water in endemic areas where such practices are common often become the hosts of one or more of the parasites as well as various viruses and bacteria.

In other instances, the careless disposal of feces and its access to flies, which may mechanically transmit some of the infective stages, also serve to disseminate the parasites. Poor personal hygiene of food handlers may also aid and abet the spread of parasites.

The third manner in which parasitic infections may be spread is that normally associated with more sophisticated societies. In highly developed urban areas, elaborate sewer systems transport waste products to a central disposal area, and the spread of fecally transmitted parasitic organisms, even in the presence of many infected individuals, is prevented. Occasionally the befouling of drinking water with sewage does occur, and outbreaks of parasitism take place. Two documented instances of water-borne amebiasis in the United States are the Chicago Hotel epidemic which occurred in 1933 (McCoy *et al.,* 1936) and the South Bend, Indiana, woodworking plant outbreak which occurred in 1950 (LeMaistre *et al.,* 1956). Both of these epidemics took place in highly developed urban areas with adequate sewage disposal systems. The individuals in these outbreaks were infected when drinking ostensibly clean water which had been contaminated with sewage.

A. Protozoan Parasites (Schemes 9, 10, 11; pp. 212, 213, and 214)

Entamoeba histolytica, Giardia lamblia, and *Dientamoeba fragilis* are three potentially pathogenic protozoan parasites of man which are acquired by ingestion of stages passed from the body in feces. The protozoan cyst is usually the infective stage, but *D. fragilis* is passed in the trophic form. These three protozoa and their nonpathogenic counterparts require no period of embryonation or development outside the body. Cyst forms liberated in feces are infective, and an immediate fecal–oral transmission is possible. All three forms often exist as nonpathogens and elicit

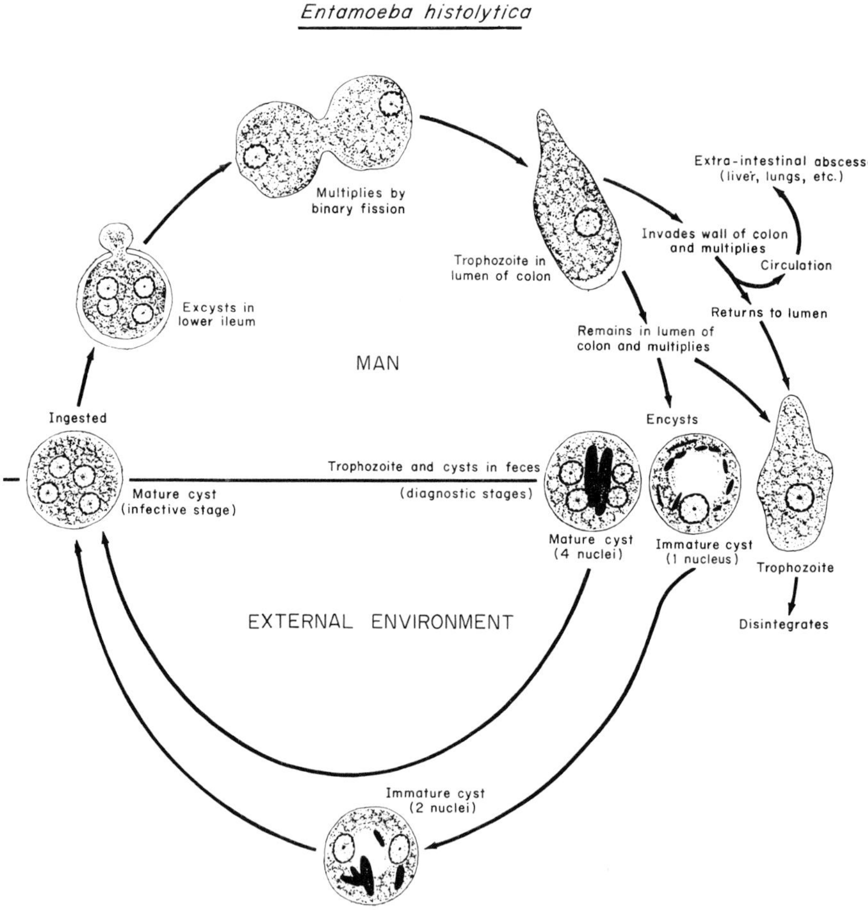

Entamoeba histolytica

SCHEME 9. Life cycle of _Entamoeba histolytica_.

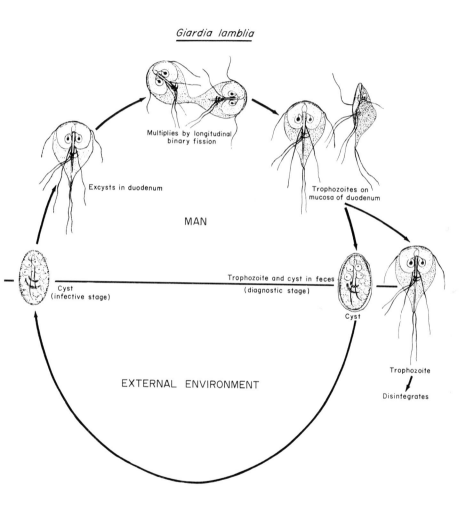

SCHEME 10. Life cycle of *Giardia lamblia*.

Dientamoeba fragilis

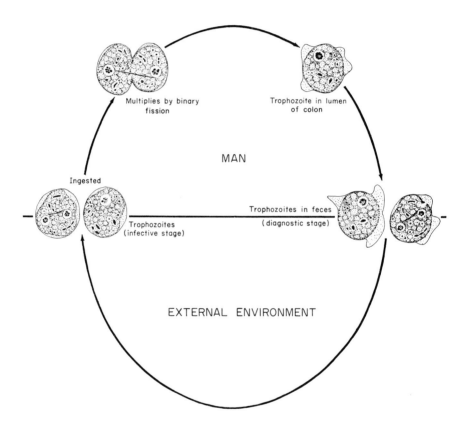

SCHEME 11. Life cycle of *Dientamoeba fragilis.*

no symptoms. On the other hand, symptomatic infections do occur. *Entamoeba histolytica* may cause severe disease, as seen in classical amebic dysentery, or even death, as when the parasite invades extraintestinal tissues such as the liver, lungs, or brain. Clinical infections with *G. lamblia* are often associated with diarrhea, constipation, steatorrhea, and gastrointestinal pains. Although *G. lamblia* does not invade the tissues as does *E. histolytica,* it apparently can cause symptoms by blocking absorptive functions of the upper intestine, particularly the absorption of fats. Infections with *D. fragilis* can be symptomatic (Kean and Malloch, 1966) causing diarrhea and epigastric distress, although the organism is not a tissue invader per se.

B. Helminth Parasites (Schemes 12 and 13; pp. 216 and 217)

Ascaris lumbricoides and *Trichuris trichiura* are likewise transmitted by ingestion of fecally polluted foods or water. These nematode parasites differ from the fecally transmitted protozoa in two ways: (1) The eggs of the two worms require a period of several days in the soil to allow the infective larvae to develop within the eggs. Although under such conditions the direct feces-to-mouth transmission is not possible, this is counterbalanced by the fact that the eggs are very resistant to environmental changes. In warm, humid, tropical areas, eggs in polluted soil or those in "night soil" used to fertilize vegetables are infective for a long period of time. (2) No multiplication occurs within the body of the host. The number of organisms parasitizing man depends on the number of eggs ingested. Thus the "worm burden" is constant for a particular infection, although the severity of the infection is generally correlated with the number of worms present.

Ascaris lumbricoides may cause symptoms in infected persons and is capable of inducing pathologic processes. After hatching from the ingested eggs, the larvae migrate in their normal life cycle. In some infections, severe lung involvement can result in an "ascaris pneumonitis." Larvae wandering through the body may also invade the brain, eye, or kidney where they can initiate severe tissue reactions during the process of being destroyed. The adult parasite, after the period of larval migration, lives in the small intestine and causes colicky pain, gastrointestinal upset, diarrhea, and allergic reactions to toxic products liberated by living or disintegrating adults. There is the added hazard of pathology resulting from the wanderings of a single adult *Ascaris* into the pancreatic duct, bile duct, liver, appendix, and other areas, as well as the possibility of intestinal perforation with subsequent peritonitis. Large numbers of

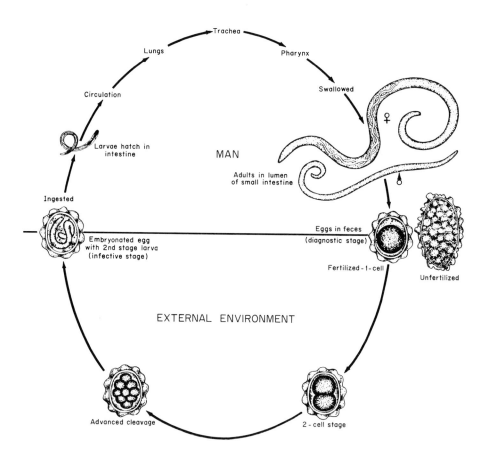

SCHEME 12. Life cycle of *Ascaris lumbricoides*.

Trichuris trichiura

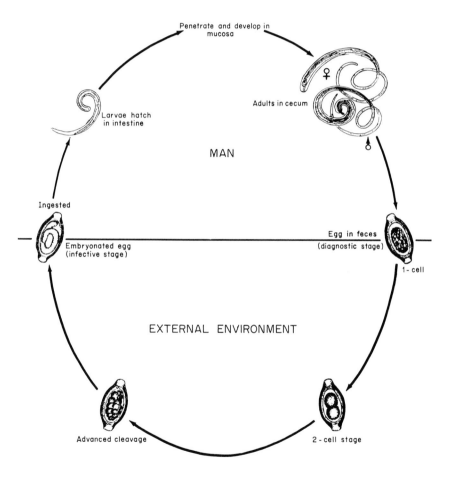

Penetrate and develop in
mucosa

Adults in cecum

Larvae hatch
in intestine

MAN

Ingested

Egg in feces
(diagnostic stage)

Embryonated egg
(infective stage)

1- cell

EXTERNAL ENVIRONMENT

Advanced cleavage

2 - cell stage

SCHEME 13. Life cycle of Trichuris trichiura.

adult worms may cause intestinal blockage which can be alleviated only by surgery.

Trichuris trichiura worms inhabit the cecal area of the large bowel, although in heavy infections they may be found in the upper colon and appendix. Symptoms are generally dependent on the magnitude of the worm burden and may include abdominal pain, constipation, flatulence, loss of appetite, vomiting, and symptoms of systemic intoxication. In severe infections, especially in children, diarrhea with blood-streaked stools and rectal prolapse can occur. Worms threaded into the appendiceal mucosa may open portals of entry for pyogenic bacteria and cause inflammatory processes.

C. Laboratory Methods

Since fecally transmitted protozoan and helminth parasites have infective stages which are microscopic in size, it is impossible to detect them grossly as contaminants of vegetables. Foods suspected of being fecally polluted should be cooked. Immersing them in such solutions as potassium permanganate is not effective (Most, 1964).

In endemic areas one also runs the risk of acquiring fecally transmitted viruses and bacteria, and measures taken for destruction of food-borne protozoa and helminths, other than cooking, may induce a false sense of security.

Diagnosis of human infections with the fecally transmitted helminths and protozoa can be made by utilizing the stool examination techniques previously noted. In persons having soft or diarrhetic stools, the trophic forms of the protozoa may predominate. In such cases, if fresh specimens can be examined, Kohn's one-step combination fixative-stain provides a simple method for obtaining stained slides for the identification of either cystic or trophic forms (Gleason and Healy, 1965). When immediate examination cannot be made, the use of polyvinyl alcohol (PVA) preservation and subsequent Wheatly trichrome stain will aid in the detection of the more fragile trophozoites (Healy, 1964).

The use of a direct wet mount, formalin–ether concentration, and stained slide enables one to make a thorough diagnostic search for helminth eggs, larvae, protozoan cysts, and trophozoites of all the intestinal parasites including those of food-borne origin.

VII. SUMMARY

The aforementioned parasites constitute the majority of those important as food-borne entities. Some, like *Trichinella spiralis,* are well known and of long historical significance; others, such as *Anisakis,* have only recently become recognized as medically important food-borne organisms. Space does not permit the listing of other food-borne parasites which are of rare occurrence and are the subject of individual case reports in the literature. They may be found listed in textbooks of parasitology such as those by Faust and Russell (1964) and Belding (1965). Many of the helminths — cestodes, trematodes, and nematodes — are normally parasites of domestic or wild mammals and birds. Their infective stages are often found in foods not ordinarily ingested by man either for sustenance or as a part of occult practices such as warding off the spirits, improving virility, or acting as medicinals. These rare food-borne organisms are acquired by the ingestion of all manner of uncooked plant and animal life, grass, weeds, beetles, grubs, cockroaches, snakes, frogs, fish, salamanders, and the myriad of organisms which form an essential chain in the often-complicated life cycle of a parasite.

All the food-borne parasites share one common epidemiologic feature; that is, in acquiring them, man either consumes the infective stages raw or ingests them in inadequately cooked or frozen food wherein they are still viable.

ACKNOWLEDGMENTS

The life cycle charts contained herein are taken from the following publications:

Melvin, D. M., Brooke, M. M., and Sadun, E. H. (1965). Life Cycle Charts of Common Intestinal Helminths of Man. *U. S. Public Health Service Publ.* **1235.**

Brooke, M. M., and Melvin, D. M. (1965). Life Cycle Charts of Common Intestinal Protozoa of Man. *U. S. Public Health Service Publ.* **1140.**

Melvin, D. M., Brooke, M. M., and Healy, G. R. (1965). Life Cycle Charts of Common Blood and Tissue Parasites of Man. *U. S. Public Health Service Publ.* **1234.**

REFERENCES

Acha, P. N., and Aguilar, F. J. (1964). *Am. J. Trop. Med. Hyg.* **13,** 48.

Africa, C. M., Garcia, E. Y., and DeLeon, W. (1935). *Philippine J. Public Health,* **2,** 1.

Alicata, J. (1965). *Advan. Parasitol.* **3,** 223.

Alicata, J., and Brown, R. W. (1962). *Can. J. Zool.* **40,** 755.

Ameel, D. J. (1934). *Am. J. Hyg.* **19,** 279.

Asami, K., Watanuki, T., Sakai, H., Imano, H., and Okamoto, R. (1965). *Am. J. Trop. Med. Hyg.* **14,** 119.

Ash, L. R. (1962). *J. Parasitol.* **48,** 298.

Ashby, B. S., Appleton, P. J., and Dawson, I. (1964). *Brit. Med. J.* **i,** 1141.

Beaver, P. C., and Rosen, L. (1964). *Am. J. Trop. Med. Hyg.* **13,** 589.

Beckett, J. V., and Gallicchio, V. (1966). *J. Parasitol.* **52,** 511.

Belding, D. L. (1965). "Textbook of Clinical Parasitology." Appleton-Century-Crofts, New York.

Biagi, F., Navarrete, F., Pina, A., Santiago, A. M., and Tapia, L. (1961). *Rev. Med. Hosp. Gen. (Mex.)* **25**, 501.

Bisgard, G. E., and Lewis, R. E. (1964). *J. Am. Vet. Med. Assoc.* **144**, 501.

Bonsdorf, B. von. (1956). *Exptl. Parasitol.* **5**, 207.

Bovornkitti, S., and Tandhanand, S. (1959). *Diseases Chest* **35**, 1.

Brown, H. W. (1960). *Clin. Pharmacol. Therap.* **1**, 87.

Buck, A. A., Sadun, E. H., Lieske, H., and Lee, B. K. (1958). *Z. Tropenmed. Parasitol.* **9**, 310.

Burrows, R. B. (1965). "Microscopic Diagnosis of the Parasites of Man." Yale University Press, New Haven.

Cathie, I. A. B. (1955). *Lancet* **i**, 149.

Cohen, B. (1962). *Southern Med. J.* **55**, 46.

Corkum, K. C. (1966). *J. Parasitol.* **52**, 444.

Cuckler, A. C., Egerton, J. R., and Alicata, J. E. (1965). *J. Parasitol* **51**, 392.

Dent, J. H. (1957). *J. Am. Med. Assoc.* **164**, 401.

Desmonts, G. (1960). *In* "Human Toxoplasmosis" p. 112 (J. C. Siim, ed.). Munksgaard, Copenhagen.

Eyles, D. E. (1960). *In* "Human Toxoplasmosis" p. 127 (J. C. Siim, ed.). Munksgaard, Copenhagen.

Facey, R. V., and Marsden, P. D. (1960). *Brit. Med. J.* **ii**, 619.

Faust, E. C., and Russell, P. F. (1964). "Craig and Faust's Clinical Parasitology." Lea and Febiger, Philadelphia.

Feldman, H. A. (1958). *Pediatrics* **22**, 559.

Frenkel, J. I., and Jacobs, L. (1958). *A.M.A. Arch. Ophthalmol.* **59**, 260.

Fulton, J. D. (1965). *Trans. Roy. Soc. Trop. Med. Hyg.* **59**, 694.

Fulton, J. D., and Fulton, F. (1965). *Nature* **205**, 776.

Gleason, N. N., and Healy, G. R. (1965). *Am. J. Clin. Pathol.* **43**, 494.

Goldman, M., Gordon, M. A., and Carver, R. K. (1962). *Am. J. Clin. Pathol.* **37**, 541.

Gould, S. E. (1945). "Trichinosis." Thomas, Springfield.

Habegger, H. (1953). "Le reservoir biologique animale et sa relation avec l'infection toxoplasmique humaine." Ambilly-Annamasse, Imprimerie Franco-Suisse, Geneva.

Haining, R. B., and Haining, R. G. (1960). *J. Am. Med. Assoc.* **172**, 2036.

Hartley, W. J., and Kater, J. C. (1963). *Res. Vet. Sci.* **4**, 326.

Healy, G. R. (1964). *Am. J. Gastroenterol.* **42**, 191.

Horio, S. R., and Alicata, J. E. (1961). *Hawaii Med. J.* **21**, 139.

Jacobs, L. (1963). *Ann. Rev. Microbiol.* **17**, 429.

Jacobs, L., and Lunde, M. N. (1957). *J. Parasitol.* **43**, 308.

Jacobs, L., Remington, J. S., and Melton, M. L. (1960). *J. Parasitol.* **46**, 11.

Jacobs, L., Moyle, G. G., and Bis, R. R. (1963). *Am. J. Vet. Res.* **24**, 673.

Jindrak, K., and Alicata, J. E. (1965). *Ann. Trop. Med. Parasitol.* **59**, 294.

Kagan, I. G. (1960). *J. Infect. Diseases* **107**, 65.

Kagan, I. G. (1965). *Am. J. Public Health* **55**, (11), 1820.

Kean, B. H., and Malloch, C. L. (1966). *Am. J. Digest. Diseases* **11**, 735.

Kelen, A. E., Ayllon-Leindl, L., and Labzoffsky, N. A. (1962). *Can. J. Microbiol.* **8**, 545.

Ketsuwan, P., and Pradatsundarasar, A. (1966). *Am. J. Trop. Med. Hyg.* **15**, 50.

Klemme, W. (1966). *Arch. Neurol.* **15**, 185.

Kozar, Z. (1962). *In* "Proc. 1st Intern. Conf. Trichinellosis" (Z. Kozar, ed.), p. 15.

LeMaistre, C. A., Sappenfield, R., Culbertson, C., Carter, F. R., Offutt, A., Black, H., and Brooke, M. M. (1956). *Am. J. Hyg.* **64**, 30.

Lewis, W. P., and Kessel, J. F. (1961). *Arch. Ophthalmol.* **66**, 49.

Linaweaver, P. G. (1966). *Military Med.* **131,** 579.

Lombardo, L., and Mateos, J. H. (1961). *Neurology* **11,** 824.

Lucker, J. T., and Douvres, F. W. (1960). *Proc. Helminthol. Soc. Wash.* **27,** 110.

Mackerras, M. J., and Sandars, D. F. (1955). *Australian J. Zool.* **3,** 1.

Markell, E. K., and Haber, S. L. (1964). *Am. J. Med.* **37,** 491.

Maynard, J. E., and Kagan, I. G. (1963). *Practitioner* **191,** 622.

Maynard, J. E., and Pauls, F. P. (1962). *Am. J. Hyg.* **76,** 252.

McAllister, R. A. (1964). *Can. J. Comp. Med. Vet. Sci.* **28,** 53.

McCoy, G. W., Hardy, A. V., Gorman, A. E., Bundesen, H. N., Connolly, J. I., and Rawlings, I. D. (1936). *Natl. Inst. Health Bull.* **166.**

McIntosh, A., and Miller, D. (1960). *Am. J. Vet. Res.* **21,** 169.

Miller, D. (1956). *Public Health Rept. (U.S.)* **71,** 1239.

Miyazaki, I. (1960). *Exptl. Parasitol.* **9,** 338.

Moore, J. T. (1915). *Am. J. Trop. Disease Prevent. Med.* **2,** 518.

Most, H. (1964). "Health Hints for the Tropics." American Society of Tropical Medicine and Hygiene, Bethesda, Maryland.

Mueller, J. F., Hart, E. P., and Walsh, W. P. (1963a). *J. Parasitol.* **49,** 294.

Mueller, J. F., Hart, E. P., and Walsh, W. P. (1963b). *N. Y. State J. Med.* **63,** 715.

Nieto, D. (1956). *Neurology* **6,** 725.

Nishimura, K., and Yogore, G. (1965). *J. Parasitol.* **51,** 928.

Nyberg, W., Grasbeck, R., Saarni, M., and von Bonsdorff, B. (1961). *Am. J. Clin. Nutr.* **9,** 606.

Oterdoom, H. J. (1961). *Trop. Geograph. Med.* **13,** 97.

Perkins, E. S. (1961). "Uveitis and Toxoplasmosis." Little, Brown, Boston.

Piekarski, G., Saathoff, M., and Schafer, E. (1961). *Zentr. Bakteriol. Parasitenk., Abt. 1. Orig.* **181,** 407.

Prommindaroj, K., Leelawongs, N., and Pradatsundarasar, A. (1962). *Am. J. Trop. Med. Hyg.* **11,** 759.

Rawal, B. D. (1959). *Trans. Roy. Soc. Trop. Med. Hyg.* **53,** 61.

Read, C. P. (1952). *J. Parasitol.* **38,** 29.

Remington, J. S., Jacobs, L., and Kaufman, H. E. (1960). *New Engl. J. Med.* **262,** 180.

Ritchie, L. S. (1948). *Bull. U.S. Army Dept.* **8,** 326.

Roselle, H. A., Schwartz, D. T., and Geer, F. G. (1965). *New Engl. J. Med.* **272,** 304.

Rosen, S. W., and Kiefer, E. D. (1958). *J. Am. Med. Assoc.* **167,** 2065.

Rosen, L., Chappell, R., Laqueur, G. L., Wallace, G. D., and Weinstein, P. P. (1962). *J. Am. Med. Assoc.* **179,** 620.

Round, M. C. (1961). *J. Hyg.* **59,** 505.

Saarni, M., Nyberg, W., Grasbeck, R., and von Bonsdorff, B. (1963). *Acta Med. Scand.* **173,** 147.

Sabin, A. B., and Feldman, H. A. (1948). *Science* **108,** 660.

Sadun, E. H. (1955). *Am. J. Hyg.* **62,** 81.

Sadun, E., and Maiphoom, C. (1953). *Am. J. Trop. Med. Hyg.* **2,** 1070.

Sadun, E. H., Buck, A. A., and Walton, B. C. (1959). *Military Med.* **124,** 187.

Sawada, T., Takei, K., Williams, J. E., and Moose, J. W. (1965). *Exptl. Parasitol.* **17,** 340.

Scholtens, R. G., Kagan, I. G., Quist, K. D., and Norman, L. (1966). *Am. J. Epidemiol.* **83,** 489.

Short, R. B., and Lewis, A. L. (1964). *J. Parasitol.* **50,** 800.

Siim, J. C. (1960). *In* "Human Toxoplasmosis" p. 53 (J. C. Siim, ed.). Munksgaard, Copenhagen.

Siim, J. C. (1961). *Surv. Ophthalmol.* **6,** 781.

Siim, J. C., and Lind, K. (1960). *Acta Pathol. Microbiol. Scand.* **50,** 445.

Silverman, P. H., and Griffiths, R. B. (1955). *Ann. Trop. Med. Parasitol.* **49,** 436.

Spencer, F. M., and Monroe, L. S. (1961). "The Color Atlas of Intestinal Parasites." Thomas, Springfield.

Stemmermann, G. N. (1953). *Am. J. Pathol.* **29,** 731.

Stiles, C. W. (1908). *Hyg. Lab. Bull.* **40,** 7.

Strauss, W. G. (1962). *J. Am. Med. Assoc.* **179,** 290.

Suessenguth, M. A., Schnurrenberger, P. R., Bauer, A., Wentworth, F. H., and Masterson, R. A. (1965). *Am. J. Vet. Res.* **26,** 1298.

Summers, W. A., and Weinstein, P. P. (1943). *Am. J. Trop. Med.* **23,** 363.

Swartzwelder, J. C., Beaver, P. C., and Hood, M. W. (1964). *Am. J. Trop. Med. Hyg.* **13,** 43.

U. S. Department of Agriculture, Agricultural Research Service, Meat Inspection Division, (1960). Regulations governing the meat inspection of the United States Department of Agriculture, Edition of 1 June 1959, Washington, D. C. U. S. Government Printing Office, 242 pp.

U. S. Public Health Service (1966). *Morbidity Mortality Ann. Suppl.* **14,** 53.

Van Thiel, P. H. (1962). *Parasitology* **52,** 16.

Van Thiel, P. H., Kuipers, F. C., and Roskam, R. Th. (1960). *Trop. Geograph. Med.* **2,** 97.

Walls, K. W., Taraska, J. J., and Goldman, M. (1963). *J. Parasitol.* **49,** 930.

Walton, B. C., and Walls, K. W. (1964). *Am. J. Trop. Med. Hyg.* **13,** 530.

Weinman, D., and Chandler, A. H. (1956). *J. Am. Med. Assoc.* **161,** 229.

White, J. C., Sweet, W. H., and Richardson, E. P., Jr. (1957). *New Engl. J. Med.* **256,** 479.

World Health Organization (1962). Technical Report Series #241. Joint FAO/WHO Expert Committee on Meat Hygiene.

Wright, W. H., Kerr, K. B., and Jacobs, L. (1943). *Public Health Rept. (U.S.)* **58,** 1293.

Wykoff, D. E., and Winn, M. M. (1965). *J. Parasitol.* **51,** 207.

Yokogawa, M. (1965). *Advan. Parasitol.* **3,** 99.

Yokogawa, M., and Yoshimura, H. (1965). *Am. J. Trop. Med. Hyg.* **14,** 770.

Yokogawa, M., Iwasaki, M., Shigeyasu, M., Hirose, H., Okura, T., and Tsuji, M. (1963). *Am. J. Trop. Med. Hyg.* **12,** 859.

Yokogawa, S., Cort, W. W., and Yokogawa, M. (1960). *Exptl. Parasitol.* **10,** 81.

Yoshimura, H. (1965). *J. Parasitol.* **51,** 961.

Yun, D. J. (1960). *J. Pediat.* **56,** 736.

Zimmerman, W. J., and Brandly, P. (1965). *Public Health Rept. (U.S.)* **80,** (12), 1061.

Zimmerman, W. J., Schwarte, L. H., and Biester, H. E. (1961). *J. Parasitol.* **47,** 429.

CHAPTER VI | **INFECTIONS DUE TO**

MISCELLANEOUS MICROORGANISMS

Frank L. Bryan

I. INTRODUCTION

From 1957 to 1962, no specific etiological agent was determined for about half of the reported outbreaks of food-borne disease in the United States (Dauer, 1952–1961; Dauer and Davids, 1959, 1960; Dauer and Sylvester, 1954–1957; Bryan, 1962; Food Protection Committee, 1964). In England and Wales, about one-fourth of all outbreaks and cases of food-borne disease were caused by unidentified agents (Vernon, 1965). According to Nikodémusz et al. (1964a), the etiological agent is not identified in a large percentage of outbreaks occurring in most countries. For instance, the causative agent was not determined in 75% of the reported cases of food poisoning in Hungary, 49% in Italy, 32% in France, 26% in East Germany, and 20% in West Germany. Inadequate investigation, lack of suitable samples, a difference in classification schemes, and inadequate laboratory support have contributed to this situation; but inability to detect some agents by conventional laboratory procedures is also partly responsible (Food Protection Committee, 1964).

The food-borne diseases that arouse contemporary interest and those that cause the highest morbidity have been discussed in detail by the other authors. However, there are other food-borne diseases that are caused by various organisms of different kinds and character. For discussion purposes these organisms may be categorized in the following ways:

1. Those organisms for which there is no definite proof of food-borne transmission, although the organisms are sometimes reported as etiological agents in outbreaks because large numbers are recovered in the incriminated foods.

2. Those organisms which have been historically recognized as milkborne.

3. Those organisms which are usually transmitted by modes other than food but are occasionally food-borne.

4. Those organisms whose roles as agents of food-borne illness are little known.

5. Those organisms which are transmitted to individuals who handle food products during processing operations.

There is obviously considerable overlapping among these categories, and factors such as time and place will move individual organisms from category to category. The reason for the classification is to emphasize the relative importance of each organism as an etiological agent of food-borne-disease.

II. ORGANISMS IN WHICH PROOF IS INCONCLUSIVE

There has been a tendency to accord undue significance to the mere presence of large numbers of a particular organism in a food thought to have caused illness (Wilson and Miles, 1964). A factor which contributes to this situation, as pointed out by Dack (1947), is that samples are usually collected many hours after the suspected meal was eaten. The food may have been stored in a manner which subjected it to contamination, and the food may have been left at room temperature in the meantime. During the interval between serving and sampling, bacterial multiplication occurs, and rapidly growing cultures may outgrow the causative organism. Dolman (1943) showed that this phenomenon can occur and may completely mask the causative organism. Many of the organisms considered to be causes of nonspecific bacterial food poisoning are those that live as saprophytes outside the body and are found as commensals and contaminants of foods. Under favorable conditions they multiply and may be found in large numbers in foods implicated in outbreaks. Certain foods may contain millions of organisms per gram without causing illness. When the organism in question constitutes part of the normal fecal flora (such as *Streptococcus, Proteus, Klebsiella, Providencia, Citrobacter, Actinomyces, Pseudomonas,* and *Aeromonas*), it becomes extremely difficult to evaluate its part in intestinal disease. However, repeated isolation of a particular strain or species from foods to which outbreaks have been attributed does cast suspicion on the organism, and its concurrent recovery from fecal specimens of patients and food workers strengthens the hypothesis. But it must be kept in mind that it would be expected to find the organisms which are isolated from heavily contaminated foods, whether pathogenic or not, in vomitus and feces of patients. Before final proof of pathogenicity of organism can be obtained, tests complying with Koch's postulates must be performed. Although animal experimentation may give additional support to the organism's role in food poisoning, laboratory animals have given inconsistent results with the nonspecific bacteria. Thus, for conclusive proof, the disease must be produced in human volunteers. In the case of the organisms discussed under this heading, reports of their incrimination as causative agents of food-borne disease appear in the literature, but proof of their role in causing illness is lacking.

A. Enterococci

Enterococci have been implicated as etiological agents of food-borne illness. In most cases, the mere isolation (usually in high numbers) of enterococci from suspect food has been the grounds for this implication; but a few volunteer studies have also been made.

1. CLASSIFICATION

Before commencing a discussion of the reported incidence and experimental studies of enterococcal food-borne disease, it is necessary to define the enterococci. Throughout the years, classification of the species within the group D streptococci — those that possess the group D antigen according to Lancefield (1933) — has been in a continuous state of flux. According to the most recent review of the taxonomy of this group, Deibel (1964) classified *Streptococcus bovis, Streptococcus equinus, Streptococcus faecalis* (including the former *Streptococcus liquefaciens* and *Streptococcus zymogenes*), and *Streptococcus faecium* (and its variety *durans*) within this group. For purposes of his review, Deibel (1964) reserved the term enterococci to connote specifically those group D streptococci which follow Sherman's criteria (Sherman, 1937). According to these criteria, enterococci are differentiated from other streptococci by their ability to grow in the presence of 6.5% NaCl; to grow at pH 9.6, at temperatures of 10°C and 45°C; and to withstand a temperature of 60°C for 30 minutes. However, in this regard, Deibel (1964) stated that these tests are marginal in nature, and the occurrence of strains that fail to give a positive result under one or more of these criteria is to be expected. Thus, *S. faecalis* and *S. faecium* comprise the enterococci.

Enterococci may be hemolytic or nonhemolytic. Alpha-hemolysis, a greenish zone around colonies on blood agar due to oxidation of hemoglobin to green choleglobins, is observed in some enterococci. But a few strains, such as *S. faecalis* variety *zymogenes* (on horse blood agar) and *S. faecium* var. *durans,* produce beta-hemolysis. This reaction is identified by a clear zone around colonies due to lysis of red blood cells. Most strains of *S. faecalis* are nonhemolytic. Differentiation of the enterococci, *S. faecalis* and *S. faecium,* is made on the basis of tellurite sensitivity, tetrazolium reduction, susceptibility to neuraminidase, nutritional differences, fermentation patterns, and serological type (Shattock, 1962; Deibel, 1964).

2. ASSOCIATED SYMPTOMS

The illness attributed to the enterococci has been described as being milder than staphylococcal intoxication (Dack, 1956), and symptoms include nausea, colic pain, diarrhea, and, in some cases, vomiting. Incubation periods as short as 2 hours (Buchbinder *et al.,* 1948) and as long as 36 hours (Dack et al., 1949) have been reported; however, the average range is 6 to 12 hours (Jensen, 1957). Shattock (1962) remarked that in cases with short incubation times vomiting was the dominant symptom, and in cases with longer incubation times the dominant symptom was diarrhea. The former symptom resembles staphylococcal intoxication, and the latter resembles salmonellosis or *Clostridium perfringens* and *Bacillus cereus* food poisoning.

3. REPORTED INCIDENCE

In regard to enterococcal food poisoning, Linden et al. (1926) isolated streptococci from cheese that was incriminated in two human outbreaks. When cats were fed cultures of these streptococci, symptoms of food poisoning resulted. Years later, one of the preserved cultures was identified as *S. faecalis* by Sherman et al. (1943) and later changed to *S. faecium* (Deibel and Silliker, 1963; Hartman et al., 1965). Nelson (1928) failed to produce illness in monkeys when he fed them chocolate cream pie (and a large number of streptococci that was isolated from the pie) that was implicated in an outbreak. Jordan and Burrows (1934) used streptococci that were implicated in an outbreak in a feeding test on monkeys. Filtrates repeatedly produced symptoms in some monkeys but not in others. Strains listed as *Streptococcus viridans* (alpha-hemolytic) and *Streptococcus haemolyticus* (beta-hemolytic) also produced symptoms in monkeys. Further reports of enterococci isolation from foods implicated as a cause of food-borne outbreaks have been made (Cary et al., 1931, 1938; Dack, 1943; Getting et al., 1944; Fabian, 1947; Buchbinder et al., 1948; Moore, 1948; Dack et al., 1949; Dack, 1956; Dauer and Sylvester, 1956; Fujiwara et al., 1956; Hayashi, 1960; Brown, 1962; Todd, 1962; and Lüönd and Gasser, 1964). Feig (1950) listed 34 outbreaks of gastroenteritis attributed to streptococci during a three-year period (1945–1947). However, an important aspect of the above consideration is that 62% of the outbreaks showed the presence of other bacteria. Of 21 outbreaks with mixed contamination, 17 were with staphylococci and 2 with salmonellae.

4. HUMAN VOLUNTEER STUDIES

Evidence to support the hypothesis that the enterococci cause food-borne disease is meager. Both animal and human feeding studies have been performed. Moore (1955) and Hartman et al. (1965) reviewed the animal studies. However, animal experimentations tell us little, since the reaction of experimental animals may not be the same as the reaction of man. To learn anything decisive about the role of enterococci in causing

illness, human volunteers must be fed substrates containing large numbers of the organisms or their metabolic products. Meyer (1953) stated
that, in studies with humans, individual susceptibility must be accom-
modated by the use of large numbers of volunteers, the concentration of
toxic metabolites at the different growth phases must be established, and
the interval between isolation of the organisms from the suspected food
and the test on volunteers must be short. As will be noted in the following
review, few studies have conformed to these criteria.

The first human volunteer study was conducted by Cary et al. (1931).
They reported an outbreak involving 75 (of 182) boys. The patients developed gastroenteritis 4 to 12 hours after eating Vienna sausage. Large
numbers of alpha-hemolytic streptococci were isolated from the leftover
sausage. Cans from the same lot, some of which had defective seams,
were examined, and a "similar" organism was isolated. Five hours after
a laboratory worker ate half a sausage from each of four cans (which had
been incubated for 4 days) he experienced nausea, vomiting, epigastric
pain, and exhaustion. The sample ingested by the volunteer probably
represented a sample similar to that consumed by the patients, although
incubation may have altered the numbers of contaminating organisms
or metabolic products. It should be pointed out, however, that the
contaminated cans of sausage probably had contained streptococci and
perhaps other organisms for some weeks before they were consumed.
(Thus, the flora may have changed during this period.) Another volunteer,
5 hours after drinking 40 ml. of a 5-day Berkefeld filtered beef heart culture that contained an estimated 10,800 cells, showed symptoms of belch-
ing, epigastric pain, tenesmus, cold perspiration, low fever, and marked
exhaustion. The consumption of 50 ml of a bacteria-free filtrate prepared
from a similar culture caused no ill effects in a third volunteer. Cats fed a
48-hour milk culture of the streptococci developed soft stools, but mice,
guinea pigs, rabbits, and monkeys showed no symptoms. The organism
used in these experiments was not identified in accordance with the
classification of enterococci now employed and was referred to as a green
pleomorphic streptococcus.

Subsequently, Cary et al. (1938) isolated enormous numbers of two
biochemical types of alpha-hemolytic streptococci from beef croquettes
(made from leftover beef stew, round of beef, and Swedish meat balls)
two days after an institutional outbreak involving 117 of 208 people. The
main symptom reported was diarrhea, although abdominal cramps,
nausea, and vomiting, in order of decreasing incidence, occurred. Incubation periods were 3 to 18 hours and averaged 12 hours. Five of seven
subjects felt abdominal distress after taking broth cultures containing

living streptococci one or more times, but no signs of illness resulted when seven volunteers (giving a history of not being subject to gastrointestinal disturbance) drank milk containing 20 ml of a filtrate from a culture grown in a veal-infusion broth. These experiments indicated that filtrates from the particular strains were not responsible for illness. One drawback of this investigation was the time lag between the outbreak and the isolation of the enterococci in the laboratory. Other limitations of the study, pointed out by Moore (1948), are that veal broth is a poor medium for toxin production, and that toxin content may be lowered by Seitz filtration.

Dolman (1943) failed to produce symptoms in volunteers who were fed bacteria-free filtrates or meat pies in which enterococci were grown. The pies contained over 1.8×10^9 *S. viridans* (alpha-hemolytic) per gram. Topley (1947) fed four cultures of enterococci (three from eggs implicated in outbreaks) to six volunteers, and no illness resulted.

A study which more closely followed the criteria of Meyer (1953) was carried out by Osler *et al.* (1948). Four strains of *S. faecalis* (one strain isolated from evaporated milk by Buchbinder *et al.* (1948) and three strains isolated from human feces within two months prior to this study) were used in a feeding study. Two of the four strains grown in milk, custard, or egg salad at 37°C for 5 hours produced symptoms of afebrile gastric or intestinal disturbance (including nausea, vomiting, diarrhea, dizziness, and abdominal cramps) in six and possibly seven of 26 volunteers after an incubation period of 2½ to 10 hours. Over 2.2×10^{10} organisms were consumed by subjects who gave no history of recent gastroenteritis. Some of those who showed no symptoms ingested larger doses than others who became ill. When the same strains were grown for 20 hours, symptoms did not occur in 21 volunteers who ingested from 2.5 to 50 ml of the culture. Only four complained of slight nausea. Controls eating the same food without the organisms were unaffected.

Moore (1948) isolated a strain of streptococcus (S1129) from a chocolate pudding that was implicated in an outbreak involving 153 cases among 309 children. A volunteer was fed broth cultures, bacteria-free filtrates of cultures grown in pudding, and chocolate pudding that had been inoculated with the streptococcus (S1129). Symptoms of acute gastroenteritis similar to those reported in the outbreak developed 3 to 6 hours following ingestion. No symptoms occurred in the volunteer following the consumption of a suspension of the isolated streptococci which were washed in Ringer's solution. Although the organism used in this study was demonstrated to be an alpha-hemolytic streptococcus, it neither possessed group D antigen nor grew in bile or 6.5% NaCl. The organism was also destroyed by heating to 60°C for 30 minutes. These tests would exclude this organism from the enterococcus group as defined by Breed *et al.* (1957) and Deibel (1964).

Sherman (1937) and Gale (1944) suggested that tyramine released as a result of decarboxylation of tyrosine by *S. faecalis* was responsible for illness. To evaluate this hypothesis, Dack *et al.* (1949) fed six volunteers milk containing 0.3 to 1 gm of tyramine monohydrochloride. They also fed ten volunteers cheese that was made with a starter strain of *S. faecalis,* containing large numbers of viable enterococci as well as appreciable quantities of tyramine. No illness was observed in the subjects. Also, when a culture of the starter strain was fed to ten volunteers, symptoms did not occur. Shattock (1949) stated that one of the most active tyrosine decarboxylase-producing strains of *S. faecalis* that she had encountered was isolated in large numbers from cheese which was consumed without ill effects. These two studies, and the fact that all *S. faecalis* strains produce tyramine, indicate that tyramine is not associated with food-borne illness. Dack *et al.* (1949) also reported that three of four volunteers developed mild diarrhea of short duration 32 to 60 hours after drinking broth cultures of *S. faecalis* variety *liquefaciens.* This organism was isolated from turkey à la king a short time prior to the volunteer study. Subsequently, Niven (1955), one of the investigators, questioned the significance of the mild diarrhea suffered by the subjects. Meals not under the control of the investigators were consumed by the subjects prior to their illness, and in one case a volunteer was subjected to pressures of an all-day academic examination. When large numbers (up to 317 billion of a strain of *S. faecalis* that were isolated from turkey meat which had been incriminated in an outbreak) were fed to seven volunteers, no effects were observed. Negative results were also obtained when billions of cells of the strain isolated by Buchbinder *et al.* (1948) were fed to three volunteers.

Streptococcus zymogenes (S. faecalis) was isolated from fish sausage that was implicated in an outbreak of gastroenteritis (Fujiwara *et al., 1956).* When boiled filtrates of cultures of this organism were fed to volunteers, nausea, vomiting, and diarrhea resulted.

To clarify the role of enterococci as etiological agents of food poisoning, Deibel and Silliker (1963) performed a study involving 23 enterococcus strains. Nine of these strains had been implicated previously in food-borne disease outbreaks (two were recent isolates). Six were from human feces, four from plant sources, three from foods, and one from an unknown source. Fourteen were *S. faecalis,* seven were *S. faecium,* and two were *S. faecium* variety *durans.* Each of these strains, after having been grown for 24 hours at 37°C in sterile milk or on sterile ham slices, were fed to volunteers. In addition, six strains of *S. faecalis* variety *liquefaciens* were consumed after complete liquefaction of gelatin. Strains of *S. faecalis* were also consumed after they were grown in media (arginine, gluconate, malate, and pyruvate) which altered their energy metabolism. The end products associated with these products differ both qualitatively and quantitatively from those that were grown in medium containing glucose. Cultures of one strain of *S. faecium,* the same strain that was studied by Osler *et al.* (1948), were inoculated into multiple tubes containing sterile milk. After incubation at 2, 4, 6, 8, 16, and 24 hours, cultures were frozen and subsequently ingested at weekly intervals by three volunteers. Approximately 1 to 10 billion cells of *S. faecalis* and *S. faecium* strains were disrupted by sonic treatment and fed to three volunteers. The investigators (Deibel and Silliker, 1963) reported that in no instance were any of the above conditions of growth conducive to producing food-poisoning symptoms in the volunteers. Guthof (1957) associated increased numbers of proteolytic enterococci in the feces of individuals with various pathological conditions, but since illness was not produced by gelatin-hydrolyzing enterococci or by end products of their activity, this hypothesis must be discounted. Deibel and Silliker (1963) pointed out shortcomings of their own experiment—the possible loss of toxicity due to repeated laboratory transfer of all but two of the strains and possible immunity due to the repeated challenge of a few individuals.

Lüönd and Gasser (1964) described an outbreak and the subsequent investigation which shed some new light on the role of enterococci as an etiological agent of food-borne disease. Following a meal, 36 of 40 individuals developed symptoms of vomiting, nausea, abdominal pain, and diarrhea There was no fever. The incubation period was remarkably constant in all patients, 19 to 22 hours. In most cases the illness lasted 3 to 4 days. Stool samples were negative for salmonellae, shigellae and staphylococci. A meat pie containing aspic was also negative for these common food-borne pathogens, but streptococci were isolated. The pie filling without aspic contained 25 million enterococci per gram, and the aspic contained 240 million enterococci per gram. These were identified microscopically, biochemically, and serologically as group D streptococci. *Streptococcus bovis, S. equinus,* and *S. faecalis* were ruled out, and the investigators felt that the isolated enterococci were *S. faecium* or its variety *durans.* As later stated by Lüönd (1965), in reply to an inquiry by Dack, the leftover piece of pie was cut up into seven slices; five slices were sent to the laboratory and examined at once, and two slices were eaten by the manager on the same day. Lüönd and Gasser (1964) stated that the manager developed the same symptoms as the patients after a corresponding incubation period. It was felt by the investigators that the aspic solution was contaminated by the workers' hands or kitchen equipment. After the aspic was placed in the pies, the pies were held at room temperature for 15 to 20 hours, allowing time for the development of enormous numbers of the incriminated organisms. The leftover pie, three-quarters of a cut of a large pie, was reported (Lüönd, 1965) to have been kept in a freezer at approximately −6°C until the moment of sample collection three days later. Thus, the slices surveyed in the laboratory were representative of the pie consumed by the patients. Because of the refrigeration practices employed, there is little likelihood that there was appreciable growth of the organisms in the remnants of the pie eaten by the patients.

To summarize the human volunteer studies, abdominal distress or illness has resulted after investigators fed living cultures of enterococci or similar organisms to volunteers (Carey et al., 1931, 1938; Osler et al., 1948; Lüönd and Gasser, 1964), but negative results have been observed when cultures were fed to volunteers (Topley, 1947; Osler et al., 1948; Deibel and Silliker, 1963). Moore (1948) and Fujiwara et al. (1956) reported that streptococcal filtrates were toxigenic; however, Cary et al. (1938) and Dolman (1943) obtained negative results with filtrates. Hartman et al. (1965) felt that certain conditions may be necessary for certain strains of enterococci to cause food-borne disease. They suggest further study of the effect of incubation temperature, of the variations of pH of the culture medium, of differences of freshly isolated and stock cultures, and of the effect of mixed cultures and synergistic action.

5. Isolations from Foods

Shattock (1962) and Deibel (1964) stated that, if group D streptococci cause food poisoning, it is surprising that streptococcal food poisoning does not occur more frequently because these organisms often occur (sometimes in large numbers) in foodstuffs. Examples include: cheese fermented by an enterococcus culture (Dahlberg and Kosikowsky, 1948; Dack et al., 1949); commercially frozen fruits, fruit juices, and vegetables (Hucker et al., 1952; Kaplan and Appleman, 1952; Larkin et al., 1955a, b); ready-to-eat meats (Floyd and Blakemore, 1954); frozen seafood (Larkin et al., 1956; Raj et al., 1961); turkeys (Wilkerson et al., 1961); spray-dried whole egg powder (Solowey and Watson, 1951); and raw and pasteurized milk (White and Sherman, 1944). Mattick and Shattock (1943) stated that English hard cheese containing from 10^4 to 10^7 enterococci did not produce illness, and Buttiaux (1956) reported that ham which contained more than 3 million *S. faecalis* was eaten without ill effects. Topley (1947) recovered alpha-hemolytic streptococci from five samples of eggs associated with food poisoning, but he also found alpha-hemolytic streptococci, frequently in high numbers, in 21 of 28 routine samples. There was no obvious difference in the streptococcal flora of the samples associated with outbreaks and the control samples. The enterococci are well adapted to grow under a wide range of environmental conditions and to survive conditions which would be destructive to many bacteria (Sherman, 1937; Deibel, 1964).

Surgalla *et al.* (1944) observed that alpha-hemolytic streptococci experimentally inoculated into canned foods were able to grow, and these organisms survived for at least 30 days at 37°C in canned foods of low acid content, semi-acid content, and high acid content, and in seafoods. In most cases they demonstrated growth for 60 days at 22°C and 37°C. No growth occurred in sugar-preserved canned peaches, a highly acid product.

6. Sources of Enterococci

The natural habitat of group D streptococci appears to be the alimentary tract of man and other animals. Some studies (Cooper and Ramadan, 1955; Bartley and Slanetz, 1960; Kenner *et al.,* 1960) indicated that *S. faecalis* is the dominant species in humans; however, other investigations (Buttiaux, 1958; Kjellander, 1960; Guthof, 1957) suggest that *S. faecium* is the dominant species. Shattock (1962) felt that this difference was geographical and must be associated with diet. In pigs, cows, sheep, and turkeys the dominant species are *S. bovis, S. equinus,* and *S. faecium* (Kjellander, 1960; Kenner *et al.,* 1960; Raibaud *et al.,* 1961; Wilkerson *et al.,* 1961); and in chickens the dominant species is *S. faecalis* (Bartley and Slanetz, 1960).

Kenner *et al.* (1960) found that the median density of streptococci in millions per gram of moist feces was 1.3 for cows, 3.0 for human beings, 3.4 for fowl, 38.0 for sheep, and 84.0 for pigs. Enterococci densities were 0.16, 2.29, 2.10, 9.42, and 8.4 million per gram, respectively. The enterococcus and *Streptococcus salivarius* groups accounted for 92.6% of the streptococcal population.

7. Challenge

The mechanism causing illness associated with the consumption of large numbers of enterococci is unknown. In fact, convincing evidence of enterococcal food-borne disease is lacking. These gaps in our knowledge, therefore, impose a challenge on all professional public health workers and researchers to collect samples and carry out volunteer studies in a manner which will settle the dilemma. When conducting a study to identify the role of enterococci in causing food-borne disease, the following procedures must be used: (1) Obtain a complete history of food preparation and handling, including time and temperatures used in cooking and holding the foods after preparation; (2) obtain samples of the same foods that were eaten by the afflicted group, preferably from a source which has been held in a manner in which enterococci counts would not appreciably change from the time of consumption to the time the sample was collected; (3) maintain the samples in a manner in which organisms will not multiply before laboratory work has begun; (4) attempt to isolate other etiological agents, such as *Clostridium perfringens,* staphylococcal enterotoxin, members of the Enterobacteriaceae, and rule these out as causes; (5) perform biochemical and serological tests to identify the species and strains and compare strains recovered from the patients with strains recovered from the incriminated food; (6) perform feeding tests with the freshly isolated cultures, using appropriate controls, on a significant number of volunteers; and (7) evaluate cultures that have been grown for varying periods of time in substrates of the same or similar composition as the incriminated food.

B. *Proteus*

Proteus species are gram-negative, motile bacteria which conform to the definition of the family Enterobacteriaceae (Edwards and Ewing, 1962). They decompose urea rapidly and deaminate phenylalanine. At present, four species are included in the genus: *Proteus vulgaris, Proteus mirabilis, Proteus morganii,* and *Proteus rettgeri.* Only the first two species produce hydrogen sulfide in abundant quantities, liquefy gelatin, and swarm on moist agar.

1. OUTBREAKS ASCRIBED TO *Proteus*

Bengtson (1919), Savage (1920), Tanner (1933), Tanner and Tanner (1953), Jordan and Burrows (1935), and Barnes (1944) discussed outbreaks of food poisoning in which members of the genus *Proteus* were suspected, but they concluded that the probability of these organisms being the etiological agent was not convincing. Many of these outbreaks occurred before the role of staphylococci or *Clostridium perfringens* as etiological agents was known. Incrimination of *Proteus* has usually been based on the presence of large numbers of the organisms in implicated foods, although occasionally *Proteus* has been found in the stools of patients with symptoms of gastroenteritis. For instance, Klieman *et al.* (1942) found that *Proteus* was the predominant organism in smoked white fish, the food supposed to have been the cause of an outbreak of gastroenteritis involving 34 people. Rustigian and Stuart (1943) recovered *P. rettgeri* (*P. entericus* 33111) from patients with gastroenteritis and from food handlers. Two documented outbreaks attributed to *Proteus* have been described by Cooper *et al.* (1941) and Cherry *et al.* (1946).

Cooper *et al.* (1941) isolated *P. vulgaris,* a rather rare *Proteus* species, from fecal specimens of ten cases. Salmonellae and staphylococci were not found. Symptoms of vomiting, diarrhea, collapse, cyanosis, and sleepiness occurred 3 to 5 hours after eating brawn. The brawn was prepared from meat which was pickled in a brine bath. Following what seemed to be inadequate cooking and cooling, the brawn sat on a serving table for as long as 11 hours prior to consumption. In the subsequent investigation, *Escherichia coli* and organisms of the *P. vulgaris* "type" were recovered from the brine. Pursuant to this investigation, Cooper *et al.* (1941) examined feces from 537 patients with diarrhea due to causes other than *Proteus* and recovered *Proteus* from 15% of the samples, but of these only 0.6% were *P. vulgaris.* In the outbreak just described, other *Proteus* species were not found as frequently in the stools of patients as were *P. vulgaris* strains. Later, Cherry *et al.* (1946) recovered large numbers of *P. mirabilis* from vomitus and feces of victims of food poisoning as well as from suspected (boned) baked ham. The ham also yielded approximately 12,000 staphylococci per gram. Nausea, vomiting (almost continuous emisis), diarrhea (an average of 8.9 watery stools), abdominal cramps, and peripheral collapse were reported. Only 3 of the 27 patients had a fever; the others had normal or subnormal temperatures. The incubation period was 3 hours, and the duration of illness was 40 hours. Although Cherry *et al.* (1946) did not feel that the isolation of the relatively low numbers of staphylococci was significant, the symptoms and incubation periods for both of these outbreaks (Cooper *et al.,* 1941; Cherry *et al.,* 1946) are quite similar to the clinical picture associated with staphylococcal intoxication.

2. ANIMAL AND HUMAN VOLUNTEER STUDIES

Jordan and Burrows (1935) fed monkeys filtrates from cultures of several *Proteus* strains grown in veal infusion agar containing starch. Vomiting resulted. When these filtrates were repeatedly fed to human volunteers, no deleterious effects were observed. Hunter and Dack (1938) failed to confirm these results when they did not reproduce the vomiting

reaction in monkeys with filtrates from *Proteus* strains. Cooper *et al.*
(1941) produced illness in kittens when he fed them filtrates from a
Proteus culture which was isolated from a food implicated in an outbreak.

Dolman (1943) prepared filtrates from three *P. vulgaris* strains isolated
from pressed beef that was incriminated in an outbreak. No significant
reactions occurred when 12 volunteers ingested from 5 to 20 ml of the
filtrate. Meat pies were inoculated with *P. vulgaris* and incubated for 8
hours at 37°C. The number of ogranisms present in each pie exceeded
1½ billion per gram. When these pies were consumed by volunteers, no
ill effects followed. The subjects also reported that the flavor of the pies
was not altered by the treatment.

Dolman (1943) advanced a hypothesis that the "toxin" type of food
poisoning outbreak, frequently reported as of unknown origin or due to
Proteus, Streptococcus, or *E. coli,* is caused by enterotoxigenic staphylococci which may be masked, overgrown, or even extinct when the
suspect food is examined in the laboratory. Sometimes staphylococci
have been found in the suspected food, but in numbers too small to appear
significant (Cherry *et al.,* 1946). However, Dolman (1943) felt that the
isolated staphylococci may be survivors of larger numbers initially
present. He stated that the decline in numbers could result from exhaustion of the essential nutrients in the foodstuffs. Such exhaustion would
usually be accelerated by competition with another organism. This
hypothesis was supported by an experiment in which three persons became ill after they had consumed a saline extract of wieners that had been
inoculated with enterotoxigenic staphylococci and subsequently by *P.
vulgaris.* At the time of consumption, the wieners were decomposing and
the *Proteus* counts were 30 to 50 billion organisms per gram. *Proteus*
outnumbered the staphylococci a hundredfold. Earlier in the experiment,
when *Proteus* counts were in this same range, no staphylococci were
observed. Large amounts of extract from wieners equally infected with
the same strain of *P. vulgaris,* but not inoculated with staphylococci, were
eaten without ill effects.

Kendereski (1961) fed two volunteers meat which had been infected
with two strains of *Proteus,* then left at room temperature for 48 hours,
and finally boiled in water for 30 minutes. No illness resulted.

3. MULTIPLE FACTORS

Fujiwara *et al.* (1965) observed the coexistence of *Proteus morganii* in
cases of food poisoning due to *Vibrio parahaemolyticus.* Monkeys that
were resistant when fed with only the vibrio species manifested symptoms
of enteritis, diarrhea, and vomiting when they were fed 50 ml of a mixture

of the vibrio and the *Proteus* strain. However, when these monkeys were fed 50 ml of *P. morganii* only, *V. parahaemolyticus* only, or a *Pseudomonas* species only, no symptoms were observed.

4. FISH POISONINGS

Within the past four years, there have been two reports from California on scombroid fish poisoning (Listick and Condit, 1964; Cooper *et al.,* 1964). In the first outbreak, symptoms reported by victims were diarrhea, headache, abdominal cramps, nausea, biting taste, flushing, and vomiting. In the second report, symptoms included headache, urticaria, edema, rash, dizziness, a tingling sensation, and nausea. In these outbreaks, illness occurred between 15 minutes and 3 hours after eating fish. Both investigations revealed grossly inadequate cooling practices; the fish (Spanish mackerel and albacore) remained for several hours at temperatures which would support bacterial incubation. In the outbreak reported by Listick and Condit (1964), *P. morganii* and *P. vulgaris* were isolated from four frozen, uncooked mackerel which were caught at the same time as the incriminated fish. Halstead (1957) stated that scombroid fish poisoning can be attributed to bacterial action of certain strains of *Proteus* resulting in elaboration of a histamine-like substance, saurine (Kawataba *et al.,* 1955). As an editorial note following the report by Cooper *et al.* (1964), Farber stated that even though most fish implicated in California food poisoning outbreaks were scombroids, all showed bacterial spoilage involving *Pseudomonas, Achromobacter,* and *Proteus* organisms. In studying food poisoning cases associated with marine products, and exhibiting allergy-like symptoms, Kawabata *et al.* (1955) hypothesized that *Achromobacter histaminium,* which produces large amounts of histamine, is a synergist.

Obviously, we lack a great deal of knowledge about ichthyosarcotoxism (fish poisoning) as well as the role, if any, that common spoilage organisms play in producing illness. Overgrowth of these organisms may have masked the presence of the real culprits.

C. *Providencia*

The *Providencia* genus is a well-defined group of Enterobacteriaceae which resemble *Proteus morganii* and *Proteus rettgeri* in certain respects. They do not hydrolyze urea or form hydrogen sulfide. Bacteria of this group have been isolated from stools in sporadic cases and in small outbreaks of diarrhea in man, but they occur also in feces of normal individuals (Ewing *et al.,* 1954). The importance of certain serotypes of this genus has yet to be investigated and evaluated. *Providencia* have been recovered from food handlers and from patients with gastroenteritis (Stuart *et al.*, 1946; Galton *et al.*, 1947; Neter, 1956). Plass (1947) recovered a member of the genus *Providencia* from patients with diarrhea and from fricasseed chicken, the incriminated food in an outbreak. The patients vomited, were nauseated, had abdominal cramps, and had watery diarrhea. Mild shock was reported in a few cases. The shortest incubation period was 2 hours, and the duration of illness was 24 hours. Stools were examined for staphylococci, but none were found. In addition to *Providencia, Proteus rettgeri* and *Proteus mirabilis* were recovered from 12 of 16 hospitalized patients. Whether the *Providencia,* the various *Proteus* species, some organism that was overlooked, or a combination of the mixed flora caused the outbreak could not be definitely established.

D. *Escherichia coli*

Escherichia coli, a common lactose-fermenting fecal organism, belongs to the family Enterobacteriaceae (Edwards and Ewing, 1962). Within the antigenic analysis of this group there are 146 O (somatic) antigens, 88 K (capsular or surface) antigens, and 49 H (flagella) antigens. A few serotypes have been shown to be pathogenic for humans. Those serotypes that are frequently mentioned in the literature on infantile diarrhea belong to nine OB groups: O26:K60(B6), O55:K59(B5), O86a:K61(B7), O111:K58(B4), O119:K69(B14), O125:K70(B15), O126:K71(B16), O127:K63(B8), and O128:K67(B12). Six additional OB or OL groups and two additional O groups have less frequently been mentioned as causes of infantile diarrhea: O18:K76(B20), O18:K77(B21), O20, O28:K73(B18), O44:K74(L), O86, O112:K68(B13), and O124: K72(B17). Complete antigenic composition for the enteropathogenic *E. coli* (E.E.C.) are given by Ewing *et al.* (1957, 1963). Most of the gastroenteritis due to E.E.C. has occurred in infants; several serotypes are accepted as the cause of infantile diarrhea. These serotypes appear to be quite host-specific. Contact appears to be the most common mode of transmission.

Feces containing large numbers of E.E.C. are the main sources of contamination. Patients ill with E.E.C. enteritis frequently excrete enormous numbers of the organisms. *Escherichia coli* may be rapidly spread by contaminated hands, gowns, and possibly fomites (Rogers and Koegler, 1951). The organism may survive for several days or weeks in dust and on furniture (Rogers, 1951; Hutchison, 1957). Since E.E.C. may be present in the upper respiratory tract of infants with enteritis, air-borne transmission must be considered. Carrier rates in asymptomatic contacts of positive index cases have ranged from 4.8 to 10% (Kessner *et al.*, 1962). During a period of one year, Gamble and Rowson (1957) isolated E.E.C from feces of 1 to 2% of adults and from 20% of hospitalized children. In a survey of approximately 25,000 stools from normal children between birth and 4 years of age, enteropathogenic *E. coli* were recovered from 2.42% of the specimens.

A review of the epidemiology of enteropathogenic *E. coli* has been made by Herweg *et al.* (1956) and by Neter (1959). Cattle may harbor *E. coli* of a type similar to those that cause human infection (Orskov, 1951; Charter, 1956; Dunne *et al.*, 1956), and milk may contain pathogenic serotypes. Thomson (1956) recovered serotypes with similar O groups to the E.E.C. from 1% of farm milk samples. He also observed that 13% of chickens surveyed were excreting *E. coli* with the same O groups that are associated with human disease, but only a few were of the H groups incriminated in human illness. Although the serotypes have been isolated from human foods and food-producing animals, there is little evidence to incriminate these sources in human infection. In a common-source epidemic involving 47 children at a school, Hobbs *et al.* (1949) reported the recovery of a paracolon organism 411, which was later identified as *E. coli* O124 (Taylor, 1955). Illness with symptoms of headache, malaise, fever, diarrhea, and vomiting occurred 14 to 48 hours after consumption of a meal. Duration of the illness did not exceed 48 hours. Salmon was incriminated, although this food item was not available for laboratory testing. Callao and Henares (1957) and Costin *et al.* (1964) have made reports of food poisoning in adults in which *E. coli* O26 and *E. coli* O86:B7:H34, respectively, were isolated. In the report of Costin *et al.* the incubation period was 7 to 12 hours, and the same serotype was recovered from a substitute coffee drink and from the patients' stools. Serological tests showed relevant rises in titer against *E. coli* in the patients. Staphylococci, clostridia, and other Enterobacteriaceae were sought during the investigation, but none were found. In Poland, Ulewicz (1956) found *E. coli* O26:K60(B) in a suspect food, in almost pure culture in the stools of nine patients who ate the food, and in the stools of two kitchen personnel.

Human volunteer studies have been performed with some of the enteropathogenic *E. coli*. Hobbs *et al.* (1949) fed two human volunteers four drops of a young broth culture of a paracolon organism (*E. coli* O124). One volunteer developed abdominal discomfort, fever, chills, pains in limbs and head, nausea, and diarrhea, 22 hours later. The second volunteer complained of abdominal discomfort and lack of appetite after 3 days. The ingested organism was recovered from specimens of feces from both volunteers. The occurrence of accidental laboratory infection suggests that low numbers can produce infection. Neter and Shumway (1950) fed a 2-month-old infant 100 million viable *E. coli* O111 cells, and the child developed diarrhea and lost weight within 24 hours; within 48 hours large numbers of the organism were recovered from the infant's throat, nasopharynx, and feces. Kirby *et al.* (1950) observed mild symptoms of diarrhea and abdominal discomfort in four of six adults 12 hours after these volunteers were fed one strain of *E. coli* O111:B4. Each of the volunteers ingested 2 billion organisms. Another strain of *E. coli* O111:B4 given to three volunteers did not produce illness. Six men who ingested fecal *E. coli* (not serologically related to O111:B4) did not become ill. Symptoms similar to acute bacterial food poisoning were produced in adult volunteers when they were fed more than 1 million cells of *E. coli* O111:B4 in milk by Ferguson and June (1952). When fewer cells were consumed, very mild illness or no effect was observed. It took between 6 and 9 billion organisms to produce severe or moderate reactions in most of the volunteers. The symptoms varied from nausea and cramps to violent diarrhea, cramps, either nausea or vomiting, or both. A temperature elevation was also noted. When 9 billion cells were ingested, the onset of symptoms came on the average of 10 hours after feeding. Five hours was the shortest recorded incubation period. Symptoms persisted to some degree for the first 24 hours. A longer incubation period, averaging 12 hours, occurred when 6 billion E.E.C. organisms were fed to volunteers. Mild symptoms occurred within 72 hours after the ingestion of approximately ½ billion cells. Slight illness was also produced in a few volunteers when they consumed 7 million E.E.C. No illness resulted in the control groups or in individuals who ingested nonpathogenic *E. coli* which had been isolated from a normal infant. Similar results were obtained when June *et al.* (1953) fed similar concentrations of three strains of *E. coli* O55:B5 to human volunteers. However, the ingestion of killed strains of E.E.C. produced no ill effects in volunteers. These and similar feeding studies have been reviewed by Ferguson (1956).

From these observations it is conceivable that certain serotypes of *E. coli* may produce food-borne disease. More attention should be paid to the complete serological typing of *E. coli* isolated from suspected food and from the related cases (Taylor, 1955; Ewing 1956, 1963). The procedures for sample collection and the laboratory diagnosis of E.E.C. are discussed by Ewing (1963).

E. *Citrobacter*

The *Citrobacter* genus of Enterobacteriaceae is composed of cultures previously classified as *Escherichia freundii*. The group includes a part of the species *Paracolobactrum intermedium,* those designated as the "Bethesda-Ballerup" group. *Citrobacter* differs in its biochemical reactions from *Escherichia coli* in many respects and more closely resembles the *Salmonella* group (Edwards and Ewing, 1962). The biochemical and serological reactions for this group are discussed by West and Edwards (1954).

A few members of the genus *Citrobacter* have been recovered from the stools of patients with diarrhea, and in a few outbreaks these organisms have been recovered from patients and food handlers and, in one instance, from an implicated food (West and Edwards, 1954; Barnes and Cherry, 1946). However, their role in enteric infections still remains to be proved. *Citrobacter* are also recovered from the intestinal tract of apparently normal persons and animals. Thus, unless the organisms are examined serologically the question concerning their pathogenicity will not be answered. *Citrobacter* (a Bethesda serotype) was recovered from several of 52 patients suffering from a mild diarrhea (average of 4.8 stools), abdominal cramps, nausea, and vomiting. The incubation period was 12 hours, and the duration of the illness was 12 hours. Corn pudding was implicated, but no samples were available for laboratory study (Barnes and Cherry, 1946). Edwards *et al.* (1948) mentioned four outbreaks in which *Citrobacter* was recovered from the stools of patients. In one outbreak the same serotype was recovered from 9 of 12 cases. In another instance four *Citrobacter* of the same serotype were recovered from four patients, and two of another serotype were isolated from food handlers. In the third outbreak two *Citrobacter* were found in the stools of two cases. *Citrobacter* was also isolated from spoiled milk and from a person who drank the milk and became ill.

Dizon *et al.* (1962) concluded that Bethesda-Ballerup organisms were responsible for an outbreak of gastroenteritis involving over 75 people. The symptoms in descending order of frequency were diarrhea, abdominal pain, fever, chills, headache, tenesmus, vomiting, nausea, dizziness, and fainting. No foods were left for examination, and the source was not determined. Two of ten food handlers were found to be carriers of Bethesda-Ballerup strains, and one was a carrier of *Shigella*. Some of the patients showed positive agglutination reactions for the Bethesda-Ballerup group. *Staphylococcus aureus* was isolated from four stool specimens from six patients. However, the symptoms and incubation period, 1 to 48 hours (average 13 hours), were not characteristic of staphylococcal intoxication.

F. *Klebsiella*

Klebsiellae are gram negative, capsulated, nonmotile organisms which conform to the definition of the family Enterobacteriaceae. Their biochemical reactions are discussed by Fife *et al.* (1965).

Horvath *et al.* (1964) reviewed a food-borne outbreak which affected 190 pupils and workmen who ate at a boarding school. The symptoms experienced by some of the patients were headache, dizziness, and nausea. All had violent abdominal pain and five to six evacuations of watery stools. Fever was not reported. Most of the patients recovered in 6 to 8 hours, and all within 24 hours. The investigators felt that a food served at the noon meal was responsible, since about 30 students on an all-day excursion escaped the illness and all those eating lunch, including nonboarding students who ate only this meal, became ill. This sets the incubation period for the outbreak at 10 to 15 hours. Seventy percent of 41 stool samples contained pure cultures of *Klebsiella* (serotypes not given), and the remainder contained a small number of other intestinal bacteria as well as *Klebsiella*. Antibiotics were not used in treatment. The same strains were recovered from pea soup, from beef cooked in rice, and from a knife. The soup and the beef in rice yielded 10^6 and 10^7 klebsiellae per gram, respectively. From the soup, 10^3 *Staphylococcus aureus* organisms per gram and a small number of *Bacillus cereus* were isolated, and from the beef in rice approximately 10^4 clostridia were also found. It must be pointed out that the foods were not properly stored in the school prior to sampling by the investigators; thus, the great number of klebsiellae cultured from the foods cannot establish these organisms as the causative agent. Nevertheless, it is interesting that klebsiellae were the predominant flora in the fecal samples, and thus it appeared that these organisms were ingested with the food. The investigation did not determine the manner in which the food was contaminated or the source of the organism.

G. *Enterobacter*

The genus *Enterobacter* consists of four species: *Enterobacter cloacae* (formerly *Cloaca* and *Aerobacter cloacae*), *Enterobacter aerogenes* (formerly *Aerobacter aerogenes* B), *Enterobacter liquefaciens* (formerly *Aerobacter* C), and *Enterobacter alvei* (formerly the *Hafnia* group which included bacteria previously designated as *Paracolobacter aerobacter*,

Paracolobacter aerogenoides, and intermediate "paracolon" bacteria). The classification, serology, and biochemical reactions of this genus are discussed by Edwards and Ewing (1962) and Fife *et al.* (1965). In the literature, a number of years ago, two reports of food-borne disease were attributed to these organisms. They described members of the cloacae-aerogenes group as the etiological agents (Buchanan and Megrail, 1929; Gilbert *et al.,* 1932). Custard cream puffs were the responsible food in both outbreaks, and symptoms and incubation periods were consistent with staphylococcal intoxication. Stuart and Rustigian (1943) discussed an outbreak of mild gastroenteritis caused by *Aerobacter* 30011 *(E. alvei)* among 28 hospital patients and staff. Milk was incriminated, and these organisms were recovered from the milk. At the same time as the hospital outbreak, cases occurred in the city, and the same organism was isolated from cream pastry. In discussing outbreaks due to *E. alvei* *(Aerobacter* 32011) they mentioned that a careful search uncovered a food handler carrying small numbers of the organism. Large numbers were recovered from the patients. Kerrin (1947) recovered *Bacillus asiaticum (E. alvei)* from a fatal infection resembling acute food poisoning. The patient became ill 4 hours after he ate some stew. Since the food was discarded, there was no opportunity for laboratory confirmation.

H. *Pseudomonas*

Pseudomonas species are gram-negative rods that usually produce a water-soluble pigment which diffuses through media. Both motile (polar flagella) and nonmotile species are known. Pseudomonads are common water and soil organisms. Of approximately 100 species, only two are known to be pathogenic to man. *Pseudomonas aeruginosa* is a common contaminant of the skin. This organism usually produces a bluish-green pigment, and in infections a blue-green purulent exudate may be observed. Barnes (1944) stated that *P. aeruginosa* has been encountered frequently enough in patients with gastroenteritis to cast suspicion on it as an enteric pathogen. *Pseudomonas aeruginosa* has been isolated on numerous occasions from stools of adult patients with gastroenteritis (Anusz, 1966).

The Kansas State Board of Health (1946) and Ensign and Hunter (1946) reviewed an epidemic involving 409 cases of acute enteritis. In the adult population the disease ran a mild to moderate course from one to several days with symptoms of diarrhea, cramps, nausea, and vomiting predominating. The disease was more severe among children. In newborn infants, there was extreme prostration, explosive diarrhea with 10 to 20 thin watery stools per day, rapid dehydration, and cyanosis. Nine cases terminated fatally. *Pseudomonas aeruginosa* was isolated from many of the patients' stools and was recovered from autopsy specimens. The community milk supply was associated with the cases on epidemiological grounds, and during an investigation of the incriminated plant *P. aeruginosa* was recovered from a rag used to wrap a pipe joint. The drippings from the rag had fallen into the milk.

In Egypt, Moustafa *et al.* (1948) reported that 30 of 125 food-borne outbreaks were attributed to *B. pyocyanea (P. aeruginosa)*. When they fed living and dead cultures of these organisms to kittens, the animals developed diarrhea, and some vomited. Carter (1954) gave an account of a food poisoning in Glasgow in which *P. aeruginosa* was isolated from rabbit meat and from patients. Gentzgow has related the story about an outbreak of gastroenteritis in military personnel. Illness occurred a short time after breakfast consisting of pancakes. During the subsequent investigation, a gelatinous mass was observed in the syrup. When the syrup was examined bacteriologically, a pure culture of *P. aeruginosa* was found (Ewing, 1966).

Bongkrek (flat white cakes covered with white mold and wrapped in banana or other large leaves) is made by fermenting pressed or grated coconut with the fungus *Rhizopus oryzae*. In Java, this product is eaten by millions of people without harm, but occasionally a species of bacteria, *Pseudomonas cocovenenans,* overgrows the fungus and produces a heat-stable, unsaturated fatty acid (bongkrek acid). Small quantities of this toxin cause hypoglycemia and death in humans and experimental animals. Bongkrek acid is not produced in soybean or peanut press cakes by *P. cocovenenans;* thus, tempeh or ontjom have not caused illness. Leaves from a certain variety of oxalis cause the pH of bongkrek to drop to 5.5, which inhibits the growth of *P. cocovenenans* but not *R. oryzae;* however, the use of these leaves have not been readily accepted by manufacturers of bongkrek (van Veen, 1966).

I. Bacillus subtilis

Bacillus subtilis is a straight or slightly curved cylindrical gram-positive rod with round ends. It is found singly and in chains. Spores are oval and centrally located. The organism is aerobic and motile, with peritrichic flagella.

Meyer (1951) hypothecated the cause of two outbreaks of gastroenteritis to members of the *B. mesentericus–B. subtilis* group, but the evidence is weak in both incidents. In the first outbreak large numbers of these organisms were found in incriminated fish which were reported to be very soft. Other common food poisoning organisms were not isolated. In the second outbreak both staphylococci and members of the *Bacillus* group were recovered. Nikodémusz *et al.* (1964*b*) attributed an outbreak to *B. subtilis viscosus.* One million of these organisms per gram were recovered from pickled fish obtained from the same lot incriminated in the outbreak.

Tong *et al.* (1962) reviewed an outbreak of gastroenteritis involving 161 people. Symptoms, in order of magnitude, were diarrhea, abdominal cramps, nausea, prostration, and vomiting. Fever was absent. The average incubation period was 10.7 hours, and the average duration of illness was 16 hours. Gram-positive, sporulating bacilli (1.2 × 10⁷ per gram) were isolated from the incriminated food, roast turkey. *Clostridium perfringens,* coagulase-positive straphylococci, and salmonellae were not recovered. Five of twelve stool specimens submitted by the patients also yielded small numbers of the same gram-positive bacillus. The organism was very similar to *B. subtilis,* except that good growth occurred when the organism was cultured under anaerobic conditions. During the preparation of the suspect meal, seven 24-pound turkeys were thawed in a refrigerator and under cold water. Since the cooking facilities could only hold three turkeys at a time, the birds were cooked a day or two in advance of the meal. Three turkeys were cooked at 204.4°C (400°F) for approximately 4 hours, and then the oven was cracked open until the next morning. The next day, a similar procedure was carried out on the other turkeys. Later the meat was refrigerated, sliced, and placed on a steam table. During the investigation, the investigators purchased two turkeys from the same wholesaler that supplied the incriminated birds and subjected them to the same precooking process. The turkeys were inoculated with *B. subtilis,* cooked, and held in a manner similar to the procedures used for preparing the banquet. This study showed that *B. subtilis* spores survived the heat treatment. In fact, the spores were recovered from an 18-pound turkey cooked at 176.7°C (350°F) for 8 hours. The study also demonstrated that the organisms grew rapidly at 43°C (the temperature of the steam table). *Bacillus subtilis* food poisoning is mentioned also in Chapter 4, Sections II,C, and III,D.

J. Actinomyces

One hour after a breakfast of hominy, canned fish roe, and corn muffins, two women developed diarrhea, vomiting, and abdominal cramps, and were prostrated (Chinn, 1939). The victims reported that the roe tasted unusual and burned their mouths. From a sample of the roe, secured a short time after the foods were consumed, large numbers of (branching, pleomorphic, budding forms) actinomyces-like organisms were recovered. Neither anaerobic growth nor chemicals were detected. The isolated actinomyces were killed after heating for 30 minutes at 65°C. Animal tests were performed. Rabbits lost from 10 to 15% weight, and diarrhea occurred in one for several days when fed a killed culture of the organism. A rabbit also died after inoculation with a killed culture, and another rabbit died after inoculation with a living culture. No reaction occurred when cats were fed cultures of this organism or when animals were fed sterile filtrates. The evidence for actinomyces as an etiological agent in this outbreak is weak (Tanner and Tanner, 1953).

K. Phospholipase C-Producing Bacteria

Lewis and MacFarlane (1953) demonstrated the formation of lecithinase C by *Clostridium oedematiens* and *Clostridium sordelli (bifermentans)*. This latter organism was reported to be responsible for an outbreak of food poisoning involving approximately 75 people (Duncan, 1944). Diarrhea occurred 6 to 7 hours after eating of a meat and potato pie which reportedly smelled or tasted so bad that only small portions were eaten. Vomiting was not reported. A sample of the liquid portion of the pie contained an almost pure culture of *C. bifermentans*. After the potatoes had been peeled, traces of soil remained on them. The potatoes were then hand-sliced. *Clostridium bifermentans* could be isolated at this stage. Partially cooked potatoes were added to the cooked meat, and the ingredients were covered with a layer of dough. The pies were cooked for 15 minutes and then placed in an insulated container where the food remained warm until it was served 6 to 7 hours later. Stools were not obtained in this investigation. *Clostridium bifermentans* food poisoning is mentioned also in Chapter 4, Sections I,A, II,B, and II,C.

MacGaughey and Chu (1948) and Colmer (1948) have shown that *Bacillus mycoides, Bacillus albolactis, Bacillus praussnitzii,* and certain strains of *Bacillus anthracis* also form phospholipase C. Nygren (1962) stated that probably many other bacteria possess the capacity to form phospholipases. If the hypothesis that phosphorylcholine, an end product of the hydrolysis of lecithin by phospholipase C, is the toxic factor in food poisoning caused by *Clostridium perfringens* and *Bacillus cereus,* then these other organisms producing phospholipase C may also be responsible for food-borne illness. Dack (1964a), however, cast doubt on this hypothesis when he reported that a volunteer in his laboratory consumed

100 mg and then 500 mg of phosphorylcholine with no ill effects. *Clostridium perfringens* and *B. cereus* food poisoning is discussed in detail in Chapter 4.

L. Candida

Candida albicans is an oval, budding, yeast-like fungus which produces both blastospores and pseudomycelium.

Species of *Candida* have repeatedly been found in the normal mouth, skin, food, and fecal material. In fact, they can be found in about 15% of feces from normal individuals. Suspicion has been cast on *Candida albicans* as a cause of intestinal conditions, but proof is lacking (Skinner, 1947). Floyd and Blakemore (1954) recovered *Candida albicans* from 28.9% of ready-to-eat meat samples obtained from butcher shops in Cairo, Egypt.

III. MILK-BORNE ORGANISMS

The organisms discussed in this category are often transmitted by means other than milk, but this category was established because of the historical importance of milk as a source of epidemics. Milk-borne outbreaks due to these organisms rarely occur now in technically developed countries. In many parts of the world, however, raw milk, raw milk products, and occasionally other foods are still important vehicles of these organisms.

Milk may become contaminated from one of three sources: the cow may have a bovine infection, and the causative ogranisms may reach the milk (brucellosis, tuberculosis); the cow may be infected directly or indirectly by human contamination of the udder (staphylococcal and streptococcal infections); or the milk may become contaminated after it leaves the cow (typhoid, diphtheria, streptococcal infections).

A. Brucella

Brucellae are small gram-negative coccobacilli occurring singly, in pairs, or in short chains. Brucellosis in man as well as in animals may be caused by any one of the three species of *Brucella*. In general, *Brucella abortus* is predominant in cattle, *Brucella suis* in swine, and *Brucella melitensis* in goats and sheep. Other animals (horses, mules, cats, dogs, chickens, rabbits, deers, and buffaloes) are also susceptible and thus may serve as reservoirs. Brucellosis is a classic example of a zoonosis, and human cases almost always result from contact with infected animals or from ingestion of their products.

1. EPIDEMIOLOGY

In the United States, close contact with an infected animal is a much more common route of transmission than is ingestion of a contaminated food; however, in most other parts of the world the reverse is true — goat's milk and fresh goat's cheese remain the major source of human brucellosis. Dalrymple-Champneys (1960) lists 74 countries that have creditably reported human brucellosis. As an example of the current situation in the United States, 411 human cases were reported during 1964. The downward trend in the number of cases of human brucellosis, apparent since 1947, has leveled off at just a little over 400 cases since 1962. In 1964, 47% of the cases with epidemiologic follow-up occurred in nonfarm, animal industry employees. Farmers accounted for 17%, and veterinarians, a small occupational group, accounted for 5% of the cases. The occupational risk is apparent. The majority of cases in which the source was determined were traced to swine. Seventy-five percent of the cases occurred in individuals, mostly males, between 20 and 50 years of age. There were 27 cases, mostly children, in which the illness was attributed to the consumption of raw milk (U. S. Public Health Service, 1965a).

Castaneda et al. (1942) attributed most brucellosis in Mexico to the consumption of products made from goat's milk. Later, Angelini (1946) reported that the principal source of brucellosis in Mexico City was cheese. The percentages that he attributed to contacts and vehicles were as follows: animals — 0.1%; patients — 5.5%; ice cream — 1.7%; un-pasteurized cream — 12.7%; raw milk — cow's 10.0%, goat's 7.3%; fresh cheese — goat's 42.7%, cow's 28.2%. Persons employed in domestic work (housekeepers and cooks) constituted the largest group of infected persons (44.6%). Angelini suggested that infection may have been acquired during procuring, preparing, or tasting raw foods.

2. SYMPTOMS

In general, the incubation period for brucellosis is from 1 to 3 weeks, although longer periods of time have been recorded (Hughes, 1897; Hardy et al., 1938). The onset may be sudden or gradual, depending on the type of illness. Symptoms include weakness, chills, sweats, anorexia, and aches and pains. Fever usually occurs. In the case of a drawn-out course of infection due to B. melitensis the temperature curve undulates — the reason for coining the term "undulant fever." However, this response is not generally observed in cases of brucellosis caused by *B. abortus* (Spink, 1956). A diurnal variation in temperature is usually seen—one with the temperature normal or slightly elevated in the morning and then rising in the late afternoon or evening. Weakness and disability are often prolonged for a period of years, and, although mortality is low, the loss in productive capacity of the afflicted is tremendous.

3. TRANSMISSION BY FOODS

On the Island of Malta in 1887, *Brucella melitensis* was first isolated by Bruce from spleens of fatal cases of Malta fever. To carry out investigations on the disease, experimental animals were needed. Goats, practically the only animal available on the island, were subjected to agglutination tests before studies were initiated. Strong agglutination reactions were observed in five of six goats obtained from two herds. Later it was discovered that about 40% of the 20,000 goats on the island had positive agglutination reactions. After the conclusion was reached that goats were the reservoir of infection, the Royal Society of London, Mediterranean Fever Commission (1905–1907), recommended that goat's milk and its products be prohibited in governmental establishments. As a result, when the recommendation was put into effect, the incidence of Malta fever (brucellosis) was markedly reduced in the military, but the incidence remained high in the civilian population. Since this classical observation, numerous reports of brucellosis outbreaks have been traced to raw milk supplies. Examples of these can be found in articles by Farbar and Mathews (1929), Simpson (1930), Hasseltine and Knight (1931), Cameron and Wells (1934), Hall and Learmonth (1934), Elkington *et al.* (1940), Steele and Hastings (1948), Magoffin *et al.* (1949), Anderson (1950), and Barrow and Peel (1965). In addition, outbreaks caused by *B. suis* were traced to cow's milk (Beattie and Rice, 1934; Horning, 1935; Borts *et al.,* 1943). Besides these outbreaks, the recovery of *Brucellae* from milk has been reported by several workers (Fabyan, 1913; McDiarmid, 1960; Stableforth *et al.,* 1960). Goat's cheese has been responsible for outbreaks due to *B. melitensis* (Fabian, 1947). However, no proved cases of human brucellosis due to *B. abortus* have been traced to cheese, ice cream, butter, or contaminated water (Spink, 1956).

Morales-Otero (1929, 1930) demonstrated in human volunteers that infection occurred more readily through the abraded skin than through the gastrointestinal tract, and that much larger doses of *B. abortus* were required to produce illness than with *B. melitensis*. Orr and Huddleson (1929) observed a group of 500 people exposed to *B. abortus* through their milk supply and found that only 1.4% gave evidence of infection.

Signs and symptoms of acute brucellosis appeared in 11 of 16 volunteers receiving a viable vaccine derived from *B. melitensis* and in 2 of 16 individuals receiving a viable vaccine derived from *B. abortus* (Spink *et al.*, 1962).

There is a possibility that brucellosis may be transmitted by ingestion of uncooked or insufficiently cooked meat, such as sausage; however, evidence to support this hypothesis is weak and indefinite. Löffler (1931) described a case of brucellosis that occurred in a slaughterhouse worker who ingested raw swine blood. Levine (1943) reported 17 cases of brucellosis in packing house workers; 9 had eaten partially cooked meat or sausages, but 7 denied any direct contact with meat. In some cases the meat had been grilled on steam radiators and was raw in the center. *Brucella abortus* was isolated from 3 workers and *B. suis* from the rest. In both of these reports the possibility of infection through skin cannot be completely excluded. A more incriminating incident, occurring in Argentina, was reported by Molinelli *et al.* (1948). Eight of ten persons contracted brucellosis after a picnic meal in which goat's meat was consumed half-cooked.

Dalrymple-Champneys (1960) stated that he knew of no evidence suggesting *Brucella* infection from meat in the United Kingdom. He also stated that food may become contaminated with *Brucella* by excreta, dust, and other fomites. The possibility of such a situation is intensified in countries where humans live in intimate contact with animals that harbor *Brucella;* however, during an investigation it would be almost impossible to determine the source of infection — whether it be manure, animal contact, contaminated food or milk, or any of other source.

In regard to transmission of brucellosis by the oral route, the Joint FAO/WHO Expert Committee on Brucellosis (1951) stated: "Viable brucella may be present in the viscera, flesh, and lymph-nodes of infected carcasses for periods in excess of one month. ... Human cases of disease have on occasion been traced to meats and meat products which were not properly cooked. Wild greens used in salad, contaminated with sheep and goat urine and faeces, have been found to produce human infection."

4. SURVIVAL

According to Horrocks (1905, 1906), *B. melitensis* remain viable in sterile tap water for 37 days, in sterile sea water for 25 days, in dried soil for 43 days, in damp soil for 72 days, in unsterilized garden soil containing manure for 20 days, in sterile milk for 17 days, on cloth and fabrics for 5 to 78 days, and exposed to the sun, for only a few hours. *Brucella melitensis* in ice cream and in goat's cheeses was not destroyed when the foods were processed (Eyre *et al.,* 1907). Stiles (1945) recovered *B. melitensis* from cheese manufactured from unpasteurized goat's milk. *Brucella abortus* can be shed in milk by healthy-appearing cattle for months or years (Fabyan, 1912; U.S. Bureau of Animal Industry, 1912). The vast majority of the organisms were in the cream fraction (Carpenter and Boak, 1928; Fitch and Bishop, 1933). These workers were able to recover *B. abortus* from inoculated butter for up to 142 days. Fitch and Bishop (1933) reported the isolation of *B. abortus* from butter, buttermilk, and ice cream that had been prepared from naturally contaminated milk. During the same year, Thompson (1933) reported the isolation of *B. abortus* from ice cream after one month's storage below freezing. Kuzdas and Morse (1954) observed that *B. abortus* remained viable in tap water for 10 days at 25°C and for 57 days at 8°C, in manure for 385 days at 8°C, and in unpasteurized milk for a few hours at 37°C and for 48 hours at 8°C. Freezing or near freezing permitted survival for periods over 824 days. Kuzdas and Morse showed also that changes in pH and variations in temperature greatly affected the survival of the organisms. Viable *B. melitensis* have been isolated from hams after curing in covered pickle brine for 21 days, but no recoveries were made after the smoking process. *Brucella suis* was recovered from hog carcasses after 21 days of refrigeration (Hutchings *et al.,* 1951). According to Pagnini (1962), *B. abortus* survived in sausage for 175 days.

Gilman *et al.* (1946) found that approximately 1500 *B. abortus* per gram survived for 6 months in cheddar cheese but not for a year when stored at 4.4°C. When milk taken from reacting cows and containing 700 to 800 *B. abortus* cells per milliliter was made into cheddar cheese, all the cheeses were positive after 3 months, some after 6 months, but none after a year. However, as a result of conducting studies on commercial cheese, some of which was made from milk contaminated with *B. abortus,* Gilman *et al.* (1946) failed to recover *Brucella* when the cheese was tested from 41 to 85 days after manufacture. They concluded that an aging period of 60 days was reasonable assurance against the presence of viable *B. abortus* in commercially prepared cheddar cheese. Stiles (1945) isolated *B. melitensis* from 16 specimens of cheese (feta, yellow cream, and Romano) made from raw goat's milk from infected herds. The age of the specimens of cheese was estimated to be between 38 and 100 days. Harris (1950), after reviewing the literature on cheese-borne infections and *Brucella* survival studies, suggested that cheese should be (1) prepared from pasteurized milk, (2) subjected to heat in the pasteurization range in its manufacture, or (3) aged for 6 months or more. *Brucella* succumbs in 1 hour at 55°C and in 3 minutes at 62° to 63°C (Sartwell, 1965).

B. Mycobacteria

Mycobacteria are gram-positive, slender rods, although other forms (short swollen rods, granular, beaded, filamentous, and branching) have been described. The organisms are nonmotile, aerobic, and produce no endospores. They have a unique resistance to drying, to chemical disinfectants, and to other adverse environmental influences because of their waxy or fatty composition. Mycobacteria are acid-fast; that is, having once been stained, they resist subsequent decolorization by mineral acids. Growth of mycobacteria is relatively slow compared to the more common bacterial species. The generation time for *Mycobacteria tuberculosis,* for instance, is 18 hours or longer; but the time for other species such as *Mycobacteria avium* and certain atypical strains may be somewhat less. Tuberculosis is generally transmitted by droplet nuclei, but there is record of milk-borne transmission of *Mycobacteria bovis.*

1. BOVINE TUBERCULOSIS

Mycobacteria bovis is as dangerous to man as is *M. tuberculosis,* but alimentary infection is less dangerous than inhalation infection (Torning, 1965). The probability of transmission of milk-borne tuberculosis is influenced by the incidence of infection in cows, by the presence and quantity of *M. bovis* in milk, by the milk-drinking habits of the populace, and by the application of the practice of pasteurizing or boiling milk. Tubercle bacilli may reach milk from infected udders, from animal or human respiratory contamination, or from feces. Dried fecal material clinging to the udder, tail, and flanks of cows are a major source of contamination of milk (Steele and Ranney, 1958).

Alimentary infection with tubercle bacilli may cause bone and joint tuberculosis or lymph node enlargement in the cervical or the mesenteric region. Several investigators have reported that milk was responsible for the transmission of tuberculosis (Kober, 1903; Park and Krumwiede, 1910; Fraser, 1912; Mitchell, 1914; Park, 1927; Boer, 1933; Chang, 1933; Griffith and Summers, 1933; Price, 1934, 1938; Griffith and Smith, 1935; Munro and Walker, 1935; Griffith, 1937; Mushatt, 1940). As an example of a milk-borne epidemic of tuberculosis, involving a total of 56 infected persons, Stahl (1939) found 29 positive reactors among a group of 32 children who had drunk raw milk from a single dairy. At the dairy there was a cow that had clinical mastitis. This cow's milk contained tubercle bacilli. Postmortem examination of the cow showed recent tuberculosis in one lung and pleura. Tuberculosis foci were also observed in a number of mesenteric nodes and in the udder. In a group of 102 children who had not consumed this cow's milk, only eight reactors were found. Later, van Zwanenberg *et al.* (1956) reported seven cases of tuberculosis of the cervical lymph nodes. Twelve additional cases were found during a tuberculin survey in the same community. All the cases gave a history of having drunk raw milk from the same dairy. Two relatively recent incidents of milk-borne tuberculosis in England are cited by Black and Sutherland (1961). In all these reports, strong circumstantial evidence incriminates milk, but proof in specific cases is rare.

In the United States, incidents of alimentary tuberculosis have not occurred during the past few decades. However, in certain other countries such reports are more common. Birkenfeld (1955) stated that, in 1952, 40,000 inhabitants of Germany had tuberculosis of the bovine type; more than 1000 of these illnesses resulted in death. In Egypt, Ibrahim *et al.* (1955) found that 12% of 147 tuberculosis patients were infected with *M. bovis*. These represented cases of pulmonary, cervical, and bone and joint tuberculosis. In France, Verge and Paraf (1956) reported that 40% of the *M. bovis* infection gave rise to tuberculosis of cervical lymph nodes, 36% to abdominal tuberculosis, and 11% to pulmonary tuberculosis.

There is little evidence of the transmission of tuberculosis from meat or other foods. M'Fadyean (1890) found that, when meat from tuberculous animals was fed to rabbits and guinea pigs, some of the animals developed tuberculosis. In an abattoir, Lillengen (1945) found an average of 71% of utensils, 100% of wiping cloths, and 68% of floor samples after cold water flushing to be contaminated with tubercle bacilli (total number of samples not specified). Brown and Petroff (1919) reported that flies which fed on tuberculous sputum became contaminated with tubercle bacilli. These contaminated flies were allowed to crawl over food. However, when susceptible guinea pigs consumed this food, they did not develop tuberculosis. Shrader (1939) felt that there may be some danger from eating certain organs such as liver, lungs, mesentery, udder, intestines, and other parts that are often the seats of serious infection and so may contain massive numbers of tubercle bacilli. However, ingestion of meat is not the normal route for transmission of tuberculosis — apparently only heavily contaminated tissues are a risk. Francis (1958) gives a comprehensive review of bovine tuberculosis.

Veterinary inspection, pasteurization of milk, and tuberculosis eradication in animals have done much to reduce bovine tuberculosis in many parts of the world. The WHO Expert Committee on Tuberculosis (1960) summarized the current status of this disease by stating: "Bovine tuberculosis is probably not of great importance at present because in most parts of the world milk is routinely boiled before consumption, and in other areas pasteurization of milk and control of bovine tuberculosis infection have been put into extensive practice. However, in countries where tuberculosis among cattle is a problem, it is also a menace to man, and in these countries complete control of tuberculosis cannot be achieved unless attention is also paid to the reduction, or preferably eradication, of tuberculosis in cattle."

2. OTHER MYCOBACTERIA

Human tuberculosis or diseases very similar to tuberculosis are caused not only by *M. tuberculosis, M. bovis,* and (rarely) *M. avium,* but also by other kinds of mycobacteria. Reports of progressive pulmonary tuberculosis in man caused by a group of slow-growing, nonchromogenic, acid-fast bacilli which differ from *M. tuberculosis* are cited by Crow *et al.* (1957) and Scammon *et al.* (1963). These organisms resemble the avian type and are classified as group III (nonphotochromogens, Battey type) by Runyon (1959). Group I (Runyon) are the photochromogenic *Mycobacteria kansasii,* group II are the scotochromogens, and group IV are rapid growers. All may be pathogenic. Neither the source of these organisms nor the probability of their being food-borne agents of tuberculosis-like illness is known; however, certain relationships have been established. Aspiration of food or lipoidal material has occurred in some but not all cases of group IV infections. Representatives of the group III organisms have been isolated from swine by Scammon *et al.* (1963). Chapman *et al.* (1965) found representatives of Runyon's groups II, III, and IV in 261 of 770 samples of raw milk from tank trucks. The source of these organisms in milk was not determined, but study is being continued. Certain group III organisms have been shown by Scammon *et al.* (1963) and by Chapman *et al.* (1965) to survive 60°C for 30 minutes. Harrington and Karlson (1965) heated tubes containing 5 ml of skim milk inoculated with millions of cells of one of 195 strains of mycobacteria at 62.8°C for 30 minutes and at 71.7°C for 15 seconds. Small numbers of a few strains of the unclassified mycobacteria survived the lower temperature process and less regularly the higher temperature process. However, the survivors failed to resist pasteurization on subsequent trials. Moureau *et al.* (1960) found that paratuberculosis strains sometimes resisted temperatures of 90°C.

Kubica *et al.* (1963) recovered representatives of Runyon groups of mycobacteria from soil, but these organisms differed slightly in biochemical properties from those strains isolated from disease processes of man. Nevertheless, these soil-borne organisms have the potential for contaminating root and leafy vegetables.

C. Beta-Hemolytic Streptococci

The streptococcus group includes all bacteria which are spherical and occur in chains of various lengths. All species are nonsporulating and gram-positive. They have microaerophilic preferences. The streptococci are divided into groups on the basis of a precipitation reaction that uses an integral part of the cell as the antigen. This antigen is a polysaccharide substance known as C substance. Group A, which is typified by *Streptococcus pyogenes,* is composed of the more virulent human strains. This group of streptococci can be further divided into over 40 antigenic types by type-specific protein antigens (M, T, and R).

Streptococcus pyogenes produces deep colonies in blood agar plates, and a wide zone of complete hemolysis can be observed around the sharply differentiated margins of the colony. Two soluble hemolysins are produced by *S. pyogenes.* One, streptolysin O, is oxygen-labile and thus detected only under anaerobic conditions; the other, streptolysin S, is not oxygen-labile and is responsible for the lysis that occurs on the surface blood agar plates. Some strains of *S. pyogenes* produce an erythrogenic toxin which is responsible for the characteristic skin lesions of scarlet fever. *Streptococcus pyogenes* is the causative agent of septic sore throat and scarlet fever as well as a number of other pyogenic and septicemic infections in man and animals. This organism is also a common cause of mastitis in dairy cattle.

Outbreaks of scarlet fever and septic sore throat were numerous prior to the advent of milk pasteurization. In the United States, these outbreaks have been reviewed by Trask (1908), Scamman (1929), White (1929), Davis (1932), Stebbins *et al.* (1937), Frank (1940), Fuchs (1941), Dublin *et al.* (1943), Getting *et al.* (1944), Evans (1946), Dauer (1952), and Tanner and Tanner (1953); in England by Wilson (1933); and in Canada by Murray (1936). Several outbreaks were traced to the diseased udder of a cow. In some cases it was presumed that udders became infected from human sources. In at least one case milk was infected by a human carrier. The milk was bottled and capped by a worker who was suffering from an acute sore throat. One of the largest milk-borne epidemics, involving over 10,000 cases, occurred in Chicago and was described by Capps and Miller (1912). Eighty-seven percent of a group of 622 patients that were interviewed were consumers of "pasteurized" milk from a single dairy. At the same time as the human outbreak, mastitis occurred in a herd of cows which supplied milk to the incriminated dairy. Sore throats were prevalent among farmers and milkers that supplied milk to the dairy. Pasteurization records revealed a remarkable degree of failure to maintain proper temperatures. During this epidemic period Davis (1912) recovered *S. pyogenes* from the milk. Outbreaks involving dried milk powder have been reported by Allen and Baer (1944) and by Purvis and Morris (1946). Canned milk which was diluted with water the evening before serving was responsible for three outbreaks involving 835 cases of tonsillitis and scarlet fever (Taylor and McDonald, 1959).

Outbreaks of scarlet fever or septic sore throat transmitted by food agents other than milk have been described by several investigators. Scamman *et al.* (1927) incriminated lobster salad, Bloomfield and Rantz (1943) an undetermined food, but milk was ruled out, Getting *et al.* (1943) ham, Goodale and Lambrakis (1944) potato salad, Commission on Acute Respiratory Diseases (1945) eggs, Reekie (1948) ice cream café, Dauer (1953) stew, Boissard and Fry (1955) custard, Dauer and Sylvester (1956) egg salad, Dauer (1958) egg salad, Farber and Korff (1958) egg salad, Ottie and Ritzerfeld (1960) rice pudding, and Mosher *et al.* (1966) shrimp salad. The causative agent of the outbreaks was *S. pyogenes*, and in all cases the foodstuffs were allowed to stand at room temperature for several hours between their preparation and consumption.

In the outbreak reviewed by Getting *et al.* (1943) and Foley *et al.* (1943), group A, Griffin type 2, hemolytic streptococci were recovered from patients, from ground ham, and from women who prepared the ham. Gastrointestinal symptoms as well as typical symptoms associated with scarlet fever and septic sore throat were reported by patients. Of the people who ate some of the ham immediately after it was cooked, none became ill; but of 116 who ate the ground ham 36 hours later, 82 became ill. Of 20 who consumed leftover ham, 19 also became ill. Large, 18- to 20-pound hams were cooked by two women in their homes for a church luncheon. One of these women was in the pre-eruptive stage of scarlet fever, and it was believed that she contaminated the ham. The hams were cooked for 4 hours and allowed to stand in water for 12 hours. The next day the hams were sliced, boned, ground, and made into sandwiches. After standing overnight at room temperature, the sandwiches were served late in the afternoon on the next day. Thus, there appeared to be ample opportunity for contamination as well as for bacterial multiplication.

At Fort Bragg, North Carolina, an extensive study was made of a foodborne epidemic of tonsillitis and pharyngitis caused by beta-hemolytic streptococcus type 5 (Commission on Acute Respiratory Diseases, 1945). The situation leading to the outbreak was that boiled eggs were sliced by hand and held at room temperature 10 hours before serving. This outbreak, which hospitalized 100 patients, was characterized by an explosive onset with a median incubation period of 38 hours (12 to 72 hours). A typical single-source, common-event epidemic pattern was noted. The primary attack rate was 42%, and the secondary attack rate was 30% (one-half of this latter group were cases, and the other half were carriers). The reported symptoms were: sore throat 100%, feverishness 99%, malaise 80%, hoarseness 53%, nasal symptoms 69%, cough 47%, epistaxis 37%, and earache 30%.

D. Corynebacteria

Corynebacterium diphtheriae is a gram-positive, nonmotile, non-spore forming, and noncapsulated bacillus possessing meta-chromatic granules. Slightly branching forms and club-shaped cells are also found. The organisms are usually arranged in a palisade. Three fairly stable forms are known: gravis, intermedius, and mitis. Toxigenic strains of *C. diphtheriae* carry a lysogenic phage (Freeman, 1951; Barksdale *et al.*, 1960).

The diphtheria organism is essentially an obligate parasite of man. Its source is a clinical case, a convalescent carrier, or a person with a sub-clinical infection. Since the usual habitat of the *C. diphtheriae* is the upper respiratory tract, droplet- and air-borne infection are the most common modes of transmission. Hands and fomites may also play a role in trans-mission. Diphtheria epidemics have been epidemiologically associated with raw milk supplies and have been summarized by Trask (1908), McCoy *et al.* (1917), Armstrong and Parron (1927), and Goldie and Maddock (1943). In the United States, during the years 1919–1948 there were only 11 milk-borne epidemics on record and none thereafter (Hull, 1963). In these outbreaks suspected milk came from herds where cows occasionally were found to have a superficial ulcer on the udder or from farms where dairy workers were carriers or were suffering from diphtheria. *Corynebacterium diphtheriae* were recovered from milk (Trask, 1908; Marshall, 1907; McSweeney and Morgan, 1928) and were shown to proliferate in milk (Eyre, 1899). McCoy *et al.* (1917) and Bloch (1938) reported outbreaks of diphtheria associated with ice cream, and Bolton (1918) found that *C. diphtheriae* could survive freezing and frozen storage for at least 5 days and still be pathogenic for guinea pigs.

During the investigation of four cases of diphtheria in a South African village, Pfeiffer and Viljoen (1945) discovered diphtheritic mastitis in two cows in the area. The combined efforts of an immunization program, slaughter of the infected cows, and advice to the residents to boil their milk prevented subsequent occurrences of cases of diphtheria.

MacKenzie (1947) reported a milk-borne outbreak of diphtheria involving 35 children in a Melbourne orphanage and a few cases in nearby towns. One dairy farmer supplied milk to four dairies; three of these dairies supplied raw milk to all the cases, including those at the orphanage. The farmer's wife, who helped with the milking, had a small ulcer just inside a nostril. The ulcer, which would occasionally break out and discharge for a few days, had been present for 8 years since she had had diphtheria. *Corynebacterium diphtheriae* was isolated from the ulcer.

IV. ORGANISMS USUALLY TRANSMITTED BY MEANS OTHER THAN FOOD

The following organisms are usually transmitted by means other than the ingestion of foods. Some are important food-borne pathogens; others have been conveyed by foods only on rare occasions.

A. *Shigella*

Shigellae are members of the family Enterobacteriaceae. The genus *Shigella* bears a resemblance to *Escherichia* but differs in that members of the former genus are nonmotile, are anaerogenic, and do not decarboxylate lysine. Edwards and Ewing (1962) discuss the biochemical and serological characteristics of the genus. The *Shigella* group is classified into four subgroups: A, *Shigella dysenteriae;* B, *Shigella flexneri;* C, *Shigella boydii;* and D, *Shigella sonnei*. Organisms formerly described as *Bacillus* or *Shigella alkalescens* and *dispar* now constitute the alkalescens-dispar group which are variants of *Escherichia coli*.

The disease produced by shigellae is known as bacillary dysentery or shigellosis. It is characterized by an incubation period of 1 to 7 days (usually less than 4 days), by symptoms of diarrhea and fever, and often by cramps, vomiting, and tenesmus. Stools may contain blood, mucus, or pus (Gordon, 1965).

1. EPIDEMIOLOGY

Outbreaks of food-borne shigellosis are relatively infrequent compared to those of salmonellosis, but the extent of the problem is poorly defined. Careful epidemiological investigation in food-borne outbreaks of shigellosis usually reveals recent diarrheal illness among food handlers, and frequently shigellae are recovered from these workers (Dauer, 1952–

1961). Food-borne shigellosis is characterized by high attack rates, common-source epidemic patterns, and short incubation periods (7 to 36 hours).

In the United States, since the initiation of the shigella surveillance program in 1964, enough data have been accumulated to give an approximation of the relative frequencies of the more common serotypes. *Shigella sonnei* (39%) was the most frequently isolated serotype. In order of frequency, *S. flexneri* 2a, 2b, 3a, 6, and 4a were the next most common serotypes recovered. These six serotypes accounted for about 90% of all isolations (U.S. Public Health Service, 1965b). These figures include all isolations from institutional outbreaks and differ from those collected by the Enteric Bacteriology Unit, Communicable Disease Center (Ewing, 1966). According to the data of the Enteric Bacteriology Unit, when only foci of infections are considered, *S. flexneri* 2a is the most common serotype in the United States as well as the world over (with few exceptions). A number of *S. flexneri* serotypes as well as *S. boydii* and *S. dysenteriae* serotypes have rarely been recovered during the surveillance period. According to the surveillance reports, *S. flexneri* serotypes are the dominant reported shigellae in Southern United States. About equal numbers of *S. flexneri* and *S. sonnei* are found in the North. The seasonal distribution shows an increase beginning in May or June and a peak in September or October, at which time the isolations were approximately double that of the winter isolations. Since many *Shigella* isolations are reported a month or two after the onset of illness, the actual seasonal pattern may be shifted back by this interval (U.S. Public Health Service, 1965b). As in the case of salmonellosis, approximately 70% of the isolations are made from infants and children under ten years of age. Animal isolations are rare; however, recoveries have been made from other primates and a guinea pig. Other nonhuman isolations have also been made from frozen egg albumen, checked eggs, cattle feed containing turkey droppings and feathers, and a farm environment. These no doubt represent human contamination.

In England and Wales, *S. sonnei* infection is very common; 20,000 to 30,000 cases are reported annually. Other *Shigella* serotypes are rare. Twenty years ago *S. flexneri* was the serotype most frequently encountered, but since that time there has been a large increase in *S.sonnei*. The pattern of *S. sonnei* infection before 1945 suggested that most illnesses were essentially food-borne in origin; however, since that time the incidence of infection has greatly increased, but few proved food-borne outbreaks have been reported (Taylor, 1957, 1960).

2. FOOD-BORNE OUTBREAKS

Since the advent of the shigella surveillance program in the United States, a few outbreaks of food-borne shigellosis have been reported. Beans were incriminated in an outbreak involving 40 cases. The beans had been kept at ambient temperature for several hours before serving and were reported to be atypically gray and bubbling. Among the 40 cases, the median incubation period was 7¾ hours. The illness lasted more than one month in some persons. *Shigella flexneri* 1a was the responsible serotype (Foster, 1964). Three outbreaks were attributed to potato salad (Condit, 1964; Lyons and Matsuura, 1965; Legters, 1965) and one to tuna salad (McCroan and Walker, 1965). In all cases infected food handlers were found, and inadequate refrigeration of food prevailed. Two institutional outbreaks were reported which gave a common-source epidemic pattern; however, the exact food items were not identified (Anderson, 1965; Gratch, 1965). Berry (1966) reported a common-source epidemic of *S. flexneri* 2a which was attributed to macaroni salad or rice, or both. An epidemic of shigellosis and pharyngitis involving over 250 college students and cafeteria employees was reported by Mosher *et al.* (1966). Two pathogens, beta-hemolytic streptococci and *S. flexneri* (serotype not reported), were responsible. Food histories incriminated shrimp salad. Frozen shrimp was thawed in warm water the night before preparation, and the next morning shrimp salad was prepared and mixed by hand. The salad was refrigerated in large containers, and it was determined that the temperature just beneath the surface layer of the salad was 18°C (65°F) or higher. Enteropathogenic and indicator organisms were recovered from the shrimp salad and from frozen shrimp of the same brand used by the cafeteria, but no *S. flexneri* were found. Fifteen (12%) of the food handlers of the cafeteria harbored *S. flexneri*.

Sporadic reports of food-borne shigellosis appear in the literature. Kinloch (1922) incriminated milk, Bowes (1938) milk, Faulds (1943) milk, Savage (1949) milk, Gorman (1950) stewed apples, Stott *et al.* (1953) probably cress, Tucker *et al.* (1954) improperly pasteurized milk, Blair (1956) sour milk, Ehrenkranz *et al.* (1958) turkey salad, Dauer and Davids (1959) tossed salad, and Kaiser and Williams (1962) potato salad. The incubation period medians ranged from 7 hours (Blair, 1956) to 36 hours (Kaiser and Williams, 1962). In reviewing outbreaks of shigellosis involving 20 to 14,000 persons aboard ships in World War II, Meyers (1946) incriminated food handlers, contaminated foods eaten ashore, and the use of polluted sea water to wash vegetable-peeling machines. Back-siphonage into the water supply may have also accounted for some of the morbidity.

3. TRANSMISSION

Shigellae are transmitted by means of personal contact, fomites, flies, and water, as well as by food. The ease with which shigellae are spread from person to person by contact suggests that the minimum infective dose must be low, although there is little quantitative evidence other than laboratory accidents in support of this statement. Studies on feeding man and monkeys have been unfruitful.

When Shaughnessy et al. (1946) fed five strains of S. flexneri to volunteer prisoners, illness resulted from three strains after 1 to 2 billion cells were consumed (mild variable symptoms), from another strain after the ingestion of from 100 million to 10 billion cells, and from the ingestion of 625 million cells of a mixture of four strains. In this study, when the dosage of living organisms was sufficient to produce moderate or severe dysentery, symptoms of nausea and vomiting or cramps began 12 to 24 hours after the challenge of the organism. Diarrhea commenced 18 to 24 hours after ingestion. When mild or questionable dysentery resulted, the incubation period was longer, 36 to 48 hours. In evaluating these results, it must be kept in mind that the group studied were mature and healthy men from environments with the possibility of high exposure rates. Thus, because of the group studied and the manner in which the experiment was conducted, this study probably has no bearing on natural infection.

Human carriers are frequently encountered when outbreaks are investigated. Cruickshank and Swyer (1940) and Watt et al. (1942) observed that S. sonnei persisted in feces of convalescents for as long as 10 weeks. In a survey of the spread of shigellosis from primary cases to family members, Ross (1957) noted only a few secondary infections, although several convalescents remained positive. Watt and Hardy (1945) observed in the 1930's and 1940's that the shigella infection rate was 3% in Georgia, 11% in New Mexico, 0.1% in New York, and 4% in Puerto Rico. However, these rates may have changed in subsequent years. Separate surveys in English children under 5 years of age have shown the carrier rate for Shigella to be 0.4% (Spicer, 1959a) and 2.42% (Galbraith, 1965), respectively.

4. RESISTANCE OF THE ORGANISM

Gorman (1950) observed that S. sonnei survived in stewed apples (pH 3.2) for a week at 20°C but died off in 24 hours at 37°C. Spicer (1959b) found that a fraction of S. sonnei could survive for 7 to 10 days on cotton thread under favorable conditions of relative humidity and temperature. This same organism was shown to survive in cheese stored at 4°C for 19 to 72 days (Kubinstein, 1964).

In naturally infected feces kept alkaline and prevented from drying, shigellae remained alive for days, but in feces that were allowed to become acid through growth of other bacteria, shigellae perished — often in a few hours (Wilson and Miles, 1964). The main factors influencing survival are temperature, pH, and other bacteria.

Shigella flexneri 2a survived in feces for approximately 120 days when held at −20°C, 60 days at 0.5°C, and 12 days at 25°C; a *S. sonnei* strain survived at −20°C for 60 days and at 0.5°C and 25°C for over 30 days (Nakamura and Taylor, 1965). At 25°C, a *S. sonnei* strain and *S. flexneri* 2a persisted in flour and milk for over 170 days; in eggs, clams, and shrimp for over 50 days; in oysters for over 30 days; in egg whites for over 10 days; and in orange juice, tomato juice, cooking oil, root beer, and ginger ale for relatively short periods. Longer survival often occurred at lower temperatures, −20°C and 0.5°C (Taylor and Nakamura, 1964).

B. *Vibrio cholerae*

Vibrio cholerae is a short, curved, gram-negative rod which occurs singly or in chains of short spiral or S-shaped form. In old cultures these bacteria are thickened and irregularly clubbed; granular shapes and spiral filaments also occur. *Vibrio cholerae* is actively motile by means of one polar flagellum. The organism is aerobic and grows best in an alkaline medium. A hemolytic variety of *V. cholerae* is known as *V. cholerae* biotype *El Tor,* or El Tor vibrio. Serologically the organisms are identical, but there are slight biochemical differences (De, 1961).

Pathologically, cholera is essentially a local disease characterized by an acute enteritis. The vibrios grow on the epithelial cells lining the mucosa and reach enormous numbers. Following an incubation period of 2 to 3 days, a purging, nonoffensive diarrhea occurs. This consists of a continuous discharge of pints of watery fluid containing mucus. As a result, dehydration is marked. Mesenteric lymph nodes and blood are not infected. It is generally assumed that the organisms liberate an endotoxin on autolysis which is responsible for the irritating effect on the intestinal tract; however, the toxin has not been demonstrated. According to Gangarosa *et al.* (1960), the chief constituent of the rice-water stools is mucus; the epithelial desquamation is·a postmorten phenomenon.

1. Epidemiology

Between 1817 and 1923 six pandemics of cholera occurred; however, since this latter date the disease has been restricted to India and other countries in the East. Cholera has been endemic in the Indo-Pakistan subcontinent for centuries. The Ganges basin is the seed bed of cholera. The incidence of the disease is apparently on the decrease (Pollitzer, 1959). This decline has been ascribed to improvements in environmental sanitation. Generally, cholera is a highly communicable disease which is transmitted primarily by person-to-person contact; under such circumstances the attack rate is low and the disease appears sporadically. But when a vehicle such as water or food is heavily contaminated, explosive epidemics may occur. There is strong presumptive evidence that the degree of clinical illness resulting is directly related to the size of the inoculum. Any food ordinarily consumed in the raw state can serve as a vehicle. Pollitzer (1959) stated that food and drink may become dangerous vehicles of cholera infection in four ways: (1) The use of fresh "night soil" as fertilizer for vegetables, such as lettuce, which are either consumed raw or inadequately cooked. Support of this premise is given by Takano et al. (1926), although Teng (1965) felt that it was doubtful that "night soil" played a significant role in the spread of the El Tor vibrio. (2) The use of contaminated water in cold drinks, such as lemonades, and in foods, such as salads; or the use of contaminated water for cleaning fruits and vegetables which are consumed without cooking. (3) Although the validity is doubtful and the reports rare, claims have been made that cholera patients handling foodstuffs may transmit the organisms. (4) Foods which are consumed without further cooking may be contaminated by storing them in contaminated containers, by sprinkling or freshening them with contaminated water, or by exposing them to flies. In addition, Pollitzer (1959) cited that fish and shellfish can become infected from the water in which they live, and that once the organisms have gained entry in or on fish they are apt to survive for a considerable length of time and may possibly multiply. Raw fish forms a staple in the diet of people from many countries, and shellfish are commonly consumed raw. Felsenfeld (1965a) pointed out that there are many food handlers and housewives among cholera victims, but workers dealing with "night soil" seldom become infected. Teng (1965) cited two outbreaks that were related to spread from restaurants. In the first outbreak, water from a well was polluted by one of the employees who was excreting the El Tor biotype. This water was consumed by the patients only after it had been boiled for the preparation of tea; thus, foods appeared to be a more feasible direct source than water. The water, however, could have contaminated foodstuffs and kitchen utensils. In the second outbreak, 32% of the asymptomatic restaurant staff were found to be excreting vibrios. As previously mentioned, the role of carriers in the transmission of cholera is ill-defined. Traditionally, the case, especially when in the incubation period, has been considered the most important source of infection. Gilmour (1952) examined 113 convalescent cases and found that 71.6% were negative after the first week, 89.3% after the second week, and 98.1% after 3 weeks. The maximum period of intermittent excretion was 25 days. Similar results have been demonstrated in other studies. Dizon (1965) noted that a few convalescent carriers continued to excrete El Tor vibrios for months, one for as long as 15 months.

The four cardinal principles of control as described by MacKenzie (1965) are to isolate, treat, and render noninfectious the index case; to place under surveillance and appropriate management the contacts of the index case; to apply, where relevant, measures of active immunization on as comprehensive a community basis as possible; and to enforce environmental sanitation requirements which will protect common sources of water and foodstuffs from contamination.

2. ROLE OF CONTAMINATED FOOD IN CAUSING ILLNESS

There have been several outbreaks of cholera in which epidemiological evidence has implicated food, although the vibrios causing the illness were not recovered from the incriminated foodstuffs.

Donitz (1886) described a cholera outbreak which occurred in Tokyo. The disease first raged in a coastal village through which most of the fish destined for Tokyo passed. Fishermen delivering the fish to Tokyo developed the disease en route. Then the epidemic struck Tokyo. A similar occurrence was reviewed by Takano et al. (1926). Severe cases of gastroenteritis occurred in a seaport prior to the distribution of fish from that area to Tokyo. An epidemic of cholera subsequently occurred in Tokyo. The epidemic was completely checked within two weeks after fish from this port was denied entry into Tokyo. Pollitzer (1959) reviewed two epidemics which resulted from oysters that were considered unwholesome or were obtained from beds contaminated by sewage. He also discussed Philippine and Indian epidemics, in which fish, shellfish, and sun-dried fish may have been involved. On the basis of investigations of Krishnan (1953) and Pillay et al. (1954), Pollitzer (1959) stated that so far there is no convincing evidence to support the hypothesis of Pandit and Hora (1951) that fish plays a particularly important role as reservoirs of infection by maintaining cholera endemicity. Joseph (1962) investigated a Philippine outbreak of cholera El Tor in which a heavy distribution of cases occurred along a seacoast road. There was a history of consumption of raw seafood or dried fish, particularly among the early cases; however, no one food or beverage was shown to have accounted for the majority of the cases. An explosive outbreak of cholera El Tor, portraying a common-source epidemic pattern, was described by Dizon (1962). From interviews of 85 cases, it became apparent that a large proportion of the cases had eaten small raw shrimps (nipons). These shrimp were caught in shallow water during low tide in the vicinity of the cities where the outbreak occurred and were sold on the same day. Untreated sewage from one city flowed directly into the sea. Although the El Tor vibrios were not recovered from the shrimp, the investigator suggested that the infection could have been spread from city to city by sea currents or by migration of contaminated shrimp.

Sticker (1912) reported that Hankin (1898) incriminated cucumbers as the vehicle of some cholera cases in India. *Vibrio cholerae* was isolated from the cucumbers. Pollitzer (1959) mentioned that in China cut melons have played a role in the transmission of cholera.

3. SURVIVAL IN FOODS

The resistance of the cholera vibrios to environmental influences is not great; however, these organisms may survive in foods long enough to transmit cholera to susceptible individuals. The following survival periods in food and water are largely based on laboratory experiments, and thus they should be viewed with caution. In nature it is apparently rare to get positive results from samples of water or foodstuffs which can be related to infection.

Pollitzer (1959) and Felsenfeld (1965a, b) reviewed the literature on the survival of *V. cholerae* in foods. Studies have shown that these organisms may survive for a week or more on meat and meat products (Uffelman, 1892; Takano *et al.,* 1926; Lal and Yacob, 1926).

Vibrio cholerae can survive on fish for a few days. Longer periods were noted when fish was stored at low temperatures (Uffelman, 1892; Friedrich, 1893; Takano *et al.,* 1926; Pesigan, 1965). Clams and oysters kept in contaminated sea water rapidly took up the vibrio in their gastrointestinal tract. Under natural conditions, clams and oysters rapidly purged themselves of contamination when under the influence of unpolluted water; thus a continuum of *V. cholerae* to water is essential for the recovery of the vibrios from shellfish. Cholera vibrios survived in the shellfish for 1½ months at 0° to 5°C and for 15 to 20 days at 22°C. Vibrios inoculated on shelled oysters increased to a maximum after 3 days at 20°C, and disappeared in a week. On boiled oysters and clams *V. cholerae* survived for 20 days (Pollitzer, 1959). In frozen shrimp and fish the *El Tor* biotype survived for more than 3 weeks (Pesigan, 1965).

Kitasato (1889) was reported by Pollitzer (1959) as stating that *V. cholerae* survived in milk until the milk became strongly acid, 1½ to 3 days in raw milk, and 2 to 3 weeks in sterile milk (variation depended on storage temperature as well as pH). Survival of vibrio added to sour milk was only 1 hour (Shousha, 1948). Felsenfeld (1965a) and Pesigan (1965) found that vibrios survived in milk and milk products for 2 to 4 weeks.

Pollack (1912) stressed that persistence of an adequate amount of moisture was apt to promote a prolonged survival of vibrios on green vegetables. *Vibrio cholerae* survived for 5 days on vegetables, for 3 days on dates, and for 6 hours in the interior of dates, but it did not survive in honey (Gohar and Makkawi, 1948). Felsenfeld (1965a) observed the survival of vibrios for less than 2 weeks on raw undamaged fruits and vegetables stored in a refrigerator, and for about 1 week after storage at room temperature. Pesigan (1965) observed similar results. Longer survival times were observed on slices of cantaloupe, the inside of water-melons, cooked carrots, sliced and cooked eggplant, slices of melon, cocoa, and cooked tapioca.

In Pakistan and India, rice served for evening meals is often covered with water and left at ambient temperatures until the next day. This form of rice is known as "panta bhat," or water rice. When these conditions were simulated in the laboratory and the rice was inoculated with *V. cholerae,* the vibrios frequently multiplied a thousandfold or more (Benenson *et al.,* 1965).

Vibrio cholerae quickly disappears from acid foods, but survives for at least a few days in salt solutions and on sugar, bread, cakes, cooked noodles, cereals, soft desserts, sweets, condiments, coffee, and chocolate (Felsenfeld, 1965b; Pesigan, 1965).

Pesigan (1965) observed that El Tor vibrios survived for 4 hours on frying pans, porcelain plates, mortar and pestle, and drinking glasses, for 24 hours on spoons, forks, and wooden chopping blocks, and for 48 hours on kitchen knives. Scrubbing with ordinary soap and water for at least 1 minute eliminated the vibrios on all the above-mentioned utensils except the chopping block.

Vibrios do not multiply in water, but they survive from a few days to two weeks, depending on temperature, pH, salt content, organic materials, sunlight, degree of other bacterial contamination, and other factors (Pollitzer, 1959; Felsenfeld, 1963). Longer survival has been observed in ice cubes (Felsenfeld, 1965a).

El Tor vibrios survived somewhat longer than classical cholera organisms (Felsenfeld, 1965a, b). The *El Tor* biotype survived in food substrates longer when the samples were kept at refrigerator temperature than when the foods were kept at room temperature.

C. *Bacillus anthracis*

Anthrax is caused by *Bacillus anthracis,* a large, nonmotile, capsulated, gram-positive rod which grows in long chains and forms equatorial, ellipsoidal, and nonbulging spores. The vegetative forms are relatively fragile, but the spores are highly resistant to external influences.

Intestinal anthrax is rare but frequently fatal. All the 42 cases reported by Zaporozhchenko (1961) terminated fatally between 29 and 115 hours after onset. The intestinal form of the illness is characterized by an incubation period of 2 to 3 days, a high temperature, general weakness, malaise, headache, insomnia, abdominal pains, and vomiting of bile and blood.

Hutyra (1908) attributed 11 cases of intestinal anthrax to the ingestion of sausage prepared from meat of an animal which had died of anthrax. Barykin *et al.* (1929) reported 27 cases of intestinal anthrax, and Solowieff (1930) attributed 30 deaths to intestinal anthrax caused by eating infected sausage. Frainas (1932) reported an outbreak of anthrax which occurred in the Philippines. After the death of a sick water buffalo, several people skinned and butchered the animal before it could be buried by the owners. Three to five days later, cutaneous and intestinal anthrax occurred in all but 10 of the individuals who consumed the meat. Many of the patients developed carbuncles. The intestinal form of the disease manifested itself through abdominal pains, fever, dizziness, and headache. Feces in some cases were coated with tarry blood. Constipation was common. Seventeen deaths occurred.

After a group of 30 people ate the meat of a partially cooked calf which had died of anthrax, one individual developed a malignant pustule. Examination of the place of slaughter revealed the anthrax bacillus (Sinai, 1933).

There is very little evidence of milk-borne anthrax. Eichhorn (1932) described a case of intestinal anthrax in an individual who consumed milk from a cow with a malignant pustule on its udder. An animal with anthrax either ceases to lactate or gives milk that is bloody, yellowish, or visibly abnormal and thus is rejected by most potential consumers, so there is little opportunity for milk-borne transmission of the agent. Stein (1947a) cited two reports in which *B. anthracis* was recovered from the milk of infected animals; however, he stated that this organism was present in the blood in great numbers only just before death, and it rarely occurred in milk secretions.

Steele (1953) reported that Edelmann *et al.* pointed out that many people have eaten meat from infected animals without ill effects, and explained this phenomenon by the rapid destruction of the bacilli in the stomach and intestines. However, spores would not be destroyed by gastric juice. Steele (1953) suggested that an effective meat inspection program is the best method of preventing intestinal anthrax. He stated that there are no records of human anthrax having been caused by infected meat in the United States, although he cited an anthrax epizootic which occurred in Haiti in 1943 and resulted in a number of human cases of intestinal anthrax. These cases were attributed to ingestion of meat from animals that were afflicted with anthrax.

Between the years 1946 and 1952, Zaporozhchenko (1961) listed 42 cases of intestinal anthrax that occurred in Russia. In 17 of these cases illness developed after consumption of meat from animals which had died of anthrax. In 24 cases, illness was caused by contaminated hands, which in turn contaminated food during eating. In the other case an infection occurred after a man's hands were contaminated with blood during the slaughter of an animal which was dying of anthrax. Zaporozhchenko also cited four outbreaks of anthrax. Three of these were small, involving 5, 7, and 20 cases, of which 3, 2, and 4, respectively, were of the intestinal variety. In all the cases, infection resulted from consumption of meat from a cow that had died of anthrax. In the fourth outbreak 26 cases were reported, including eight cases of the intestinal form. *Bacillus anthracis* was isolated from a dead sheep's viscera. However, before the results of the laboratory investigation were known, the meat was sold and consumed by the patients. Bacteriological investigation confirmed the diagnosis of intestinal anthrax in all cases. In an isolated case of anthrax, a woman, making sausage, blew through the opening of a portion of the gut from a cow which had died of (then) unknown causes. While blowing, she swallowed part of the intestinal contents. *Bacillus anthracis* was recovered from the cow and, during autopsy, from the woman.

During the investigation of another woman's death from intestinal anthrax, Ghossain (1961) found that the woman had eaten raw goat meat 4 days prior to her illness. Just prior to slaughter, the animal was reported to be sick and dying. About 20 people were reported to have partaken of the meat. One person had a buccal anthrax lesion; another told of mild intestinal disturbances, but fecal cultures were negative.

Plants grown in soil contaminated with anthrax spores may have the organisms on the outside but not on the inside (Morris and Riley, 1926). When guinea pigs died from being inoculated with *B. anthracis* and the carcasses were held at 25° to 37°C, these organisms were rapidly destroyed by anaerobic organisms in the unopened carcasses within 3 days. At least 4 weeks were required before the organism could no longer be detected when the carcasses were exposed to 5° to 10°C. When the carcasses were opened before decomposition set in, the vegetative forms sporulated and the spores persisted for months in blood, muscle, liver, bone, and spleen. Anthrax spores survived for 9 to 10 years in frozen storage (Stein, 1947b). Stein and Rogers (1945) observed that all of 43 strains of *B. anthracis* were destroyed by vigorous boiling for 3 to 5 minutes.

D. *Francisella*

Francisella tularensis (Pasteurella tularensis) is a gram-negative, non-motile, noncapsulated, pleomorphic organism. Under optimal conditions of cultivation this organism displays a constant range of pleomorphism, and all strains were alike regardless of virulence, age, or source. Discrete units as small as 300 to 350 mμ produced tularemia in mice, and the cultures from the mice showed the complete gamut of morphologic units (Foshay and Hesselbrock, 1945). The organism is destroyed by heating for 10 minutes at 56° to 58°C.

Meyer (1965) stated that no other infection has such a variety of modes of transmission. These include: direct contact with animals (rabbits, hares, rodents, sheep, and game birds), bites of bloodsucking arthropods (ticks, deer flies, certain other flies, mosquitoes, and rarely lice, mites, and fleas), ingestion of insufficiently cooked wild rabbit meat, ingestion of contaminated drinking water, and laboratory infections.

There are a few reports in the literature in which the ingestion of in-adequately cooked rabbit meat has been incriminated in the transmission of tularemia. However, in many of these reports the rabbits were also handled by some of the patients; thus, entry through the intact skin was possible. Several of these outbreaks involved cases with no ulcers or skin involvement. Freese *et al.* (1926) described four cases in a family that ate an undercooked rabbit which was killed by a dog, and Crawford (1932) reported an instance in which seven members of a family developed tularemia following the ingestion of a rabbit. Beck and Merkel (1935) and Amoss and Sprunt (1936) also reported cases that followed the consumption of a partially cooked rabbit. In a review of the epidemiology of tularemia, Ayres and Feemster (1948) cited three epidemics in which rodent contaminated food may have played a role in transmission. They also listed another case where handling or eating a wild rabbit was given as the mode of transmission. They commented that dogs and cats are susceptible and have been known to contract the disease by eating raw meat of sick, wild rabbits. In a review of tularemia from 1939 through

October 1949, Foshay (1950) made no mention of the relationship of disease transmission by rabbit meat.

Symptoms described during the outbreaks in which rabbit meat was the suspected vehicle have included nausea, vomiting, abdominal pains, diarrhea, fever (39.4° to 40.6°C), chills, headache, thirst, convulsions, and delirium. Incubation periods were from 8 to 24 hours. Several (12 of 20) of the cases in the previously mentioned outbreaks terminated fatally. Cecil and Loeb (1959) included most of the above-mentioned symptoms and added abscesses on the roof of the mouth and ulcers in the pharynx, nasopharynx, and submaxillary and anterior cervical lymph nodes in cases of oral and abdominal tularemia transmitted by ingestion of contaminated water or insufficiently cooked wild rabbit meat.

As a follow-up to the outbreak reported by Freese et al. (1926), an experimental rabbit that died of tularemia was skinned and parted into three pieces for frying. These pieces were rolled in graham flour and fried with grease in a pan over a hot gas flame for 10 minutes. When the meat was thought to be sufficiently cooked, as evidenced by a brown crust, the pieces were sliced. Successive layers of muscle appeared white until near the bone where some red strands of muscle were seen, surrounded by red juice. The red muscle was injected into four guinea pigs, and the red juice was injected subcutaneously into four pigs. All the experimental animals died with typical lesions of tularemia. *Francisella tularensis* retained its virulence for 6 to 18 months in rabbit tissue which was kept frozen at −14°C (Rosenthal, 1950).

Although the domestic rabbit is susceptible to experimental tularemia, rabbits raised in rabbitries in the United States have very rarely been found infected and, therefore, may be handled and eaten with apparent safety (Francis, 1925; Jellison and Parker, 1945). The domestic rabbit, *Oryctolagus,* as a reservoir or source of human infection is negligible or unknown.

E. *Streptobacillus moniliformis*

Streptobacillus moniliformis is a gram-negative, pleomorphic organism 2 to 15 μ in length. In the chains in which the streptobacilli are linked, swollen bodies are interspersed among the bacillary forms. The major source of the organism is the nasopharynx of rats.

Place et al. (1926) studied an epidemic of erythema arthriticum epidemicum (Haverhill fever) in which 86 cases occurred in 39 families, all living in a small area. Parker and Hudson (1926) isolated the *Streptobacillis moniliformis* from 11 of 17 patients studied and demonstrated serum antibodies in other patients. All the cases were users of milk from a small dairy. When the milk was pasteurized, the epidemic stopped. Although the source of the infection was not definitely known, it was assumed that rats were involved. A cow which exhibited a healed lesion on one teat suggestive of a rat bite showed serum antibodies for *Streptobacillus moniliformis*. Later, Place and Sutton (1934) compared this outbreak to a previous outbreak of undetermined etiology involving 600 residents of Chester, Pennsylvania. Ninety-two percent of the patients used raw milk from a single source. Other outbreaks of this nature have not appeared in the literature. Oeding and Pederson (1950) cited a single case of Haverhill fever in a waiter in which milk or food contact was possible.

The onset of Haverhill fever, following an incubation period of 1 to 5 days, is abrupt and is accompanied by chills, fever, vomiting, headache, and rather severe pains in the back and joints. Within the first 2 days a maculopapular rash usually develops. One or more joints may become swollen, red, and painful, and acute arthritis is one of the prominent and persistent symptoms. In the case of a food-borne infection, according to Dubos and Hirsch (1965), the portal of entry of the organisms is thought to be the intestine, since gastrointestinal symptoms are common. However, the rapid loss of viability of streptobacillus cultures in a slightly acid environment casts serious doubt on this view. The oral region or the throat might also be considered as a portal of entry; throat soreness is a frequent symptom.

F. Nitrate-Reducing Bacteria

Nitrite poisoning (methemoglobinemia) has been reported to occur in children after they ate spinach (Holscher and Natzschka, 1964; Simon, 1966). A high nitrate content was found in the spinach; its source was attributed to excessive fertilization. Storage at room temperature after preparation of the spinach allowed time for bacterial reduction of nitrates. Many common microorganisms (including the Enterobacteriaceae, staphylococci, *Pseudomonas, B. subtilis,* and *C. perfringens*) reduce nitrates.

V. UNKNOWN ROLE

Little is known about the role of the following organisms as agents of food-borne disease. However, their association with animals used for food and food products causes concern.

A. Listeria

Listeria monocytogenes is a small, motile (at 25°C), gram-positive, nonsporeforming, and noncapsulating rod which is aerobic to micro-aerophilic. It is widely distributed in nature and has been recovered from at least 35 mammalian species and 17 fowl species, and from trout, ticks, and crustaceans (Gray, 1963). In a number of animals L. monocytogenes produces an infection which is characterized by a monocytosis. The most common manifestation of infection with this organism is meningitis. Other manifestations include septicemia, abortion, localized external or internal abscesses, endocarditis, conjunctivitis, and pharyngitis (Gray, 1962a). A possible role of L. monocytogenes in cerebral damage (Seeliger, 1962), in mental retardation (Hood, 1962), in habitual abortion (Potel, 1962), and in infectious mononucleosis (Nyfeldt, 1962) has been suggested.

Although food-borne transmission of listeriosis has not been confirmed in humans, the possibility of contaminated milk, whether raw or pasteurized, as a source of infection cannot be overlooked. Potel (1953) related a case in which a woman gave premature birth to twins with granulomatosis after she ingested milk from a cow with atypical mastitis. Listeria monocytogenes was isolated from the infants and from the milk. This organism has been isolated from raw milk samples in Europe (Wramby, 1944; Potel, 1953). Gray et al. (1955, 1956) suggested the oral route in the transmission of infection to pregnant rabbits and goats. Abortion, stillbirth, or early death of the newborn resulted. Listeriosis has been observed in cattle (Smith et al., 1955), and udder infections have been described (De Vries and Strikworda, 1956).

Donker-Voet (1962) and Kampelmacher (1962) described a cow which, from all four quarters, continuously shed L. monocytogenes in her milk. The milk appeared completely normal, although there was a mastitis. Seeliger (1961) mentioned that in several cases of listeriosis which occurred in Halle, Germany, the consumption of raw milk was suspected as the cause, and in other cases suspicion was cast on cream, sour milk, and cottage cheese.

Gray (1958) and Seeliger (1961) pointed out the possible relationship between listeriosis in fowl and in man. Recoveries of L. monocytogenes have been made from poultry found in the immediate vicinity of human cases; however, isolations of the causative agent have not been reported from eggs or egg products. Urbach and Schabinski (1955) observed that L. monocytogenes in infected eggs increased 500,000 times when incubated at room temperature for 6 to 10 days. These organisms also increased in number after storage in a refrigerator. When the eggs, incubated at room temperature, were fed to guinea pigs after the pigs had undergone a period of starvation, a fatal listeria septicemia occurred. Although listeriosis in chickens is frequent, there are thus far no indications that eggs or egg products are involved in the transmission of the disease (Gray, 1958; Kampelmacher, 1962).

Although listeriosis is seen in slaughtered animals, meat and meat products have not been proved a source of infection. Gudkova *et al.* (1958) found *L. monocytogenes* in the viscera of pigs used for food on a farm where there had been several cases of mononucleosis due to *L. monocytogenes.*

Potel (1951) reported that *L. monocytogenes* survived 80°C for 5 minutes and cited a study by Ozgen in which he reported that this organism survived 100°C for 15 seconds. In another study *L. monocytogenes* survived heating at 63°C for 5 minutes but not for 10 minutes (Donker-Voet, 1962). Gray *et al.* (1956) cited two studies that indicated that *L. monocytogenes* survived pasteurization. Bearns and Girard (1958) found that, if over 5×10^4 listeriae per milliliter were present in milk or water prior to pasteurization (61.7°C for 35 minutes), organisms of this kind were recovered. The survivors in the pasteurized milk reached 10^8 cells per milliliter after 48 hours at 22°C. With this number of organisms present in the milk, no gross changes or off-odors were detected. Bearns and Girard (1958) and Bojsen-Moeller (1962) observed that *L. monocytogenes* grew well at 4° to 6°C. When eggs infected with *L. monocytogenes* were fried so that the whites were congealed but the yolks were soft, *Listeria* were isolated from them (Urbach and Schabinski, 1955). Stenberg and Hammainen (1955) observed that *L. monocytogenes* could survive most salting processes used in meat processing. This organism can survive also in wood shavings, rabbit ration pellets, fodder, dry straw, animal feces, manure, and soil for several weeks, and in some cases for years (Gray, 1963). *Listeria monocytogenes* has the ability to withstand repeated freezing and thawing. It withstands direct sunlight and ultraviolet light (Lehnert, 1960). Gray (1962b) reported that *L. monocytogenes* may remain viable and pathogenic for mice for at least 15 months in naturally contaminated silage extracts.

After reviewing the role of animal products as a source of infection of *Listeria*, Kampelmacher (1962) concluded that a relationship can be suspected but has not yet been proved. At this time no more can be added to this conclusion.

B. *Erysipelothrix*

Erysipelothrix are gram-positive, nonsporulating, noncapsulating, aerobic rods which are members of the family Corynebacteriaceae. They are similar to *Listeria*. *Erysipelothrix* are resistant to salting, pickling, and smoking. These organisms will survive in meat that has undergone such treatment for 1 to 3 months (Reed, 1965). They survive for relatively long periods in putrefying meat and in water. Erysipeloid, the disease caused by these organisms, is primarily an occupational disease in individuals handling meat, poultry, shellfish, fish, and bone buttons. In man, inoculation almost invariably takes place through broken skin, but in swine other portals of entry are possible. Rare reports of food-borne transmission have been made. For instance, Fiessinger and Brovet (1934) reported a case of *Erysipelothrix* infection in a man a few hours after he consumed a meal of salt pork. Symptoms of malaise, generalized pruritus, erysipeloid eruption localized in the ears and cheeks, and anemia occurred. The patient died about 2 months after the onset of illness.

C. *Leptospira*

Leptospires are spirochets which are distinguished from *Treponema* and *Borrelia* by a fine coiling of their primary spirals. When leptospires are at rest their ends usually appear hooked; when in motion, these organisms take the form of a curved letter, such as a C, O, or S. Over 100 serotypes of *Leptospira* are known. Leptospires are widely distributed in animals throughout the world. Infection caused by these organisms localizes in tissues and organs, especially the kidneys. Symptoms frequently associated with an infection are fever, headache, chills, malaise, vomiting, and muscular aches.

Leptospires in meat, organs, milk containing blood clots, and urine of an infected animal can infect humans (Bernkopf *et al.*, 1947). The main portal of entry for these organisms is the skin, especially when it has been cut or abraded, or has become sodden from long contact with water. The mucous membranes of the eyes, nose, and throat are also penetrable (Van der Hoeden, 1958).

During the acute phase of leptospirosis in lactating animals, leptospires may be shed in milk (Little and Baker, 1950). In this regard, however, Kirschner and Maguire (1955) and Kirschner *et al.* (1957) reported that milk has a leptospirocidal effect. Their experiments showed that in undiluted milk leptospires lost motility after 1 hour and disappeared within 3 hours. The effect was observed in milk after storage for 2 months at 40°C, after pasteurization, and after heating for 5 minutes at 80°C. However, boiling destroyed the antileptospiral property. Bernkopf *et al.* (1947) and Gsell (1952) found active leptospires present in inoculated sterile milk 3 days after refrigerated storage. When milk and tap water were mixed (1 : 40 to 1 : 80), the leptospires survived for 60 days (Kirschner and Maguire, 1955). Dolman (1957) referred to a case of leptospirosis which occurred in a man 8 days after he consumed ham that had been kept in a cellar. This ham reportedly showed evidence of rat gnaws and, presumptively, was polluted with rodent urine. Galton *et al.* (1958) commented that there have been no proved cases of leptospirosis that could be traced to drinking infected milk. They also stated that the pH of the stomach is usually such that ingested organisms would be rapidly destroyed. Infection could, however, presumably occur in portions of the alimentary tract anterior to the stomach.

D. *Pasteurella pseudotuberculosis*

Meyer (1953) cited three reports of cases of this rare and frequently fatal infection. All except one of the reported infections gave rise to abdominal symptoms. This supported, but did not prove, a hypothesis that the infection is acquired via the digestive tract. The same relationship was discussed by Seeliger (1960). Also, children operated on for appendicitis have yielded cultures of *Pasteurella pseudotuberculosis* (Knapp, 1956). As long as only single cases and no epidemics occur, it will be difficult if not impossible to uncover the mode of transmission. The possible modes suggested by Knapp were: direct contact with infected animals or their excreta; eating contaminated meat or other foodstuffs; and drinking contaminated water or milk. Daniels (1961) reported the isolation of *P. pseudotuberculosis* from the feces of guinea pigs, a canary, and a human with symptoms similar to appendicitis.

DeSmet and Van Ussel (1966) reported an outbreak of gastroenteritis possibly caused by *P. pseudotuberculosis*. Two forms of human infection are known. One is a generalized or septic form which is often fatal; the other is a localized, benign form — mesenteric adenitis. The authors described a patient who manifested definite icterus, fever, chills, headaches, anorexia, and abdominal pain. The index case and six other persons remained ill for more than a month after the initial attacks of nausea, vomiting, abdominal pains, diarrhea, and fever. The outbreak occurred 24 to 36 hours after a group of persons ate leftovers from a wedding dinner. The leftovers had been kept overnight in the cellar of a farmhouse, where they could have become contaminated because of the presence of mice, rats, and a cat. The food had not caused illness among others who had eaten the dinner the day before. None of the food was available for examination, but the period of storage to which the leftovers were subjected in the cellar was adequate for bacterial multiplication. Most of the affected persons recovered within five days. Serological examinations of those who experienced the gastroenteritis were made at times ranging from one to three months after the group shared the incriminated meal. The index case and 9 of 14 others who had become ill yielded positive agglutination reactions for *P. pseudotuberculosis* strain II. An examination of rodents from the farm three months after the outbreak did not disclose that any were positive for *P. pseudotuberculosis*.

VI. CONTACT INFECTION

Individuals who handle food products during processing operations may develop infection or become carriers of many of the organisms that have been discussed in this chapter. Enterococci, members of the family Enterobacteriaceae that infect animals, and other enteric organisms that originate with animal intestinal contents may be spread to muscle or viscera and then to workers' hands. Once on the hands these organisms may be ingested and the carrier state or illness may result. As previously mentioned, *Brucella, Erysipelothrix, Leptospira, Francisella tularensis,* and *Bacillus anthracis* are more commonly transmitted by contact of meat or viscera than by ingestion. Rich (1944) mentioned that handling tissues of slaughtered tuberculous animals can lead to cutaneous infection.

Individuals may be exposed to air-borne infection during processing of animals or storing of foodstuffs. For instance, brucellosis (Hendricks *et al.,* 1962), tuberculosis, and fungal infections, such as cryptococcosis and histoplasmosis (Dack, 1964*b*), may be transmitted by the air-borne route. Pulmonary aspergillosis was once an occupational disease in France among workers who chewed up grain for squabs. Engaging in this practice, the workers were exposed to *Aspergillus* spores in moldy grain (Bridges, 1963).

This category is mentioned only briefly in this chapter, but transmission by contact of raw foods or by inhalation of dust or aerosols associated with food processing should also be considered as food-borne diseases, in the broad sense of the term.

REFERENCES

Allen, R. F., and Baer, L. S. (1944). *J. Am. Med. Assoc.* **124,** 1191.
Amoss, H. L., and Sprunt, D. H. (1936). *J. Am. Med. Assoc.* **106,** 1078.
Anderson, F. M. (1950). *In* "Brucellosis," pp. 220–224. American Association for the Advancement of Science, Washington, D. C.

Anderson, N. W. (1965). *Shigella Surveillance Report* 7, 4. U. S. Public Health Service, Communicable Disease Center, Atlanta, Georgia.

Angelini, A. (1946). "Epidemiologic Characteristics of Brucellosis in Mexico City." First Inter-American Congress on Brucellosis, Mexico City.

Anusz, Z. (1966). *Epidemiol. Rev.* **20**, 38.

Armstrong, C., and Parron, T. (1927). "Further Studies on the Importance of Milk and Milk Products as a Factor in the Causation of Outbreaks of Disease in the United States." Supplement No. 62 to the Public Health Report, U. S. Government Printing Office, Washington, D. C.

Ayres, J. C., and Feemster, R. F. (1948). *New Engl. J. Med.* **238**, 187.

Barksdale, W. L., Garmise, L., and Horibata, K. (1960). *Ann. N. Y. Acad. Sci.* **88**, 1093.

Barnes, L. A. (1944). *U. S. Naval Med. Bull.* **43**, 707.

Barnes, L. A., and Cherry, W. B. (1946). *Am. J. Public Health* **36**, 481.

Barrow, G. I., and Peel, M. (1965). *Monthly Bull. Min. Health Public Health Lab. Serv.* **24**, 21.

Bartley, C. H., and Slanetz, L. W. (1960). *Am. J. Public Health* **50**, 1545.

Barykin, Vygodschikov, and Sazhina. (1929). *Hyg. Epidemiol.* **23**. Cited in Tanner and Tanner, 1953.

Bearns, R. E., and Girard, K. F. (1958). *Can. J. Microbiol.* **4**, 55.

Beattie, C. P., and Rice, R. M. (1934). *J. Am. Med. Assoc.* **102**, 1670.

Beck, H. G., and Merkel, W. C. (1935). *Southern Med. J.* **28**, 422.

Benenson, A. S., Ahmad, S. Z., and Oseasohn, R. O. (1965). *In* "Proceedings of the Cholera Research Symposium," pp. 332–336. *U. S. Public Health Serv. Publ.* **1328**.

Bengtson, I. A. (1919). *J. Infect. Diseases* **24**, 428.

Bernkopf, H., Olitzki, L., and Stuczynski, L. A. (1947). *J. Infect. Diseases* **80**, 53.

Berry, R. B. (1966). *Shigella Surveillance Report* 9, 9. U. S. Public Health Service, Communicable Disease Center, Atlanta, Georgia.

Birkenfeld, M. (1955). *Tuberkulosearzt* **9**, 517.

Black, J. M., and Sutherland, I. B. (1961). *Brit. Med. J.* i, 1732.

Blair, M. R. (1956). *S. African Med. J.* **30**, 1144.

Bloch, E. (1938). *Lancet* ii, 837.

Bloomfield, A. L., and Rantz, L. A. (1943). *J. Am. Med. Assoc.* **121**, 315.

Boer, H. D. (1933). *Maandschr. Kindergeneesk.* **2**, 337.

Boissard, J. M., and Fry, R. M. (1955). *J. Appl. Bacteriol.* **18**, 478.

Bojsen-Moeller, J. (1962). *In* "Second Symposium on Listeric Infection" (M. L. Gray, ed.), pp. 169–172. Montana State College, Bozeman, Montana.

Bolten, J. (1918). *Public Health Rept. (U. S.)* **33**, 163.

Borts, I. H., Harris, D. M., Joynt, M. F., Jennings, J. R., and Jordan, C. F. (1943). *J. Am. Med. Assoc.* **121**, 319.

Bowes, G. K. (1938). *Brit. Med. J.* i, 1092.

Breed, R. S., Murray, E. D., and Smith, N. E. (1957). "Bergey's Manual of Determinative Bacteriology," 7th ed. Williams & Wilkins, Baltimore, Maryland.

Bridges, C. H. (1963). *In* "Diseases Transmitted from Animals to Man" (T. G. Hull, ed.), pp. 453–507. Thomas, Springfield, Illinois.

Brown, D. E. (1962). *Morbidity Mortality Weekly Rept.* **11**, 154. U. S. Public Health Service, Communicable Disease Center, Atlanta, Georgia.

Brown, L., and Petroff, S. A. (1919). "The occurrence of tubercle bacilli outside the body in a sanatorium and health resort." Contributions to Medical and Biological Research, Dedicated to Sir W. Osler, Vol. I, p. 359. Hoeber, New York.

Bryan, F. L. (1962). "Food-Borne diseases." (Mimeo.) U. S. Public Health Service, Communicable Disease Center, Atlanta, Georgia.

Buchanan, E. B., and Megrail, E. (1929). *J. Infect. Diseases* **44,** 235.

Buchbinder, L., Osler, A. G., and Steffen, G. I. (1948). *Public Health Rept. (U.S.)* **63,** 109.

Buttiaux, R. (1956). *Rev. Med. Liege* **11,** 521.

Buttiaux, R. (1958). *Ann. Inst. Pasteur* **94,** 778.

Callao, V., and Henares, M. (1957). *Rev. Sanidad Hig. Publica* **31,** 1.

Cameron, W. R., and Wells, M. (1934). *Southern Med. J.* **27,** 907.

Capps, J. A., and Miller, J. L. (1912). *J. Am. Med. Assoc.* **58,** 1848.

Carpenter, C. M., and Boak, R. (1928). *Am. J. Public Health* **18,** 743.

Carter, H. S. (1954). *Glasgow Med. J.* **35,** 244.

Carey, W. E., Dack, G. M., and Myers, E. (1931). *Proc. Soc. Exptl. Biol. Med.* **29,** 214.

Carey, W. E., Dack, G. M., and Davison, E. (1938). *J. Infect. Diseases* **62,** 88.

Castaneda, M., Toval, R., and Velez, R. (1942). *J. Infect. Diseases* **70,** 97.

Cecil, R. L., and Loeb, R. F. (eds.) (1959). "A Textbook of Medicine." Saunders, Philadelphia, Pennsylvania.

Chang, C. S. (1933). *New Engl. J. Med.* **209,** 690.

Chapman, J. S., Bernard, J. S., and Speight, M. (1965). *Rev. Respirat. Diseases,* **91,** 351.

Charter, R. E. (1956). *Brit. Med. J.* **ii,** 339.

Cherry, W. B., Lentz, P. L., and Barnes, L. A. (1946). *Am. J. Public Health* **36,** 484.

Chinn, B. D. (1939). *Food Res.* **4,** 239.

Colmer, A. R. (1948). *J. Bacteriol.* **55,** 777.

Commission on Acute Respiratory Diseases (1945). *Bull. Johns Hopkins Hosp.* **77,** 143.

Condit, P. K. (1964). *Shigella Surveillance Report* **4,** 6. U. S. Public Health Service, Communicable Disease Center, Atlanta, Georgia.

Cooper, K. E., and Ramadan, F. M. (1955). *J. Gen. Microbiol.* **12,** 180.

Cooper, K. E., Davies, J., and Wiseman, J. (1941). *J. Pathol. Bacteriol.* **52,** 91.

Cooper, M., Hasman, D., and Condit, P. K. (1964). *Morbidity Mortality Weekly Rept.* **13,** 166. U. S. Public Health Service, Communicable Disease Center, Atlanta, Georgia.

Costin, I. D., Voiculescu, D., and Gorcea, V. (1964). *Pathol. Microbiol. (Basle)* **27,** 68. Cited in *Bull. Hyg.* **40,** 363 (1965).

Crawford, M. (1932). *J. Am. Med. Assoc.* **99,** 1497.

Crow, H. E., King, C. T., Smith, C. E., Corpe, R. F., and Stergus, I. (1957). *Am. Rev. Tuberc. Pulmonary Diseases* **75,** 199.

Cruickshank, R., and Swyer, R. (1940). *Lancet* **239,** 803.

Dack, G. M. (1943). "Food Poisoning." University of Chicago Press, Chicago, Illinois.

Dack, G. M. (1947). *Am. J. Public Health* **37,** 360.

Dack, G. M. (1956). "Food Poisoning," 3rd ed. University of Chicago Press, Chicago, Illinois.

Dack, G. M. (1964a). *Food Technol.* **18,** 1904.

Dack, G. M. (1964b). "Annual Report 1964." Food Research Institute, University of Chicago, Chicago, Illinois.

Dack, G. M., Niven, C. F., Kirsner, J. B., and Marshall, H. (1949). *J. Infect. Diseases* **85,** 131.

Dahlberg, A. C., and Kosikowsky, F. V. (1948). *J. Dairy Sci.* **31,** 275.

Dalrymple-Champneys, W. (1960). "Brucella Infection and Undulant Fever in Man." Oxford University Press, London.

Daniels, J. J. (1961). *Brit. Med. J.* **ii,** 997.

Dauer, C. C. (1952). *Public Health Rept. (U. S.)* **67,** 1089.

Dauer, C. C. (1953). *Public Health Rept. (U. S.)* **68,** 696.

Dauer, C. C. (1958). *Public Health Rept. (U. S.)* **73,** 681.

Dauer, C. C. (1959). *J. Milk Food Technol.* **22,** 332.

Dauer, C. C. (1961). *Public Health Rept. (U. S.)* **76,** 915.

Dauer, C. C., and Davids, D. J. (1959). *Public Health Rept. (U. S.)* **74,** 715.

Dauer, C. C., and Davids, D. J. (1960). *Public Health Rept. (U. S.)* **75,** 1025.

Dauer, C. C., and Sylvester, G. (1954). *Public Health Rept. (U. S.)* **69,** 538.

Dauer, C. C., and Sylvester, G. (1955). *Public Health Rept. (U. S.)* **70,** 536.

Dauer, C. C., and Sylvester, G. (1956). *Public Health Rept. (U. S.)* **71,** 797.

Dauer, C. C., and Sylvester, G. (1957). *Public Health Rept. (U. S.)* **72,** 735.

Davis, D. J. (1912). *J. Am. Med. Assoc.* **58,** 1852.

Davis, D. J. (1932). *J. Bacteriol.* **23,** 87.

De, S. N. (1961). "Cholera: Its Pathology and Pathogenesis." Oliver & Boyd, Edinburgh.

Deibel, R. H. (1964). *Bacteriol. Rev.* **28,** 330.

Deibel, R. H., and Silliker, J. H. (1963). *J. Bacteriol.* **85,** 827.

DeSmet, P., and Van Ussel, E. (1966). *Acta Gastroenterol. Belg.* **29,** 341.

DeVries, J., and Strikworda, R. (1956). *Zentr. Bakteriol. Parasitenk. Abt. I. Orig.* **167,** 229.

Dizon, J. J. (1962). "Epidemiological Studies on Cholera El Tor" EPI-63-5-2. U. S. Public Health Service, Communicable Disease Center, Atlanta, Georgia.

Dizon, J. J. (1965). *In* "Proceedings of the Cholera Research Symposium," pp. 322–326. *U. S. Public Health Serv. Publ.* **1328.**

Dizon, J. J., Alvero, M. G., Rustia, F. S., and Tamayo, J. F. (1962). *J. Philippine Med. Assoc.* **38,** 313. Cited in *Bull. Hyg.* **38,** 158 (1963).

Dolman, C. E. (1943). *Can. J. Public Health* **34,** 205.

Dolman, C. E. (1957). *In* "Meat Hygiene," pp. 11–108. World Health Organization, Geneva.

Donitz, W. (1886). *Z. Hyg.* **1,** 405.

Donker-Voet, J. (1962). *In* "Second Symposium on Listeric Infection" (M. L. Gray, ed.), pp. 133–139. Montana State College, Bozeman, Montana.

Dublin, T., Rogers, E. F. H., Perkins, J. E., and Graves, F. W. (1943). *Am. J. Public Health* **33,** 157.

Dubos, R. J., and Hirsch, J. G. (1965). "Bacterial and Mycotic Infections of Man." Lippincott, Philadelphia, Pennsylvania.

Duncan, J. T. (1944). *Monthly Bull. Mint. Health Emergency Public Health Lab. Serv.* **3,** 61.

Dunne, H. W., Glantz, P. J., Hokanson, J. F., and Bortree, A. L. (1956). *Ann. N.Y. Acad. Sci.* **66,** 129.

Edwards, P. R., and Ewing, W. H. (1962). "Identification of Enterobacteriaceae." Burgess, Minneapolis, Minnesota.

Edwards, P. R., West, M. G., and Bruner, D. W. (1948). *J. Bacteriol.* **55,** 711.

Ehrenkranz, N. J., Takos, M. J., Hoffert, W. R., and Riemer, F. (1958). *New Engl. J. Med.* **259,** 375.

Eichhorn, A. (1932). *Lederle Vet. Bull.* **1,** 3. Cited in Steele, J. H. (1953).

Elkington, G. W., Wilson, G. S., Taylor, J., and Fulton, F. (1940). *Brit. Med. J.* **1,** 477.

Ensign, P. R., and Hunter, C. A. (1946). *J. Pediat.* **29,** 620.

Evans, A. C. (1946). *J. Infect. Diseases* **78,** 18.

280 FRANK L. BRYAN

Ewing, W. H. (1956). Ann. N.Y. Acad. Sci. 66, 61.

Ewing, W. H. (1963). "Isolation and identification of Escherichia coli Serotypes Associated with Diarrheal Diseases." U. S. Public Health Service, Communicable Disease Center, Atlanta, Georgia.

Ewing, W. H. (1966). Personal communication.

Ewing, W. H., Tanner, K. E., and Dennard, D. A. (1954). J. Infect. Diseases 94, 134.

Ewing, W. H., Tatum, H. W., and Davis, B. R. (1957). Public Health Lab. 15, 118.

Ewing, W. H., Davis, B. R., and Montague, J. S. (1963). "Studies on the Occurrence of Escherichia coli Serotypes Associated with Diarrheal Diseases." U.S. Public Health Service, Communicable Disease Center, Atlanta, Georgia.

Eyre, J. (1899). Brit. Med. J. ii, 586.

Eyre, J. W. H., McNaught, J. G., Kennedy, J. C., and Zammit, T. I. (1907). In "Reports of the Royal Society of London, Mediterranean Fever Commission," Part VI. Harrison & Sons, London. Cited in Spink, W. W. (1956).

Fabian, F. W. (1947). Am. J. Public Health 37, 987.

Fabyan, M. (1912). J. Med. Res. 26, 441.

Fabyan, M. (1913). J. Med. Res. 28, 85.

Farbar, M. E., and Mathews, F. B. (1929). Ann. Internal Med. 2, 875.

Farber, R. E., and Korff, F. A. (1958). Public Health Rept. (U.S.) 73, 203.

Faulds, J. S. (1943). Monthly Bull. Min. Health Emergency Public Health Lab. Serv. 2, 143.

Feig, M. (1950). Am. J. Public Health 40, 1372.

Felsenfeld, O. (1963). Bull. World Health Organ. 28, 289.

Felsenfeld, O. (1965a). "Review of Recent Trends in Research and Control of Cholera." World Health Organization, Geneva.

Felsenfeld, O. (1965b). In "Proceedings of the Cholera Research Symposium," p. 313. U. S. Public Health Serv. Publ. 1328.

Ferguson, W. W. (1956). Ann. N.Y. Acad. Sci. 66, 71.

Ferguson, W. W., and June R. C. (1952). Am. J. Hyg. 55, 155.

Fiessinger, N., and Brovet, G. (1934). Presse Med. 42, 889. Cited in Dolman, C. E. (1957).

Fife, M. A., Ewing, W. H., and Davis, B. R. (1965). "The Biochemical Reactions of the Tribe Klebsielleae." U. S. Public Health Service, Communicable Disease Center, Atlanta, Georgia.

Fitch, C. P., and Bishop, L. M. (1933). Proc. Soc. Exptl. Biol. Med. 30, 1205.

Floyd, T. M., and Blakemore, C. F. (1954). J. Infect. Diseases 94, 30.

Foley, G. E., Wheeler, S. M., and Getting, V. A. (1943). Am. J. Hyg. 38, 250.

Food Protection Committee of the Food and Nutrition Board (1964). "An Evaluation of Public Health Hazards from Microbiological Contamination of Foods." Natl. Acad. Sci.-Natl. Res. Council Publ. 1195.

Foshay, L. (1950). Ann. Rev. Microbiol. 4, 313.

Foshay, L., and Hesselbrock, W. (1945). J. Bacteriol. 49, 233.

Foster, S. (1964). Shigella Surveillance Report 1, 5–6. U. S. Public Health Service, Communicable Disease Center, Atlanta, Georgia.

Frainas, E. C. (1932). Bur. Animal Ind. Gaz. 2, (No. 3).

Francis, E. (1925). J. Am. Med. Assoc. 85, 378.

Francis, J. (1958). "Tuberculosis in Animals and Man. A Study in Comparative Pathology." Cassell, London.

Frank, L. C. (1940). Public Health Rept. (U. S.) 55, 1373.

Fraser, J. (1912). J. Exptl. Med. 16, 432.

Freeman, V. J. (1951). *J. Bacteriol.* **61,** 675.

Freese, H. L., Lake, G. C., and Francis, E. (1926). *Public Health Rept. (U. S.)* **41,** 369.

Friedrich, A. (1893). *Arb. Gesundh. (Berlin)* **8,** 465.

Fuchs, A. W. (1941). *Public Health Rept. (U.S.)* **56,** 2277.

Fujiwara, K., Sekiya, T., and Bamba, K. (1956). *Japan. J. Bacteriol.* **11,** 411.

Fujiwara, K., Tsuchiya, Y., and Miyaji, M. (1965). *Ann. Rept. Inst. Food Microbiol., (Chiba Univ.)* **18,** 5.

Galbraith, N. S. (1965). *Monthly Bull. Min. Health Public Health Lab. Serv.* **24,** 376.

Gale, E. F. (1944). *Brit. Med. J.* **i,** 631.

Galton, M. M., Hess, M. E., and Collins, P. (1947). *J. Bacteriol.* **53,** 649.

Galton, M. M., Menges, R. W., and Steele, J. H. (1958). *Ann. N. Y. Acad. Sci.* **70,** 427.

Gamble, D. R., and Rowson, K. E. K. (1957). *Lancet* **ii,** 619.

Gangarosa, E. J., Beisel, W. R., Benyajati, C., Sprinz, H., and Prapont, P. (1960). *Am. J. Trop. Med. Hyg.* **9,** 125.

Getting, V. A., Wheeler, S. W., and Foley, G. E. (1943). *Am. J. Public Health* **33,** 1217.

Getting, V. A., Rubenstein, A. D., and Foley, G. E. (1944). *Am. J. Public Health* **34,** 833.

Ghossain, A. (1961). *Mem. Acad. Chir.* **87,** 289.

Gilbert, R., Coleman, M. B., and Laviano, A. B. (1932). *Am. J. Public Health* **22,** 721.

Gilman, H. L., Dahlberg, A. C., and Marquardt, J. C. (1946). *J. Dairy Sci.* **29,** 71.

Gilmour, C. C. B. (1952). *Bull. World Health Organ.* **7,** 343.

Gohar, M. A., and Makkawi, M. (1948). *J. Trop. Med. Hyg.* **51,** 95.

Goldie, W., and Maddock, C. G. (1943). *Lancet* **i,** 285.

Goodale, R. H., and Lambrakis, J. S. (1944). *U. S. Naval Med. Bull.* **43,** 1277.

Gordon, J. E. (ed.) (1965). "Control of Communicable Diseases in Man," 10th ed. American Public Health Association, New York, N. Y.

Gorman, J. (1950). *Med. Officer* **83,** 241. Cited in *Bull. Hyg.* **25,** 1104 (1950).

Gratch, I. F. (1965). *Shigella Surveillance Report* **6,** 5; **7,** 4, U. S. Public Health Service, Communicable Disease Center, Atlanta, Georgia.

Gray, M. L. (1958). *Avian Diseases* **2,** 296.

Gray, M. L. (1962a). *In* "Second Symposium on Listeric Infection" (M. L. Gray, ed.), pp. 36–41. Montana State College, Bozeman, Montana.

Gray, M. L. (1962b). *In* "Second Symposium on Listeric Infection" (M. L. Gray, ed.), pp. 85–92. Montana State College, Bozeman, Montana.

Gray, M. L. (1963). *Am. J. Public Health* **53,** 554.

Gray, M. L., Singh, C., and Thorp, F., Jr. (1955). *Proc. Soc. Exptl. Biol. Med.* **89,** 169.

Gray, M. L., Singh, C., and Thorp, F., Jr. (1956). *Am. J. Vet. Res.* **17,** 510.

Griffith, A. S. (1937). *Tubercle* **18,** 529.

Griffith, A. S., and Smith, J. (1935). *Lancet* **2,** 1339.

Griffith, A. S., and Summers, G. J. (1933). *Lancet* **1,** 875.

Gsell, O. (1952). *Med. Sci. Publ. Army Med. Serv. Grad. School, Walter Reed Army Med. Center* **1,** 34.

Gudkova, E. I., Mironova, K. A., Kus'minskii, A. S., and Geine, G. O. (1958). *Zh. Mikrobiol. Epidemiol. i Immunobiol.* **29,** 1373.

Guthof, O. (1957). *Zentr. Bakteriol. Parasitenk. Abt. I. Orig.* **70,** 327.

Hall, I. C., and Learmonth, R. (1934). *J. Infect. Diseases* **55,** 184.

Halstead, B. H. (1957). *In* "Conference on Shellfish Toxicology," pp. 37–76. U. S. Public Health Service, Washington, D. C.

Hankin, E. H. (1898). *Brit. Med. J.* **i,** 205.

Hardy, A. V., Frant, S., and Kroll, M. M. (1938). *Public Health Rept. (U. S.)* **53,** 796.

Harrington, R., Jr., and Karlson, A. G. (1965). *Appl. Microbiol.* **13**, 494.

Harris, H. J. (1950). "Brucellosis: Clinical and Subclinical," 2nd ed. Hoeber, New York.

Hartman, P. A., Reinbold, G. W., and Saraswat, D. S. (1965). *J. Milk Food Technol.* **28**, 344.

Hasseltine, H. E., and Knight, I. W. (1931). *Public Health Rept. (U. S.)* **46**, 2291.

Hayashi, K. (1960). *Endemic Disease Bull. Nagasaki Univ. (Japan)* **2**, 181.

Hendricks, S. L., Borts, I. H., Heeren, R. H., Hansler, W. J., and Held, J. R. (1962). *Am. J. Public Health* **52**, 1166.

Herweg, J. C., Middelkamp, J. N., and Thornton, H. K. (1956). *J. Pediat.* **49**, 629.

Hobbs, B. C., Thomas, M. E. M., and Taylor, J. (1949). *Lancet* **ii**, 530.

Holscher, P. M., and Natzshka, J. (1964). *Deut. Med. Wochschr.* **89**, 1751.

Hood, M. (1962). *In* "Second Symposium on Listeric Infection" (M. L. Gray, ed.), pp. 340–345. Montana State College, Bozeman, Montana.

Horning, B. G. (1935). *J. Am. Med. Assoc.* **105**, 1978.

Horrocks, W. H. (1905). *Proc. Roy. Soc.* **B,76**, 510.

Horrocks, W. H. (1906). *In* "Reports of the Commission on Mediterranean Fever," Part 4, pp. 27. Harrison and Sons, London.

Horvath, I., Hanny, I., and Pethes, A. (1964). *Zentr. Bakteriol. Parasitenk. Abt. I. Orig.* **193**, 191.

Hucker, G. J., Brooks, R. F., and Emery, A. J. (1952). *Food Technol.* **6**, 147.

Hughes, M. L. (1897). "Mediterranean, Malta or Undulant Fever," Macmillan, London.

Hull, T. G. (1963). "Disease Transmitted from Animals to Man," 5th ed. Thomas, Springfield, Illinois.

Hunter, F. R., and Dack, G. M. (1938). *J. Infect. Diseases* **63**, 346.

Hutchings, L. M., McCullough, N. B., Donham, C. R., Eisele, C. W., and Bunnell, D. E. (1951). *Public Health Rept. (U. S.)* **66**, 1402.

Hutchison, R. I. (1957). *J. Hyg.* **55**, 27.

Hutyra, F. (1908). *Ztschrf. Fleisch-u. Milch-hyg. (Berlin)* **19**, 84.

Ibrahim, A., Issa, A. A., and Gohar, M. A. (1955). *J. Egypt. Med. Assoc.* **38**, 260.

Jellison, W. L., and Parker, R. R. (1945). *Am. J. Trop. Med. Hyg.* **25**, 349.

Jensen, L. B. (1957). *J. Forensic Sci.* **2**, 355.

Joint FAO/WHO Expert Committee on Brucellosis: (1951). "Report on First Session." World Health Organ. Tech. Rept. Ser. **37**.

Jordan, E. O., and Burrows, W. (1934). *J. Infect. Diseases* **55**, 363.

Jordan, E. O., and Burrows, W. (1935). *J. Infect. Diseases* **57**, 121.

Joseph, P. R. (1962). Epi-62-35-3. (Mimeo.) U.S. Public Health Service, Communicable Disease Center, Atlanta, Georgia.

June, R. C., Ferguson, W. W., and Worfel, M. T. (1953). *Am. J. Hyg.* **57**, 222.

Kaiser, R. L., and Williams, L. D. (1962). *Penn. Med. J.* **65**, 351.

Kampelmacher, E. H. (1962). *In* "Second Symposium on Listeric Infection" (M. L. Gray, ed.), pp. 146–151. Montana State College, Bozeman, Montana.

Kansas State Board of Health (1946). "Final Report: Acute Enteritis Epidemic, Great Bend, Kansas." Kansas State Board of Health, Topeka, Kansas.

Kaplan, M. T., and Appleman, M. D. (1952). *Food Technol.* **6**, 167.

Kawabata, T., Ishizaka, K., and Miura, T. (1955). *Japan. J. Med. Sci. Biol.* **8**, 487.

Kendereski, S. (1961). *Higijena (Belgrade)* **13**, 39. Cited in *Bull. Hyg.* **37**, 30, (1961).

Kenner, B. A., Clark, H. F., and Kabler, P. W. (1960). *Am. J. Public Health* **50**, 1553.

Kerrin, J. C. (1947). *Monthly Bull. Min. Health Public Health Lab. Serv.* **6**, 112.

Kessner, D. M., Shaughnessy, H. J., Googins, J., Rasmussen, C. M., Rose, N. J., Marshall, A. L., Andelman, S. L., Hall, J. B., and Rosenbloom, P. J. (1962). *Am. J. Hyg.* **76**, 27.

Kinloch, J. P. (1922). *J. Hyg.* **21**, 451.

Kirby, A. C., Hall, E. G., and Coackley, W. (1950). *Lancet* **ii**, 201.

Kirschner, L., and Maguire, T. (1955). *New Zealand Med. J.* **54**, 560.

Kirschner, L., Maguire, T., and Bertaud, W. S. (1957). *Brit. J. Exptl. Pathol.* **38**, 357.

Kitasato, S. (1889). *Z. Hyg.* **5**, 494.

Kjellander, J. (1960). *Acta Pathol. Microbiol. Scand.* **48**, Suppl. **136**, 1.

Klieman, I., Frant, S., and Abrahamson, A. E. (1942). *Am. J. Public Health* **32**, 151.

Knapp, W. (1956). *New Engl. J. Med.* **259**, 776.

Kober, G. M. (1903). "The Transmission of Bovine Tuberculosis by Milk with a Tabulation of Eighty-six Cases." Transactions Association of American Physicians, Georgetown University, Washington, D. C.

Krishnan, K. V. (1953). *In* "Indian Council of Medical Research, Cholera Advisory Board, Technical Report 1952." New Delhi, India.

Kubica, G. P., Beam, R. E., and Palmer, J. W. (1963). *Am. Rev. Respirat. Diseases* **88**, 718.

Kubinstein, S. S. (1964). *Presse Med.* **72**, 2115. Cited in *Excepta Med.* **4203** (1965).

Kuzdas, C. D., and Morse, E. V. (1954). *Cornell Vet.* **44**, 216.

Lal, R. B., and Yacob, M. (1926). *Indian J. Med. Res.* **14**, 245.

Lancefield, R. C. (1933). *J. Exptl. Med.* **57**, 571.

Larkin, E. P., Litsky, W., and Fuller, J. E. (1955a). *Appl. Microbiol.* **3**, 98.

Larkin, E. P., Litsky, W., and Fuller, J. E. (1955b). *Appl. Microbiol.* **3**, 102.

Larkin, E. P., Litsky, W., and Fuller, J. E. (1956). *Am. J. Public Health* **46**, 464.

Legters, L. J. (1965). *Shigella Surveillance Report* **5**, 6. U. S. Public Health Service, Communicable Disease Center, Atlanta, Georgia.

Lehnert, C. (1960). *Zentr. Bakteriol. Parasitenk. Abt. I. Orig.* **180**, 350.

Levine, M. G. (1943). *J. Ind. Hyg.* **25**, 451.

Lewis, G. M., and MacFarlane, M. G. (1953). *Biochem. J.* **54**, 138.

Lilleengen, K. (1945). *Svenska Jägareförb meddel.* **10**, 1. Cited in *Vet. Bull.* **18**, 319 (1948).

Linden, B. A., Turner, W. R., and Thom, C. (1926). *Public Health Rept. (U.S.)* **41**, 1647.

Listick, F. A., and Condit, P. K. (1964). *Morbidity Mortality Weekly Rept.* **13**, 30. U. S. Public Health Service, Communicable Disease Center, Atlanta, Georgia.

Little, R. B., and Baker, J. A. (1950). *J. Am. Vet. Med. Assoc.* **116**, 105.

Löffler, W. (1931). *Schweiz. Med. Wochschr.* **61**, 968.

Lüönd, H. (1965). Letter to G. M. Dack, Dec. 8, 1965. Chemisches Laboratorium der stadt Zurich, Zurich, Switzerland.

Lüönd, H., and Gasser, H. (1964). *Mitt Gebiete Lebensm. Hyg.* **55**, 144.

Lyons, W. F., and Matsuura, H. T. (1965). *Shigella Surveillance Report* **5**, 5. U. S. Public Health Service, Communicable Disease Center, Atlanta, Georgia.

MacGaughey, C. A., and Chu, H. P. (1948). *J. Gen. Microbiol.* **2**, 334.

MacKenzie, D. J. M. (1965). *In* "Proceedings of the Cholera Research Symposium," pp. 341–346. U. S. Public Health Serv. Publ. 1328.

MacKenzie, E. F. (1947). *Health Bull. (Melbourne)* **23**, 2395.

Magoffin, R. L., Kabler, P., Spink, W. W., and Fleming, D. S. (1949). *Public Health Rept. (U. S.)* **64**, 1021.

Marshall, W. E. (1907). *J. Hyg.* **7**, 32.

Mattick, A. T. R., and Shattock, P. M. F. (1943). *Monthly Bull. Emergency Public Health Lab. Serv.* **2**, 73.

McCoy, G. W., Bolten, J., and Bernstein, H. S. (1917). *Public Health Rept. (U.S.)* **32,** 1787.

McCroan, J. E., and Walker, R. J. (1965). *Shigella Surveillance Report* **5,** 6. U. S. Public Health Service, Communicable Disease Center, Atlanta, Georgia.

McDiarmid, A. (1960). *Vet Record* **72,** 423.

McSweeney, C. P., and Morgan, W. P. (1928). *Lancet* **ii,** 1201.

Meyer, K. F. (1953). *New Engl. J. Med.* **249,** 765, 804, 843.

Meyer, K. F. (1965). *In* "Bacterial and Mycotic Infections of Man" (R. J. Dubos and J. G. Hirsch, eds.), 4th ed., pp. 681–697. Lippincott, Philadelphia, Pennsylvania.

Meyer, R. (1951). *Ztschr. Hyg. Infektionskr.* **133,** 211. Cited in *Bull. Hyg.* **27,** 583 (1952).

Meyers, W. A. (1946). *U. S. Navy Bu. Med. News Letter* **7,** 1853.

M'Fadyean, J. (1890). *Ann. Rept. Vet. Dept. (London)* **13,** 19.

Mitchell, A. P. (1914). *Brit. Med. J.* **i,** 125.

Molinelli, E. A., Basso, G., and Miyara, S. (1948). "Brucelosis Humana en la Republic Argentina, Pt. I. Introduccion," *Inter-Am. Reunion Brucelosis,* 1946.

Moore, B. (1948). *Monthly Bull. Min. Health Public Health Lab. Serv.* **7,** 136.

Moore, B. (1955). *J. Appl. Bacteriol.* **18,** 606.

Morales-Otero, P. (1929). *Puerto Rico J. Public Health Trop. Med.* **5,** 144.

Morales-Otero, P. (1930). *Puerto Rico J. Public Health Trop. Med.* **6,** 3.

Morris, H., and Riley, K. (1926). *Louisiana State Univ. Agric. Expt. Sta. Bull.* **196,** 4.

Mosher, W. E., Elsea, W. R., Markellin, V., and Lennon, R. G. (1966). *Shigella Surveillance Report* **9,** 6. U. S. Public Health Service, Communicable Disease Center, Atlanta, Georgia.

Moureau, M. H., Bretey, J., and Roy, D. (1960). *Ann. Inst. Pasteur* **99,** 586.

Moustafa, M. N. E. D., Elyan, A., and Gohar, M. A. (1948). *J. Egypt. Med. Assoc.* **31,** 556.

Munro, W. T., and Walker, G. (1935). *Lancet* **i,** 252.

Murray, R. H. (1936). *Can. J. Public Health* **27,** 555.

Mushatt, C. (1940). *J. Hyg.* **40,** 396.

Nakamura, M., and Taylor, B. C. (1965). *Health Lab. Sci.* **2,** 220.

Nelson, N. A. (1928). *New Engl. J. Med.* **199,** 1145.

Neter, E. (1956). *Bacteriol. Rev.* **20,** 272.

Neter, E. (1959). *J. Pediat.* **55,** 223.

Neter, E., and Shumway, C. N. (1950). *Proc. Soc. Exptl. Biol. Med.* **75,** 504.

Nikodémusz, I., Bouquet, D., and Novotny, T. (1964a). *Nepegezsegugy Budapest* **45,** 337. Cited in *Bull. Hyg.* **40,** 603 (1965).

Nikodémusz, I., Bojan, M., Hoch, V., Kiss, M., and Kiss, P. (1964b). "Yearbook Institute of Nutrition 1963." Budapest.

Niven, C. F., Jr. (1955). *Ann. Inst. Pasteur* **7,** 120.

Nyfeldt, A. (1962). *In* "Second Symposium on Listeric Infections" (M. L. Gray, ed.), pp. 335–337. Montana State College, Bozeman, Montana.

Nygren, B. (1962). *Acta Pathol. Microbiol. Scand. Suppl.* **160,** 1.

Oeding, P., and Pederson, H. (1950). *Acta Pathol. Microbiol. Scand.* **27,** 436.

Orr, P. F., and Huddleson, I. F. (1929). *Am. J. Public Health* **34,** 38.

Orskov, F. (1951). *Acta Pathol. Microbiol. Scand.* **29,** 373.

Osler, A. G., Buchbinder, L., and Steffen, G. I. (1948). *Proc. Soc. Exptl. Biol. Med.* **67,** 456.

Ottie, H. J., and Ritzerfeld, W. (1960). *Deut. Med. Wochschr.* **85,** 1625. Cited in *Bull. Hyg.* **36,** 205 (1961).

Pagnini, P. (1962). *Riv. Ital. Igiene* **22,** 154. Cited in *Excepta Med.* 2002 (1963).

Pandit, C. G., and Hora, S. L. (1951). *Indian J. Med. Sci.* **5,** 343.

Park, W. H. (1927). *Am. Rev. Tuberc.* **15,** 399.

Park, W. H., and Krumwiede, C., Jr. (1910). *J. Med. Res.* **23,** 205.

Parker, F., Jr., and Hudson, N. P. (1926). *Am. J. Pathol.* **2,** 357.

Pesigan, T. P. (1965). *In* "Proceedings of the Cholera Research Symposium." *U. S. Public Health Serv. Publ.* **1328.**

Pfeiffer, D. H., and Viljoen, N. F. (1945). *S. African Vet. Med. Assoc. J.* **16,** 148.

Pillay, T. V. R., Dutta, S. N., and Rajagopal, S. (1954). *Alumni Assoc. Bull. All-India Inst. Hyg. Public Health* **1,** 27.

Place, E. H., and Sutton, L. E. (1934). *Arch. Internal Med.* **54,** 659.

Place, E. H., Sutton, L. E., and Willner, O. (1926). *Boston Med. Surg. J.* **194,** 285.

Plass, H. F. R. (1947). *J. Lab. Clin. Med.* **32,** 886.

Pollak, F. (1912). *Zentr. Bakteriol. Parasitenk. Abt. I. Orig.* **66,** 491.

Pollitzer, R. (1959). "Cholera." World Health Organization, Geneva.

Potel, J. (1951). *Zentra Bakteriol. Parasitenk. Abt. I. Orig.* **156,** 490.

Potel, J. (1953). *Wiss. Z. Univ. Halle,* **3,** 341.

Potel, J. (1962). *In* "Second Symposium on Listeric Infection" (M. L. Gray, ed.), pp. 323–324. Montana State College, Bozeman, Montana.

Price, R. M. (1934). *Can. J. Public Health* **25,** 13.

Price, R. M. (1938). *Can. J. Public Health* **29,** 251.

Purvis, J. D., and Morris, G. C. (1946). *U. S. Naval Med. Bull.* **46,** 613.

Raibaud, P., Caulet, M., Galpin, J. V., and Mocquot, G. (1961). *J. Appl. Bacteriol.* **24,** 285.

Raj, H., Wiebe, W. J., and Liston, J. (1961). *Appl. Microbiol.* **9,** 295.

Reed, R. G. (1965). *In* "Bacterial and Mycotic Infections of Man" (R. J. Dubos and J. G. Hirsch, eds.). Lippincott, Philadelphia, Pennsylvania.

Reekie, A. G. (1948). *Health Bull. (Scot.)* **6,** 65. Cited (1947) in *Bull. Hyg.* **24,** 117.

Rich, A. R. (1944). "The Pathogenesis of Tuberculosis," 2nd ed. Thomas, Springfield, Illinois.

Rogers, K. B. (1951). *J. Hyg.* **49,** 140.

Rogers, K. B., and Koegler, S. J. (1951). *J. Hyg.* **49,** 152.

Rosenthal, J. W. (1950). *New Orleans Med. Surg. J.* **102,** 558.

Ross, A. I. (1957). *Monthly Bull. Min. Health Public Health Lab. Serv.* **16,** 174.

Royal Society of London, Mediterranean Fever Commission. (1905–1907). "Reports." Harrison & Sons, London. Cited (1956) in Spink, W. W.

Runyon, E. H. (1959). *Med. Clin. N. Am.* **43,** 273.

Rustigian, R., and Stuart, C. A. (1943). *Proc. Soc. Exptl. Biol. Med.* **53,** 241.

Sartwell, P. E. (ed.) (1965). "Preventive Medicine and Hygiene, Maxcy and Rosenau," 9th ed. Appleton-Century-Crofts, New York.

Savage, W. G. (1920). "Food Poisoning & Food Infection." Cambridge University Press, Cambridge.

Savage, W. (1949). *Brit. J. Soc. Med.* **3,** 45.

Scamman, C. L. (1929). *Am. J. Public Health* **19,** 1339.

Scamman, C. L., Lombard, H. L., Beckler, E. A., and Lawson, G. M. (1927). *Am. J. Public Health* **17,** 311.

Scammon, L. A., Pickett, M. J., Froman, S., and Will, D. W. (1963). *Am. Rev. Respirat. Diseases* **87,** 97.

Seeliger, H. P. R. (1960). *Bull. World Health Organ.* **22,** 469.

Seeliger, H. P. R. (1961). "Listeriosis." Hafner, New York.

Seeliger, H. P. R. (1962). *In* "Second Symposium on Listeric Infection" (M. L. Gray, ed.), pp. 338–339. Montana State College, Bozeman, Montana.

Shattock, P. M. F. (1949). *Proc. 12th Intern. Dairy Congr. Stockholm 1949,* **2,** 598.

Shattock, P. M. (1962). *In* "Chemical and Biological Hazards in Food" (J. C. Ayres, H. E. Snyder, and H. W. Walker, eds.), pp. 303–319. Iowa State University Press, Ames, Iowa.

Shaughnessy, H. J., Olsson, R. C., Bass, K., Friewer, F., and Levinson, S. O. (1946). *J. Am. Med. Assoc.* **132,** 362.

Sherman, J. M. (1937). *Bacteriol. Rev.* **1,** 3.

Sherman, J. M., Smiley, K. L., and Niven, C. F. (1943). *J. Dairy Sci.* **26,** 321.

Shousha, A. T. (1948). *Bull. World Health Organ.* **1,** 353.

Shrader, J. H. (1939). "Food Control." Wiley, New York.

Simon, C. (1966). *Lancet* **i,** 872.

Simpson, W. M. (1930). *Ann. Internal Med.* **4,** 238.

Sinai, G. J. (1933). *J. Am. Med. Assoc.* **101,** 782.

Skinner, C. E. (1947). *Bacteriol. Rev.* **11,** 227.

Smith, R. E., Reynolds, I. M., and Bennett, R. A. (1955). *J. Am. Vet. Med. Assoc.* **126,** 106.

Solowey, M., and Watson, A. J. (1951). *Food Res.* **16,** 187.

Solowieff, N. (1930). *Arch. Hyg.* **104,** 132.

Spicer, C. C. (1959a). *Monthly Bull. Min. Health Public Health Lab. Serv.* **18,** 86.

Spicer, C. C. (1959b). *J. Hyg.* **57,** 210.

Spink, W. W. (1956). "The Nature of Brucellosis." University of Minnesota Press, Minneapolis, Minnesota.

Spink, W. W., Hall, J. W., III, Finstad, J., and Mallet, E. (1962). *Bull World Health Organ.* **26,** 409.

Stableforth, A. W., McDiarmid, A., and Bothwell, R. W. (1960). *Vet. Record* **72,** 419.

Stahl, S. (1939). *Am. J. Public Health* **29,** 1154.

Stebbins, E. L., Ingraham, H. S., and Reed, E. A. (1937). *Am. J. Public Health* **27,** 1259.

Steele, J. H. (1953). *Advan. Vet. Sci.* **1,** 329.

Steele, J. H., and Hastings, J. W. (1948). *Public Health Rept. (U. S.)* **63,** 144.

Steele, J. H., and Ranney, A. F. (1958). *Am. Rev. Tuberc. Pulmonary Diseases* **77,** 908.

Stein, C. D. (1947a). *Ann. N. Y. Acad. Sci.* **48,** 507.

Stein, C. D. (1947b). *Vet. Med.* **42,** 13.

Stein, C. D., and Rogers, H. (1945). *Vet. Med.* **40,** 406

Stenberg, H., and Hammainen, T. (1955). *Nord. Veterinarmed.* **7,** 853.

Sticker, G. (1912). Abhandlungen aus der Senchengeschichte und Senchenlehre. II. Band: Die Cholera, Giessen.

Stiles, G. W. (1945). *Rocky Mt. Med. J.* **42,** 18.

Stott, L. B., Cruickshank, B., Robertson, M., Tyser, P. A., James, A., and Clarke, R. H. (1953). *Monthly Bull. Min. Health Public Health Lab. Serv.* **12,** 224.

Stuart, C. A., and Rustigian, R. (1943). *Am. J. Public Health* **33,** 1323.

Stuart, C. A., Wheeler, K. M., and McGann, V. (1946). *J. Bacteriol.* **52,** 431.

Surgalla, M., Segalove, M., and Dack, G. M. (1944). *Food Res.* **9,** 112.

Takano, R., Ohtsubo, I., and Inouye, Z. (1926). "Studies of Cholera in Japan." *League of Nations Publ.* **C.H. 515.**

Tanner, F. W. (1933). "Food-Borne Infection and Intoxication." Twin City Printing Co., Champaign, Illinois.

Tanner, F. W., and Tanner, L. P. (1953). "Food-Borne Infection and Intoxication," 2nd ed. Garrard Press, Champaign, Illinois.

Taylor, B. C., and Nakamura, M. (1964). *J. Hyg.* **62**, 303.

Taylor, I. (1957). *Proc. Roy. Soc. Med.* **50**, 31.

Taylor, J. (1955). *J. Appl. Bacteriol.* **18**, 596.

Taylor, J. (1960). *Bull. World Health Organ.* **23**, 763.

Taylor, P. J., and McDonald, M. A. (1959). *Lancet* **i**, 330.

Teng, P. H. (1965). *In* "Proceedings of the Cholera Research Symposium," pp. 328–332. *U. S. Public Health Serv. Publ.* **1328.**

Thompson, R. (1933). *Can. Med. Assoc. J.* **29**, 9. Cited in Spink, W. W. (1956).

Thomson, S. (1956). *J. Hyg.* **54**, 311.

Todd, J. C. (1962). *Morbidity Mortality Weekly Rept.* **11**, 122. U. S. Public Health Service, Communicable Disease Center, Atlanta, Georgia.

Tong, J. L., Engle, H. M., Cullyford, J. S., Shimp, D. J., and Love, C. E. (1962). *Am. J. Public Health* **52**, 976.

Topley, E. (1947). *Med. Res. Council Spec. Rept. Ser.* **260**, 53.

Torning, K. (1965). *Diseases Chest* **47**, 241.

Trask, J. W. (1908). *Hyg. Lab. Bull.* **41**, 21.

Tucker, C. B., Fulkerson, G. C., and Neudecker, R. M. (1954). *Public Health Rept. (U. S.)* **69**, 432.

Uffelman, J. (1892). *Klin. Wochschr.* **29**, 1209.

Ulewicz, K. (1956). *Przegl. Epidol.* **10**, 341.

Urbach, H., and Schabinski, G. (1955). *Z. Hyg.* **141**, 239.

U.S. Bureau of Animal Industry (1912). "The Bacterium of Contagious Abortion of Cattle Demonstrated To Occur in Milk." Circular 198.

U. S. Public Health Service (1965*a*). "Zoonoses Surveillance 6. Brucellosis." Communicable Disease Center, Atlanta, Georgia.

U.S. Public Health Service (1965*b*). "Shigella Surveillance Summary." Communicable Disease Center, Atlanta, Georgia.

Van der Hoeden, J. (1958). *Advan. Vet. Sci.* **4**, 277.

van Veen, A. G. (1966). *In* "Toxicants Occurring Naturally in Foods." *Natl. Acad. Sci.– Natl. Res. Council Publ.* **1354.**

van Zwanenberg, D. F., Stewart, C. J., Harding, K. M., and Gray, S. T. G. (1956). *Brit. Med. J.* **i**, 1464.

Verge, J., and Paraf, A. (1956). *Hospital* (Paris) **44**, 222.

Vernon, E. (1965). *Monthly Bull. Min. Health Public Health Lab. Serv.* **24**, 321.

Watt, J., and Hardy, A. V. (1945). *Public Health Rept. (U. S.)* **60**, 261.

Watt, J., Hardy, A. V., and Decapito, J. M. (1942). *Public Health Rept. (U.S.)* **57**, 524.

West, M. G., and Edwards, P. R. (1954). "The Bethesda-Ballerup Group of Paracolon Bacteria." *U. S. Public Health Serv., Monograph* **22.**

White, B. (1929). *New Engl. J. Med.* **200**, 797.

White, J. C., and Sherman, J. M. (1944). *J. Bacteriol.* **48**, 262.

Wilkerson, W. B., Ayres, J. C., and Kraft, A. A. (1961). *Food Technol.* **15**, 286.

Wilson, G. S. (1933). *Lancet* **ii**, 829.

Wilson, G. S., and Miles, A. A. (1964). "Topley and Wilson's Principles of Bacteriology and Immunity," Vols. 1 and 2. Williams & Wilkins, Baltimore, Maryland. *World Health Organ. Tech. Rept. Ser.* 195.

World Health Organization Expert Committee on Tuberculosis. (1960). "Seventh Report."

Wramby, G. O. (1944). *Skand. Vet. Tidskr.* **34**, 278. Cited in Bearns, R. E. and Girard, K. F. (1958).

Zaporozhchenko, A. Y. (1961). *Zh. Mikrobiol. Epidemiol. Immunobiol.* **32**, 41.

PART II **INTOXICATIONS OF MICROBIAL ORIGIN**

CHAPTER VII | **BOTULISM —**

TYPES A, B, AND F

Hans Riemann

I. GENERAL INTRODUCTION

Clostridium botulinum types A and B were the first serological types recognized as causing human botulism. Type F was recognized only recently; the first outbreak was observed in Denmark in 1958, and the second in California in 1966.

Of the three remaining types—C, D, and E—only the latter plays a role in human botulism. It will be discussed in the second part of this chapter. Types C and D are responsible for serious outbreaks among animals. These types will not be dealt with here except for a brief discussion of animal botulism.

Botulism is a highly lethal food poisoning which generally occurs as small-scale scattered outbreaks in the human population but often as large-scale outbreaks in animal populations. It is caused by the heat-labile toxins of *C. botulinum.* All serotypes form spores and grow rapidly under anaerobic conditions in various organic materials. The powerful toxins are formed during the growth period and cause intoxications when ingested by man or animals.

The high mortality and the dramatic occurrence of botulism have made it one of the most feared food intoxications, and the danger of botulism has been a deciding factor in the formulation of food processing techniques.

Human botulism is rare, and it may seem to occupy a place out of proportion to its frequency as a cause of death (Meyer, 1956). But the potential danger of botulism as a major public health hazard should not be overlooked. *Clostridium botulinum* spores are widespread in nature, and botulism is a persistent hazard in some areas, as illustrated by the almost endemic occurrence of botulism among Eskimos in Alaska, the Canadian Northwest Territories, and Labrador (Dolman, 1965). Outbreaks of fish borne botulism in the United States in 1963 demonstrated that spores may easily gain access to commercially processed foods and cause quite large outbreaks if the food is not properly processed or handled. These recent outbreaks have caused a renewal of interest in research on botulism in the United States (Oscheroff *et al.,* 1964) and in other countries.

Human botulism is almost invariably caused by food which has been inadequately preserved, stored for some time, and then consumed cold or without sufficient heating. Experience shows that it may take only something slightly unusual to produce an outbreak of botulism, such as defects in can seaming operations or a change in packaging technology.

The symptoms of botulism vary only a little for the various types, but the mortality seems to be highest for type A.

II. HISTORY

The history of botulism has been treated in two excellent reviews (Meyer, 1956; Dolman, 1965), and only a summary will be given here.

Botulism, or sausage poisoning (*botulus,* latin word for sausage), was recognized as a disease entity in the first part of the nineteenth century when Justinus Kerner in Wurttemberg, Germany, collected data for two monographs on 230 cases of sausage poisoning. As a result of his work "sausage poisoning," or "Kerner's disease," was made reportable, and warnings were issued against consumption of spoiled sausages. In Wurttenburg smoked blood sausages were associated with sausage poisoning, and the way the sausages were used is rather typical for food items causing botulism. The sausages were smoked (this process does not destroy *C. botulinum* spores) and then stored under conditions that permit multiplication and toxin production. The sausages were eaten without heating.

Actually, blood sausages became suspect much earlier. Emperor Leo VI of Byzantium forbade the eating of blood sausage because of its harmfulness to health.

Despite the reporting and the warnings after Kerner's studies, about 2000 cases of botulism with a case mortality rate of 30% occurred in Europe between the middle of the nineteenth century and the beginning of the twentieth century. Around 1870 the disease became known as botulism because of the frequent implication of sausages.

In 1895 the organism that causes botulism was isolated and described by Van Ermengem during the investigation of a food poisoning outbreak in Belgium. This outbreak involved 34 members of a musical society who had eaten a meal containing raw salted ham. Twenty to thirty-six hours later most of the musicians developed a neuroparalytic syndrome, and three died within a week. Van Ermengem isolated an anaerobic, sporeforming organism from the ham and from the spleen of one of the victims. Filtrates of cultures of this organism, when injected into laboratory animals, produced characteristic and often fatal paralyses. Van Ermengem likened this outbreak of food poisoning to sausage poisoning and proposed the name *Bacillus botulinus* for the causative organism.

For a long time it was believed that botulism was provoked only by foods containing animal protein. However, in 1904 wax bean salad caused an outbreak involving 12 cases and 11 deaths in Darmstadt, Germany. A salad prepared from home-canned beans and mayonnaise was implicated in another outbreak involving 12 persons and 1 death in Stanford, California, in 1913.

In the five-year period from 1918 to 1922, 83 outbreaks with 297 cases and a case fatality rate of 62.3% were reported for California alone. Underprocessed vegetables, often home-canned, were implicated in most of these outbreaks. This increase in the incidence of botulism was related to an increase in home-preserved food during and after World War I.

Intensive studies of the causative organism and of the heat resistance of its spores made it possible to formulate heat processes that safeguard against botulism. Since 1925 commercial canners have practically controlled the disease in their field.

A total of 477 outbreaks, with 1281 cases and a case fatality rate of 65%, was reported for the United States in the period from 1899 to 1949. Eighty percent of the outbreaks in which the agent was verified were caused by type A, and 18% by type B.

A paralytic disease ichthyism, associated with the consumption of raw, usually salted fish, has been known in Russia since 1880. The causative agent, *Bacillus ichthyismi,* was isolated in 1914 and recognized as a variety of, or even as identical with, *Bacillus botulinus.*

A total of 163 outbreaks, with 1283 cases and a case fatality rate of 36%, was reported for the U.S.S.R. from 1880 to 1939. Most of the outbreaks seem to have been caused by type A. Sixty-two outbreaks were reported in the period 1920–1932. Six hundred and seventy-four persons were affected, of whom 172 died (25%). From 1958 to 1964, 95 outbreaks were reported. Three were caused by commercially produced foods, and 92 by home-preserved foods. Type A was isolated in 19 outbreaks, type B in 17. Two cases were caused by type C (Matveev *et al.,* 1967).

Four hundred and thirty-four outbreaks, with 1294 cases and a case fatality rate of 13.8%, have been recorded for Germany during the first half of this century. The lower fatality rate, compared to figures for the United States and the U.S.S.R., can be explained by the fact that most of the outbreaks in Germany were caused by the less-lethal type B.

Only 24 authentic outbreaks took place in France before 1940, but 500 outbreaks with more than 1000 cases were reported during World War II. One and one-half percent of the verified outbreaks were caused by type A, and 98.5% by type B. Most of the cases were caused by stored products of clandestinely slaughtered pigs. The meat was often given some kind of heat-treatment before being consumed, and this accounts for the low case fatality rate, 1.5%.

Small group outbreaks involving less than 40 cases have been reported from Argentina, Austria, Hungary, Belgium, Czechslovakia, Great Britain, The Netherlands, Switzerland, Yugoslavia, and Canada. One hundred and eight cases were reported in Scandinavia in 1901–1963: 2% caused by type A, 48% by type B, 5% by type F, 22% by unknown types, and the remaining by type E (Skulberg, 1964). A few cases of botulism have been recorded from India and Australia.

Two outbreaks of type F botulism have been reported. Both outbreaks were small (Møller and Scheibel, 1960; Conduit and Renteln, 1966).

III. INCIDENCE, FOODS INVOLVED, AND CLINICAL MANIFESTATIONS

A. Incidence

Type B is a more frequent cause of botulism in Europe than type A, whereas the opposite is the situation in North America. Both types, however, are presently of less importance than type E in northern countries. Type E has accounted for 46% of outbreaks of botulism in Japan, Canada, and Scandinavia in the twentieth century (Dolman, 1964). This may be a reflection of temperature requirements, since type E is able to develop at lower temperatures than A and B.

The incidence of botulism in the United States is about 20 cases per year, with a tendency toward decline (Fig. 1). The increase in 1963 was due to type E botulinum toxin in smoked, vacuum-packed fish.

B. Foods Involved

The types of foods involved in botulism vary according to eating habits in different regions, and it seems that almost any type of food with a pH above 4.5 can support growth and toxin formation (Oscheroff et al., 1964; Dack, 1956, 1964). Table I summarizes the types of food which have been involved in outbreaks of botulism in the United States. The majority of outbreaks were caused by preserved vegetables. Home-canned string beans, corn, beans, spinach, and asparagus account for more than 50% of the total number of outbreaks caused by vegetables. The dominating role of home-canned foods emphasizes the importance of a proper preservation technique in the prevention of botulism. Home-canned foods are often processed in open jars with a maximum temperature of 100°C, and such a temperature can be selective for C. botulinum because of the high resistance of its spores. The canning industry has adopted nonselective heat processes involving retort temperatures of 121°C or higher to ensure killing of all botulinum spores.

Fruits, pickles, and similar acid products have occasionally been reported as vehicles for botulism. This is surprising, since it seems established that C. botulinum is inhibited at pH values below 4.5. The growth

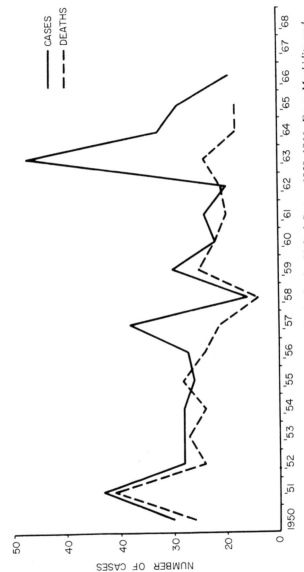

FIG. 1. Botulism—reported cases and deaths in the United States, 1950–1966. From *Morbidity and Mortality Annual Supplement, 1966.*

TABLE I

Foods Involved in Outbreaks of Botulism in the United States between 1899 and 1967, Verified Type E Outbreaks Excluded[a]

Food	Number of times involved	Home-preserved
Vegetables	389	357
Meat	41	34
Milk and milk products	7	5
Fish and seafood	30	21
Fruit and pickles	32	31

[a] From Dack (1956, 1964).

and toxin formation by *C. botulinum* in acid foods has been discussed by Ingram and Robinson (1951), who pointed out that in some cases a shift in pH had occurred as a result of growth of a fungus, *Mycoderma*. In other cases only the average pH was recorded, and it is known that a considerable variation between individual jars may be expected. It is also possible that different varieties of fruit and different degrees of ripeness might result in different pH values in the canned product. Furthermore, time is required for acid to diffuse out of the foods and decrease the pH of the syrup; this equilibration will generally take place during cooking, but it depends on the cooking temperature, which may be as low as 70°C in home canning.

Another factor which may be of importance is that the botulinum toxin is more stable under acid conditions, and once formed it may remain active much longer in acid foods than in neutral foods.

The effect of acidity on *C. botulinum* is discussed in more detail in the chapter on food processing.

Home-canned vegetables seem to have been the most common cause of types A and B botulism in the United States and Canada, but meat and meat foods apparently play a more important role in Europe. Historically, botulism is related to blood and liver sausages and other made-up meats. Pork, beef, and poultry products are still the most important vehicles for botulism in Germany. Since World War II, pickled fish has also been involved. In France mainly pork products but also other meat products and preserved vegetables have been involved. A few fish-borne type B outbreaks have been reported. Meat products and preserved (pickled or salted) fish have each been responsible for 50% of the cases in the Scandinavian countries (Skulberg, 1964). Various meat products have caused most outbreaks in England and Belgium.

Marine foods, mainly fish, have been responsible for most outbreaks in Japan and the U.S.S.R. Home-processed fish accounted for 57% and home-preserved ham for 15% of the outbreaks in the U.S.S.R. from 1958 to 1964. Type A was mainly associated with fish, and type B with ham. The mortality rate was 23% in fish-borne outbreaks (including type E) and 4.8% in outbreaks caused by ham (Matveev et al., 1967). In Australia, as in North America, home-canned vegetables have been the most frequently botulinum-contamined foods (Skulberg, 1961; Meyer, 1956). The two reported outbreaks of type F botulism were both meat-borne. The first was caused by home-made liver paste, and one of the three affected persons died (Møller and Scheibel, 1960). Home-made jerky (dry cured smoked meat) prepared from venison was implicated in the outbreak in California (Conduit and Renteln, 1966).

Although the types of foods causing botulism in different countries may largely be a function of the types of foods which are traditionally preserved, different patterns of distribution of types of C. botulinum and their spores could conceivably play a role. For example, animal carriers could be a more important source in Europe than in North America. However, little is known about animal carriers or about other aspects of C. botulinum ecology.

The type of food which is likely to cause botulism also depends on the degree of organoleptic changes which are caused by the growth of C. botulinum and the degree of spoilage which is accepted by the consumer. When C. botulinum types A and B grow in low-acid foods of bland taste, a foul and rancid odor (butyric acid, etc.) is generally observed. The taste of home-canned pears and apricots in which C. botulinum had grown has been described as "acrid," "sharp," "strong," "bitter," or "peculiar" (Dack, 1956). Spoilage was also noted in outbreaks of sausage poisoning described by Kerner.

However, signs of danger in the form of spoilage have not prevented botulism. Many people do not eat foods which have a disagreeable odor or bad taste, but there are degrees of tolerance among individuals. Uncertainty may exist in the case of unfamiliar foods, since certain foods are normally "spoiled," as, for example, some types of cheeses, and some individuals will eat foods even when they are aware that there is a sign of spoilage. In foods which are low in protein or are more acid, or are heavily spiced or fermented, the off-odor is more difficult to recognize. Spoilage odors are also less pronounced in cold dishes, and botulism is typically caused by foods which are consumed without final heating.

C. Clinical Manifestations

1. INCUBATION

The typical symptoms of botulism usually appear within 12 to 36 hours, but the onset may vary from 4 hours to 4 days in extreme cases. Gastrointestinal upset often precedes the typical symptoms, occurring within 12 to 24 hours. There does not seem to be any difference in incubation time among the different types of botulism.

2. SYMPTOMS

The symptoms produced by the different botulinum toxins are similar. The typical symptoms involve the nervous system, but they are often preceded by nausea, vomiting, and digestive disturbance in the form of burning and distress in the abdomen. Pain or tenderness in the abdomen is usually absent, but diarrhea may occur. Constipation often develops in the later stages of the disease.

Nausea and vomiting seem to be related to the amount of food consumed and the degree of spoilage in the food (Dack, 1956). The gastrointestinal upset is usually followed by feelings of lassitude and dizziness and vertigo, often accompanied by dryness of the mouth and tongue and pharyngeal pain.

Typical neurological symptoms develop most rapidly in severe cases. Blurring of vision and double vision occur as a result of paralysis of extraocular muscles, accompanied by blepharoptosis, mydriasis, nystagmus, and loss of reflex to light stimulation. Difficulty in swallowing occurs, followed by difficulty in speech. Paralysis of pharyngeal muscles occurs in fatal cases. There is also weakness of muscle groups in the neck. Death is usually due to respiratory failure. The temperature is normal, and there is no impairment of the patient's mental abilities.

In contrast to type E botulism, urine retention is usually not observed in patients with type A or B botulism.

3. DURATION

The duration of illness in fatal cases is usually 3 to 6 days after ingestion of the toxic food. In a series of 173 fatal cases, 15% died within 48 hours, 60% within 3 to 6 days, and one patient died after 26 days (Dack, 1956).

IV. ECOLOGY

A. Distribution of Spores

Clostridium botulinum has been commonly found in soil. From about 1% to 25% or more of several thousand soil samples taken in North America, Australia, Hawaii, China, Sweden, England, and Germany were positive during earlier studies by K. F. Meyer and his colleagues (Meyer, 1956). In the United States, type A was most frequent west of the Rocky Mountains. Botulinum spores were found less frequently in soil from states on the Atlantic seaboard and relatively rarely in samples from the Great Plains and Mississippi Valley. Strains found in the Mississippi Valley and around the Great Lakes were mainly type B.

Fifteen percent of soil samples from New York State were found to contain *C. botulinum* in 1946. Eighty percent of the strains were type A, 15% were B, and 5% were A and B. In 1950, 0.7% of soil samples from Georgia were found to contain *C. botulinum*. Type B but not type A was found in soil samples from England, Denmark, Belgium, The Netherlands, and Switzerland.

Clostridium botulinum type A was isolated from soil in Sweden in 1949 and in France in 1953 (Scholtens and Cohen, 1964; Smith, 1955). Occurrence of *C. botulinum* spores may be related to the nature of the soil. In Meyer and Dubowsky's survey of California soil samples in 1922, the proportion of type A to type B was 5–1 in cultivated soil and 38–1 in virgin soil. Twenty-seven percent of samples from cultivated soil and 58% of samples from virgin soil were toxic.

During a recent examination of more than 4000 soil samples in the Soviet Union, 10.5% were found to contain *C. botulinum* (Kravtchenko and Shiskulina, 1967). Type E was dominating (62% of the samples), type B was found in 28%, type A in 8.3%, and type C in 2.1% of the samples. Type D was isolated from only one sample. The sampling was done in five geographic areas, and soil contamination seemed to be highest in the most southern regions, as shown in Table II. No relation was found between number of positive soil samples and the density of human population, cultivation of soil, cattle raising, or other possible factors of contamination, but the type of soil seemed to have influence. Silt soils yielded a high percentage of positive *C. botulinum* samples, but very few positive samples were found in sandy soil. Only 4 of 103 samples of water yielded *C. botulinum* cultures. The 4 positive samples were from the Azow Sea.

It should be pointed out that *C. botulinum* is difficult to isolate from

TABLE II
DISTRIBUTION OF *Clostridium botulinum* SPORES IN VARIOUS AREAS OF THE U.S.S.R.[a]

Area	Percent positive samples	*C. botulinum* types isolated
European, north of 52°N	2.5 (0.95–11.8)	B and E
European, south of 52°N	12.7 (7.1–23)	A,B,C,D,E
Middle Asia	14.2 (3.2–46.5)	A,B,C,E
Western Asia Novosibirsk	1.1	
Taishet	6.7	
Obluyche, Vyazemsk, and Southern Primarye	13.8 (5–25)	

[a] From Kravtchenko and Shiskulina (1967).

mixed soil cultures. Toxin production in mixed soil cultures may also be erratic. Therefore, negative results do not prove that the organism is absent, and *C. botulinum* may be much more frequent than studies indicate.

Clostridium botulinum type E has in the past been associated with water and marine environments, whereas types A, B, C, etc., have been regarded as terrestrial types. More recent studies have shown, however, that types A and B may be present and even outnumber type E in marine environments. Both types have been isolated from fish and shrimps from the Gulf of Mexico and Gulf of Darien (Carroll *et al.,* 1966). Types A and B and also type F have been detected in samples of crabs, fish, and mud from the Pacific Coast of the United States (Craig and Pilcher, 1966; Eklund and Paysky, 1967).

There does not seem to be any reason to believe that *C. botulinum* types have a preference for a marine environment. Their presence in this environment may reflect ability of the spores to survive in water.

Although there are indications that *C. botulinum* can multiply in certain types of soil, it has not been demonstrated how important this factor is in the ecology of *C. botulinum*. It seems likely that the distribution of *C. botulinum* in nature is related to multiplication of the organism in rich organic matter, such as animal carrions. This raises the question of the importance of animals in the ecology of *C. botulinum*.

B. Animal Botulism

Animal species differ in their susceptibility to botulinum toxin, and the types that cause botulism in susceptible animals differ among animals and locations (Table III).

TABLE III
BOTULISM IN MAN AND ANIMALS[a]

Type	Principal victims	Common vehicle	Highest incidence
A	Man, chicken	Canned vegetables, meat, and fish	Western United States
B	Man, horse	Preserved meat, silage, forage	Eastern United States, Europe
Ca	Wild birds	Rotten vegetables, fly larvae	North and South America, South Africa, Australia
Cb	Cattle sheep, horse, mink	Silage, carrion	North America, Europe, Australia
D	Cattle	Carrion	South Africa
D	Cattle	Carrion	South Africa
E	Man, mink, wild bird	Fish, marine foods	North America, Northern Europe, Japan
F	Man	Preserved meat	United States, Denmark

[a] From McCroan *et al.* (1964).

The most resistant species seems to be the turkey buzzard, which has been reported to survive injection of 10^5 pigeon lethal doses of botulinum toxin.

Dogs, cats, and pigs are relatively resistant. Dogs are susceptible to injected botulinum toxin types A and B, but no verified natural cases have been reported (Smith, 1957). The mechanism of resistance in these animals is not known.

The animals most commonly affected by botulism in the United States are chickens (limberneck) and ducks (alkali disease, western duck disease), but botulism also occurs in mink, horses, and cattle (loin disease, forage poisoning) (McCroan *et al.,* 1964).

Mass intoxication, killing hundreds of thousands of water birds, was observed in 1932 at the northern shore of Great Salt Lake in Utah. Such incidents occur annually on areas of lakes, ponds, and mud flats in the United States, Canada, Mexico, Argentina, Uruguay, Australia, South Africa, and Germany (Meyer, 1956). Outbreaks have also been observed in Sweden and Denmark (Müller, 1967). The causative organism is *C. botulinum* type C, which seems to multiply in stagnant water, especially if the water is alkaline and covered with algae, which helps to maintain anaerobic conditions. Putrifying larvae of flies *(Lucilia)* and water beetles are also infected and serve as toxin reservoirs. Type C botulism associated with stagnant water has also been observed in turkeys (Smith, 1957).

In 1963 and 1964, large-scale death of water birds (gulls and loons) was observed in October–December along the shore of Lake Michigan. Botulinum type E toxin was detected in 50 to 100% of the dead birds (Kaufman and Fay, 1964; Fay *et al.,* 1965). It is not known whether the birds died from botulism or from some other disease, but fledgling ring-billed gulls were found to be sensitive to high doses of type E toxin. Twenty percent died after an oral dose of 60,000, and 70% died after 140,000 mouse LD_{50}. The surviving birds did not seem to develop resistance (Kaufman *et al.,* 1967).

Later studies have indicated that 4- to 5-week old birds which have been hatched and kept in an environment free of type E toxin may be sensitive to 2000 mouse lethal doses. It was also found that smaller doses induce resistance. Some birds may have an innate resistance to type E toxin (Kaufman and Crecelius, 1967).

Mallards and California gulls did not respond to 3.2 to 3.8 million MLD of type E toxin in another series of experiments, but E toxin increased the lethal effect of C toxin (Jensen and Gritman, 1967). These studies suggest that birds may be significant in relation to the ecology of type E as well as type C.

Cattle botulism (lamziekte) occurs in areas in Africa where the shortage of phosphorus in plants coincides with the presence of *C. botulinum* type D. Cattle die from eating carrions which they do to satisfy their hunger for phosphorus. Decaying tortoises seem to be the most toxic carrions, and the toxin is preserved for a period of time in dried tortoise carrions. In some districts every carrion is toxic. In spite of this, human cases have not been verified, although the local population consumes cattle carrions. This confirms the fact that humans are rather resistant to type D toxin.

The toxin can be demonstrated in the blood in 60% of the poisoned cattle and may persist for 6 days (Smith, 1957). *Clostridium botulinum* spores are present in the cattle that die from lamziekte or from other diseases and also in 8% of the healthy cattle living in infected areas. Soil near carrions contains spores, but the spore load in random samples is generally low (Meyer, 1956). *Clostridium botulinum* type D has been found only in Africa except for one isolation in the U.S.S.R.

Mass poisoning has also been reported in sheep suffering from nutritional deficiencies (lack of protein) (Smith, 1957). One hundred thousand sheep died in Western Australia in 1932–1933. The feeds implicated in such outbreaks are grain, harvesting refuse, silage, peanut straw, rotten potatoes, soiled melon, and carrions of cattle, sheep, and rabbit (Meyer, 1956). Spontaneous recovery in cattle and sheep is 30% to 50%.

Horses are very sensitive to botulinum toxins types, A, B, C, and D. The feed involved is generally fermented feeds or feed contaminated with rat or cat carrions (Meyer, 1956; Smith, 1957).

Animal carriers have been found in areas where animal botulism is relatively rare. Spores of types A and B have been isolated from liver and intestines of cattle which died from hemoglobinuria in Nebraska and from beef liver in a slaughterhouse in Paris. Catgut manufactured from Spanish sheep intestines has been found to contain *C. botulinum* spores, and spores have been isolated from bone marrow and intestines of healthy dogs (Meyer, 1956). Cats have been reported to carry *C. botulinum* type C spores (Smith, 1957).

These observations suggest that animals may be of importance in the ecology of *C. botulinum*. However, no epidemiological relationship between animal and human botulism has been demonstrated, and animals rarely get type A botulism, which is a common human type.

In a recent survey of more than 2000 samples of raw poultry, pork, and beef from packing plants distributed over the United States and Canada, *C. botulinum* (type C) was found in only one sample. Twenty thousand cultures of other clostridia were isolated during this survey. This observation suggests that if *C. botulinum* is present in healthy animals there is very little carryover into meat, and *C. botulinum* spores in raw meats are generally greatly outnumbered by other clostridia (Greenberg *et al.,* 1966).

V. MORPHOLOGY AND PHYSIOLOGY

A. Description and Classification

The genus *Clostridium* comprises more than one hundred species of anaerobic bacteria which have been divided into nine subgroups by Prevot and Fredette (1966). *Clostridium botulinum* species form subgroup No. 6. The following brief description of this group is based on Prevot and Fredette's monograph. A comprehensive discussion of clostridia has been published by Prevot *et al.* (1967).

1. *Clostridium botulinum* A (van Ermengen) Holland.

Synonym: *Bacillus botulinus* van Ermengen 1896.

Habitat: Soil, vegetables, fruit, human and animal feces, preserved meat and vegetables.

Morphology: Long rods, 0.9 to 1.2 μ by 4 to 6 μ; occur singly, in pairs, or in short chains; motile by means of 6 to 20 peritrichous flagella; oval, subterminal spore, fusiform or club-shaped; gram-positive.

Physiology: Strict anaerobe. Optimum temperature 25° to 33°C. Heat resistance quite high, but varies with the strains. Neutral red and safranin reduced.

Cultural characters: Gas and fetid odor produced.

Deep agar: Thin disk-like colonies, riddled with gaseous bubbles; some strains have fluffy colonies; gas.

Blood-deep agar: Irregular, hemolytic colonies.

Glucose broth: Turbid; gas; rancid odor; indole not formed.

Meat broth: Abundant turbidity; gas and putrid odor.

Gelatin: Liquefaction; H_2S formed.

Coagulated egg white: Digested; H_2S formed; some strains do not attack egg white.

Carbohydrates: Acid and gas from glucose, fructose, maltose, dextrin, salicin, and glycerol.

Biochemistry: Nitrites not formed from nitrates; sulfites not reduced; produces NH_3, H_2S, volatile amines, alcohols, ketones, acetylmethyl-carbonol, and acetic, butyric, and lactic acids.

Hemolysin: Appears early, disappears rapidly; has been identified as a lecithinase related to the α factor of *C. perfringens*.

2. *Clostridium botulinum* B (Leuchs) Dickson.

Physiology: Spore slightly less thermoresistant than that of type A. Proteolytic activity less marked than that of type A. Slightly less marked reducing power than type A (neutral red only is reduced).

Hemolysin: Appears early.

Biochemistry: Nitrites not formed from nitrates; sulfites not reduced; produces NH_3, H_2S, volatile amines, alcohols, ketones, and acetic, butyric, and lactic acids.

3. *Clostridium botulinum* C Bengston 1922.

Synonym: *Clostridium luciliae* Bergey *et al.* 1923.

Morphology: Similar to type A; 10 to 20 peritrichous flagella; rapidly becomes motile.

Physiology: Spore less thermoresistant than that of types A and B. Low reducing power (neutral red not reduced).

Deep agar: Irregular colonies; no gas.

Glucose broth: Flakey growth; no gas.

Gelatin: Liquefaction.

Milk: Coagulated, then digested.

Coagulated proteins: Not attacked.

Carbohydrates: Glucose, fructose, maltose, and lactose are fermented by all strains; galactose, sucrose, starch, and glycerol are feebly attacked by some strains.

Biochemistry: Nitrites not formed from nitrates; sulfites not reduced; produces NH_3, H_2S, volatile amines, ketones, and acetic, propionic, butyric, and lactic acids.

Type C may be subdivided into types C_α and C_β, which share a common toxic factor, but type C_α has a specific minor toxic factor; thus, C_α antitoxin neutralizes C_α and C_β toxins, and C_β antitoxin neutralizes C_α partly.

4. *Clostridium botulinum* D Theiler 1927.

Morphology: Similar to type A.

Physiology: Spore less thermoresistant than that of the preceding types.

Cultural characters: Small amounts of gas and fetid odor produced.

Deep agar: Small, translucent, irregular colonies.

Broth: Flaky growth.

Gelatin: No liquefaction.

Milk: Coagulated by some strains, but never digested.

Coagulated proteins: Not attacked.

Carbohydrates: Glucose, fructose, maltose, galactose, and lactose are fermented by all strains; some strains ferment, in addition, glycerol, starch, and dulcitol.

Biochemistry: Nitrites not formed from nitrates; sulfites not reduced; produces NH_3, volatile amines, alcohols, and acetic, butyric, and lactic acids.

5. *Clostridium botulinum* E Gunnison, Cummings, and Meyer 1936 (fish type). Type E is discussed in the second section of this chapter.

6. *Clostridium botulinum* F Møller and Scheibel 1960.

Morphology: Similar to that of the preceding species.

Physiology: Strictly anaerobic. Unstable and highly thermoresistant spores. Optimum temperature, 30°C.

Cultural characters: Proteolytic and saccharolytic activity similar to that of types A and B.

7. Relation between *C. botulinum* strains.

The description show that the six known species of *C. botulinum* are related. They differ from each other by the antigenic structure of their toxins and in the nature and degree of their metabolism.

Type A and B toxins are specific and neutralized only by their respective antitoxins. Type D toxin is partially neutralized by type C antitoxin. Type E antitoxin neutralizes type F toxin partly, and type F antitoxin has a similar effect on type E toxin.

The varieties described under the name of *C. parabotulinum* exhibit only differences of degree in their proteolytic activity. At the Rome Congress of 1953, the subcommittee on the nomenclature of the clostridia has recommended that it be dropped. More detailed descriptions of the characteristics of *C. botulinum* can be found in publications by Skulberg (1964) and Eklund *et al.* (1967) and in proceedings from symposia on botulism (Lewis and Cassel, 1964; Ingram and Roberts, 1967).

B. Metabolism

Clostridium botulinum A and B metabolize glucose to carbon dioxide, ethanol, volatile acids, and lactic acid; glycerol is fermented to carbon dioxide and ethanol (Clifton, 1940*a*). Pyruvic acid is fermented to acetic acid, carbon dioxide, and ethanol (Clifton, 1940*b*). There are indications that types A, B, and E utilize the Embden-Meyerhof pathway in glucose metabolism (Green *et al.*, 1967).

Amino acids are utilized in pairs by *C. botulinum* A and B. Alanine, leucine, and serine are oxidized in the presence of glycine and, more slowly, proline. Acetic acid, carbon dioxide, and ammonia are formed from proline and alanine. Proline is reduced to δ-amino-*n*-valeric acid. Acetic acid, carbon dioxide, and ammonia are also formed from alanine and glycine (Clifton, 1939, 1940*a*).

Leucine and particularly serine can be utilized when they are present alone. Serine is metabolized to ammonia, carbon dioxide, ethanol, and acetic acid (Clifton, 1940*c*). Also, ornithine is fermented as a single substrate and partially decarboxylated to putrescein; it is partially fermented to ammonia, carbon dioxide, δ-aminovaleric acid, and volatiles (acetate, propionate, butyrate, and valerate) (Mitruka and Costilow, 1966). *Clostridium botulinum* can deaminate a number of amino acids (Landgrebe and Weaver, 1964). All serotypes deaminate arginine (by arginine deaminase), asparagine, citrulline, ornithine, serine, and threonine. Alanine, cystine, glycine, histidine, lysine, proline, tryptophan, and valine were not deaminated by any type when they were tested one at a time.

The knowledge of the biochemistry of *C. botulinum* is still very limited, and it has not yet been explained why this organism is anaerobic. It seems that some of the biochemical pathways are identical with those found in aerobic organisms, and enzymes regarded as typically "aerobic" have been found in *C. botulinum* (Hobbs and Anderson, 1967). The chemical composition of *C. botulinum* has not been extensively studied. It has been found that type A cells contain fatty acids with 12 to 22 carbon atoms, and several long-chain unsaturated acids are also present (McCollough *et al.*, 1966). The significance of these findings is unknown.

C. Growth and Toxin Production

In many studies of the growth of *C. botulinum* a spore inoculum has been used. This makes it impossible to decide whether failure of growth is due to inhibition of germination or to growth inhibition. In some studies presence of toxin has been the only parameter used to estimate growth; here again it is not possible to decide whether absence of toxin is due to failure of germination, to growth inhibition, to failure of toxin production, or to instability of the toxin.

1. Nutritional Requirements

Clostridium botulinum types A and B can grow and form toxin in a large variety of media provided that pH, temperature, water activity, etc., are not inhibitory. This has been illustrated by the many types of food which can cause botulism.

Synthetic media have been formulated for vegetative growth of *C. botulinum* A and B (Campbell and Frank, 1956; Elberg and Meyer, 1939).

Type A requires nine amino acids, thiamine, and *p*-aminobenzoic acid (PAB) for growth in a mineral salts medium. Type B requires five amino acids and no thiamine or PAB.

A simple medium which permits outgrowth (development of a germinated spore into a vegetative cell) of *C. botulinum* A consists of eight amino acids and thiamine dissolved in phosphate buffer. Some amino acids were consumed and others were produced during outgrowth, but there was no net loss of amino nitrogen. This suggests that A spores contain all enzymes required for outgrowth (National Canners Association, 1966a). None of the synthetic media support toxin production as well as do complex organic media, and media such as corn steep medium and trypticase–yeast extract media which are very rich in organic materials seem to be the most favorable for toxin production (Sterne and Wentzel, 1950; Duff et al., 1957a,b; Bonventre, 1957).

Toxin synthesis has not been studied at the molecular level. The effect of a number of factors on toxin production in complex media has not been clarified, but it has been observed that carbohydrates are of importance. Clostridium botulinum A and B grow at almost equal rates when a number of different carbohydrates are added to a laboratory medium. However, the highest yield of cells and toxin was obtained when glucose or maltose was added. These sugars also increased cell lysis. Even 0.1% glucose was effective, and glycerol also gave good yields of cells and toxin (Bonventre and Kempe, 1959a).

In other media, 1% added glucose has been found to increase cell population but to delay lysis. Glucose has also been found to induce production of forespores but can be replaced by 1% arginine (Bowers and Williams, 1963). Glucose and yeast extract were required for maximum toxin production by C. botulinum type F in a laboratory medium (Holdeman, 1967).

Growth and toxin production by C. botulinum A and B is usually accompanied by putrid spoilage, but the degree of spoilage varies in different types of toxic foods. In chopped beef and pork, spoilage is noticeable before toxin can be detected (Greenberg et al., 1958). In some vegetables, in spiced food, and in cured meats, spoilage may not be detected at a time when active amounts of toxin have been produced. In cured meats toxin may be produced at salt concentrations where spoilage cannot be detected even after prolonged incubation; this phenomenon is discussed in the chapter on food processing.

2. EFFECT OF OXYGEN

Clostridium botulinum is an obligate anaerobe, and atmospheric oxygen has an inhibitory or direct toxic effect. The organism may be susceptible to hydrogen peroxide formed as a metabolic product by cells exposed to oxygen. Other peroxides formed by oxidation of compounds in the medium may be toxic.

Growth of anaerobes in laboratory media and in food is related to the oxidation-reduction potential. The highest potential which permits growth depends on the number of cells present, since cells reduce the potential in their immediate vicinity. Incubation in an anaerobic container is, therefore, not always required to produce growth. Thus, *C. botulinum* A and B spores initiated growth after 10 to 20 days in skim milk which was stored without anaerobic precautions (Kaufman and Brilland, 1964), and toxin was produced when smoked fish inoculated on the skin with type A spores was incubated with an excess of air for 8 days at 30°C (Thatcher *et al.*, 1962). In these cases the food itself produced a sufficiently low oxidation-reduction potential. Contaminating oxygen-consuming microorganisms may have improved the anaerobic conditions on the surface of the smoked fish.

The effect of other microorganisms on growth and toxin production of *C. botulinum* is discussed in the chapter on food processing.

3. EFFECT OF TEMPERATURE

The temperature range for growth and toxin production by types A and B extends from 10° to 12.5°C, to 47.5° to 50°C, and the amount of toxin produced in laboratory media shows little dependence on temperature in a range from 20° to 25°C (Ohye and Scott, 1953; Presscott and Tanner, 1938; Starin, 1926; Kaufman and Brilland, 1964). When type F was cultivated in a laboratory medium at 30°C, the highest level of toxin was reached after 48 to 72 hours (Holdeman, 1967). Type F also produces toxin at 4°C (Watts, 1967).

4. EFFECT OF pH

In controlled experiments no toxin production has been observed in foods with pH lower than 4.6, and the limiting pH varies from pH 4.6 to 5.3, depending on the type of food. In otherwise adequate laboratory media, growth may be inhibited at pH values of 5.3 or higher. Toxin production seems to be independent of pH within the growth range, but the stability of the toxin is better at the low pH values.

5. EFFECT OF MOISTURE

Growth and toxin production are inhibited as water in the food becomes less available—that is, when the water activity decreases. The water activity can be reduced by drying or by addition of sugar or salt. *Clostridium botulinum* A and B are unable to grow or produce toxin at a water activity below 0.93 to 0.94, which corresponds to 50% sucrose or 10% NaCl.

Thus it is the availability of water and not the percentage of water in the food which is of importance. This has not always been realized, and in some of the earlier work on the effect of drying on growth and toxin production the limiting water content is reported as percentage of moisture. The effects of temperature, pH, water activity, curing salts, etc., on toxin production in foods are discussed in the chapter on food processing.

D. Germination

An important property of *C. botulinum* is the ability to form spores which are so resistant that they can survive not only heat-treatment and irradiation but also drying, freezing, and exposure to chemical agents for long periods of time. The spores will germinate and grow in foods and other media when conditions become favorable. The germination of spores has been reviewed recently (Riemann, 1963; Sussman and Halvorson, 1966), and general information about bacterial spores can be found in the proceedings of spore symposia (Halvorson, 1957, 1961; Campbell and Halvorson, 1965). A brief account of the germination will be given here, since spore germination is important in relation to botulism.

1. CHANGES IN THE SPORES DURING GERMINATION

Spores consist of cytoplasm which contains nuclear bodies. The cytoplasm is surrounded by a delicate membrane, the spore wall. The spore wall is generally surrounded by at least two further layers, the cortex and the spore coat, which are in close contact with one another. The cortex consists of mucopeptide containing muramic acid and diaminopimelic acid. The spore coat is rich in cystine, which probably protects against radiation damage. Spores contain 5 to 15% dipicolinic acid (DPA) on a dry-weight basis. This compound is not present in vegetative cells. The calcium content in spores (1 to 4% of dry weight) is much higher than in vegetative cells and seems to be important for the heat resistance of spores. Spores can take up almost 70% water and low-molecular-weight solutes but do not seem to be osmotically active. Structural parts of spores have been visualized as a reversible sol–gel system, which keeps the essential components in a more or less insoluble state. Clostridial spores are about 1.0 μ in diameter and 1.6 μ in length. They have a high refractive index — as dehydrated protein — and look bright when observed in a phase-contrast microscope. They can be stained only by using a long staining time or by applying acid or hot staining solutions.

The first sign of germination is a dramatic loss of heat resistance, and germination of spores is conventionally defined as loss of heat resistance. A germinated spore is killed in a few minutes at 60°C. The first visible signs of germination are loss of refractibility and increase in size of the spore. These changes may take place in minutes, but it usually takes 1 to 1½ hours before the new vegetative cell is formed. Germinated spores are stainable with ordinary bacteriological stains.

During germination a considerable amount of organic matter, the so-called germination exudate, is released from the spores. The exudate contains polypeptide, amino acids, calcium, and DPA. During germination 85% to 90% of the DPA is released. At the same time, enzymes, which were masked in the dormant spore, become active.

Relatively few studies have been made of germination as distinguished from vegetative growth. In many experiments vegetative growth has been used as an indication of germination. Although it is true that vegetative growth from a spore inoculum proves that at least one spore has germinated, the absence of vegetative growth does not tell whether germination or multiplication was inhibited. In the following paragraph, only germination will be discussed.

2. RATE OF GERMINATION

Studies of the kinetics of spore germination in laboratory media have shown that the logarithm of nongerminated spores decreases linearly with time; that is, germination behaves like a first-order reaction. Germination is preceded by a lag period which increases if the spores are exposed to aeration and decreases with increasing number of spores (Wynne and Foster, 1948a). The biological interpretation of this exponential mode of germination is not known. Normal or log-normal distributions of germination times have been observed in experiments designed so that long incubation times could be used (Riemann, 1957; Schmidt, 1959). This means that some of the spores germinate in a comparatively short time but a small fraction may remain inactive (dormant) for prolonged periods. Generally, 1 to 24 hours of incubation in appropriate laboratory media results in 90% germination or more, but mean germination times of 1 month have been found for *C. botulinum* A in laboratory media that support good growth. Mean germination times of 2 and 3 months have been observed for related clostridial spores inoculated in ham and heated for 1 hour at 100°C or irradiated with 1 megarad (gamma irradiation) (Riemann, 1957, 1960). Delayed germination (dormancy) was noticed long ago. Dormant periods from 53 days to 5½ years have been reported and were discussed by Wynne and Foster (1948b). Dormancy seems to be partially a function of the environment, and sometimes quite rapid germination can be obtained in appropriate media. Dormancy also depends on the properties of the spores. It can be reduced by heat-shocking the spores, and it may be prolonged by severe heat-treatment or irradiation.

3. GERMINATION REQUIREMENTS

The mechanism that triggers germination is still unknown. Many different chemical compounds or physical treatments have been reported to cause or to stimulate germination, but it is not possible to discern any clear principles.

Clostridium botulinum A has been reported to germinate in a mixture of three amino acids; DL-alanine, L-arginine, and DL-phenylalanine, but these results have not always been reproducible. L-Alanine alone has been reported to induce germination of *C. botulinum* A at high pH (National Canners Association, 1966a). Carbon dioxide is required for rapid germination of *C. botulinum* A in laboratory media and can be replaced with oxalacetic acid and partially with oxalic acid, fumaric acid, and succinic acid. Yeast extract could also replace CO_2 (Wynne and Foster, 1948c). The function of CO_2 in the germination process is unknown.

Certain chemicals which cannot be regarded as nutrients, such as surface-active agents, *n*-alkyl primary amines, phenol, hydrogen peroxide, and chelating agents, have been reported to germinate different aerobic and anaerobic spores. None of these except caramelized glucose have been demonstrated to be active with *C. botulinum* spores.

Mechanical treatment such as grinding with glass powder causes germinal changes. Ultrasonics have a similar effect (Sehgal and Grecz, 1967). It has been suggested that this treatment breaks a permeability barrier with the result that the spores become hydrated and excrete DPA and calcium.

Normally spores germinate best in neutral solutions and fail to do so in an acid environment. Germination generally proceeds most rapidly at 30° to 40°C and is very slow at low temperatures. Studies with botulinum spores have indicated that there is an Arrhenius relationship between the velocity constant and the incubation temperature (Wynne and Foster, 1948a,b). Germination may not be completely inhibited at refrigeration temperatures. Ninety-nine percent germination of clostridial spores has been observed in 2 weeks at 4°C. This temperature is too low for vegetative growth, and under such conditions the germinated spores often die.

Clostridium botulinum spores were also observed to undergo germination-like changes at temperatures above the growth range when incubated in a glucose solution that had been autoclaved under alkaline conditions (caramelized glucose). The nature of this process is unknown.

4. GERMINATION INHIBITORS

Moisture, pH, and curing salts have a considerable influence on germination. The effect of these factors is discussed in the chapter on food processing.

Antimetabolites which interfere with vegetative growth have generally little or no effect on germination. 2,4-Dinitrophenol, cyanide, acid, fluoride, and iodoacetate have no effect on germination of *C. botulinum*. The situation is the same with antibiotics; germination proceeds at normal rates, but the emerging vegetative cell is rapidly killed if it is sensitive to the antibiotic. This destruction may take place at a very early stage of germination even before the spores have lost their refractibility (Scheibel and Lennart-Petersen, 1958). The only universal germination inhibitors seem to be salts of heavy metals such as zinc, copper, nickel, and mercury, and their effect can be reversed by chelating agents. Unsaturated rancid fatty acids such as oleic acid (sodium oleate), 100 μg/ml, inhibit germination of *C. botulinum* completely in laboratory media (Foster and Wynne, 1948; Halvorson, 1958).

5. EFFECT OF HEAT SHOCK AND AGING

It is commonly observed that heat shock accelerates germination and may be necessary for germination even in the presence of sufficient nutrients. Generally, the more heat-resistant the spores are, the more severe the heat shock required; most mesophilic spores such as *C. botulinum* germinate optimally after a heat shock of 85°C for 10 to 20 minutes. *Clostridium botulinum* type E spores are killed by such a treatment but are activated at 60°C in 13 minutes (Schmidt, 1964). It is not known how heat shock activates spores. It has been suggested that it increases the permeability of spores. Heat shock also releases small amounts of calcium and dipicolinic acid from spores. The suspending medium may influence the result of heat-shocking. Cysteine has been found to increase the effect on *C. botulinum* spores. Spores which were heat-shocked in 5% yeast extract changed so much during the heating that they become stainable with ordinary bacteriological stains (Treadwell *et al.*, 1958). The effect of heat-shocking is also influenced by the medium in which the spores were formed. *Clostridium botulinum* B spores did not respond to heat-shocking when they had been produced in a medium containing 0.4% glucose (Roberts, 1967). The effect of heat-shocking is temporary. If the spores are stored for some time, a second heat shock is required to bring about optimal germination.

During storage, spores may undergo changes that make them more like germinated spores. These changes occur most rapidly if spores are stored in aqueous solution and at elevated temperatures. The result is often that germination requirements become less exacting. Prolonged storage may eventually lead to death of the spores; that is, they become unable to form colonies.

VI. NATURAL VARIATION

Variations in colony form and toxigenicity of types A and B were observed by Meyer and others 40 years ago, and Skulberg (1964) has recently reviewed the earlier literature in a study of cultural variation in a number of *C. botulinum* strains. Skulberg observed smooth, rough, and intermediate colony forms of types A and B on blood agar plates. The smooth colonies were stable, and cultures forming smooth colonies seldom gave rise to variants. Smooth forms were usually outgrown by others. Some of the intermediate colony forms were unstable and changed into a rough form, into other intermediate forms, or into the smooth form. The rough form produced colonies which generally were much larger than the intermediate and smooth forms.

The production of toxins in types A and B was related to the colony form. Strains that grew as smooth colonies generally produced the largest amounts of toxin. Rough and intermediate forms produced little (1 to 10 MLD/ml) or no toxin and were termed hypotoxigenic. In cultures of these strains toxin could be detected during the first 12 to 36 hours of incubation but not 12 to 20 hours later.

Hypotoxigenic strains formed spores more readily than toxigenic strains and were less sensitive to oxygen.

The various strains of types A and B produced different degrees of hemolysis on blood agar, and hemolysis seemed to be a strain characteristic rather than related to various colony forms.

There were no differences in fermentation pattern between toxigenic and hypotoxigenic strains, but the latter tended to produce less gas.

The hypotoxigenic strains of types A and B were found to be more proteolytic than the toxigenic forms of the same type. These variations in type were confirmed in a later publication (Skulberg and Hausken, 1965). The mechanism involved in these variations is unknown at present, but Skulberg observed that changes in colony form and toxigenicity seemed to be induced when sterile filtrates from hypotoxigenic cultures or cultures of other clostridial species were added to *C. botulinum* types A or B cultures. Nontoxic type B strains which could be stained with fluorescent antisera prepared against a toxic strain were found by Georgala and Boothroyd (1967). Holdeman (1967) observed that different colony types were produced by type F strains but found no consistent correlation between toxigenicity and colony type.

The existence of hypotoxigenic and atoxic botulinum strains suggests that potential toxin-producing strains may be more common in nature than is indicated by surveys.

VII. RESISTANCE OF SPORES

Clostridium botulinum spores A, B, and F have a high resistance to heat and ionizing radiation. This is discussed in more details in the chapter on food processing. The resistance is variable and depends on the strain and on the medium in which the spores have been produced. Sugiyama (1951) found that reduction of Fe^{++} and Ca^{++} ions in the sporulation medium reduced heat resistance, whereas the addition of long-chain fatty acids increased resistance. Calcium is known to be important for the heat resistance of most other sporeformers. It was also reported that spores produced at 37°C were more resistant than spores produced at 41°C.

Vinton *et al.* (1947) found that spores of a nontoxigenic clostridium related to *C. botulinum* were three times as resistant when they were produced in cooked meat as when they were produced in raw meat. Lund (Halvorson, 1955) found that similar spores could be made heat-sensitive simply by exposure to raw meat extract. The phenomenon may not apply to all spores, since it has been reported that *C. botulinum* B spores have the same resistance in raw and cooked ground beef (Kempe *et al.*, 1957a,b).

The age of spores may also play a role in heat resistance, but the aging effect seems to be dependent on other, unknown factors, since different experiments have given different results (Esty and Meyer, 1922; Sommer, 1930). *Clostridium botulinum* spores are also resistant to chemicals, but chlorine is effective if properly used. Available chlorine (4½ ppm) in phosphate buffer, pH 6.5, has been found to cause 99.9% kill in 3 to 8 minutes at room temperature (Ito *et al.*, 1967).

VIII. TOXIN

A. Nature of Botulinal Toxins

The toxins of *C. botulinum* are classified as exotoxins, although they are apparently produced in the cytoplasm (Skulberg, 1964; Holdeman, 1967) and not released until the cells lyse (Bonventre and Kempe, 1960; Skulberg, 1964). The toxins are simple proteins composed only of amino acids. They possess hemagglutinating properties and seem to be serologically homogeneous.

The toxins are antigenically specific; that is, they are neutralized by their specific antitoxin, and little or no cross reaction takes place.

Botulinal toxins are very potent. One milligram of 98% pure A and B toxin is sufficient to kill several million mice by intraperitoneal injection.

B. Biochemistry

The A and B toxins have been obtained in a more or less purified form (Schantz, 1964). The general procedure is to precipitate the toxin from the supernatant of a suitable culture medium such as 1.2 to 2.5% trypticase containing yeast extract, glucose, and a reducing agent (Harrell *et al.,* 1964). Precipitation is carried out with $(NH_4)_2SO_4$, acid, or alcohol and is followed by extraction with dilute $CaCl_2$ solution. These procedures are repeated a number of times. Some success has been accomplished in purification on DEAE cellulose.

The purification progress is checked by assays of the toxic potency per milligram of nitrogen in the preparation. Biological assay is carried out in mice, and the LD_{50} is defined as the dose that will kill 50% of the mice. This dose is determined from a graph where the percentages of mice killed are plotted on probit log dose paper against the dose. A precision of ±25% can be obtained when at least three dilutions of the toxic preparation are used and 10 mice are injected with each dilution. A more rapid method to check purification progress is based on the relationship between the time from challenge to death and the size of the dose. The error can be kept within ±40% if 50 mice are used in the assay.

Criteria for purity are behavior as a single substance in the toxin–antitoxin reaction, in the ultracentrifuge, and in electrophoresis, and the ratio of absorption at 278 mμ and 260 mμ. One difficulty is that unexplained potentiation to the extent of 100- to 1000-fold may occur during assay procedures. Proteinaceous materials such as serum, meat broth, gelatin, and peptones have been reported to cause this potentiation (Schantz, 1964).

The properties of purified botulinum toxins have been reviewed by Lamanna (1959) and by Schantz (1964). Most of the work has been carried out on type A because it was the first one available in sufficient quantities in purified form. The molecular weight based on diffusion and sedimentation characteristics has been calculated to 900,000. Similar values for molecular size were found when untreated toxic cultures, types A through F, were subjected to ultracentrifugation in tubes with cellulose sponges (Schantz and Spero, 1967). Estimations of molecular weight of type A toxin based on inactivation by ionizing irradiation gave values of 1.6×10^6; possibly the toxin was present as a dimer (Skulberg, 1961). Much smaller toxic fractions have been obtained by various procedures (see below).

The A toxin has an isoelectric point of 5.6; it contains 16.2% nitrogen and is composed of 7754 residues of 19 different amino acids. Cysteine is the limiting amino acid (20 residues), and calculation of molecular weight based on one residue of cysteine gives a value of 4500.

Attempts to explain toxicity on the basis of the chemical composition of the molecule have not been fruitful. An active group composed of amino acids may exist. So far, evidence indicates that toxicity is due to a particular conformation of the molecule (coiling, folding). Boroff (1964) and Boroff and Fitzgerald (1958) have reported experiments which indicate that tryptophan plays a role in toxicity. Certain tryptophan derivatives can protect against the *in vivo* action of toxin (Boroff and Suhadalnik, 1961). However, toxicity may sometimes be lost without loss of the fluorescence which is due to tryptophan (Boroff and Suhadalnik, 1961; Schantz, 1964). Thus toxicity and fluorescence may depend on different bonds or structures.

The large A toxin molecule can dissociate into smaller units. Wagman and Bateman (1951, 1953) and Wagman (1954) discovered that toxin which was dissolved in buffers at pH 6.5 to 8.0 and ionic strength 0.13 gradually dissociated into slowly sedimenting units with molecular weights from 40,000 to 100,000. Wagman (1963) obtained toxic dialyzable units with a molecular weight of 3800 after peptic digestion of A toxin which had previously been treated in alkali at a pH near 9. Riesen *et al.,* quoted by Schantz (1964), found dissociation into small toxic units after treatment of type A toxin with 0.1 *N* HCl for 4 hours.

Type A toxin has also been separated into a fraction responsible for hemagglutination and a toxic fraction with a molecular weight of 10,000 to 20,000 (Schantz, 1964; Cammack, 1957). Heckley *et al.* (1960) found that the molecular size of A toxin in the lymph of orally poisoned rats was much smaller than that fed to the animals. Boroff *et al.* (1967*a*) obtained two apparently homogeneous fractions by DEAE Sephadex chromatography of crystalline A toxin. One fraction had a molecular weight of 158,000. It contained most of the toxicity and 20% of the protein. The other fraction had a molecular weight of 250,000.

Type B toxin has the same toxicity on a weight basis as type A. The molecular weight of type B has in some experiments been found to be similar to that of type A. Under other conditions a molecular weight of 60,000 has been found (Smith, 1955; Schantz, 1964); apparently type B toxin can also dissociate into smaller units. Boroff *et al.* (1967*b*) reported isolation of a smaller component with increased specific activity. Type F toxin has not yet been studied in purified form.

C. Activation and Stability

1. ACTIVATION

The well-known activation of type E toxin with trypsin has also been observed with types A, B, and F (Bonventre and Kempe, 1959*b*; Skulberg, 1964; Holdeman, 1967) especially with young cultures. Prolonged treatment with trypsin, chymotrypsin, or ficin destroyed the toxicity of type A and B toxins, while pepsin and papain had little or no effect (Meyer and Lamanna, 1959). Bacterial proteinases have an activating and — during prolonged action — a deactivating effect on A and B toxins. This effect has been observed for enzymes produced by *C. botulinum* types A and B and enzymes produced by *Bacillus cereus* and *Bacillus subtilis* (Meyer and Lamanna, 1959; Skulberg, 1964). Filtrates of *Streptococcus faecalis* cultures have also shown an activating effect, whereas filtrates from *Bacillus sphaericus, C. sporogenes,* and *Escherichia coli* were without effect, except that the first two increased lysis of *C. botulinum* cells, and they may in this way facilitate toxin release. Five percent gastric mucin can cause a threefold increase in the activity of botulinum toxin (Boor *et al.,* 1955; Schantz, 1964).

2. STABILITY OF BOTULINUM TOXIN

Distortion of the A toxin molecule by spreading it on a large surface or by treatment with 6 M urea at room temperature causes complete loss of toxicity. Photo-oxidation in the presence of methylene blue also destroys toxicity. Reaction of free amino groups with ketone causes detoxification, 43% of the toxicity being lost when 5% of the free amino groups have reacted; 98% loss was found when 19% of the amino groups had reacted

(Schantz, 1964). In contrast to the heat resistance of botulinum spores, toxin is readily destroyed by heat. The ease with which toxins are inactivated suggests that chemical groupings free to react are responsible for toxicity (Bonventre and Kempe, 1959c).

Purified toxin becomes inactive in a few minutes at 50°C and pH 6.9 but can be kept for an hour at 40°C at the same pH without loss in activity. Fifty percent of the toxicity was lost in 15 minutes at room temperature at pH values of 10.79, while no detoxification took place in 3 hours at pH 10.58. Ninety percent of the toxicity was lost in a few minutes at pH 11.12 (Schantz, 1964).

However, mild warming is not sufficient to eliminate the toxins in foods. Type A toxin in foods is inactivated in about 6 minutes at 80°C, in 18 minutes at 72°C, and in 1½ hours at 65°C. Type B toxin has a similar resistance (Dack, 1956). Stabilizing factors influence the speed with which the toxins are destroyed, and their stability is highest at acid pH (Bonventre and Kempe, 1959b; Dack, 1956). *Clostridium botulinum* type A spores may contain toxin. The amount of toxin within the spores represents about 0.01% of the toxin present in the sporangium which contains the spore, or about 500 molecules of toxin per spore. The spore toxin is heat-resistant and may tolerate 5 to 60 minutes of heating at 90°C without decrease in toxicity. Ninety percent of the spore toxin is released when spores are stored for a month at 2° to 4°C in aqueous suspension. The released toxin is heat-sensitive (Grecz and Lin, 1967).

The botulinum toxins are very radioresistant. A 10^3-fold reduction of the potency of purified A toxin in phosphate buffer required 0.34 Mrad, and addition of 5% trypticase to the solution increased the dose requirement 100-fold. Unpurified toxin in cheese was reduced 50-fold by 7.3 Mrads. The same degree of destruction was brought about in broth by 4.9 Mrads (Wagenaar and Dack, 1960). When dialyzed botulinum A and B cultures were irradiated in buffers with 0.2% added gelatin, the doses required for 90% destruction at 15°C varied between 0.62 and 3.10 Mrads. Freezing at −75°C caused fivefold increase in the radiostability of B toxin (Skulberg, 1964).

No antibiotics have been shown to affect botulinum toxins. One-tenth percent $KMnO_4$ or 0.025% bromine destroyed toxin in aqueous solution in a few minutes, but 10% H_2O_2 did not cause detoxification in 1 hour at 18°C. Crystal violet at a concentration of 0.05% caused destruction in 1 hour at 18°C, but aldehyde reagents were without effect (Bellinger *et al.,* 1951). Treatment with formaldehyde causes loss of toxicity and is used in the preparation of vaccines (toxoid).

Toxins may be very stable in foods. Titers of A toxin in cheese showed little or no decline during 2 months of storage at 30°C (Wagenaar and Dack, 1958).

D. Pharmacological Action

Toxin absorbed from the digestive tract is transported via lymph and blood and perhaps also within the nerves. Botulinum toxins all act the same way. Their action is specifically directed to nerves, especially the peripheral nerves. The impulse conduction in peripheral nerves remains normal, and so does the postsynaptic excitability of muscle fibers. The site of action seems to be a presynaptic block of the release of acetylcholine in the cholinergic nerve system, although the release mechanism itself may not be damaged (Brooks, 1964).

Botulinum toxin is unique in causing a presynaptic block. Curare, which has a somewhat similar effect, acts by desensitizing cholineceptive postsynaptic areas (Brooks, 1964).

It has been reported that botulinum toxin also acts on sympathetic nerve transmission (Whaler, 1967). However, rather high doses are required. Seventeen thousand LD_{50} per milliliter immediately caused a 72% blockage of adrenergic response in the isolated rabbit sinoatrial node, but this effect was not progressive and was immediately reversible (Vincenzi, 1967).

No structural changes have been observed in botulinum-poisoned nerve muscle preparations, but chronically poisoned muscles undergo the usual signs of denervation, and recovery of ganglia and muscles takes months (Brooks, 1964).

Different animals vary considerably in their susceptibility to the various types of toxin. By intraperitoneal injection rats are 25 times as resistant to A toxin as mice are, but 10,000 times as resistant to B toxin. These differences are unexplained. The toxic dose for rats is correlated with body weight, but this is not the case with mice (Lamanna and Hart, 1967). The reason for this difference is unknown.

IX. TOXOIDS

The production of toxoids has been discussed by Cardella (1964), who describes the procedures for preparation of purified aluminum phosphate-adsorbed, univalent, divalent, and pentavalent botulinum toxoids for immunization of man. These preparations are well tolerated and elicit satisfactory antitoxin responses in man.

Three injections of separate lots of pentavalent-type ABCDE toxoids produced immune responses to each toxin in a considerable proportion of individuals. A booster injection one year after the initial injection markedly increased the antitoxin titers, and measurable antitoxin was found in 86 to 100% of the individuals.

The toxoids were also antigenic to mouse, rabbit, and guinea pig and offered a high level of resistance to challenge with toxin administered intraperitoneally, orally, or via the respiratory tract.

X. ANTITOXIN

Horses produce from 10 to several hundred antitoxin units per milliliter of blood serum. One unit of antitoxin A or B neutralized 10^4 mouse LD_{50} of toxin. One unit of antitoxin E neutralizes 10^3 mouse LD_{50} of toxin.

The production in horses and the use of diagnostic and therapeutic antitoxins have been discussed recently (Harrell *et al.*, 1964; Cooper, 1964). The diagnostic sera are important in laboratory investigation of botulism. Therapeutic sera have value in the treatment of humans who have been exposed to botulinum toxin. It is important to inject antitoxin sera early, preferably before symptoms develop. Little or no effect can be expected in patients who have already developed symptoms; however, human type E botulism seems to respond more to antitoxin treatments than type A or B (see the section on type E botulism).

Botulinum toxin is probably seldom formed in the food after ingestion, but damage may continue as a result of gradual resorption of toxin as long as it remains in the intestinal tract. Good results from administering antitoxin through a duodenal tube have been reported (Minervin, 1967). Antitoxins for human use can be partially despeciated by pepsin treatment to reduce allergic reactions to horse serum. The pepsin treatment causes some loss in antitoxic activity.

XI. LABORATORY DIAGNOSIS

Laboratory diagnosis is not always easy. Demonstration of toxin in the remains of suspected food is of great importance. Toxin should also be looked for in necropsy material and if possible in stomach content and blood samples of patients. Details of laboratory technique are described in the chapter on laboratory methods.

XII. TREATMENT AND PREVENTION

A. Treatment

Treatment of botulism is unsatisfactory at best. Antitoxin is the only known specific therapeutic agent for botulism and should be used even when the disease is advanced.

B. Prevention

1. GENERAL RECOMMENDATIONS

Botulinum spores are widely distributed in nature, and almost any kind of food may become contaminated. Clean handling of food and careful cleaning of vegetables is important, especially if they are going to be preserved and stored for some time. Generally home canning and home preservation of foods, except freezing, is not advisable, as the many outbreaks of botulism from home-preserved foods have demonstrated. Home-preserved foods should preferably be heated to boiling immediately before serving in order to destroy botulinum toxin that might be present.

The canning industry has for many years kept botulism under control except for a few mishaps. However, the risk of botulism exists in commercially produced foods and may increase when changes are introduced in processing, packaging, or distribution if such changes are not carefully tested.

2. VACCINATION

As already mentioned, toxoids have been produced which can be effectively used for vaccination. However, the overall incidence of botulism has so far been so low that only people who are highly exposed to *C. botulinum,* such as laboratory workers, are vaccinated.

REFERENCES

Bellinger, H., Kornlein, M., and Lembke, A. (1951). *Zentr. Bakteriol. Parasitenk. Abt. I. Orig.* **156,** 430.
Bonventre, P. F. (1957). Ph.D. Thesis, University of Michigan.
Bonventre, P. F., and Kempe, L. L. (1959a). *Appl. Microbiol.* **7,** 372.
Bonventre, P. F., and Kempe, L. L. (1959b). *Appl. Microbiol.* **7,** 374.
Bonventre, P. F., and Kempe, L. L. (1959c). *J. Bacteriol.* **78,** 892.
Bonventre, P. F., and Kempe, L. L. (1960). *J. Bacteriol.* **79,** 18.

Boor, A. K., Tressolt, H. B. T., and Schantz, E. J. (1955). *Proc. Soc. Exptl. Biol. Med.* **89,** 270.

Boroff, D. A. *In* "Botulism 1964" (K. Lewis and K. Cassel, Jr., eds.), p. 103. U. S. Department of Health, Education, and Welfare, Public Health Service.

Boroff, D. A., and Fitzgerald, J. E. (1958). *Nature* **181,** 751.

Boroff, D. A., and Suhadalnik, R. J. (1961). *Bacteriol. Proc.* p. 137.

Boroff, D. A., DasGupta, B. R., and Fleck, U. (1967a). *In* "Botulism 1966" (M. Ingram and T. A. Roberts, eds.), pp. 278–295. Chapman and Hall, London.

Boroff, D. A., DasGupta, B. R., and Fleck, U. S. (1967b). *Bacteriol. Proc.* p. 5.

Bowers, L. E., and Williams, O. B. (1963). *J. Bacteriol.* **85,** 1175.

Brooks, V. B. (1964). *In* "Botulism 1964" (K. Lewis and K. Cassel, Jr., eds.), pp. 105–111. U. S. Department of Health, Education, and Welfare, Public Health Service.

Cammack, K. A. (1957). *Biochem. J.* **67,** 30.

Campbell, L. L., and Frank, H. A. (1956). *J. Bacteriol.* **71,** 267.

Campbell, L. L., and Halvorson, H. O. (eds.) (1965). "Spores III." American Society for Microbiology, Ann Arbor, Michigan.

Cardella, M. A. (1964). *In* "Botulism 1964" (K. Lewis and K. Cassel, Jr., eds.). U. S. Department of Health, Education, and Welfare, Public Health Service.

Carroll, B. J., Garrett, E. S., Reese, Gladys B., and Ward, B. Q. (1966). *Appl. Microbiol.* **14,** 837.

Clifton, C. E. (1939). *Proc. Soc. Exptl. Biol. Med.* **40,** 338.

Clifton, C. E. (1940a). *J. Bacteriol.* **39,** 485.

Clifton, C. E. (1940b). *Proc. Soc. Exptl. Biol. Med.* **43,** 588.

Clifton, C. E. (1940c). *Proc. Soc. Exptl. Biol. Med.* **43,** 588–590.

Conduit, P. K., and Renteln, H. A. (1966). *Morbidity Mortality Weekly Rpt.* **15,** 349.

Cooper, M. S. (1964). *In* "Botulism 1964" (K. Lewis and K. Cassel, Jr., eds.), pp. 147–164. U. S. Department of Health, Education, and Welfare, Public Health Service.

Craig, J. M., and Pilcher, K. S. (1966). *Science* **153,** 311.

Dack, G. M. (1956). "Food Poisoning." University of Chicago Press.

Dack, G. M. (1964). *In* "Botulism 1964" (K. Lewis and K. Cassel, Jr., eds.), pp. 33–38. U. S. Department of Health, Education, and Welfare, Public Health Service.

Dolman, C. E. (1964). *In* "Botulism 1964" (K. Lewis and K. Cassel, Jr., eds.), pp. 5–33. U. S. Department of Health, Education, and Welfare, Public Health Service.

Duff, J. T., Wright, G., Klener, J., Moore, D. E., and Bibler, R. H. (1957a). *J. Bacteriol.* **73,** 42.

Duff, J. T., Klener, J., Bibler, R. H., Moore, D. E., Gottfrued, E., and Wright, G. G. (1957b). *J. Bacteriol.* **73,** 597.

Eklund, M. W., and Paysky, F. (1967). *In* "Botulism 1966" (M. Ingram and T. A. Roberts, eds.), pp. 47–53. Chapman and Hall, London.

Eklund, M. W., Paysky, F. T., and Wieler, D. I. (1967). *Appl. Microbiol.* **15,** 1316.

Elberg, S. S., and Meyer, K. F. (1939). *J. Bacteriol.* **37,** 429.

Esty, J. R., and Meyer, K. F. (1922). *J. Infect. Diseases* **31,** 650.

Fay, L. D., Kaufman, O. W., and Ryel, L. A. (1965). Mass mortality of water-birds in Lake Michigan 1963–64. Publication No. 13, Great Lakes Research Division, The University of Michigan, 1965.

Foster, J. W., and Wynne, E. S. (1948). *J. Bacteriol.* **55,** 495.

Georgala, D. L., and Boothroyd, Margery (1967). *In* "Botulism 1966" (M. Ingram and T. A. Roberts, eds.), p. 494. Chapman and Hall, London.

Green, J. H., George, I., and Litsky, W. (1967). *Bacteriol. Proc.,* p. 6.

Greenburg, R. A., Silliker, J. N., Nank, W. K., and Schmidt, C. F. (1958). *Food Res.* **23,** 656.

Greenburg, R. A., Tompkin, R. B., Bladel, B. O., Kittaka, R. S., and Anellis, A. (1966). *Appl. Microbiol.* **14,** 789.

Grez, N., and Lin, C. A. (1967). *In* "Botulism 1966" (M. Ingram and T. A. Roberts, eds.), pp. 302–322. Chapman and Hall, London.

Halvorson, H. O. (1955). *Ann. Inst. Pasteur* **7,** 53.

Halvorson, H. O. (ed.) (1957). "Spores." Publication No. 5, American Institute of Biological Sciences, Washington, D. C.

Halvorson, H. O. (1958). "The Physiology of Bacterial Spores." Technical University of Trarelheim, Norway.

Halvorson, H. O. (ed.) (1961). "Spores II." Burgess Publishing Company, Minneapolis, Minnesota.

Harrell, W. K., Green, J. H., and Winn, J. F. (1964). *In* "Botulism 1964" (K. Lewis and K. Cassel, Jr., eds.), pp. 147–164. U. S. Department of Health, Education, and Welfare, Public Health Service.

Heckley, R. J., Hildebrand, G. J., and Lamanna, C. (1960). *J. Exptl. Med.* **111,** 745.

Hobbs, G., and Anderson, A. W. (1967). *In* "Botulism 1966" (M. Ingram and T. A. Roberts, eds.), pp. 417–424. Chapman and Hall, London.

Holdeman, Lillian L. (1967). *In* "Botulism 1966" (M. Ingram and T. A. Roberts, eds.), pp. 176–184. Chapman and Hall, London.

Ingram, M., and Roberts, T. A. (eds.) (1967). "Botulism 1966." Chapman and Hall, London.

Ingram, M., and Robinson, R. H. M. (1951). *Soc. Appl. Bacteriol. Proc.* **14,** 62.

Ito, K. A., Seslar, Donna J., Mercer, W. A., and Meyer, K. F. (1967). *In* "Botulism 1966" (M. Ingram and T. A. Roberts, eds.), pp. 108–122. Chapman and Hall, London.

Jensen, W. I., and Gritman, R. B. (1967). *In* "Botulism 1966" (M. Ingram and T. A. Roberts, eds.), pp. 407–413. Chapman and Hall, London.

Kaufman, O. W., and Brilland, A. R. (1964). *Am. J. Publ. Health* **54,** 1514.

Kaufman, O. W., and Crecelius, E. M. (1967). *Vet. Res.* **128,** 1857.

Kaufman, O. W., and Fay, L. D. (1964). *Quart. Bull. Mich. Agr. Expt. Sta.* **47,** No. 2, p. 236.

Kaufman, O. W., Solomon, H. M., and Monheimer, R. H. (1967). *In* "Botulism 1966" (M. Ingram and T. A. Roberts, eds.), pp. 400–406. Chapman and Hall, London.

Kempe, L. L., Graikoski, J. T., and Bonventre, P. F. (1957a). *Appl. Microbiol.* **5,** 292.

Kempe, L. L., Graikoski, J. T., and Bonventre, P. F. (1957b). *Appl. Microbiol.* **6,** 261.

Kravtchenko, A. T., and Shiskulina, L. M. (1967). *In* "Botulism 1966" (M. Ingram and T. A. Roberts, eds.), pp. 13–19. Chapman and Hall, London.

Lamanna, C. (1959). *Science* **130,** 767.

Lamanna, C., and Hart, E. R. (1967). *In* "Botulism 1966" (M. Ingram and T. A. Roberts, eds.), pp. 370–377. Chapman and Hall, London.

Landgrebe, J. C., and Weaver, R. H. (1964). *Bacteriol. Proc.* 103.

Lewis, K., and Cassel, K., Jr. (eds.) (1964). "Botulism 1964." U. S. Department of Health, Education, and Welfare, Public Health Service.

Matveev, K. I., Nefedjev, N. P., Bulatova, T. I., and Sokolov, I. S. (1967). *In* "Botulism 1966" (M. Ingram and T. A. Roberts, eds.), pp. 1–10. Chapman and Hall, London.

McCollough, M. L., Anderson, A. W., and Elliker, P. R. (1966). *Bacteriol. Proc.* p. 76.

McCroan, J. E., McKinley, T. W., Brim, A., and Henning, W. C. (1964). *Public Health Rept.* (U.S.) **79**, 997.

Meyer, E. A., and Lamanna, C. (1959). *J. Bacteriol.* **78**, 175.

Meyer, K. F. (1956). *Bull. World Health Organ.* **15**, 281.

Minervin, S. M. (1967). *In* "Botulism 1966" (M. Ingram and T. A. Roberts, eds.), pp. 336–345. Chapman and Hall, London.

Mitruka, B. M., and Costilow, R. N. (1966). *Bacteriol. Proc.* p. 102.

Møller, V., and Scheibel, I. (1960). *Acta Pathol. Microbiol. Scand.* **48**, 80.

Müller, J. (1967). Medlamblad for den danske Dyrlaegeforening **50**, 887.

National Canners Association (1966a). Research Information, March 1966. No. 112.

National Canners Association (1966b). Research Information, November 1966. No. 120.

Ohye, D. F., and Scott, W. J. (1953). *Australian J. Biol. Sci.* **6**, 178.

Oscheroff, B. J., Slocum, G. G., and Decker, W. M. (1964). *Public Health Rept. (U.S.)* **79**, 871.

Presscott, S. C., and Tanner, F. W. (1938). *Food Res.* **3**, 189.

Prevot, A. R., Turpin, A., and Kaiser, P. (1967). "Les bacteries anaerobies." Dund, Paris.

Prevot, A. R., and Fredette, V. (1966). "Manual for the Classification and Determination of the Anaerobic Bacteria." Lea and Febiger, Philadelphia.

Riemann, H. (1957). *J. Appl. Bacteriol.* **20**, 404.

Riemann, H. (1960). *Nord. Vet. Med.* **12**, 86.

Riemann, H. (1963). "Germination of Bacterial Spores with Chelators." Copenhagen, 1963.

Roberts, T. A. (1967). Personal communication.

Schantz, E. J. (1964). *In* "Botulism 1964" (K. Lewis and K. Kassel, Jr., eds.), pp. 91–103. U. S. Department of Health, Education, and Welfare, Public Health Service.

Schantz, E. J., and Spero, L. (1967). *In* "Botulism 1966" (M. Ingram and T. A. Roberts, eds.), pp. 296–301. Chapman and Hall, London.

Scheibel, J., and Lennant-Petersen, O. (1958). *Acta Pathol. Microbiol. Scand.* **44**, 222.

Schmidt, C. F. (1959). *Bacteriol. Proc.* p. 10.

Schmidt, C. F. (1964). *In* "Botulism 1964" (K. Lewis and K. Cassel, Jr., eds.), p. 69. U. S. Department of Health, Education, and Welfare, Public Health Service.

Scholtens, R. G., and Cohen, D. B. (1964). "Scientific Proceedings of the 101st Annual Meeting of the American Veterinary Medical Association, 1964," p. 224.

Sehgal, L. R., and Grecz, N. (1967). *Bacteriol. Proc.* p. 5.

Skulberg, A. (1961). *Nord. Vet. Med.* **13**, 84.

Skulberg, A. (1964). "Studies on the Formation of Toxin by *Clostridium botulinum.*" A/S Kaare Grytting, Orkunger, Norway.

Skulberg, A., and Hausken, O. W. (1965). *J. Appl. Bacteriol.* **28**, 83.

Smith, L. D. S. (1955). "Introduction to the Pathogenic Anaerobes," pp. 109–112. University of Chicago Press.

Smith, L. D. S. (1957). *Advan. Vet. Sci.* **3**, 363.

Sommer, E. W. (1930). *J. Infect. Diseases* **46**, 85.

Starin, W. A. (1926). *J. Infect. Diseases* **38**, 106.

Sterne, M., and Wentzel, L. M. (1950). *J. Immunol.* **65**, 175.

Sugiyama, H. (1951). *J. Bacteriol.* **62**, 81.

Sussman, A. S., and Halvorson, H. O. (1966). "Spores, Their Dormancy and Germination." Harper and Row, New York.

Thatcher, F. S., Robinson, J., and Erdman, I. (1962). *J. Appl. Bacteriol.* **25**, 120.

Treadwell, P. E., Jann, G. J., and Salle, A. J. (1958). *J. Bacteriol.* **76**, 549.

Vincenzi, F. F. (1967). *Nature* **213**, 394.

Vinton, C., Martin, S., and Gross, C. E. (1947). *Food Res.* **12**, 184.
Wagenaar, R. O., and Dack, G. M. (1958). *J. Dairy Sci.* **41**, 1182, 1191, 1196.
Wagenaar, R. O., and Dack, G. M. (1960). *Food Res.* **25**, 279.
Wagman, J. (1954). *Arch. Biochem. Biophys.* **50**, 104.
Wagman, J. (1963). *Arch. Biochem. Biophys.* **100**, 414.
Wagman, J., and Bateman, J. B. (1951). *Arch. Biochem. Biophys.* **31**, 424.
Wagman, J., and Bateman, J. B. (1953). *Arch. Biochem. Biophys.* **45**, 375.
Watts, Nancy W. (1967). *In* "Botulism 1966" (M. Ingram and T. A. Roberts, eds.), pp. 158–168. Chapman and Hall, London.
Whaler, B. C. (1967). *In* "Botulism 1966" (M. Ingram and T. A. Roberts, eds.), pp. 377–387. Chapman and Hall, London.
Wynne, E. S., and Foster, J. W. (1948a). *J. Bacteriol.* **55**, 69.
Wynne, E. S., and Foster, J. W. (1948b). *J. Bacteriol.* **55**, 61.
Wynne, E. S., and Foster, J. W. (1948c). *J. Bacteriol.* **55**, 331.

CHAPTER VIII | BOTULISM — TYPE E

Genji Sakaguchi

I. INTRODUCTION

Clostridium botulinum type E represents one of six types of the species, distinguished by the toxin neutralization reaction. The symptoms of type E botulism in man and animals are essentially the same as those of

329

other types. There are, however, some characteristic features of type E botulism, such as foodstuffs implicated, physiology of the causative organisms, heat lability of the spores, and activation phenomenon of the toxin.

Type E spores have been demonstrated in aquatic as well as in terrestrial specimens in different areas of the world. The foodstuffs implicated in type E botulism have mainly been uncooked or partially cooked, preserved fish or sea mammals. Such foodstuffs have generally been home-prepared ones, but recent outbreaks due to commercially prepared foodstuffs have stimulated the present revival of interest in botulism.

History of Type E Botulism

Gunnison *et al.* (1936) proposed for the first time the designation of *Clostridium botulinum* type E for the cultures that had been isolated from the intestines and muscles of sturgeon in the U.S.S.R.

Two outbreaks of type E human botulism occurred in New York State in 1932 and 1934. The first outbreak involving three cases with one death was due to smoked salmon from Canada; the second one involving three cases with one death was due to canned sprats from Germany (Hazen, 1938).

Between the 1941 outbreak in San Francisco (Geiger, 1941) and the 1960 outbreak in Minneapolis (Kautter, 1964), no type E outbreak occurred on the mainland of the United States. During and after these two decades, type E outbreaks have occurred in Canada, Alaska, Japan, Denmark, Sweden, Norway, the U.S.S.R., and Poland.

II. TYPE E BOTULISM

A. Incidence and Mortality

From 1932 to 1964, there were at least 96 bacteriologically proved outbreaks in eight different countries, involving 449 cases with 135 deaths, the fatality rate being 30.1%. The total outbreaks in each country are listed in Table I. Of the total, 50 were in Japan, 29 in North America, and 17 in Europe. Total outbreaks in various countries were presented by Dolman (1964). Further information was provided by Forfang and Skulberg (1964), by Meisel *et al.* (1964), and also through personal communications from Dr. H. Iida of Hokkaido Institute of Public Health, Sapporo, from Dr. H. O. Pedersen of Frederiksberg Municipal Depart-

TABLE I
INCIDENCE OF TYPE E BOTULISM IN DIFFERENT COUNTRIES[a]

Place		Year	Outbreaks	Cases	Deaths
Japan:	Hokkaido	1951–1964	31	225	42
	Tohoku area	1953–1963	19	82	35
United States:	Alaska	1950–1962	8	22	8
	Other states	1932–1963	8	37	15
Canada		1944–1963	13	36	22
Poland		1962–1963	5	23	5
U.S.S.R.		1938–1961	4	5	4
Denmark		1951–1964	4	9	2
Sweden		1960–1962	3	5	1
Norway		1963	1	5	1
Total			96	449	135

[a] Case fatality rate: 30.1%

ment of Health, Copenhagen, and from Dr. H. Meisel of State Institute of Hygiene, Warsaw. Outbreaks have occurred only in the countries located in the northern latitude of the Northern Hemisphere, where inhabitants traditionally eat uncooked, preserved fish or sea mammals.

In addition to those listed in Table I, many additional cases and deaths with characteristic symptoms due to consumption of uncooked or partially cooked fish or sea mammals are recorded in various countries (Dolman, 1964). No doubt a large part of these cases were type E botulism.

B. Foodstuffs Involved

1. GENERAL CONSIDERATIONS

Foodstuffs involved in type E botulism in the different countries are listed in Table II. Almost all the outbreaks were due to saltwater or freshwater fish, fish eggs, or sea mammals. The exceptions were the outbreak in the United States in 1941 due to canned mushroom sauce and the two outbreaks in the U.S.S.R. in 1961 due to pickled ham and pickled red beets. All the foodstuffs implicated were uncooked or lightly cooked products, except the three outbreaks in the United States in 1934, 1941, and 1963, due, respectively, to commercially canned sprats, mushroom sauce, and tuna fish; the one in Japan in 1956 due to commercially canned mackerel; and the four outbreaks in Poland in 1962 due to canned herring. In the canned tuna fish outbreak in Detroit, Michigan, type E toxin was

TABLE II
FOODSTUFFS INVOLVED AND GEOGRAPHICAL AREAS FOR OUTBREAKS OF TYPE E BOTULISM

Area	Foodstuff	Number of outbreaks	Area	Foodstuff	Number of outbreaks
Japan	"Izushi" of:		United States— mainland	Smoked fish (salmon 1; whitefish 2)	3
	Sole	15		Canned fish (sprats, tuna)	2
	"Hata-hata" (*Arctoscopus japonicus*)	11		Salmon eggs	1
	Horse mackerel	3		Cisco, vacuum-packed	1
	Mackerel pike	3		Canned mushroom sauce	1
	Mackerel	2		Canned herring	4
	Sardine	2		Smoked cod	1
	Herring	1		Smoked herring	1
	Salmon	1		Salted sea bass	1
	Other saltwater fish	7		Pickled ham	1
	Freshwater fish	3		Pickled beets	1
	"Kirikomi" (of sole, herring, "onago")	3	Poland	Pickled fish (herring, trout)	2
	Trout eggs	1	U.S.S.R.	Salted herring	1
	Canned mackerel	1		Uncooked trout	1
Canada	Salmon eggs	5	Denmark	Herring (pickled, rollmops)	2
	Seal flippers (utjak)	3	Sweden	Unknown	1
	Pickled fish (herring, trout)	2	Norway	Fermented fresh-water trout	1
	Salted herring	1			
	Canned salmon	1			
	Beluga flippers (muktuk)	1			
United States— Alaska	Beluga flippers (muktuk)	4			
	Salmon eggs	3			
	Whale fluke	1			

found in cans from the same lot which had faulty seals. The product was apparently contaminated after retorting (Dack, 1964).

The food items listed in Table II may look uncommon, but they are characteristic of each area. For instance, Canada and Alaska, where Eskimos and Indians have been the main victims, had outbreaks resulting from consumption of salmon eggs, beluga flippers, and seal flippers; Japan, from "izushi" of various fishes; Scandinavian countries, from pickled or other types of preserved fish. Outbreaks of type E botulism appear to be closely associated with the peculiar eating habits of local inhabitants.

There seem to be some common characteristics of these foodstuffs: (1) The foods are fish or sea mammals. (2) The fish are seldom cooked, since cooking destroys the flavor and acceptability of the products. (3) The seafoods are kept standing for a considerable period to remove blood and to allow fermentation or ripening. (4) The processed foods are usually acidic.

2. SALMON EGG CHEESE OR STINK EGGS

Salmon eggs are often eaten in the form of cheese or stink eggs by Eskimos and Indians. Salmon roe are put in a gunnysack and kept in a running stream overnight to remove blood and other foreign substances. The eggs are smoked for about 12 hours, crushed, and placed in a large wooden container, where they are kept standing for a few weeks to allow fermentation. When it is no longer gummy and no longer sticks to the teeth, salmon egg cheese is ready to eat (Dolman *et al.*, 1955).

3. BELUGA FLIPPERS, OR "MUKTUK"

The skin and underlying blubber or beluga flippers are cut into chunks or strips. They are dried outdoors on a rack or over a pole for a few days, then taken into a hut, put in brine, and kept standing for several weeks to mature. The preparation is sliced into small pieces to serve (Dolman, 1964).

4. SEAL FLIPPERS, OR "UTJAK"

Seal flippers are kept in a hut for several days in drums of seal oil, until the skin falls loose, and then the food is ready to eat (Dolman, 1964).

5. PICKLED HERRING

Herring are gutted, the heads and tails chopped off, and the bones removed. The fillets are put into a pickling solution consisting of vinegar and water for a day or so and then soaked in oil and spices for an additional few days (Pedersen, 1955).

6. "Izushi"

"Izushi" is a traditional home-made food in northern Japan. Any kind of fish, either freshwater or saltwater fish, can be used. The fish are gutted, the heads and tails are removed, and the fish are soaked in water for 4 to 7 days to remove blood. Large fish are sliced. They are packed in a wooden tub with cooked rice, malted rice or "koji," and cut-up vegetables, such as carrots, cabbages, and radishes. Small amounts of vinegar, salt, sometimes "sake," and red pepper are added. The lid of the tub is weighted with heavy stones to compress the solids. After standing for a few weeks to allow lactic fermentation to take place, the contents are ready to eat. The preparation usually has a pH value of 4 to 5 (Nakamura *et al.*, 1956).

7. "Kirikomi"

Fish are prepared in the same way as for "izushi." After the blood is removed by soaking in water, the thinly sliced fish are packed in a wooden tub with malted rice and salt. This mixture is kept standing for about 2 weeks for ripening.

C. Clinical Manifestations

1. Incubation Period

Symptoms of botulism usually develop within 12 to 36 hours; for type E, shorter incubation periods have often been reported. In the Mashike outbreak in Hokkaido in 1957, among 35 patients, 14 developed symptoms within 16 to 20 hours, 8 within 11 to 15 hours, and 7 within 6 to 10 hours (Iida *et al.*, 1958b). In the Toyotomi outbreak in 1962, from a total of 55 patients, 17 developed symptoms within 16 to 20 hours, 8 within 11 to 15 hours, 8 within 21 to 25 hours, 8 within 26 to 30 hours, 5 in less than 5 hours, and the remaining 9, all slightly ill, within 31 hours or longer (Iida *et al.*, 1964).

2. Symptoms

The symptoms of type E botulism are usually indistinguishable from those of type A or B botulism (see Chapter 7, Section II, C.). There is no retention of urine in type A or B botulism (Dack, 1956), although this has been recorded in type E cases occasionally. In the Mashike outbreak, retention of urine was recorded in 8 cases, or 23% (Iida *et al.*, 1958b).

The same was also recorded in the United States (Rogers *et al.*, 1964), together with severe nausea and vomiting, and pharyngeal erythema, which
had not been stressed in type A outbreaks.

3. DURATION

In the Mashike outbreak, all the 9 fatal cases died within 60 hours
after the development of symptoms, 2 of them within as short a period as
10 hours. Such rapid death was occasionally observed in other outbreaks
(Iida *et al.*, 1958*b*). Dolman (1960) described several cases in which death
occurred in less than 24 hours and stated that there had been no fatalities
later than the third day. From a total of 24 patients hospitalized in the
Toyotomi outbreak, 1 died in 28 hours, and the other 23, who received
antitoxin injections, were discharged in 3 to 35 days, the average being
20.3 days (Iida *et al.*, 1964).

The shorter incubation period and the rapid course of type E botulism
are demonstrated also in laboratory animals. The mice inoculated intraperitoneally with type E toxin of 1 MLD or so usually die within 24 hours.

III. ECOLOGY: DISTRIBUTION OF SPORES

Type E spores have been demonstrated in both aquatic and terrestrial
specimens: soil, lakeshore mud, coastal sand, sea-bottom sludge, fish,
potato peels, etc. (Occurrence of type E botulism in animals is discussed
in Chapter 7, Section III, B.)

Nakamura *et al.* (1956) collected a total of about 2300 samples of soil,
lakeshore mud, and coastal sand in Hokkaido and demonstrated type E
toxin in 82 cultures (3.6%). Kanzawa (1960) demonstrated type E toxin
in 30 of 315 soil samples (9.5%) taken along the River Ishikari in
Hokkaido from the mouth to 50 km upstream. He also demonstrated
type E toxin in 7 cultures from 30 (23%) samples of coastal sand.

Dead, floating fish in Hachiro Lagoon in Akita near the endemic area
yielded type E cultures in high frequency: 10 of 12 (83%) in one survey,
and 11 of 12 (92%) in another, in comparison with only 2 of 500 (0.4%)
from fresh fish (Kodama *et al.*, 1964). It is of interest that of 33 "izushi"
samples — 14 home-made and 19 commercial preparations — 4 samples
(12%), 3 home-made and 1 commercial, yielded toxigenic type E cultures.

Kobayashi (1961) isolated 2 type E cultures from mud samples collected at the Central Fish Market, Tokyo. Of 756 soil samples collected
in Yamaguchi Prefecture, where no outbreak has been reported, Yamagata (1963) isolated 3 type E strains.

One hundred and eighty-three sea-bottom samples from British Columbia coastal inlets yielded 5 type E cultures (2.7%). Twenty-seven samples taken from more shallow bottoms along the northern British Columbia coast yielded 7 strains (26%). From 47 surface sand samples collected near the outflow of Lake Pitt, type E toxin was demonstrated in 11 (23%). Type E spores were negative in 20 samples taken from or near Vancouver beaches, in 80 silt samples from Hooper Bay, and in 142 ocean-bottom samples from the Bering Sea and Bering Strait near the endemic areas in Alaska (Dolman and Iida, 1963).

Of 34 soil samples from fish markets, canals, and harbors in Copenhagen, 16 (47%) yielded type E toxin, as did 1 of 4 sea mud samples from Greenland (25%) (Pedersen, 1955). A total of 118 bottom samples from the Baltic Sea, the Sound Sea, the straits of Kattegat and Skagerrak, and major watercourses was examined by Johansen (1963). All the cultures were positive for type E toxin. Of 167 shore samples of Sweden's coast, major and minor lakes, and watercourses, 118 (71%) were positive. A total of 94 fish caught in the Sound Sea, the southern Baltic Sea, and the strait of Kattegat were tested, and type E toxin was demonstrated in 43 (46%). Of 87 inland soil samples of various origins, 71 (82%) yielded type E toxin. More surprising is the fact that, of 40 samples of peels of potatoes from different districts, including Israel, 27 (68%) were type E toxigenic.

In the United States, 604 samples of water, mud, fish, and other substances from Lake Michigan and Lake Superior were tested by Bott *et al.* (1964). Mouse lethal toxin was shown in 227 samples (38%), of which type E toxin was shown in 24 samples (2 of 8 water samples, 1 of 20 mud samples, 16 of 82 fish samples, 3 of 4 swab samples of fishing boats, and 2 of 9 other samples). Ward and Carroll (1965) demonstrated the presence of type E spores in mud samples taken from estuarine waters of the Gulf of Mexico.

Although the percentage of detected organisms of a certain species in large numbers of samples may vary according to the method employed, no doubt type E spores are distributed widely not only in endemic areas but also in areas where no outbreak has occurred. The findings in different laboratories show that type E spores are terrestrial in origin and are carried to the sea or lakes by streams (Kanzawa, 1960; Johansen, 1963; Dolman and Iida, 1963). It may also be true that certain areas are more heavily contaminated than others. Dolman and Iida (1963) described possible nidification by type E spores. This nidification or replenishing of the spots with type E spores may be accomplished by the steady transportation by streams and by multiplication of the organisms in fish bodies as indicated by much higher frequency of detection in dead, floating fish in Hachiro Lagoon. The same mechanism of replenishment might be possible in the carcasses of land animals (Kanzawa, 1960).

IV. MORPHOLOGY AND PHYSIOLOGY

A. Morphology

Clostridium botulinum type E is a gram-positive, rod-shaped, spore-forming, anaerobic organism producing a type-specific toxin. The rod has long peritrichous flagella and is actively motile. The size of the rod is 0.8 to 1 μ by 4 to 6 μ, with round ends. Spores are oval and sub-terminal and may or may not distend the sporangium.

B. Biochemical Properties

Gunnison *et al.* (1936) reported that the Russian strains were non-hemolytic, while weak hemolytic activity was reported by other workers.

All type E strains ferment glucose, fructose, maltose, sucrose, sorbitol, and mannose, with production of acid and gas; all the strains do not ferment lactose, raffinose, mannitol, rhamnose, galactose, or dulcitol.

Coagulated egg white or serum, meat or liver particles, casein, etc., are not digested. Positive liquefaction of gelatin was described (Gunnison *et al.*, 1936; Hazen, 1937; Dolman *et al.*, 1947, 1950). All 22 type E strains isolated in Japan were negative for gelatin liquefaction (Nakamura *et al.*, 1956). All 10 strains originated in Denmark did not liquefy gelatin (Pedersen, 1955).

Type E organisms can be differentiated from types A and B in the fermenting of sucrose and mannose and in their failure to liquefy gelatin, in addition to toxin neutralization tests.

C. Germination, Development, and Outgrowth

1. GENERAL CONSIDERATIONS

The spores are found in foodstuffs, environmental specimens, and cultures, but little is known about sporulation. Schmidt *et al.* (1962) utilized trypticase-glucose-peptone broth for the production of type E spores. The medium supported production of 50 to 130 million spores per milliliter in one day at 28°C.

General acidity of the foodstuffs involved and the relationship between type E organisms and fish or sea mammals may seem to suggest acid and salt tolerance of the organisms. The available information, however, does not support this view. It has been shown that type E organisms have a psychrophilic property; the spores can germinate and multiply at 3° to 4°C. This property may play an important role in pathogenesis of type E botulism.

2. EFFECT OF ENVIRONMENT

a. *Physical.* A slight toxin production by a type E strain at 6°C was reported in macerated herring (Dolman *et al.,* 1950). Ohye and Scott (1957) demonstrated positive growth by 10 strains in cooked meat medium at 5°C. Type E toxin of 160 MLD/gm was demonstrated after trypsin treatment of sole inoculated with the spores and kept at 4°C for 21 days (Sakaguchi *et al.,* 1960). Schmidt *et al.* (1961) demonstrated growth and toxin formation in 31 to 45 days at 3.3°C with 4 type E strains and in a beef stew medium, but no growth at 1.1°C or 2.2°C after 104 days. At 6.1°C and 9.4°C, growth and toxin production were more rapid (Schmidt *et al.,* 1962).

b. *Chemical.* Herring were inoculated with type E spores and kept at 23°C for 3 days. Portions of the toxic herring were pickled and kept longer. The pickling solution with a pH range of 4.5 to 4.8 exhibited a high toxicity for several weeks, and the organisms did multiply intermittently (Dolman *et al.,* 1950). The same author described toxin production in pickled herring at even lower pH (pH 4.0 to 4.2) (Dolman and Iida, 1963).

Tubes of brain heart infusion broth, adjusted to various pH's, were inoculated with spores of the Tenno strain and incubated at 30°C. Sufficient growth occurred in 46 hours in the tubes with an initial pH range of 5.9 to 8.6, but no growth at pH 5.5 or lower, or at pH 8.9 or higher (Table III) (Sakaguchi *et al.,* 1960).

Type E strains grew at 3.9% salt concentration but not at 4.1% salt in peptic digest liver broth at 30°C. In the same medium, a type A strain grew at 7.3% salt concentration and a type B strain at 6.3% salt at 37°C (Pedersen, 1957). Salt tolerance of 4 type E strains was examined also by Segner *et al.* (1964). They grew at 4.0% salt but not at 4.5% salt at 10°C or lower; at 15.5° to 29.5°C, growth occurred at 4.5% salt but not at 5.0% salt; at 36.6°C, growth occurred at 3.5% salt but not at 4.0% salt.

No information is available on the minimum nutritional requirements of type E organisms. Studies were reported on the outgrowth of type E spores at 7.8°C in 22 different protein hydrolyzates (5% solution with 0.4% glucose, pH 7.0) and in 6 complex media (Schmidt, 1964). The most

TABLE III

INFLUENCE OF pH ON GROWTH AND TOXIN PRODUCTION BY
Clostridium botulinum TYPE E, THE TENNO STRAIN,
IN BRAIN HEART INFUSION BROTH[a]

			Toxin potencies (MLD/ml)		
		Bacterial	Culture supernatant		
Initial pH	Final pH	protein (μg/ml)	Nonactivated	Activated	Cells Activated
5.53	5.52	0	0	0	0
5.89	5.43	430	16	4,100	4,100
6.09	5.26	535	32	8,200	16,000
6.58	5.40	599	130	16,000	16,000
6.93	5.71	631	1,000	16,000	4,100
7.38	5.76	603	4	1,000	8,200
7.76	6.05	535	4	260	130
8.13	6.39	469	2	64	260
8.57	6.77	453	0	0	130
8.85	8.68	0	0	0	0

[a] Determinations were made after incubation at 30°C for 46 hours.

rapid outgrowth was observed in peptone and in trypticase-peptone; that in all other protein hydrolyzates tested was slower. Schmidt (1964) also stated that, although sodium nitrite of 150 to 200 ppm did not affect the growth of type E organisms in trypticase glucose peptone broth at 28°C, these concentrations caused a twofold to threefold extension of growth at 7.8°C. A similar effect was observed with sodium benzoate at 1000 ppm.

D. Toxin Formation

It has been known from the beginning that type E strains are poor producers of toxin as compared with other types and that toxin formation is more constant at 30°C or lower temperatures than at 37°C, although growth occurs at either temperature (Gunnison et al., 1936; Hazen, 1937). Sometimes, type E strains do not produce toxin at all at 37°C (Pedersen, 1955).

The potency of toxin in the cell-free filtrate of a culture may be dependent on at least three different factors: (1) The amount of toxin precursor, which is synthesized in the cytoplasm of the bacterial cells (Sakaguchi and Tohyama, 1955b; Skulberg, 1964). An abundant growth is, therefore, a prerequisite for the high potency. (2) The rate of release of the precursor and self-activation (see Section VI, B, 1, c). (3) The rate of destruction of preformed precursor and toxin.

Toxin potencies in the culture supernatant and viable cell counts of the Tenno strain grown at 37°C were determined at intervals. The viable count reached the highest level in 24 hours, whereas toxin in the supernatant was not demonstrated before the second day. The highest potency was shown on the fourth day (Sakaguchi and Tohyama, 1955a). This apparently indicates that the exotoxin appears in the medium only after the death of the organisms.

Pedersen (1955) analyzed toxin formation at 30°C and at 37°C with 3 type E strains. The toxic potency of all strains reached the highest level in 4 days at 30°C, and the level did not change for 10 days or longer. At 37°C, the highest toxin potencies were attained in 2 days, but the level was maintained for only 2 or 3 days. Pedersen concluded that type E strains produce toxin at either temperature, but inactivation of toxin at 37°C is very rapid, whereas that at 30°C is negligible.

Seldom does a type E strain produce toxin of 5000 LD_{50}/ml or higher potency as determined by mouse intraperitoneal injections. An unusually potent toxin, 50,000 LD_{50}/ml, was described by Dolman (1953). This unusually potent toxin was reproducible with a mixed culture of the toxigenic (TOX) strain and the nontoxigenic, proteolytic mutant (TP) (Dolman, 1957a) (see Section III, E).

Toxin potency of type E cultures in a suitable medium reaches 3000 to 8000 LD_{50}/ml after 5 days of incubation at 30°C (Duff et al., 1956).

E. Mutants

Strains of Clostridium botulinum show a tendency to become nontoxigenic, and type E strains have this tendency. From the cultures that had become nontoxigenic, Dolman (1957a) demonstrated two kinds of mutant colonies; one formed opaque colonies, being rich in spores, and the other formed transparent, flat colonies showing proteolytic activity. The former was designated as OS phase (opaque, sporulating), and the latter as TP phase (transparent, proteolytic). These mutants were readily distinguishable from toxigenic colonies (TOX). He stated that all type E TOX cultures tended to degrade to the OS phase, and TOX ⟷ OS ⟷ TP or TOX ⟷ TP processes were noted less frequently. The OS and TP mutants are completely nontoxigenic, but either may regain toxigenicity. These mutant phases were different also in carbohydrate fermentation. The TP cultures produced acid and gas from glucose, fructose, and maltose; only traces of acid or gas from sorbitol, glycerol, and salicin; and no acid or gas from sucrose. The OS cultures produced acid but no gas from glucose, fructose, maltose, sucrose, and glycerol; and no acid or gas from sorbitol and salicin.

Strains having similar properties to *Clostridium botulinum* type E and yet being nontoxigenic have been isolated from samples of soil and fish (Pedersen, 1955; Dolman *et al.,* 1955; Nakamura *et al.,* 1956; Hobbs *et al.,* 1965).

Dolman's observation on colonial mutation was supported (Iida, 1963; Skulberg, 1964), whereas Hobbs *et al.* (1965) raised a question. They could demonstrate neither the restoration of toxigenicity by the OS cultures nor the appearance of OS or TP colonies from any of their toxic cultures. None of the OS cultures was stained with fluorescent antibody prepared against type E cell antigens (Walker and Batty, 1964), whereas all toxigenic type E strains showed fluorescence. Moreover, they demonstrated α-toxin of *Clostridium welchii* in culture filtrates of the 3 OS strains. They alluded to the possibility that the OS variants, described by Dolman, are contaminating organisms originating from soil or fish and are difficult to separate from *Clostridium botulinum* type E.

V. RESISTANCE OF ENDOSPORES

A. Heat

Type E botulism is rarely due to heat-processed foodstuffs because of the unusually low heat resistance of the spores. The spores were killed in a buffer of pH 7.4 after heating for 2 to 5 minutes at 100°C or for 6 to 40 minutes at 80°C, depending on the spore concentration (Gunnison *et al.,* 1936).

Dolman and Chang (1953) stated that the spores of certain strains occasionally withstood exposure to 100°C for 30 minutes. Graikoski and Kempe (1964) reported less resistance but occasionally found relatively heat-resistant spores that withstood exposure to 85°C for 120 minutes or to 90°C for 60 minutes.

Sakaguchi *et al.* (1954) reported that 5 million spores of a strain in 1 ml of *M*/15 phosphate buffer, pH 7.0, were killed by exposure to 75°C for 20 minutes or to 80°C for 5 minutes. Pedersen (1955) stated that most of the type E spores occurring in nature can be killed by heating for 5 minutes at 80°C. Nakamura *et al.* (1956) tested the heat resistance of the spores of 8 strains at 70°C, 80°C, and 90°C. All of them survived for 120 minutes at 70°C but were killed within 20 minutes at 80°C or within 5 minutes at 90°C. Five strains survived after heating at 80°C for 5 minutes; 2 of them survived for 10 minutes. Only one strain, the Iwanai strain, withstood heating at 90°C for 3 minutes but was killed in 5 minutes.

Schmidt (1964) obtained D values at 80°C; 1.8 for the Minneapolis strain in trypticase-glucose-peptone broth, and 2.3 in $M/15$ phosphate buffer, pH 7.0. The values lie between those for the strains Nanaimo ($D_{80} = 0.6$) and 108 ($D_{80} = 3.3$), which had been determined by Ohye and Scott (1957).

Roberts et al. (1965) tested 8 type E strains in aqueous suspension. The D values at 80°C ranged from 0.33 to 1.25, and the z values from 7.4 to 10.7. In most instances no survivors of 10^7 spores were detected after heating at 80°C for 6 minutes.

B. Radiation

Erdman et al. (1961) found that an irradiation dose of 0.20 megarad was necessary to kill 90% of type E spores in broth. Samples of canned beef stew inoculated with 10^6 spores were irradiated with various doses and stored at 18°C; growth and toxin formation were examined at intervals (Schmidt et al., 1962). The mean radiation D value for 6 strains was 0.132 megarad, about half that required for type A or B spores.

Roberts et al. (1965) showed radiation survival curves for 9 type E strains. There was a slight variation in the "shoulder" of the curves which extended to about 0.4 megarad, but after that the curve became exponential with a D value from 0.065 to 0.16 megarad. Some type E strains showed higher resistance than did some strains of types A and B. They also showed the effect of sporulation medium on radiation resistance.

VI. TOXIN

A. Immunology and Serology

1. ANTIGENIC SPECIFICITY

Botulinus toxin of each type is antigenically specific. A minor cross neutralization was recorded between the international standard for type E antitoxin and type D toxin, and type C and D antitoxins and type E toxin; no cross reaction was observed among types A, B, and E (Bowmer, 1963). A considerable cross neutralization between type E antitoxins and type F toxin was described (Møller and Scheibel, 1960; Dolman and Murakami, 1961).

2. Toxoid

In 14 to 22 days, four nonactivated and four activated toxins (see Section VI, B) became nontoxic for mice at pH 5.5 and 33°C in the presence of 0.6% formalin (Gordon *et al.,* 1957). Other preparations of activated toxin became nontoxic for guinea pigs in 18 to 25 days at pH 5.0 and 33°C in the presence of 0.6% formalin (Fiock *et al.,* 1961). These toxin preparations had been highly purified. As with type A and B toxins, type E toxin is more readily detoxified for mice than for guinea pigs. Dolman (1964) stated that toxoids prepared from trypsin-activated toxins are less effective antigens than those from the nonactivated toxins, although no significant change in antitoxin-combining power results from activation with trypsin (Dolman, 1957*a*).

3. Antitoxin

Antitoxin is essential for the identification of toxin type by neutralization reactions and is the only specific therapeutic agent available at present.

Therapeutic type E horse antitoxin was prepared in Canada by collaboration between the Connaught Medical Research Laboratories and the Defence Research Board, with the antigen prepared by Dolman and his associates. The potency of the preparation was roughly 5000 international units per 10-ml vial (Dolman and Iida, 1963). This preparation was made available in Japan, and the therapeutic effect was so impressive (see Section VIII, A) that the manufacture of type E antitoxin was started at the National Institute of Health, Tokyo. Partially purified trypsin-activated type E toxin of cell origin (Sakaguchi and Sakaguchi, 1961) was used as the antigen. Alum-precipitated toxoid was used for the primary inoculations, and the toxin for the hyperimmunization program. The preparations had 2600 to 8400 international units per vial. The potencies were determined as relative values to the international standard antitoxin by the parallel-line assay method (Kondo *et al.,* 1963). In our hands, the Canadian antitoxin was shown to possess 13,000 international units per vial.

B. Biochemistry

1. Activation Phenomenon

a. Activation by a Proteolytic Enzyme of Heterologous Organisms. In 1953, six persons in Tenno-machi, Akita, Japan, ate gilthead "izushi"; four became ill, and two died. A type E strain (the Tenno strain) was isolated from the "izushi." The strain was very poor in toxin formation,

and the victims had eaten only about 20 gm of the "izushi." The filtrate of the "izushi" contained 200 MLD of type E toxin per milliliter
for mice when tested 20 days after the outbreak. The culture medium
inoculated with "izushi," after the incubation at 37°C for 24 hours, contained 400 MLD/ml, but the isolated strain never produced toxin higher
than 20 MLD/ml at the same temperature (Sakaguchi et al., 1954). A
strain of the genus Clostridium (the No. 13 strain) was isolated from the
"izushi." Strain 13 was nontoxigenic, but when it was cultured together
with the Tenno strain, the potency was increased 100 times or more. The
potentiated toxicity was reproduced when the cells of the Tenno strain
and the filtrate of strain 13 had been incubated together at 37°C. The
active substance in the culture filtrate of strain 13 appeared to be a
proteolytic enzyme with limited range of pH (5.0 to 6.5) for the action.
No preformed toxin was detected in the disintegrated bacterial cells;
therefore the synthesis of a "toxin precursor" in the bacterial cells and
the "activation" were postulated as essential processes in converting the
inactive precursor into active type E toxin (Sakaguchi and Tohyama,
1955a,b). The high potency of the toxin in the "izushi" was explained by
this activation phenomenon.

   b. *Activation by Trypsin.* Duff et al. (1956) reported potentiated
toxicity by treating type E cultures with trypsin at a pH range of 5.5 to
6.5. The activation ratio — that is, the ratio of potency before and after
trypsin activation — ranged from 12 to 47. They demonstrated that both
forms of toxin are relatively stable. The optimum pH for trypsin activation was 5.5 to 6.5 at 37°C; at pH 7.0, activation and destruction of toxin
took place simultaneously, resulting in lower toxicities. Activation by
trypsin is completely inhibited by soybean trypsin inhibitor.

   c. *Activation by the Proteinase of Homologus Organism.* A culture of *Clostridium botulinum* type E produces toxin of low potency.
This might be an indication that the precursor has been partially activated
by an enzyme possessed by the organisms. No proteolytic enzyme acting
on casein substrate could be demonstrated in type E cultures. Skulberg
(1964), however, described proteolytic enzyme in toxigenic type E cultures. The results as shown in Table III, indicating a higher rate of self-
activation at a higher pH, may support the view of enzymic self-
activation.

   The unusually potent toxin described by Dolman (1953) was ascribed
to the activation of toxin produced by the TOX phase by the proteolytic
enzyme of the TP phase (Dolman, 1957a). Whether the TP phase represents a homologous type E strain needs to be established.

It is not known, as yet, whether the precursor is completely nontoxic or relatively nontoxic or whether a low toxicity usually detectable in type E cultures represents the partially activated toxin or the toxicity of the precursor itself.

d. Activation in the Digestive Tracts. The activation phenomenon takes place in the digestive tracts. This makes the activation phenomenon of practical importance in the pathogenesis of type E botulism.

When nontreated and trypsin-activated type E cultures, markedly different in parenteral toxicity, were administered orally to mice, there was no significant difference in toxic activity, suggesting that the nontreated culture given orally is activated in the alimentary tract before absorption (Duff *et al.,* 1956).

Activation in the digestive tract of human beings has been demonstrated. Three Indian women in Prince Rupert became ill after consumption of salmon egg cheese. All of them died. About 250 MLD of type E toxin per gram was demonstrated in the remaining salmon egg cheese. Stomach contents of the victims taken at autopsy were pH 6.2 to 6.4 and possessed toxicities of 4000, 8000, and 60,000 MLD/ml, respectively. Apparently, the toxicities had been potentiated in the digestive tract after ingestion (Dolman, 1957*b*).

2. MECHANISM OF ACTIVATION

The mechanism of activation by proteinases deserves more study. The search will explain what chemical structure of the toxin molecule is necessary for such extraordinary toxocity and what specific structure makes the precursor inactive or relatively inactive.

Type E toxin is synthesized in the cytoplasm of the cells (Skulberg, 1964) in the form of a "precursor" which may be inactive, and a certain substantial change or "activation" converts the precursor into active toxin (Sakaguchi and Tohyama, 1955*a,b;* Duff *et al.,* 1956). This is analogous to the enzymic activation of proenzymes. Trypsin acts on the precursor itself, as the purified precursor is activated similarly and the activated toxin can be purified without any loss of toxicity. The optimum pH for activation, 5.5 to 6.5, is considerably more acid than the optimum for the usual proteolytic activity of trypsin. Activation and destruction of toxin take place simultaneously at pH 7 or higher (Duff *et al.,* 1956).

Electorophoretic studies indicated that the type E toxin precursor from the bacterial cells is strongly acidic because of the presence of ribonucleic acid. Activation involves a change of charge from negative to neutral or positive due to the splitting off of a moiety containing ribonucleic acid (Sakaguchi and Sakaguchi, 1959). This finding suggests that activation results from the release of a "masking group" which covers the toxic groupings.

Gerwing et al. (1961) claimed that activation results from fragmentation of a large protoxin molecule by trypsin, releasing more toxic sites. This description was based on the finding that, in ion exchange chromatography on DEAE cellulose, nonactivated toxin was eluted in one peak and trypsin-activated toxin in two peaks. Further, they demonstrated that the nonactivated type E toxin had a sedimentation constant ($S_{20,w}$) of 5.6 Svedberg units, whereas trypsin-activated toxins did not form a boundary under identical conditions and were dialyzable (Gerwing et al., 1962).

Sakaguchi et al. (1964) demonstrated that neither the precursor nor the trypsin-activated toxin penetrates swollen gel of Sephadex G-200; therefore, fragmentation may not be associated with activation. This view has been further confirmed. Both precursor and activated toxin were sedimented to the same relative position in terms of toxic activity and protein in sucrose density gradient separation. The precursor had a sedimentation constant ($S_{20,w}$) of 11.1 to 11.9, and the trypsin-activated toxin 11.3 to 11.6 (to be published). This is understandable from the fact that dissociation of crystalline type A toxin ($S_{20,w} = 20$) into small fragments ($S_{20,w} = 5$) (Wagman and Bateman, 1953) does not involve any potentiation.

A small molecular precursor was demonstrated when the large molecular one was treated with chymotrypsin at pH 7.0, which caused partial inactivation, followed by separation by gel filtration (to be published). The presence of a small molecular precursor is another indication that fragmentation is not associated with activation.

The highly purified precursor and toxin were compared with each other by the following methods: starch electrophoresis, ion exchange chromatography, gel filtration on Sephadex G-200 with different eluants, antitoxin-combining power, antitoxin-producing power, sedimentation patterns in the ultracentrifuge, density gradient separation, destruction by urea, and destruction by such proteolytic enzymes as pepsin, chymotrypsin, and trypsin at different pH values. None of the methods demonstrated any difference between precursor and toxin, except the higher toxicity (200 to 500 times as high) possessed by the activated toxin (Sakaguchi et al., 1963; other data to be published). The results indicate that activation results from an extremely minor structural change in the molecule of the precursor through the cleavage of a certain bond, possibly removing a "masking group."

3. ISOLATION AND PURIFICATION

Purification of type E toxin, before and after tryptic activation, was attempted by Gordon et al. (1957). The whole culture was treated with 95% ethanol at $-7°C$ to make a final concentration of 25%. The supernatant was siphoned off, and the precipitate obtained by centrifugation was resuspended in a ¼ culture volume of water. It was then extracted with 0.075 M CaCl$_2$ at pH 6.0. The extract was subjected to two more alcohol precipitations at 25%. The third alcohol precipitate dissolved in 0.2 M succinate buffer, pH 5.5, showed 85,000 LD$_{50}$/mg N when started from an untreated whole culture and 19,000,000 LD$_{50}$/mgN from a trypsin-activated material. Higher purification was accomplished from activated materials than from nontreated ones.

Fiock et al. (1961) modified the methods and purified activated toxin from a whole culture. The first alcohol precipitation was replaced with ammonium sulfate precipitation at about 40% saturation. Extraction was done in the same manner, and the extract was precipitated twice in 25% ethanol. The precipitate showed a toxicity of 45,000 LD$_{50}$/mg N.

Sakaguchi and Sakaguchi (1959) attempted the purification of type E toxin precursor from the bacterial cells grown in glucose peptone medium of pH 6.3, which prevented release of the precursor into the medium (Table III). The bacterial cells were collected by centrifugation and washed with 0.05 M acetate buffer, pH 5. The precursor was extracted with 1 M acetate or 0.2 M phosphate buffer, pH 6.0, and precipitated at 40 to 50% saturation with ammonium sulfate. The precursor at this stage migrated toward the cathode when subjected to starch electrophoresis at pH 6.0; the sample activated with trypsin migrated toward the anode. This and other findings indicate that ribonucleic acid is bound with the precursor (Sakaguchi and Sakaguchi, 1959).

The treatment of precursor by pancreas ribonuclease removed almost all the ribonucleic acid, without affecting the potency. A highly purified toxin precursor, free from ribonucleic acid, was obtained by separating the ribonuclease-treated preparation on a CM-Sephadex column (Sakaguchi et al., 1964). The sample had toxicities of 510,000 and 83,000,000 LD$_{50}$/mg N before and after tryptic activation, respectively.

Purification of nonactivated toxin from cell-free culture filtrate was attempted by Gerwing et al. (1961). The Iwanai strain was grown for 5 days at 30°C in a cellophane casing with daily changes of the culture medium outside the casing. The culture in the casing was filtered through a Seitz filter. The filtrate was treated with 35% ethanol at $-15°C$. The precipitate was resuspended in 0.05 M acetate buffer, pH 6.0. This was separated by chromatography on DEAE cellulose treated with 2 M

GENJI SAKAGUCHI

sodium acetate solution. This sample had a toxicity of 580,000 MLD/mg N.

Later, they modified procedures for the purification of nonactivated type E toxin. The sterile culture filtrate obtained in the same manner was treated with 60% ammonium sulfate. The precipitate, dissolved in 0.01 M acetate buffer, pH 5.5, was subjected to chromatography on acidified DEAE cellulose. The purified nonactivated toxin contained 7,500,000 MLD/mg N (Gerwing *et al.*, 1962).

4. PHYSICAL AND CHEMICAL PROPERTIES

Type E toxin, like toxin of other types, appears to be a simple protein, being composed of amino acids only. Considerable difference has been reported in the molecular sizes.

Fiock *et al.* (1961) examined their material obtained from a whole culture in the ultracentrifuge. Two components with sedimentation constants ($S_{20,w}$) of 12.5 and 4.7 were observed; the fast-sedimenting boundary represented the major component.

The same level of molecular size was found with both the precursor and the trypsin-activated toxin derived from the bacterial cells. Neither of them penetrated swollen gel of Sephadex G-200 (Sakaguchi *et al.*, 1964). Sedimentation constants ($S_{20,w}$) of about 11.5 were calculated for both of them (to be published).

Much smaller molecular sizes were reported by Gerwing *et al.* (1962, 1964) for nonactivated and activated toxins obtained from culture filtrates. They reported a sedimentation constant ($S_{20,w}$) of 5.6 for nonactivated toxin and < 1.0 for trypsin-activated toxin in their earlier paper. Later, a sedimentation constant ($S_{20,w}$) of 1.70 and a molecular weight of 18,600 were calculated for the nonactivated toxin.

It is not surprising, however, that there is such a diversity in molecular size. Of course, molecules of smaller size will pass through the intestinal wall more easily. Similar variation has been observed with type A toxin; molecular weights of 3800 to 900,000 have been reported (Schantz, 1964).

Small-sized precursor and toxin were demonstrated by treating the large molecules by trypsin or chymotrypsin at pH 7. It was demonstrated that the culture supernatant of type E culture, usually with pH around 5, consisted mainly of large molecules. However, when it was adjusted to pH 7.0 and incubated at 37°C without the addition of any enzyme, an increased amount of small molecular substance was shown by gel filtration on Sephadex columns (to be published). Since highly purified materials do not dissociate at pH 7.0 and 37°C, it is apparent that Gerwing's material had undergone enzymic dissociation in cellophane casing culture with daily changes of culture medium, by which the reaction had been maintained above 7.

Type E toxin, like type A, is readily denatured by heat, acid, alkali, and many oxidizing agents. If foods are boiled for several minutes and thoroughly mixed during the boiling process before consumption, toxin is destroyed completely.

Type E toxin appears to be less heat-resistant than type A toxin. It was destroyed within 5 minutes at 60°C at pH 7.5 or 3.5, was most stable at pH 4.5 to 5, and differed from type A toxin in not being stabilized in solutions of high ionic strength (Ohye and Scott, 1957).

Dolman (1947a) noted that activated toxin was unstable even when stored in the refrigerator, and crude toxins prepared from pure TOX cultures were of greater stability. On the other hand, Duff et al. (1956) stated that the activated toxin is stable at pH 7.0 in the absence of trypsin. The rapid decrease in toxicity when the toxin is incubated with trypsin at pH 7.0 was ascribed to the proteolysis by trypsin. Exposure of the precursor or the activated toxin to room temperature at pH 6.0 or 7.5 for 12 hours or longer did not cause any appreciable destruction of either material (Sakaguchi et al., 1964).

No difference in stability between nonactivated and activated toxins was indicated in preparing toxoids (Gordon et al., 1957; Fiock et al., 1961),

Lyophilization of type E toxin causes partial denaturation (Gerwing et al., 1964), while freezing and thawing processes caused no significant destruction of precursor or toxin (Sakaguchi et al., 1964).

Gerwing et al. (1964) observed that, when type E toxin was dialyzed in untreated dialysis paper, up to 90% of toxic activity was destroyed by surface-active elements on the dialysis casing. This inactivation was prevented by treating the dialysis casing by boiling for 5 minutes in 0.01 M EDTA solution of pH 7.0.

Both type E toxin and precursor were easily detoxified by urea at concentrations of 2 M or higher at pH 6.0, and the rates of detoxification were the same (to be published).

Sakaguchi investigated destruction of type E toxin and its precursor by pepsin, chymotrypsin, and trypsin. Both were susceptible to these enzymes at the optimum pH values—for example, pH 2.0 for pepsin, and pH 7.0 to 8.0 for chymotrypsin and trypsin—and split into small molecular substances associated with loss of toxicity. In the experiment with chymotrypsin and trypsin at pH 7.0, products of small molecular size having toxicity were detected by gel filtration technique. At pH 6.0, neither precursor nor activated toxin was susceptible to either enzyme; trypsin caused activation only without reducing the molecular size of the precursor (to be published).

Radiation resistance of crude preparations of type E toxin was investigated by Roberts *et al.* (1965), and a radiation D value of 2.1 megarads was obtained. This is the same level of resistance as for type A toxin. Therefore, type E toxin, like types A and B, is not completely destroyed by the large doses of radiation usually applied to food (of the order < 5 megarads).

C. Susceptibility of Different Species of Animals

Many species of animals are susceptible to type E toxin. Before type E human botulism had been reported, the toxin of the Russian strains was found to be lethal to monkeys when 2500 mouse MLD was fed orally (Gunnison *et al.*, 1936). In that report resistance of chickens to 2000 guinea pig MLD was described.

Hazen (1937) reported that young chickens were highly susceptible to the toxin of the "salmon" strain but unsusceptible to that from the "German-canned sprats." The toxin from the Nanaimo strain did not affect chickens, but 500 mouse MLD of toxin from the VH strain caused paralysis in a young hen (Dolman *et al.*, 1950). Type E botulism in waterbirds is discussed in Section I of this chapter. Susceptibility of fish to type E toxin was described by Dolman (1953).

A lethal dose of type E toxin when fed to *Macaca mulatta* was about 1000 to 10,000 mouse MLD (Wagenaar *et al.*, 1953). About 1500 to 2500 mouse MLD per kilogram of weight was required to kill a monkey by oral administration. The figure is larger than 180 MLD for type B and 650 MLD for type A, but smaller than 50,000 to 75,000 MLD for type F, 100,000 to 250,000 MLD for type C_α, and 600,000 MLD for type D. These figures appear to correspond to the type distribution in human botulism (Dolman and Murakami, 1961).

Wagenaar *et al.* (1953) stated that minks were resistant to type E toxin when fed orally. A natural outbreak of botulism in mink by type E toxin was reported by Skulberg (1961). Of the 631 animals fed a ration including raw fish, 606 died; extracts from the ration or the fish were demonstrated to contain type E toxin.

In view of the frequent occurrences of type E human botulism in Japan, we examined the susceptibility of mink to the toxin. Trypsin-activated type E toxin was administered. The lethal amounts of various types of toxins in terms of mouse LD_{50} per mink by intraperitoneal injection were: type A 10^3, B 10^4, C_α 10^3, C_β 10^2, D 10^7, E 10^4, and F 10^6.

Those per oral administration were: type A 10^7, B larger than 10^7, C_α 10^5, C_β 10^4, D larger than 10^9, E 10^7, and F larger than 10^7. The results are in good relation to the frequent occurrence of type C_β botulism among mink and the rare occurrence of type E botulism; yet there is a possibility of natural outbreaks of type E and type A if highly toxigenic feed is given to mink.

No report has been found on natural outbreaks of type E botulism among horses, cattle, sheep, goats, or other larger animals.

VII. LABORATORY DIAGNOSIS

A. Toxicology

Detection of the toxin from the incriminated foodstuff or in a blood sample of the patient is the minimum evidence necessary for laboratory diagnosis.

Demonstration of toxin in the bloodstream of patients can be helpful, especially when the food sample is not available. In the Tennessee, Kentucky, and Alabama outbreaks in the fall of 1963, Rogers et al. (1964) demonstrated toxin in the blood of 5 of 6 patients with clinical symptoms from 1 to 10.5 days after the ingestion of the food, vacuum-packaged smoked whitefish. No toxin was demonstrated in the blood of 6 of 7 individuals without symptoms. It is of interest that the blood from one symptom-free consumer was demonstrated to contain type E toxin.

The food sample is macerated in a mortar with a small amount of diluent and sand to make an emulsion. The diluent may be either 0.85% saline or 0.1 M phosphate buffer, pH 6.0, with or without 0.2% gelatin. After centrifugation, the clear supernatant fluid is injected into mice intraperitoneally. A dilution of 1-5 or 1-10 is usually given to avoid nonspecific deaths. Groups of mice protected with type-specific antitoxins, generally types A, B, and E for human outbreaks, are also injected. Injection is also made into unprotected mice with the supernatant fluid heated at 80°C for 30 minutes or at 100°C for 10 minutes. Symptoms may appear within 2 to 12 hours. If the material contains botulinum toxin, the unprotected group receiving unheated supernatant fluid should die with typical symptoms; those receiving heated samples should survive. If it is type E toxin, the groups protected with type A and type B antitoxins die, and the group protected with type E antitoxin survive.

Potencies of type E toxins may vary widely, and sometimes toxin is demonstrated only through trypsinization of the sample. To accomplish activation, it is important to adjust the reaction mixture of the sample to pH 6.0 before adding trypsin (usually a crude preparation at 1.0%). The reaction mixtures are incubated at 37°C for 30 to 60 minutes.

If a fishery sample contains a very small amount of toxin or a small number of spores, detection may be favored by preliminary incubation (Slocum, 1964). This is accomplished by incubating raw fish or smoked fish at 30°C for 7 to 10 days. Free exudate or extract is injected into mice in a similar manner.

B. Bacteriology

The preliminary incubation technique, described in Section VII, A, permits testing a large number of specimens and detection of a small number of organisms in a sample. One to three grams of the test sample, with or without preliminary incubation, is inoculated into tubes of suitable enrichment medium. Cooked liver medium or cooked meat medium is often used for the purpose, since no other precautions are needed to obtain anaerobiosis than boiling the tubes for a few minutes and cooling before inoculation.

The inoculated tubes may be heated at 60°C for 30 to 60 minutes to kill heat-labile organisms. Because of the heat sensitivity of type E spores, heating at 70°C or higher should be avoided. The tubes are incubated at 30°C. Growth of type E organisms is indicated by vigorous gas production, although other organisms can often produce gas. Toxin can be detected within 2 days, but the cultures should be incubated for at least 10 days before they are considered negative. Selective growth of type E organisms may be obtained if the enrichment cultures are incubated at 10°C. The low-temperature incubation worked out successfully in the Minneapolis outbreak in 1960, but did not work in the Detroit outbreak in 1963 (Slocum, 1964).

Blood agar plates or other suitable nutrient agar plates are streaked with the enrichment cultures shown to be toxigenic. The plates are incubated in an anaerobic jar at 30°C.

Isolation of type E organisms from heavily contaminated materials is sometimes very difficult. No selective medium is as yet applicable to *Clostridium botulinum*.

An alcohol method for isolation of type E cultures has been devised (Slocum, 1964). The method consists in mixing an aliquot of toxic cultures with an equal amount of absolute ethyl alcohol and allowing the mixture to stand at room temperature for 1 hour or longer before streaking onto agar plates.

VIII. BOTULISM — TYPE E

After incubation for 2 days, plates are examined for type E colonies. On blood agar plates, type E organisms are slightly hemolytic. When egg-yolk agar plates are used, the colonies are surrounded by zones of yellow precipitate, and, on standing, clear zones will appear around the precipitation zones (Slocum, 1964).

Identification of type E organisms is solely dependent on toxin neutralization tests on isolated cultures with type-specific antiserum. Recently, a fluorescent staining method was described by Walker and Batty (1964). The method is able to differentiate three groups — types A, B, and F; types C and D; and type E. A fluorescent-labeled antiserum against type E organisms reacts only with type E organisms, whereas there is cross staining of flagella between the different types.

VIII. TREATMENT AND PREVENTION

A. Antitoxin Treatment

Antitoxin treatment in human cases of botulism, where symptoms were well developed, did very little in reducing the case fatality rate of type A or B botulism; only prophylactic use of antitoxin in those who have eaten the incriminated foodstuff and have not yet developed symptoms has been effective in preventing botulism (Dack, 1956).

In Canada, type E botulinus antitoxin was used on 4 patients, including 1 severe case, in an outbreak due to salmon eggs which occurred in Bella Coola, B. C., in 1961. None of them died. The antitoxin was also administered to 4 other patients with no bacteriological diagnosis, and all recovered (Dolman and Iida, 1963).

The Canadian antitoxin was administered to 4 patients in the outbreak at Samani, Hokkaido, in 1959, to 4 patients in Sapporo in 1960, and to 1 patient at Samani and 9 patients at Hamamasu in 1961. Of the total of 18 patients who received the Canadian antitoxin, 17 improved rapidly and only 1 died.

A large-scale outbreak occurred at Toyotomi, Hokkaido, in May 1962. A total of 55 persons developed symptoms after eating commercial products of herring "kirikomi" (see Section II, B, 7). One died, 15 were seriously ill, 26 were moderately ill, and the other 13 were slightly ill. A Japanese preparation of antitoxin was administered intramuscularly to all the seriously and moderately ill cases; none died.

From 1959 to 1964, there were 13 outbreaks in Hokkaido in which antitoxin was used, with 90 cases and only 3 deaths, the case fatality rate being only 3.3%. Two of the 3 deaths occurred before antitoxin administration. Table IV summarizes the results of type E antitoxin therapy in Hokkaido, Japan. The data may attest to the value of antitoxin therapy, although the severity of the diseases between the nontreated and the treated cases is not comparable. The similar morbidity rates in the outbreaks with or without antitoxin treatment, however, may well signify the difference in case fatality rate (Iida et al., 1964).

TABLE IV
EFFECT OF TYPE E ANTITOXIN THERAPY IN HOKKAIDO, JAPAN

Antitoxin	Number of outbreaks	Persons exposed	Cases	Deaths	Morbidity rate (%)	Fatality rate (%)
Not used	20	282	137	39	48.5	28.5
Used	13	190	90	3 [a]	47.4	3.3
Total	33	472	227	42	48.1	18.5

[a] Two died before antitoxin injection.

The effectiveness of type E antitoxin was also confirmed in the United States. In the Tennessee, Kentucky, and Alabama outbreaks in 1963, 17 persons became ill after eating vacuum-packed smoked whitefish which came from Lake Michigan. Of 17 cases, 5 died. Of the 12 survivors, 8 had been treated with type E antitoxin; 4 survived without the antitoxin. Of the persons who died, none had received the antitoxin (Rogers et al., 1964).

It may be worthwhile to note that some patients were severely ill, suffering from difficulty in breathing at the time of antitoxin administration. It may be desirable to inject antitoxin in as a large dose (2600 to 8400 international units per person were injected in Japan) and as early as possible; however, antitoxin treatment of patients even with advanced symptoms of botulism should never be given up.

B. Prevention

1. GENERAL CONSIDERATIONS

Since type E spores are widely distributed in soil, lake and sea water and mud, fish, vegetables, etc., the contamination of various foodstuffs by type E spores may be inevitable. Careful cleaning of surfaces of fish and early removal of the intestines will be of help to lessen the contamiating spores.

Type E botulinum spores are relatively heat-labile; exposure of food-stuffs to heating at 80°C for 30 minutes, at 90°C for 10 minutes, or at 100°C for 5 minutes would guarantee complete eradication of type E spores (Dolman, 1957b). Heating a foodstuff before consumption may destroy all the toxin, if any were formed. Such foodstuffs as "izushi," "kirikomi," salmon eggs, seal and whale flippers, pickled herring, etc., are traditionally eaten uncooked by Japanese, by Indians and Eskimoes in Canada and Alaska, and by Scandinavian people. For such food items, other methods of control must be considered.

Iida et al. (1958a) demonstrated that toxin is formed during soaking of the fish in fresh water to remove blood for a few days, in preparation of "izushi." Salmon egg cheese undergoes a similar treatment. It is essential to minimize the period and to lower the temperature of soaking to 5°C or lower. Treating the fish with chlortetracycline or oxytetracycline has been considered (Boyd et al., 1956; Sakaguchi et al., 1960), although it was not determined whether the use of these antibiotics impairs normal lactic fermentation, on which the taste of "izushi" depends.

Addition of vinegar in proportion of 40 ml/kg of fish to lower the re-action below pH 5.4 or a culture of Streptococcus lactis to allow quick lactic fermentation will be of value in preventing type E botulism from "izushi" (Kanzawa and Iida, 1957). Cutting fish into smaller slices will also favor faster penetration of acid.

If fish or marine mammals are to be kept for a long time under refrigeration, psychrophilic properties of type E organisms must be taken into consideration. Refrigeration at ordinary temperatures of 5° to 10°C does not always guarantee the complete absence of toxin formation; it may even be possible that the organisms would grow more selectively than at higher temperature because of the absence of competition with other putrefactive microorganisms. Therefore, refrigeration of fish at 3°C or at lower temperature should be recommended.

2. VACCINATION

Botulinus toxoid is used to protect such animals as mink, foxes, and cattle. In human beings, however, the toxoid has been used to protect only laboratory workers subject to the exposure to toxins. In such countries as Japan, where people habitually like partially preserved raw fish, protection of individuals with toxoid must be considered. Recently, there seems to be a tendency to administer type E antitoxin for a number of cases of "izushi" poisoning with no proof of botulism, thus increasing the chances for serum sickness in case botulism recurs in these individuals at a later time when they are again given antitoxin.

In our laboratory, an alum-precipitated toxoid for human use was prepared experimentally from highly purified type E toxin. The final preparation, containing 16 μg of protein nitrogen per milliliter, was injected subcutaneously twice into guinea pigs at a dose of 0.2 or 0.5 ml. Antitoxin of 14 to 22 international units/ml was demonstrated in the serum. In guinea pigs, when antitoxin potency was $\frac{1}{50}$ international unit/ml, more than 90% of the animals tolerated the challenge of type E toxin of 2000 to 10,000 LD_{50} by intraperitoneal injection. Therefore, if the antitoxin potency in blood is 10 international units/ml or higher, there would be no possibility of natural intoxication. It was also demonstrated that the toxoids prepared from either the precursor or the trypsin-activated toxin are similarly effective (to be published).

Human beings were injected with a pentavalent aluminum phosphate-precipitated toxoid including types A, B, C, D, and E. Production of type E antitoxin was demonstrated to an extent similar to the other types. It was also shown that type E toxoid was antigenic in mice, guinea pigs, and rabbits, showing high levels of resistance to the challenge of toxin administered by various routes (Cardella, 1964).

REFERENCES

Bott, T. L., Deffner, J. S., Foster, E. M., and McCoy, E. (1964). In "Botulism 1964" (K. H. Lewis and K. Cassell, Jr., eds.), pp. 221–234. U.S. Department of Health, Education, and Welfare, Public Health Service.

Bowmer, E. J. (1963). Bull. World Health Organ. 29, 701.

Boyd, J. S., Bluhm, H. M., Muirhead, C. R., and Tarr, H. L. A. (1956). Am. J. Public Health 46, 1531.

Cardella, M. A. (1964). In "Botulism 1964" (K. H. Lewis and K. Cassel, Jr., eds.), pp. 113–130. U. S. Department of Health, Education, and Welfare, Public Health Service.

Dack, G. M. (1956). "Food Poisoning," 3rd ed. University of Chicago Press.

Dack, G. M. (1964). In "Botulism 1964" (K. H. Lewis and K. Cassel, Jr., eds.), pp. 33–40. U. S. Department of Health, Education, and Welfare, Public Health Service.

Dolman, C. E. (1953). Atti VI Congr. Intern. Microbiol. Roma 4, 130.

Dolman, C. E. (1957a). Can. J. Public Health 48, 187.

Dolman, C. E. (1957b). Japan. J. Med. Sci. Biol. 10, 383.

Dolman, C. E. (1960). Arctic 13, 230.

Dolman, C. E. (1964). In "Botulism 1964" (K. H. Lewis and K. Cassel, Jr., eds.), pp. 5–32. U. S. Department of Health, Education, and Welfare, Public Health Service.

Dolman, C. E., and Chang, H. (1953). Can. J. Public Health 44, 231.

Dolman, C. E., and Iida, H. (1963). Can. J. Public Health 54, 293.

Dolman, C. E., and Kerr, D. E. (1947). Can. J. Public Health 38, 48.

Dolman, C. E., and Murakami, L. (1961). J. Infect. Diseases 109, 107.

Dolman, C. E., Chang, H., Kerr, D. E., and Shearer, A. R. (1950). Can. J. Public Health 41, 215.

Dolman, C. E., Darby, G. E., and Lane, R. F. (1955). *Can. J. Public Health* **46**, 135.

Duff, J. T., Wright, G. G., and Yarinsky, A. (1956). *J. Bacteriol.* **72**, 455.

Erdman, I. E., Thatcher, F. S., and MacQueen, K. F. (1961). *Can. J. Microbiol.* **7**, 207.

Fiock, M. A., Yarinsky, A., and Duff, J. T. (1961). *J. Bacteriol.* **82**, 66.

Forfang, K., and Skulberg, A. (1964). *Tidsskr. Norske Laegeforen.* **84**, 973.

Geiger, J. C. (1941). *J. Am. Med. Assoc.* **117**, 22.

Gerwing, J., Dolman, C. E., and Arnott, D. (1961). *J. Bacteriol.* **81**, 819.

Gerwing, J., Dolman, C. E., and Arnott, D. (1962). *J. Bacteriol.* **84**, 302.

Gerwing, J., Dolman, C. E., Reichmann, M. E., and Bains, H. S. (1964). *J. Bacteriol.* **88**, 216.

Gordon, M., Fiock, M. A., Yarinsky, A., and Duff, J. T. (1957). *J. Bacteriol.* **74**, 533.

Graikoski, J. T., and Kempe, L. L. (1964). *Bacteriol. Proc.* p. 3.

Gunnison, J. B., Cumming, J. R., and Meyer, K. F. (1936). *Proc. Soc. Exptl. Biol. Med.* **35**, 278.

Hazen, E. L. (1937). *J. Infect. Diseases* **60**, 260.

Hazen, E. L. (1938). *Science* **87**, 413.

Hobbs, G., Roberts, T. A., and Walker, P. D. (1965). *J. Appl. Bacteriol.* **28**, 147.

Iida, H. (1963). *Japan. J. Med. Sci. Biol.* **16**, 307.

    Biol. **11**, 215.

Iida, H., Nakamura, Y., Nakagawa, I., and Karashimada, T. (1958a). *Japan J. Med. Sci. Biol.* **11**, 215.

Iida, H., Karashimada, T., Nakane, M., and Saito, T. (1958b). *Hokkaido Inst. Public Health Rept.* **9**, 31 (in Japanese).

Iida, H., Kanzawa, K., Nakamura, Y., Karashimada, T., Ono, T., and Saito, T. (1964). *Hokkaido Inst. Public Health Rept.* **14**, 6 (in Japanese).

Johansen, A. (1963). *J. Appl. Bacteriol.* **26**, 43.

Kanzawa, K. (1960). *Hokkaido Inst. Public Health Rept.* **11**, 161.

Kanzawa, K., and Iida, H. (1957). *Hokkaido Inst. Public Health Rept.* **8**, 33 (in Japanese).

Kautter, D. A. (1964). *J. Food Sci.* **29**, 843.

Kobayashi, S. (1961). *Hirosaki Med. J.* **12**, 682 (in Japanese).

Kodama, E., Fujisawa, S., and Sakamoto, T. (1964). *Akita Inst. Public Health Rept.* **8**, 15 (in Japanese).

Kondo, H., Kondo, S., Murata, R., and Sakaguchi, G. (1963). *Japan J. Med. Sci. Biol.* **16**, 31.

Meisel, H., Albrycht, H., Rymkiewicz, D., Switalska, A., and Trembowler, P. (1964). *Med. Dosw. I Mikrobiol.* **16**, 193 (in Polish).

Møller, V., and Scheibel, I. (1960). *Acta Pathol. Microbiol. Scand.* **48**, 80.

Nakamura, Y., Iida, H., Saeki, K., Kanzawa, K., and Karashimada, T. (1956). *Japan. J. Med. Sci. Biol.* **9**, 45.

Ohye, D. F., and Scott, W. J. (1957). *Australian J. Biol. Sci.* **10**, 85.

Pedersen, H. O. (1955). *J. Appl. Bacteriol.* **18**, 619.

Pedersen, H. O. (1957). *Proc. 2nd Intern. Symposium Food Microbiol. Cambridge 1957* p. 289.

Roberts, T. A., Ingram, M., and Skulberg, A. (1965). *J. Appl. Bacteriol.* **28**, 125.

Rogers, D. E., Koening, G., and Spickard, A. (1964). *In* "Botulism 1964" (K. H. Lewis and K. Cassel, Jr., eds.), pp. 133–145. U. S. Department of Health, Education, and Welfare, Public Health Service.

Sakaguchi, G., and Sakaguchi, S. (1959). *J. Bacteriol.* **78**, 1.

Sakaguchi, G., and Sakaguchi, S. (1961). *Japan. J. Med. Sci. Biol.* **14,** 243.

Sakaguchi, G., and Tohyama, Y. (1955a). *Japan. J. Med. Sci. Biol.* **8,** 247.

Sakaguchi, G., and Tohyama, Y. (1955b). *Japan. J. Med. Sci. Biol.* **8,** 255.

Sakaguchi, G., Tohyama, Y., Saito, S., Fujisawa, S., and Wada, A. (1954). *Japan J. Med. Sci. Biol.* **7,** 539.

Sakaguchi, G., Sakaguchi, S., Kawabata, T., Nakamura, Y., Akano, T., and Shiromizu, K. (1960). *Japan. J. Med. Sci. Biol.* **13,** 13.

Sakaguchi, G., Sakaguchi, S., and Kondo, H. (1963). *Japan. J. Med. Sci. Biol.* **16,** 309.

Sakaguchi, G., Sakaguchi, S., and Imai, N. (1964). *J. Bacteriol.* **87,** 401.

Schantz, E. J. (1964). *In* "Botulism 1964." (K. H. Lewis and K. Cassel, Jr., eds.). pp. 91–103. U.S. Department of Health, Education, and Welfare. Public Health Service.

Schmidt, C. F. (1964). *In* "Botulism 1964" (K. H. Lewis and K. Cassel, Jr., eds.), pp. 69–82. U.S. Department of Health, Education, and Welfare, Public Health Service.

Schmidt, C. F., Lechowich, R. V., and Folinazzo, J. F. (1961). *J. Food Sci.* **26,** 626.

Schmidt, C. F., Lechowich, R. V., and Nank, W. K. (1962). *J. Food Sci.* **27,** 85.

Segner, W. P., Schmidt, C. F., and Boltz, J. K. (1964). *Bacteriol. Proc.,* p. 3.

Skulberg, A. (1961). *Nord. Vet. Med.* **13,** 87.

Skulberg, A. (1964). "Studies on the Formation of Toxin by *Clostridium botulinum.*" A/S Kaare Grytting, Organger, Norway.

Slocum, G. G. (1964). *In* "Botulism 1964" (K. H. Lewis and K. Cassel, Jr., eds.), pp. 83–88. U. S. Department of Health, Education, and Welfare, Public Health Service.

Wagenaar, R. O., Dack, G. M., and Mayer, D. P. (1953). *Am. J. Vet. Res.* **14,** 479.

Wagman, J., and Bateman, J. B. (1953). *Arch. Biochem. Biophys.* **45,** 375.

Walker, P. D., and Batty, I. (1964). *J. Appl. Bacteriol.* **27,** 140.

Ward, B. Q., and Carroll, B. J. (1965). *Appl. Microbiol.* **13,** 502.

Yamagata, H. (1963). *Yamaguchi Prefectural Res. Inst. Health, Rept.* **1,** 101 (in Japanese).

CHAPTER IX | STAPHYLOCOCCAL INTOXICATIONS

Robert Angelotti

I. INTRODUCTION

Staphylococcus aureus occupies a unique position among microbial pathogens. It lives in close association with man, and, although every strain is potentially capable of causing disease, the host-parasite relationship is relatively stable, with infection occurring only on the disruption of man's systemic defense mechanisms. Little is known of the means by which staphylococci have achieved extensive colonization of man, particularly of the nasopharynx and skin, of the evolutionary processes involved, or of the manner by which man achieves containment of the organism without apparent ill effects to himself or the bacteria. The ubiquity of staphylococci in man's environment, particularly in milk and food, makes it all the more remarkable that staphylococcal infection and, more pertinently, staphylococcal intoxication are not more prevalent. The demonstrated inability of staphylococci to compete effectively in complex ecological systems is an important contributing factor to this equilibrium (Straka and Combs, 1952; Dack and Lippitz, 1962; Peterson *et al.*, 1962*a,b*; Troeller and Frazier, 1963*a,b*). Without some such biologically operative restraining mechanism, the commensal relationship between man and this organism may never have developed.

With the sole exception of their ability to produce enterotoxin, enterotoxigenic staphylococci do not differ greatly from other members of the species. Consequently, any treatise dealing with the enterotoxigenic staphylococci must necessarily be preceded by a full account of the characteristics and activities of the numerically superior nonenterotoxigenic members.

II. HISTORY

Staphylococci were observed in pus and cultivated by Koch and Pasteur independently in 1878 and 1880. Ogston is credited with applying the name "staphylococcus" in 1881 because of the typical grapelike clusters of cocci he observed in cultures. It was Rosenbach, however, who, after careful and systematic study of the organism, obtained in 1884 pure cultures of the microorganism on solid media. Using the generic name staphylococcus, previously suggested by Ogston, he classified the group into the *albus* and *aureus* forms. By the turn of the century, other varieties of staphylococci were studied and named, and their ability to produce toxic substances was recognized. In spite of an early demonstration (in 1884) that staphylococci produced what was then thought to be a single filterable toxin having a series of effects, it was not until 1924 that Parker (1924), stimulated by the report of the streptococcal erythrogenic toxin (Dick and Dick, 1924), reinvestigated the formation and activity of staphylococcal exotoxins.

Dr. Gail M. Dack, in his chronicle of staphylococcal food poisoning (Dack, 1956), states that food poisoning by this organism predates the scientific discovery of the staphylococci. Dr. Dack's accounts of the experiences of Dr. V. C. Vaughan and Dr. G. M. Sternberg, who independently experimented with cheeses involved in food poisoning incidents, and his review of the discovery on four separate occasions—by Denys in Belgium, in cow meat, in 1894; by Owen in the United States, in dried beef, in 1907; by Barber in the Philippines, in milk, in 1914; and by Dack and his co-workers in the United States, in cream-filled sponge cakes, in 1930—that certain strains of staphylococci are capable of producing a toxic irritant to the gastrointestinal tract are invaluable to both student and laboratory investigator. Though Barber's work was classically performed and outlined particularly well a toxin type of food poisoning, little attention was paid to the report. Staphylococcal enterotoxin was not differentiated from other staphylococcal toxic factors until Dack and co-workers investigated the cause of illness in 11 people who had eaten ornately decorated Christmas cakes composed of three layers of sponge cake with a thick cream filling (Dack *et al.*, 1930; Dack, 1956). Dack and his colleagues demonstrated in human volunteers that culture filtrates of staphylococci isolated from the cakes caused mild gastroenteritis in two individuals who drank 5- and 10-ml portions of the filtrate. A third volunteer, after having consumed 25 ml of the filtrate, suffered acute gastroenteritis, which began 3 hours after ingestion. Shortly after this report, several food poisoning outbreaks were reported by others (Jordan, 1930; Jordan and Burrows, 1934) in which staphylococci produced a toxic substance in food and staphylococcal food poisoning was recognized clinically.

III. CHARACTERIZATION OF THE ORGANISM

Presently, only *Staphylococcus aureus* and *Staphylococcus epidermidis* are listed as members of the genus *Staphylococcus* in "Bergey's Manual" (Breed *et al.*, 1957). The genus is differentiated from other genera of the family *Micrococcaceae* on the basis of mannitol fermentation and coagulase production. The type species of the genus is *Staphylococcus aureus* Rosenbach, which ferments in mannitol and is coagulase-positive, whereas the second member of the genus is incapable of mannitol fermentation and does not produce coagulase. *Staphylococcus aureus* is a pathogenic species; and, although virulence may vary, all strains are potentially capable of causing disease. Cultural and biochemical reactions, such as chromogenesis, coagulase production, mannitol fermentation, production of hemolysins, antigenic structure, and gelatin liquefaction, have been used in the past to distinguish between virulent and avirulent strains. Unfortunately, none of the above reactions is correlated with certainty to pathogenicity, although coagulase production is almost always manifested by staphylococci on primary isolation from man and animals or their suppurative exudates. Coagulase production is generally accepted to be the best indicator of potential pathogenicity. On the other hand, the relationship, if one exists, between the ability to cause infectious disease and enterotoxigenicity is not fully understood. Enterotoxin production, with rare exception, is restricted to coagulase-positive strains. From a total of 263 strains of cocci isolated from clinical sites and dairy products, Thatcher and Simon (1956) obtained seven coagulase-negative strains that caused emesis in cats when they were injected intraperitoneally with 5.0 ml of boiled (30 minutes) culture filtrate. In fact, by current taxonomic criteria, certain of these cultures would necessarily be classified as members of the genus *Micrococcus*.

A. Morphology

The staphylococci typically occur as grapelike clusters approximately 0.8 to 1.0 μ in diameter. The spatial arrangement of cells in clusters is more frequently observed on solid media than in liquid, and the occurrence in broth cultures of pairs, short chains, and clusters of few cells is not uncommon. Staphylococci are nonmotile and nonsporing. Capsules may be formed in very young cultures, but they disappear within a few hours. The cells stain readily with basic aniline dyes, but less so with acid dyes. They are gram-positive in a young culture (18 to 24 hours), but tend to gram-variability as the culture ages.

In a nutritionally adequate liquid medium, *S. aureus* usually yields abundant growth in a still culture typified by heavy turbidity, sedimentation, and, in many strains, the formation of a ring of growth at the liquid–air interface. Pigment formation occurs in an agitated broth culture, but less frequently in a still culture. On solid media, such as trypticase soy agar or Staphylococcus Medium 110, *S. aureus* produces circular, entire, raised, convex colonies that are smooth and characteristically pigmented in shades varying from cream to orange. Nonpigmented colonies of *S. aureus,* as well as those that are lightly pigmented, often become pigmented or darker in color after a secondary incubation for a few days at room temperature or 30°C. (Chapman, 1947; Donnelly *et al.,* 1964).

B. Habitat

Staphylococci are ubiquitously distributed in man's environment and are found in varying numbers in air and dust, as well as in water, milk, food, feces, and sewage. The primary habitats of these organisms are the mucous membranes of the nasopharynx and the skin of man and animals. Colonization by *S. aureus* of the nasal passages of man is common, and a large proportion of normal persons carry these organisms in the nose. Williams (1963), in his review of the healthy carriage of *S. aureus*, tabulated the nasal carrier rates in different population groups as reported by numerous investigators. These data reveal that the carrier rates in normal adults not associated with hospitals are between 30 and 50%. The incidence of nasal carriage increases to approximately 60 to 80% in patients and working personnel associated with hospitals (Miles *et al.*, 1944; Rountree and Barbour, 1951; Goslings and Buchli, 1958; Williams *et al.*, 1959; Macfarlane *et al.*, 1960).

Staphylococci are isolated most frequently from the nose, but the throat and skin are also important sources. Throat carriage in healthy adults in the United States and England approximates 4 to 7% (Commission on Acute Respiratory Disease, 1949; Campbell, 1948), whereas much higher rates (40 to 70%) are reported by Scandanavian workers (Packalen and Bergqvist, 1947; Vogelsang, 1951, 1958). The percentage of skin carriage in adults varies from about 4 to 44%, depending on the skin sites examined: Carriage on the nose is consistently high (40 to 44%), whereas hand carriage varies (14 to 40%), and only 4 to 16% of individuals examined harbored staphylococci on the leg (Williams, 1946; Martin and Whitehead, 1949; Ridley, 1959).

With carrier rate frequencies of the orders described above, it is no small wonder that staphylococci are present in foods, particularly those that come in intimate contact with food handlers during processing and preparation. Except for milk and dairy products, in which staphylococci are present as a result of shedding from the bovine udder, the single most important source of *S. aureus* in foods is man; and until he is no longer directly involved in the preparation and service of food, staphylococci will remain a food-borne health hazard.

C. Metabolism

The temperature range for optimum growth of staphylococci is 35° to 37°C, although growth occurs throughout the much wider range of approximately 10° to 45°C. The organisms are facultative anaerobes that grow best in the presence of oxygen, but no growth occurs in the complete absence of CO_2 (Gladstone *et al.*, 1935). Initiation of growth is most easily accomplished in a slightly alkaline medium. In spite of the physiological complexity of staphylococci, growth does occur in chemically defined media, but in a manner inferior to that in media containing meat extracts or casein hydrolyzates.

The vitamin requirements for growth of staphylococci have not been developed in detail. Both nicotinic acid and thiamine are required (Knight, 1937a,b; Knight and McIlwain, 1938). That additional vitamins are needed to stimulate growth of some strains or initiate growth of others on continuous subculture in chemically defined media has been demonstrated: biotin (Porter and Pelczar, 1940), tryptophan (Gladstone, 1937), and pantothenic acid (Sevag and Green, 1944) act in this fashion. Surgalla (1947) was not able to demonstrate a stimulatory effect of biotin or tryptophan on the growth of S. aureus 196, an enterotoxigenic strain, but he did observe stimulation by calcium pantothenate. Riboflavin reportedly is synthesized by the organisms (O'Kane, 1941).

The nitrogen requirements are complex and vary among strains. Meat extracts are replaceable in synthetic media with a combination of 14 amino acids (Fildes et al., 1936). Those amino acids responsible for most rapid growth are arginine, phenylalanine, aspartic acid, cystine, leucine, valine, proline, and glycine (Gladstone, 1937). By contrast, Surgalla (1947) demonstrated that synthetic media composed of salts, glucose, thiamine, nicotinic acid, and 2 to 16 amino acids supported growth and enterotoxin production by S. aureus strains. The simplest medium contained only arginine and cystine in addition to the salts, glucose, and vitamins. Anaerobic growth in synthetic media of the types described by Fildes and his co-workers and Gladstone requires the presence of uracil (Richardson, 1936).

The technical difficulties associated with obtaining metal-free constituents for preparation of defined media in which metal ion requirements for growth may be studied and the ease with which these ions are substituted one for another in microbial metabolism have thwarted definition of essential mineral requirements. In spite of the technical difficulties involved, the necessity for calcium, magnesium, and potassium for the growth of S. aureus has been demonstrated (Haynes et al., 1954; Shooter and Wyatt, 1955, 1956). On the other hand, sodium is not required (Shooter and Wyatt, 1957).

Organic sulfur is required for growth and is usually obtained from cystine. Fildes and Richardson (1937) and Gladstone (1937) demonstrated that methionine and sodium dithiodiacetate could be substituted as sources.

Detailed studies of the carbohydrate metabolism of S. aureus have been few in number. It is generally assumed that glucose is metabolized by glycolysis and subsequent oxidation of pyruvic acid. Until the studies of Strasters and Winkler (1963) were reported, information concerning alternative pathways of glucose breakdown was scanty and was based primarily on limited observations by Hancock (1960) and Das and Chatterjee (1962) of an active pentose cycle. By means of Warburg experiments, enzyme studies with cell-free extracts, and experiments with radioactive substrates applied to five separate strains of different phage groups, Strasters and Winkler were able to develop the general outline of carbohydrate metabolism by *S. aureus*.

Glucose is broken down by the glycolytic system quantitatively to L(+)lactic acid under anaerobic conditions. The main reactions of oxidative assimilation of glucose are the production of CO_2 and acetic acid and a further oxidation of acetic acid by the citric acid cycle. Glucose oxidation occurs by means of the pentose cycle, which is most active in cells grown in the absence of glucose. Cells grown in either the absence or the presence of glucose are not capable of metabolizing ribose anaerobically, although ribose-grown cells primarily oxidize the sugar to CO_2 and acetic acid. Cocci grown in glucose display enhanced glycolysis, suppression of the Krebs cycle, decreased activity of the pentose cycle, decreased pyruvic acid oxidation, and inhibition of synthesis of the citric acid cycle enzymes.

D.   Effects of Physical and Chemical Agents

Staphylococci appear to be more resistant to harsh environmental conditions than many of the other nonsporulating types of microorganisms. Since they withstand desiccation well, drying on porcelain beads is a common method of preserving staphylococcal cultures. Freezing, particularly slow freezing, usually reduces viable members, but cocci can withstand frozen storage in foods for long periods. On the other hand, food poisoning strains of staphylococci have not been observed to multiply in foods with internal temperatures at or below 5.56°C (42°F) (Angelotti *et al.*, 1961*a*). The lowest temperature at which enterotoxin production has been recorded is 18°C (64.4°F) (Segalove and Dack, 1941). Although they are not as heat-resistant as micrococci, their tolerance to moist heat must be considered high for a nonsporing organism. For example, 59 minutes of exposure at 140°F (60°C) was required to reduce 1×10^7 food poisoning staphylococci per gram of custard to nondetectable levels (Angelotti *et al.*, 1961*b*). Even though they persist at elevated temperatures, they cease to multiply in foods at 46.6°C (116°F) (Angelotti *et al.*, 1961*c*).

Staphylococci grow well in the presence of high concentrations of sodium chloride. This fact has been advantageously applied in developing enrichment methods for culturing small numbers of staphylococci from

mixed cultures or from complex ecological situations such as those pre-
vailing in many foods. Luxuriant growth occurs in media containing 10%
sodium chloride. In addition to sodium chloride tolerance, they appear to
resist destruction with mecuric chloride, a 1% solution being required to
kill them in 10 minutes (Wilson and Miles, 1964). With this exception,
most other laboratory disinfectants are effective, particularly the quar-
ternary ammonium formulations.

E. Bacteriophage Typing and Lysogeny

That staphylococci are phage-typable has long been recognized and
used as an epidemiological tool in determining the relationships among
strains and the sources with which they are associated. The review and
appraisal (Munch-Petersen, 1963) of phage-typing results obtained with
staphylococci isolated from food in food intoxications revealed that the
majority involved in food poisoning belonged to phage group III, with
types 6 and 47, either alone or with others, the most common. Although
sufficient data are not available to categorize with certainty the phage
types associated with normal or wholesome foods, between one-fourth
and one-third of such isolates are lysed by phages of group III, whereas
most of the remainder are untypable. Staphylococci from cheese, butter,
buttermilk, cream, ice cream, fermented milk, and dried and condensed
milk appear to be lysed mainly by phages of groups III and IV (Munch-
Petersen, 1963). It is not possible by phage typing to determine whether
a strain of staphylococcus is capable of producing enterotoxin. Too many
food poisoning strains have been isolated that are not lysed by group III
for this system to be acceptable. In spite of the attendant dangers in
assuming such a correlation, it is interesting to note that in Great Britain
during the years 1950 to 1962 food-borne intoxications due to strains
lysed by group III ranged from 64.5 to 94.7% (Munch-Petersen, 1963).

One of the complications affecting the classification of food poisoning
strains by phage type is the factor of lysogenic conversion. That such
conversions are probably occurring continually among the staphylococci
is evidenced from the ease with which Rosendal and his colleagues
(1964) were able to demonstrate the conversion from ability to inability
to produce extrocellular "Tween"-splitting enzyme by staphylococci
that survived exposure to the filtrate of phage obtained from a strain that
did not produce the enzyme. In repeated tests they demonstrated that 84
to 87% of the cells from the receptor strain were lysed and that the remaining colonies were all lysogenic and did not produce the enzyme. Lysogenic conversion of enterotoxin production has also been accomplished.

Enterotoxigenicity has been conferred on nonenterotoxigenic staphylococci by lysogenizing the latter with temperate phage carried by an enterotoxigenic staphylococcus (Casman, 1965). This conversion appears to be achieved easily with enterotoxin A-producing strains, but has not as yet been demonstrated with strains that produce both enterotoxins A and B or enterotoxin B alone (Casman, 1965; Read and Pritchard, 1963). The majority of enterotoxigenic staphylococci appear to be lysogenic. A greater similarity exists among the temperate phages of cultures that produce enterotoxin B than among strains that produce other types of enterotoxins (Read and Pritchard, 1963).

Presently phage typing of staphylococci in relation to food poisoning appears to be restricted to identifying the human or environmental source from which the organism came, and little correlation has yet been established, at least in the United States, between enterotoxigenicity and phage type.

F. Toxigenesis

Most pathogenic strains of staphylococci when grown in suitable media and under proper conditions produce toxic substances that can be demonstrated to be hemolytic, leukocytic, dermonecrotic, and lethal to the mouse and rabbit. The hemolytic toxins or hemolysins of staphylococci are differentiated on the basis of the animal species of red blood cells against which they are active. The α-lysin is most active against rabbit red cells, but not against human red cells, and causes lysis at 37°C. The β-lysin acts on sheep and ox red cells, but not on human or rabbit cells, and causes lysis only after incubation at room temperature or in the refrigerator ("hot-cold lysin"). Two other hemolytic toxins, the γ and δ lysins, are produced. The former has lytic activity against a wide variety of animal cells, and the latter is most active against washed human and horse red cells. The α-hemolysin is also leukocidal, and part of the leukocytic effect of staphylococcal toxic filtrates is attributable to this lysin. In addition, however, a different leukocidin composed of two separate components is produced.

On primary isolation from lesions and exudates, most virulent strains of staphylococci can be shown to produce coagulase, an extracellular substance that causes coagulation of plasma. Both a free coagulase, which is liberated into the surrounding medium, and a bound coagulase, which is attached to the bacterial cell wall, are produced and may be demonstrated by means of the tube or slide coagulase tests, respectively.

Staphylococci also produce a fibrinolysin, or, more correctly, an activator (staphylokinase) of profibrinolysin that acts on human and other animal plasmas and is demonstrated by the clearing on plates of precipitated fibrin.

Other substances produced by staphylococci include hyaluronidase, deoxribonuclease, and various lipases. The best known of the latter is the lipase giving rise to opacity in egg-yolk-containing media; it is apparently fairly well correlated to coagulase production.

Lastly, certain strains of coagulase-positive staphylococci produce enterotoxin, the subject of a major portion of this chapter. Enterotoxin may be produced by strains characterized by all or only some of the above factors. Except that most enterotoxigenic strains are coagulase-positive, there appears to be no correlation between enterotoxigenicity and production of other toxins or virulence factors.

IV. STAPHYLOCOCCAL FOOD POISONING

During the short period since 1930, when Dack and his co-workers established the etiology of staphylococcal food poisoning, the disease has steadily increased in incidence to the point that today it is the most commonly reported identifiable type of food poisoning in the United States. The incidence remains high in spite of the continued efforts of Federal, state, and local health agencies to improve environmental sanitation in food processing, preparation, and service establishments and to educate food handlers to the dangers of subjecting potentially hazardous foods to conditions that permit staphylococci growth and toxin formation. Better control of staphyloccal food poisoning is possible through application of current knowledge related to food sanitation and proper heating and cooling practices, but it is doubtful that full control can ever be achieved.

A. Clinical Syndrome

Following ingestion of food containing the enterotoxin by susceptible individuals, the onset of symptoms is quite rapid and characteristic. The incubation period, usually approximately 3 hours, may vary from less than 1 hour to several hours (rarely more than 6 to 8 hours) and depends on the susceptibility of the individual and the dosage consumed. Primary symptoms are nausea, vomiting, retching, abdominal cramping, and diarrhea. These symptoms are often accompanied by headache and muscular cramps. In severe cases prostration accompanies the vomiting and diarrhea. Temperature changes are not uncommon in patients; fever is common, but temperature sometimes drops. The disease is acute and self-limiting; complete recovery within 24 to 72 hours is the rule. In severe cases requiring hospitalization, supportive treatment is directed toward relieving shock, preventing dehydration, and controlling vomiting and diarrhea.

B. Incidence

The true incidence of staphylococcal food poisoning in the United States is unknown. The total number of reported food-borne disease outbreaks occurring annually in the United States averages approximately 250 and involves about 10,000 individuals (Dauer *et al.,* 1952–1961) (see Figs. 1 and 2). These data are meaningless, however, because a vast

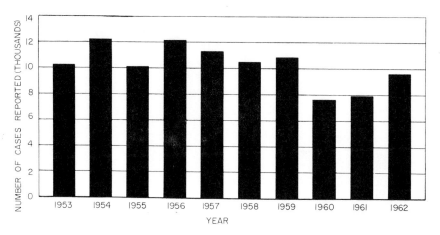

Fig. 1. Number of cases of food-borne illness reported in the United States (1953–1962).

disparity exists between population of reporting areas and numbers of outbreaks reported. These disparities are presented graphically in Figs. 3 and 4.

Within the recognized limitations of the reporting system for food-borne disease employed in this country, the incidence of staphylococcal food poisoning can be compared to that caused by other etiological agents (see Fig. 5). Staphylococci are the major identified cause of food poisoning in the United States. This is not the case, however, in other countries. In England and Wales, for instance, salmonellae were the cause of 96%, 92%, and 95% of all food poisoning incidents in the years 1961, 1962,

and 1963, respectively, compared to only 2%, 5%, and 2% for staphylococci in the same years (Report of the Public Health Laboratory Service,
1962–1964). The Netherlands, on the other hand, report little staphylococcal food poisoning but have numerous outbreaks caused by *Bacillus*

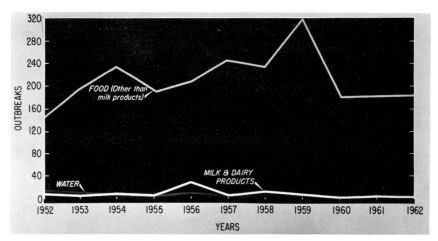

FIG. 2. Food-, milk-, and water-borne disease outbreaks reported in the United States
(1952–1962).

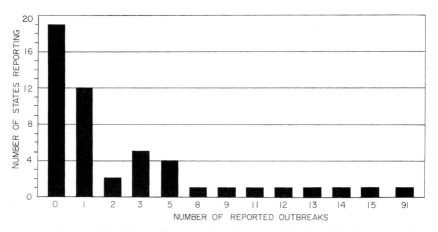

FIG. 3. Types of food-borne disease outbreaks reported in the United States (1952–
1962).

cereus. About half of all the food poisoning incidents in Japan are caused
by ingestion of fish and fish dishes contaminated with *Vibrio parahaemolyticus.* Undoubtedly staphylococcal food poisoning is related to
certain dietary preferences among nations, but it appears that in the

United States the high incidence may be correlated to the enormous number of commercially prepared, served, and catered meals consumed by the American people in public establishments.

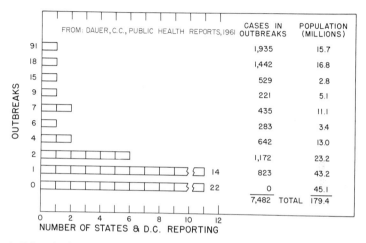

FIG. 4. Disparity between reported food-borne disease and population of reporting areas (1960).

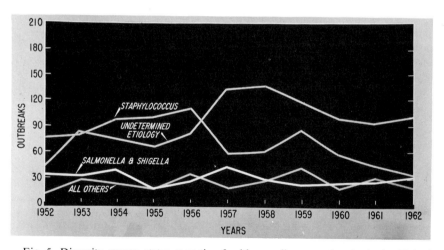

Fig. 5. Disparity among states reporting food-borne disease outbreaks in the United States (1961).

C. Human Susceptibility

Human susceptibility to staphylococcal enterotoxin is very difficult to assess quantitatively because tolerance to the toxin is displayed in varying degrees by people eating the same poisoned food and the exact quantity of food eaten by each individual is not ascertainable. The degree of individual immunity to the enterotoxin achieved through previous experience and the differences in potency of the toxins produced by various strains, as well as the differences in growth and toxin production by a given strain in various foods, all serve to further complicate the picture.

It is possible, however, to determine by means of the gel-diffusion tests the concentration of enterotoxin per gram of food involved in food poisoning incidents, and studies indicate that the concentrations may be expected to be quite low. The sensitivities of the single- and double-gel-diffusion tube tests for detecting enterotoxins is 1.0 μg/ml of antigen for the single-diffusion test and between 0.05 and 0.1 μg/ml of antigen for the double-diffusion test (Hall *et al.*, 1965). Experience has shown that, to obtain these concentrations of antigen per milliliter, it is necessary to extract the enterotoxin from the food and concentrate the extract. Data collected in the laboratories of the R. A. Taft Sanitary Engineering Center of the Public Health Service indicate that, after extraction and concentration, the minimum levels of enterotoxins that can be recovered by gel-diffusion methods after addition to foods are in the range of 0.02 to 0.05 μg/gm of food (Hall *et al.*, 1965; Read *et al.*, 1965a,b). Recovery figures are more reliable for enterotoxin B than for A because the former was highly purified, whereas the latter was not. The actual concentrations of enterotoxin A detected were probably much lower than the data indicate. These same investigators have been able, by the above techniques, to demonstrate the presence of one or both of the toxins in a number of foods that were shown epidemiologically to be responsible for staphylococcal food poisoning. If one assumes that at least the minimally detectable concentrations of toxin per gram were present in these foods and that at least 100 gm of the poisoned food was consumed, it is conceivable that the dosage necessary to cause human illness is approximately 1 to 5 μg. Casman and Bennett (1965) reported that, in ham and pastries responsible for food poisoning incidents, they found from 50 to 200 million staphylococci per gram. They also reported that the average amount of enterotoxin A produced in aerated Brain Heart Infusion* broth and measured by the slide gel-diffusion test, the sensitivity of which is also limited to 1 μg of enterotoxin per milliliter of antigen, was 2 to 4 μg/ml of broth containing 15 to 20 billion organisms. By assuming that there is a direct correlation between growth and enterotoxin produc-

---

*Mention of commercial names does not constitute endorsement by the Public Health Service.

tion and by disregarding the differences between media in supporting enterotoxin production, they calculated that the concentration of enterotoxin present in the foods was from 0.01 to 0.04 μg/gm. Again, if one assumes an average portion to be 100 gm, dosage figures of 1 to 4 μg are obtained, which agree closely with those calculated above.

Unfortunately these figures are highly speculative and are probably higher than the true emetic dose for humans because they are based on test procedures of limited sensitivity. They do serve, however, to indicate that the concentrations of enterotoxin capable of causing human illness are quite low. Until such time as human volunteer studies are performed with highly purified enterotoxin preparations, the human emetic dose will remain a matter of speculation.

V. STAPHYLOCOCCAL ENTEROTOXINS

Until fairly recently little was known about the nature of staphylococcal enterotoxin, its mode of action, or the methods by which it could be produced and assayed in the laboratory. Various investigators associated with Dr. Gail M. Dack at the Food Research Institute of the University of Chicago since the end of World War II are responsible for the accumulation of much of the basic information on enterotoxin presently available. These studies have also provided the foundation on which many investigators have built and the impetus for others to perform research on this hitherto elusive biological entity.

A. Types

Staphylococcal enterotoxins are weakly antigenic; however, antibodies to these antigens have been produced in cats, rabbits, monkeys, horses, and burros. Dolman and Wilson (1938) were the first to report on what appears to be a specific antibody to staphylococcal enterotoxin, and Surgalla and his co-workers (1953) conclusively established the antigenicity of enterotoxin and the existence of more than one immunological type. These findings subsequently were confirmed in short order by others (Thatcher and Matheson, 1955; Bergdoll *et al.*, 1959*a*; Casman, 1958, 1960). Presently, three antigenically distinct enterotoxins have been identified and designated as enterotoxins A, B, and C (Casman *et al.*, 1963*a*; Bergdoll *et al.*, 1965*a*). Preliminary data suggest that a fourth enterotoxin exists; it has been designated tentatively as enterotoxin D by Casman and his co-workers (1966).

Staphylococcus aureus 196E (ATCC 13565) is the prototype strain for enterotoxin A (Casman *et al.,* 1963*a*), and it has been extensively employed by the group at the Food Research Institute for the production and purification of this enterotoxin. This strain, however, produces relatively large amounts of β-hemolysin that may influence bioassay procedures, and for this reason the use of other strains for the production and study of enterotoxin A is desirable.

Staphylococcus aureus 243 (ATCC 14458) is the prototype strain for enterotoxin B; it was selected from the several strains known to produce enterotoxin B because it is the only strain that produces B without producing A (Casman *et al.,* 1963*a*). Before 243 was selected as the prototype strain, the more widely known S_6 strain had been used extensively as the source of crude enterotoxin B for purification and immunological studies because of the favorable yields and potency of the toxin produced. With the subsequent recognition that the S_6 strain also produced small quantities of enterotoxin A, it was discarded as the prototype strain in favor of the single toxin-producing strain 243.

Prior to the development of serological test procedures (of which more will be said later) for the detection of A and B enterotoxins, it was not possible to undertake extensive studies of the frequency of distribution of these two types of toxin-producing strains from various sources or to determine whether more than two types of enterotoxins were produced. In a survey of this type, Casman (1965) found that, of 50 strains of food poisoning origin, from the United States and Canada, 90% produced enterotoxin A, 6% produced B, and 4% produced both enterotoxins A and B.

More recently Bergdoll and his associates (1965*a*) identified enterotoxin C from two strains of staphylococci designated as 137 and 361. The former strain was isolated from a leg abscess of a patient in Albert Billings Hospital, University of Chicago, in 1933, and the latter, by Dr. Betty C. Hobbs of England from cooked chicken implicated in a food poisoning outbreak. The original designation applied to the English strain by Dr. Hobbs was No. F.4626/62. Bergdoll subsequently applied the Food Research Institute number of 361 to this same culture. The enterotoxins produced by these two strains were shown to be antigenically distinct from enterotoxins A and B by gel diffusion in Ouchterlony plates and by neutralization tests in monkeys. Strain 137 has been adopted as the prototype strain for the production of enterotoxin C and is cataloged by the American Type Culture Collection as strain 19095.

The results of Casman's survey (1965) indicate that, in the United States at least, most staphylococcal food poisoning incidents are probably caused by enterotoxin A. The importance of enterotoxin C, however, cannot be ignored. Of 50 staphylococcal cultures that were observed to be enterotoxin-positive by the monkey feeding test, 20 (although negative for enterotoxins A and B by the plate gel-diffusion test) produced the new enterotoxin. Eleven of these 20 strains were later isolated from foods implicated in five separate staphylococcal food poisoning outbreaks in Georgia, Indiana, and England (Bergdoll *et al.*, 1965*a*).

The fact that 30 of the above 50 cultures caused emesis in monkeys that was not due to enterotoxins A, B, or C, and that 11 of these 30 cultures were isolated from foods implicated in food poisoning outbreaks, indicates that there are more than three types of enterotoxin. In addition to this evidence, a number of cat-emetic substances different from enterotoxins A and B have been demonstrated in various culture supernatant fluids of staphylococci isolated from the noses of healthy carriers, clinical specimens, frozen foods, and raw milk (Casman, 1965). The role of these substances in human illness has not been determined; it is believed, however, that some will be identified as enterotoxins C and D. The identification of other enterotoxins must await the development of suitable immunological procedures, and it may be that the staphylococci will produce an array of enterotoxins even greater in number than the lethal toxins of *Clostridium botulinum*.

B. Production

Although production of enterotoxin is achieved rather easily in the laboratory by cultivating toxigenic strains in Brain Heart Infusion broth or foodstuffs, specialized media must often be used to obtain sufficient quantities for chemical and purification studies.

Large volumes of enterotoxin B may be produced in a medium containing 15 gm of pancreatic digest of casein (Mead Johnson), 1230 μg of nicotinic acid, 50μg of thiamine hydrochloride, and 2.25 gm of anhydrous glucose per liter. The medium is adjusted to pH 7.6 prior to sterilization in the autoclave. High yields are attainable by shallow culture in Roux bottles containing 200 ml of culture incubated at 37°C in a sealed incubator containing 20% carbon dioxide and 80% air; by deep culture in sealed 12-gallon bottles containing 10 liters of culture under 20% carbon dioxide and 80% air and incubated at 37°C with continuous turning of the bottle in a horizontal position; and by deep aerated culture in which 1 to 15 liters of culture is aerated with either air or a mixture of 20% carbon dioxide and 80% oxygen during incubation at 37°C. The inoculum level and incubation period vary according to the culture method employed and range from 0.5 to 70 ml of inoculum and 2 to 5 days of incubation. Young cultures (6 to 24 hours old) are usually used as an inoculum (Surgalla *et al.,* 1951).

Enterotoxin B may also be produced in a medium consisting of Hy-Case S. F. (Sheffield Chemical, Norwich, New York), 20 gm; yeast extract, 3 gm; nicotinic acid, 1200 μg; calcium panthothenate, 500 μg; and thiamine hydrochloride, 40 μg/liter. The initial reaction of the medium is pH 7.4 to 7.6. Volumes of medium (100 ml) in 500-ml Erlenmyer flasks are inoculated with 1% inoculum of rapidly metabolizing staphylococcal cells and incubated for 24 hours on a reciprocal shaker (90 strokes of 2-inch length per minute) at 37°C. The cells are removed by centrifugation, and the supernatant fluid is concentrated by dialysis against Carbowax 4000 or polyethylene glycol. The concentrate is apparently stable for long periods if maintained at -15°C (Frea *et al.,* 1963).

The medium described by Casman (1958), modified by the replacement of Casamino Acids (Difco) with N-Z Amine A (Sheffield Chemical, Norwich, New York), can be employed to produce large quantities of enterotoxin A (Casman and Bennett, 1964). The composition of the medium per liter is: ferric citrate, 0.025 gm; K_2HPO_4, 1.0 gm; KH_2PO_4, 1.0 gm; $MgSO_4 \cdot 7H_2O$, 0.2 gm; 1-cystine, 0.025 gm; sodium acetate, 7.0 gm; 1-tryptophan, 0.075 gm; N-Z Amine A, 20 gm; calcium pantothenate, 500 μg; thiamine hydrochloride, 40 μg; and nicotinic acid, 1200 μg. The medium is adjusted to pH 7.2 to 7.4 and sterilized in the autoclave. Inoculum levels are small (0.1 ml of an 18- to 24-hour-old culture for 30 to 100 ml of medium), and no special atmosphere is required. The mode of aeration is determined by the volume of culture. Rotation at 20 rpm of 30-ml quantities of the culture in 8-ounce nursing bottles (Favorite and Hammon, 1941), rocking of 100-ml quantities in Roux bottles through an arc of 20° each for 1½ seconds (Casman, 1940), and sparging of air through larger volumes (Achorn and Schwab, 1948) are all satisfactory.

The "sac-culture" method is widely used for routine production of small quantities of highly concentrated crude enterotoxins A and B for immunological studies. The procedure was originally applied to enterotoxin production by Casman and Bennett (1963) as a means of achieving a concentrated toxin free from ammonium sulfate-precipitable substances. Cellophane sacs, 14 inches in length, are prepared from 0.75-inch-diameter dialysis tubing. The open end of the sac is attached to an 8-inch length of glass tubing of sufficient diameter to admit entry of a 1-ml pipet. The glass tubing is cotton-stoppered, and the whole is placed in a Roux bottle containing 100 ml of medium. A cotton plug is placed in the

Roux bottle, and the glass tubing is permitted to extend outside the bottle. The entire assembly is sterilized in the autoclave and cooled; then 0.5 ml of an aqueous suspension of the test organism prepared from an 18- to 24-hour-old nutrient agar slant is introduced into the sac via the glass tubing. The inoculum is distributed over the interior of the sac by inflating the sac and tipping the bottle. Incubation is at 35° to 37°C for 72 hours in a horizontal position, and the growth is removed from the sac with a 1-ml pipet. To prevent dilution of the toxin, only a small volume of saline or water is used to rinse residual growth from the sac. The cells are removed from the crude toxin by high-speed centrifugation. Any of the media described above or Brain Heart Infusion broth may be employed for the production of both enterotoxins A and B by this method.

Crude enterotoxins A and B also may be produced by a reverse modification of the "sac culture," which is more easily accomplished. Double-strength Brain Heart Infusion broth (100 ml) is placed in a 48A-average-porosity dialysis bag suspended in 18 ml of phosphate-buffered distilled water in a 300-ml Erlenmyer flask. The flask is cotton-stoppered, and the assembly is sterilized in the autoclave. Growth from a 7-hour-old Brain Heart Infusion agar slant, incubated at 35°C, is harvested in 2 ml of phosphate-buffered distilled water and used to inoculate the 18 ml of buffered water surrounding the sac. The inoculated "sac-cultures" are incubated for 72 hours at 35°C on a rotary shaker at 160 excursions per minute. The cells from the fluid collected from outside the sac are removed by centrifugation. The supernatant fluid contains the crude enterotoxin (Donnelly et al., 1963; Read et al., 1965a,b). As a rule sufficient enterotoxin B is produced in "sac-culture" to be assayed directly by immunological procedures. This is not always the case with enterotoxin A, however, and additional concentration may be achieved by forced evaporation in the cold or osmotic dialysis.

Enterotoxins A and B can also be produced in foods for certain types of laboratory studies. Enterotoxin A has been produced experimentally by inoculating the surface of raw and cooked beef and pork and the surface of cooked ham with a dilution of an enterotoxin A-producing strain followed by incubation at 30°C for 72 hours (Casman et al., 1963b). Both enterotoxins A and B may be produced routinely in the laboratory in food slurries prepared by homogenizing one part food with one part distilled water and inoculating the slurry with 1 ml of an appropriate 18-hour-old trypticase soy broth culture, followed by still or shaking incubation for 24 to 48 hours at 35°C. By this procedure enterotoxins have been produced in custard, chicken à la king, chicken pie filling, ham salad, shrimp, fish, beef, and meat pies (Hall et al., 1963, 1965).

Enterotoxin C is produced in a medium consisting of 30 gm of N-Z Amine NAK (Sheffield Chemical, Norwich, New York); 30 gm of Protein Hydrolyzate Powder (Mead Johnson and Co., Evansville, Indiana); 0.01 gm of niacin; and 0.005 gm of thiamine per liter. The medium is adjusted to pH 7.6 and dispensed in 400-ml portions in 2-liter Erlenmyer flasks. The sterile medium is inoculated with 4 ml of appropriate staphylococcal culture which has been prepared by incubating the culture in 50 ml of medium in a 250-ml Erlenmyer flask at 37°C for 24 hours without shaking. Following inoculation the medium is incubated for 24 hours at 37°C on a rotary shaker operated at approximately 275 rpm. The toxin is in the fluid portion of the culture from which the cells may be removed by centrifugation; it is concentrated by dialysis for 24 hours against an equal volume of 50% aqueous Carbowax 20 M (Union Carbide Corp., New York, New York) followed by dialysis for 1 week against deionized water. The dialyzed crude enterotoxin may be lyophilized with no apparent loss of potency (Bergdoll et al., 1965a).

C. Purification and Physicochemical Properties

Twenty-one years elapsed between the demonstration by Dack and his co-workers (1930) that culture filtrates of staphylococci isolated from cream-filled sponge cakes caused gastroenteritis in volunteers who drank portions of the filtrates and the first report on the partial purification of enterotoxin (Bergdoll et al., 1951). A number of factors contributed to the length of this interval, not the least of which was the applied direction necessarily undertaken by microbiological efforts during World War II. Significant progress had to await the development of modern biochemical and analytical procedures. Although several investigators contributed to the present knowledge concerning the purification of enterotoxin, credit must be given to Dr. M. S. Bergdoll and his colleagues at the Food Research Institute for the sustained effort over several years that resulted in the first purification of an enterotoxin. Because the existence of more than one type of enterotoxin was not recognized in the years immediately following World War II, enterotoxin B, which is now known to be less frequently involved in food poisoning, was purified first. This development unwittingly, but ironically, prolonged the subsequent development of in vitro methods for detecting enterotoxin in foods involved in outbreaks.

Early workers reported that enterotoxin was nondialyzable (Davison and Dack, 1936; Minnet, 1938; Hammon, 1941), which indicated that it might be a protein. Following the development of a means of producing large volumes of crude enterotoxin (Surgalla *et al.,* 1951), Bergdoll and his associates (1951, 1952, 1956; Surgalla *et al.,* 1954) provided evidence that partially purified toxic materials contained a water-soluble protein of approximately 15,000 to 25,000 molecular weight with an isoelectric point near pH 8.5, that the toxic material was antigenic, that it contained a high percentage of lysine, and that it was trypsin-resistant. The first published report describing a method for obtaining a highly purified enterotoxin (now designated as enterotoxin B) appeared in 1959 (Bergdoll *et al.,* 1959*b*) and involved acid precipitation of the bacterial supernatant fluid, adsorpiton on alumina, alcohol precipitation, chromatography on Amberlite IRC-50 resin, and zone electrophoresis on starch. The most active preparations obtained by this method gave emesis in monkeys at approximately 1 μg of nitrogen per dose, which was equivalent to 8 to 10 mg of sample weight.

About twice the amount of enterotoxin B of apparent comparable purity to that described above can be obtained from staphylococcal culture supernatant fluids by omitting the acid precipitation and alumina adsorption steps of the above method and replacing them by direct passage of the supernatant fluids over Amberlite XE64(IRC-50). The enterotoxin from 200 ml of bacterial culture supernatant fluid is diluted with 1 volume of water and adjusted to pH 6. The toxin is removed from the fluid by passage through 1 gm of resin pretreated with 0.02 M sodium phosphate, pH 6.2 The toxin is eluted from the resin with 0.2 M sodium phosphate, pH 6.2, and precipitated with ethyl alcohol at $-10°C$. Concentration is effected by dialysis and freeze-drying (Bergdoll *et al.,* 1961).

A modification of the chromatographic method described by Bergdoll has been employed by Frea and his colleagues (1963); it involves gel filtration on a G-100 Sephadex column followed by Sephadex electrophoresis. Toxin preparations obtainable by this method are usually clean, but not as free of minor impurities as the toxins purified by the procedures of Bergdoll (Schantz *et al.,* 1965).

Because all the above methods yield only milligram quantities of toxin, detailed studies of the chemical composition of the toxin had to await the development of methods that would result in preparation of purified enterotoxin B in greater yields than were previously possible. Schantz and his co-workers reported on a large-scale purification method that entailed a series of modifications of the chromatographic method described by Bergdoll and was followed by separation of the few remaining impurities on carboxymethylcellulose. The overall yield of purified toxin obtained by their method usually amounted to 50 to 60% based on serological tests of the original culture. The freeze-dried protein derived from the purification is a snow-white fluffy powder that is highly hygroscopic and very soluble in water and salt solutions. It is a simple protein consisting only of amino acids and is particularly rich in aspartic acid and lysine, which together comprise one-third of the total weight and number of residues of the protein. It has no free sulfhydryl groups and only one disulfide bridge. Glutamic acid is the N-terminal residue, and lysine is the C-terminal residue. The structure appears to be a single polypeptide chain with one N-terminal and one C-terminal residue per mole of protein (Spero et al., 1965). The molecular weight is 35,300, the isoionic point is pH 8.55, the isoelectric point is about pH 8.6, and the partial specific volume is 0.743. Sedimentation analyses indicate the toxin preparation to be homogeneous with respect to molecular weight and density (Wagman et al., 1965). The toxin has a maximum absorption at 277 mμ with an extinction ($E_1^{1\%}{}_{cm}$) of 14.0. It has a sedimentation coefficient ($S_{20,w}$) of 2.89S (single component) and a diffusion coefficient ($D_{20,w}$) of 7.72×10^{-7} cm^2 sec^{-1}. By electrophoresis the toxin is shown to be a single component.

When assayed in rhesus monkeys by intravenous injection, this purified enterotoxin B preparation causes vomiting in 50% of the animals (effective dose, or ED_{50}) at a concentration of 0.1 μg/kg of body weight (Schantz et al., 1965).

Bergdoll and his colleagues (1965b) also reported on the physicochemical properties of enterotoxin B purified according to the method of Schantz. They used somewhat different methods from those previously reported (Schantz et al., 1965; Wagman et al., 1965) for certain of the physical measurements, but the values obtained are in fairly good agreement with those described above. Sedimentation analysis indicated no heterogeneity; and by extrapolation of the $S_{20,w}$ value obtained at different concentrations to zero quantity, a value of 2.78 ($S_{20,w}$) was obtained. The diffusion coefficient was 8.22×10^{-7} cm^2 sec^{-1} at 20°C. The intrinsic viscosity was determined as 0.038 dl per gram. The molecular weight calculated from the sedimentation, diffusion, and viscosity data was 30,650. Spectrophotometric analysis of the toxin indicated maximum absorption at 277 mμ with an extinction coefficient ($E_1^{1\%}{}_{cm}$) of 12.1.

With the exception of the value for half-cystine, which is about 20% higher than that reported by Spero et al. (1965), the results of the amino acid analysis performed by Bergdoll and his co-workers agree with those of Spero and his colleagues. From the half-cystine content of the enterotoxin and the minimal molecular weight values for the other amino acids, Bergdoll and his group calculated the molecular weight to be 30,000 $\pm$ 1000, which agrees well with the value they obtained from their sedimentation, viscosity, and diffusion data but is lower than the value of

35,300 reported by Schantz *et al.* (1965) or that of 35,380 reported by Spero *et al.* (1965). Not only did the Chicago group obtain molecular weight values significantly different from those previously reported, but by using different analytical procedures, they also observed differences in the C- and N-terminal residues. They report the N-terminal amino acid sequence to be glutamic acid–serine–aspartic acid–lysine– and the C-terminal sequence to be –leucine–tyrosine–lysine–lysine–COOH.

Although much is known of the physical and chemical properties of enterotoxin B, not much information is available on the properties of enterotoxins A and C. Enterotoxin A has been independently isolated and purified by both Schantz and Bergdoll, and detailed studies of the toxin molecule are presently under way in their laboratories. At this writing, however, insufficient data are available on the purification process and chemical properties of the toxin to permit detailed presentation. In general, the toxin appears to be of comparable molecular weight to that of enterotoxin B, but the amino acid sequence and C- and N-terminal residues are different (Bergdoll, 1965).

D. Inactivation

Enterotoxin B is the only toxin of the three so far isolated that has been sufficiently purified to permit definitive studies on its properties and conditions necessary for inactivation. Although coagulation of purified enterotoxin B occurs after 5 minutes of exposure at 100°C. less than 50% of the biological activity is lost. No change in biological activity is noted after a solution of the toxin has been heated at 60°C and pH 7.3 for as long as 16 hours. In 0.05 M phosphate buffer ranging from pH 4 to 7.3 the toxin is stable for a week or more at room temperatures, although some loss of activity occurs on longer standing. Storage at pH 10 for several days causes no loss in activity, and the freeze-dried toxin is stable for over a year when stored at 4°C. Storage at room temperature for the same period, however, results in some loss of activity. Trypsin, chymotrypsin, rennin, and papain do not destroy biological activity of the toxin, but ficin and crude protease do. Pepsin destroys activity at pH 2, but not at higher pH values (Schantz *et al.,* 1965).

Studies on the thermal resistivity of crude and purified enterotoxin B in Veronal buffer reveal the toxin to be especially resistant to heat inactivation. The double-gel-diffusion test for the assay of heated enterotoxin was used to establish the following D values (minutes) for enterotoxin at exposure temperatures of 210°, 230°, and 250°F: for crude, 64.5, 52.3, and 29.7; for purified, 23.5, 11.4, and 9.9. From the inactivation data on purified enterotoxin B, the following values were obtained:

$z = 58.3°F$; experimental activation energy = 20,700 cal/gm mole; standard enthalpy of activation at 120°C = 19,900 cal/gm mole; standard entropy of activation at 120°C = 21.4 cal/gm mole K; and standard free energy of activation at 120°C = 28,200 cal/gm mole (Read and Bradshaw, 1966).

Because purified enterotoxin A is not generally available, the conditions for inactivation have not yet been determined; however, certain assumptions may be made, based on experience with crude culture filtrates and partially purified enterotoxin A. Crude culture filtrates used in cat injection tests usually are either boiled or treated with trypsin prior to injection to destroy the hemolysins. Neither procedure results in complete inactivation of the toxin, but boiling for an hour sometimes results in as much as a 50% reduction in concentration as measured serologically. Trypsinization (crude trypsin 1–250 for 30 minutes), on the other hand, has little effect on the crude toxin. Even after prolonged storage in the refrigerator and freezer, toxic culture filtrates containing enterotoxin A can cause emesis in cats; and the partially purified material is stable for several years when lyophilized and stored in the cold.

VI. METHODS FOR THE ISOLATION AND ENUMERATION OF STAPHYLOCOCCI IN FOODS

In food samples not involved in food poisoning outbreaks, coagulase-positive staphylococci make up a relatively small part of the total food flora. Because this is so, laboratory workers have resorted to the use of selective media to prevent the overgrowth of the staphylococci during primary cultivation. Though necessary, this practice has thwarted attempts to develop quantitative detection methods for staphylococci in foods because the media used, are, to some degree, also inhibitory to the staphylococci.

Direct plating methods for staphylococci in foods have been restricted in the past primarily to the use of mannitol salt agar (Chapman, 1945) and Staphylococcus Medium 110 (Chapman, 1946), whose selective properties are mainly attributable to the sodium chloride concentration present in the media. These media require subjective judgment on colonial morphology and pigmentation. Since they have generally been found to permit the concomitant growth of many other bacterial types, testing the numerous colonies for coagulase production as a confirmatory procedure is necessary.

As a result of the difficulties associated with media whose selectivity is accomplished through the addition of sodium chloride, other formulations were tried to yield better differentiation and improved selectivity. On the first of these formulations, which contained potassium tellurite (Ludlam, 1949), staphylococci grew as black colonies. Zebovitz et al. (1955) developed a modification of Ludlam's medium for enumerating coagulase-positive staphylococci in meats, which became widely used for enumerating staphylococci in meats and other foods. Unfortunately, the black colony formation on tellurite-containing media is not restricted to coagulase-positive staphylococci; micrococci and certain *Proteus* species, particularly *P. vulgaris,* form black colonies similar to those of staphylococci.

The findings of Gillespie and Alder (1952) that most coagulase-positive staphylococci produced opacity when grown in media containing egg yolk led to the development of tellurite–egg yolk media for the purpose of maintaining the characteristic black appearance of staphylococcal colonies on tellurite agar and the high correlation of the egg yolk reaction with coagulase production (Innes, 1960).

The recent comparative evaluation of several staphylococcal plating media conducted in the author's laboratories (Crisley et al., 1965), intended to determine the relative merits of each for quantitatively recovering staphylococci from foods revealed that the tellurite–egg yolk formulations were superior to those in which either egg yolk or tellurite was absent. Such factors as food composition, types and numbers of associative flora, and concentrations of staphylococci affected recovery rates obtained with the different media. The tellurite–egg yolk formulations of Innes (1960) and Baird-Parker (1962) yielded superior results with specific classes of foods. The tellurite–polymyxin–egg yolk (TPEY) medium of Crisley et al. (1964) was less affected by food constituents and yielded good differentiation and quantification of staphylococci.

After 24 hours of incubation at 35°C, coagulase-positive staphylococci usually display one or more of the following reactions on TPEY medium: (1) a zone of precipitate around the colony; (2) a clear zone or halo around the colony with precipitate beneath colony; and (3) no halo around the colony, but precipitate beneath. On some occasions an additional 12 to 24 hours of incubation may be necessary to permit large-enough colonies to develop for these reactions to be distinct. The testing for coagulase production of five colonies with the above characteristics from plates containing between 30 and 300 colonies is employed as a confirmatory step.

The number of staphylococci usually present in normal foods is small, and TPEY medium as well as other tellurite–egg yolk formulations are surface-plating media. Quantities of food greater than that contained in a 1–8 dilution usually cannot be accurately delivered by pipet, and the largest volume of fluid that can be conveniently spread over the surface of a medium in a standard petri dish is about 0.1 ml. Consequently, the minimum range of staphylococci that can be enumerated consistently in food by such a procedure is approximately 1×10^3 to 1×10^4 per gram. Foods involved in outbreaks characteristically contain more staphylococci than this, whereas the staphylococcal counts associated with wholesome foods vary from too few to detect to a few hundred per gram. Thus, to enumerate low levels of staphylococcal contamination of this type, a most probable number (MPN) enrichment method must be used. A selective enrichment broth is employed with plating on a solid medium and confirmation by testing selected colonies for coagulase production. The most commonly employed selective enrichment broth is cooked meat medium containing 10% sodium chloride. A loopful of culture from each tube that is positive for growth after 24 hours of incubation is streaked on the surface of one of the tellurite–egg yolk plating media, and after incubation the typical staphylococcal colonies are tested for coagulase production. A mannitol–sorbic acid medium (Raj and Liston, 1961) has been proposed for testing frozen foods by the MPN method in which confirmation is obtained on plates of Staphylococcus Medium 110–egg yolk agar that are incubated at 45°C. The elevated incubation temperature reportedly increases the specificity of the medium.

As of 1968 no staphylococcal medium had been devised that permits only the growth of coagulase-positive staphylococci. Consequently, suspected staphylococcal isolates obtained from foods must be tested for coagulase production before any conclusion is drawn concerning the contamination of food with coagulase-positive staphylococci. The attention that microbiological standards for foods are presently receiving indicates that a numerical limit will be set for staphylococci in certain food categories. With the methods available presently, a test for coagulase production will have to be included in any procedure recommended for detecting and enumerating staphylococci in foods.

VII. DETECTION OF ENTEROTOXINS IN FOODS

A. Animal Assays

Until fairly recently the investigation and diagnosis of staphylococcal food poisoning employed by many laboratories depended on correlating the incubation period and symptoms of patients to the finding of coagulase-positive staphylococci in the food eaten in common. The testing for enterotoxin by assay in cats or monkeys of pure culture supernatant fluids or filtrates derived from the isolated staphylococci was performed by only a limited number of laboratories. The maintenance costs and purchase price of these animals were too expensive for routine use in most laboratories, and laboratory investigation of outbreaks usually proceeded no further than the isolation of coagulase-positive staphylococci. Many attempts have been made over the years to develop an inexpensive and rapid assay procedure for enterotoxin. In addition to cats and monkeys, frog and rabbit gut sections have been tried, as well as nematodes and tissue culture. The reliability of these tests is not yet confirmed.

The kitten test (Dolman *et al.,* 1936) has been used successfully in many laboratories for detecting enterotoxin in culture filtrates. The test as originally described called for the injection of 1 to 3 ml of filtrate in a kitten weighing 350 to 700 gm. On injection of filtrate containing enterotoxin, projectile vomiting accompanied by diarrhea, weakness, and unsteadiness usually developed in less than an hour. Davison *et al.* (1938) and Hammon (1941) reported on intravenous injection tests in kittens and cats. This latter test has been used extensively in the Milk and Food Research laboratories of the Public Health Service and at times is still employed to confirm certain serological findings. Because unidentified emetic factors for kittens and cats are known to occur in some foods, intraperitoneal and intravenous injection tests are restricted to culture fluids only.

Apart from human volunteers, the best assay animal appears to be the rhesus monkey, *Macaca mulatta* (Surgalla *et al.,* 1953). In this test incriminated food or culture fluid is fed by stomach tube, and only vomiting is acceptable as a positive reaction.

B. Serological Assays

Fortunately for those who have the responsibility of determining the etiology of food-borne disease outbreaks, animal assay procedures for enterotoxin are rapidly being replaced by serological procedures. Since 1963 a number of publications have appeared from the laboratories of the Public Health Service and the Food and Drug Administration on the detection and identification of enterotoxins A and B in foods by means of gel-diffusion methods. The first such report (Hall *et al.,* 1963) described (1) the extraction of enterotoxin B from a number of foods to which it had been added or in which enterotoxin B-producing strains of staphylococci had grown and (2) the quantitative detection of the toxin in food extracts by Oudin (Crowle, 1961) or Oakley and Fulthorpe (1953) single-

and double-gel-diffusion agar precipitin techniques and by the micro-Ouchterlony (Crowle, 1958) diffusion method. Shortly thereafter Casman and Bennett (1965) described methods for removing enterotoxins A and B from foods by separating the toxins from food extracts by absorption on carboxymethylcellulose or by filtration through Sephadex G-100 and identification of the toxins by the micro-Ouchterlony slide test. The publication by Read and his co-workers (1965a,b) appeared a month later and described extraction methods specifically adapted to milk and cheese followed by identification of enterotoxins A and B by single- and double-gel-diffusion tests. By midyear Hall et al. (1965) reviewed the status of serological methods for detecting enterotoxins in foods and included a more improved extraction and concentration procedure than the method they reported in 1963. It is very encouraging that three separate groups of workers have been able, by dissimilar methods, to extract enterotoxins A and B from foods and to quantify their concentrations by means of gel-diffusion techniques. That this has been accomplished suggests that the methods may find wide application in public health and diagnostic laboratories.

The three serological methods (single- and double-gel-diffusion tube tests and micro-Ouchterlony slide tests) depend on the use of antiserum that is specific to the toxin against which it has been produced. Although specific antiserum to enterotoxin B is easily obtainable because of the purity of the antigen, some nonspecific antibodies are found in such antisera. We have successfully produced antiserum to enterotoxin B in our laboratories (Hall et al., 1963); however, we have found that, in spite of the purity of the antigen, low dilutions of antiserum cross-react with antigens in the supernatant fluids of nonenterotoxigenic staphylococci as well as with strains that produce enterotoxin A. These cross reactions are probably due to the presence of staphylococcal antibodies already in the serum of the rabbits prior to immunization, as demonstrated by Cohen et al. (1961), and can be eliminated in most cases by diluting the antiserum to 1:60 and above.

More recently Genigeorgis and Sadler (1966a,b) applied a fluorescent antibody technique to the detection of enterotoxin B in a number of foods and culture filtrates with considerable success. This method may, in time, gain wide application because of its rapidity and sensitivity. Similarly, passive hemagglutination-inhibition techniques (Robinson and Thatcher, 1965; Brown and Brown, 1965; Morse and Mah, 1967; Johnson et al., 1967) have been used to detect enterotoxins A and B in culture filtrates and defined media, and these methods may have advantages over gel-diffusion techniques if adequate refinements can be developed for the extraction of the toxins from foods.

Most of the antisera to enterotoxin A presently in use have been produced to a partially purified antigen and have been rendered "monovalent" either by absorption (Casman, 1960) or by dilution.

1. SINGLE-DIFFUSION TUBE TEST

Each of the three test procedures has certain advantages and disadvantages; the selection of the method to be used depends on a number of factors and requires an understanding of the basic differences in the tests themselves. All these tests are agar gel-diffusion precipitin reactions and are, therefore, affected by conditions of electrolyte concentration, pH, reactant concentrations, and others as described in detail by Crowle (1961). In the single-diffusion tube test the antiserum agar occupies approximately one-half the volume of the tube, and the liquid food extract containing the antigen is layered over the solidified antiserum agar. The density of the zone of specific precipitation that forms in the antiserum agar is a function of the antibody concentration in the agar. At equivalent concentrations of antigen and antibody, the precipitate forms at the interface of antigen and antiserum. To ensure that a zone forms in the antiserum agar, dilutions of the antiserum are prepared to obtain the highest dilution of antiserum that will yield an observable zone of precipitation. The enterotoxin is a low-molecular-weight antigen and diffuses rapidly into the antiserum agar. The antibodies are insoluble in the presence of excess antigen, and consequently the zone formed is of a uniform density with a sharply defined leading edge. The width of the zone, usually measured in millimeters, is dependent on the concentration of enterotoxin and dilution of antiserum, the time and temperature of incubation, the ionic strength of the diluent, and the diameter of the tube employed. The effects of these variables on quantification of enterotoxin have been presented in some detail elsewhere (Hall *et al.*, 1965). In general, the single-gel-diffusion tube test is useful when the antigen solution contains 1 μg or more of enterotoxin per milliliter. Since this degree of enterotoxin is rarely observed in extracts of foods incriminated in outbreaks, the extract must be concentrated. On the other hand, such a concentration is readily obtained in culture filtrates and in some experimentally contaminated foods. The single-gel-diffusion tube test does not differentiate among multiple antigen-antibody systems and consequently cannot be used to resolve the number of such systems reacting in a given test. The rapidity of the test (reactions are usually measurable in less than 24 hours) allows its use as a rapid screening procedure when a truly mono-valent antiserum is employed.

2. DOUBLE-GEL-DIFFUSION TUBE TEST

To detect concentrations of enterotoxin below 1 μg/ml of antigen solution, the double-gel-diffusion tube test may be used. In this test (Hall *et al.*, 1965) the antigen diffuses downward into a layer of buffered agar, which is situated in the tube as an overlay over the antiserum agar. At the same time that the antigen is diffusing downward, the antibody is diffusing upward into the buffered agar layer; at the point at which the two reactants meet, a narrow band or line of specific precipitation is formed. The position of this line is dependent on the concentration of the reactants; that is, when antigen is in excess, the line appears closer to the interface of buffer agar and antiserum agar, whereas the reverse is true when antibody is in excess. The density of the precipitate is de-pendent on antibody concentration; but both reactants continue to feed the reaction at the position of equivalence, and prolonged incubation tends to increase the density. Experience in our laboratories indicates that, after 1 week of incubation, uniform results are obtained when the enterotoxin concentration is in the range of 0.1 to 10.0 μg/ml of antigen solution. This method is used not only as a means of detecting low con-centrations of enterotoxin but also as a confirmatory procedure in con-junction with single-gel-diffusion tube tests, particularly when equiva-lence reactions or nonspecific clouding of the antiserum agar occurs in the latter. The procedure will also resolve different antigen–antibody systems, each of which is detected as a separate line in the buffered agar layer.

1. MICRO-OUCHTERLONY SLIDE TEST

The third method of detecting enterotoxin is by means of the micro-Ouchterlony slide test described by Crowle (1958), but applied in slightly modified form (Casman and Bennett, 1965; Hall *et al.*, 1965) for assaying enterotoxin. In this test, lines of precipitation develop in a thin layer of agar gel located between a glass slide and an overlying plastic template with wells. Antiserum is placed in the center well, and known enterotoxin solutions are alternated with unknown samples in the sur-rounding outside wells. The preparation is incubated at room tempera-ture for 24 hours in a moist chamber and examined for lines of pre-cipitation. Such lines form between an antigen and its corresponding antibody, and identification of the unknown sample is readily determined by coalescence of its line of precipitation with a reference line of precipitation formed by the interaction of antibody with known enterotoxin. The limit of sensitivity of this test also is approximately 1 μg of enterotoxin per milliliter of antigen solution (Casman and Bennett, 1965); the test is more exacting in its preparation and in the proportion of antigen to antibody that is used. It is advisable to prepare all tests in duplicate to avoid failure due to poor absorption and precipitate formation; a series of doubling dilutions of each unknown sample must be tested to ensure the formation of lines of identity with the control in the gel between the wells rather than under a well, as may occur with high concentrations of enterotoxin. For these reasons, we have used the test only to confirm the identity of lines or zones of precipitation formed in double- and single-gel-diffusion tube tests, provided the enterotoxin concentrations were at least 1 μg/ml of antigen solution.

C. Other Methods

Experience has revealed that the amounts of enterotoxins in foods incriminated in outbreaks are below the levels detectable by the above tests and that extraction and concentration procedures must be applied to the foods before the toxins can be identified and assayed serologically. Several methods have been employed successfully for this purpose. Casman and Bennett (1965) performed a comparative evaluation of two separation methods based on absorption on carboxymethylcellulose columns and gel filtration through Sephadex G-100 columns followed by concentration of the eluates by dialysis against polyethylene glycol. Their findings indicated that the use of Sephadex G-100 generally resulted in the recovery of slightly more enterotoxin, but did not permit the degree of concentration possible after absorption to, and elution from, carboxymethylcellulose. The latter procedure permitted detection of similar amounts of enterotoxin and was selected by them as the method of choice for separating enterotoxin from food.

Hall et al. (1965) separated enterotoxin from foods by passage of the supernatant fluid that is obtained in the preliminary extraction step through an Amberlite CG50 resin column and concentrated the enterotoxin by dialysis against polyvinylpyrrolidone. Read and his co-workers (1965a,b) experienced difficulty in assaying enterotoxin from milk and cheese using the procedure of Hall et al. (1965), due to the interference of migrating opaque substances in dairy products that affected gel-diffusion results. To eliminate these substances, they developed separation procedures specifically adapted for milk and cheese.

Although these extraction and separation steps are somewhat laborious, they are necessary if enterotoxins are to be detected in incriminated foods. The examination in our laboratories of a number of foods involved in outbreaks has revealed that a concentration method must be applied more often than not to ensure detection of enterotoxin. These extraction and concentration procedures permit the assay by the single-gel-diffusion tube test of as little as 0.05 μg of either enterotoxin A or B per gram of food.

REFERENCES

Achorn, G. B., and Schwab, J. L. (1948). *Science* **107**, 377.
Angelotti, R., Foter, M. J., and Lewis, K. H. (1961a). *Am. J. Public Health* **51**, 76.
Angelotti, R., Foter, M. J., and Lewis, K. H. (1961b). *Appl. Microbiol.* **9**, 308.
Angelotti, R., Foter, M. J., and Lewis, K. H. (1961c). *Am. J. Public Health* **51**, 83.
Baird-Parker, A. C. (1962). *J. Appl. Bacteriol.* **5**, 12.
Bergdoll, M. S. (1956). *Ann. N.Y. Acad. Sci.* **65**, 139.
Bergdoll, M. S. (1965). Personal communication.
Bergdoll, M. S., Kadavy, J. L., Surgalla, M. J., and Dack, G. M. (1951). *Arch. Biochem. Biophys.* **33**, 259.
Bergdoll, M. S., Lavin, B., Surgalla, M. J., and Dack, G. M. (1952). *Science* **116**, 633.
Bergdoll, M. S., Surgalla, M. J., and Dack, G. M. (1959a). *J. Immunol.* **83**, 334.
Bergdoll, M. S., Sugiyama, H., and Dack, G. M. (1959b). *Arch. Biochem. Biophys.* **85**, 62.
Bergdoll, M. S., Sugiyama, H., and Dack, G. M. (1961). *J. Biochem. Microbiol. Technol. Eng.* **3**, 41.
Bergdoll, M. S., Borja, C. R., and Avena, R. M. (1965a). *J. Bacteriol.* **90**, 1481.
Bergdoll, M. S., Chu, F. S., Huang, I. Y., Rowe, C., and Shih, T. (1965b). *Arch. Biochem. Biophys.* **112**, 104.
Breed, R. S., Murray, E. G. D., and Smith, Nathan R. (1957). "Bergey's Manual of Determinative Bacteriology," 7th ed. Williams and Wilkins, Baltimore.
Brown, G. R., and Brown, C. A. (1965). *Bacteriol. Proc.,* p. 72.
Campbell, A. C. P. (1948). *J. Pathol. Bacteriol.* **60**, 157.
Casman, E. P. (1940). *J. Bacteriol.* **40**, 601.
Casman, E. P. (1958). *Public Health Rept. (U.S.)* **73**, 599.
Casman, E. P. (1960). *J. Bacteriol.* **79**, 849.
Casman, E. P. (1965). *Ann. N.Y. Acad. Sci.* **128**, 124.
Casman, E. P., and Bennett, R. W. (1963). *Am. J. Bacteriol.* **86**, 18.
Casman, E. P., and Bennett, R. W. (1964). *Appl. Microbiol.* **12**, 363.
Casman, E. P., and Bennett, R. W. (1965). *Appl. Microbiol.* **13**, 181.
Casman, E. P., Bergdoll, M. S., and Robinson, J. (1963a). *J. Bacteriol.* **85**, 715.
Casman, E. P., McCoy, D. W., and Brandly, P. J. (1963b). *Appl. Microbiol.* **11**, 498.
Casman, E. P., Bennett, R. W., and Kephart, R. E. (1966). *Bacteriol. Proc.* p. 13.
Chapman, G. H. (1945). *J. Bacteriol.* **50**, 201.
Chapman, G. H. (1946). *J. Bacteriol.* **51**, 409.
Chapman, G. H. (1947). *J. Bacteriol.* **53**, 367.
Cohen, J. O., Cowart, G. S., and Cherry, W. B. (1961). *J. Bacteriol.* **82**, 111.

Commission on Acute Respiratory Diseases. (1949). *Am. J. Hyg.* **50**, 331.

Crisley, F. D., Angelotti, R., and Foter, M. J. (1964). *Public Health Rept. (U.S.)* **79**, 369.

Crisley, F. D., Peeler, J. T., and Angelotti, R. (1965). *Appl. Microbiol.* **13**, 140.

Crowle, A. J. (1958). *J. Lab. Clin. Med.* **52**, 784.

Crowle, A. J. (1961). "Immunodiffusion," pp. 11–36. Academic Press, New York.

Dack, G. M. (1956). "Food Poisoning," University of Chicago Press.

Dack, G. M., and Lippitz, G. (1962). *Appl. Microbiol.* **10**, 472.

Dack, G. M., Cary, W. E., Woolpert, O., and Wiggers, H. J. (1930). *J. Prevent. Med.* **4**, 167.

Das, S. K., and Chatterjee, G. C. (1962). *J. Bacteriol.* **83**, 1251.

Dauer, C. C., *et al.* (1952–1961). Summary of disease outbreaks. *Public Health Rept (U.S.)* **67**, 1089 (1952); **68**, 696 (1953); **69**, 538 (1954); **70**, 536 (1955); **71**, 797 (1956); **72**, 735 (1957); **73**, 681 (1958); **74**, 715 (1959); **75**, 1025 (1960); **76**, 915 (1961).

Davison, E., and Dack, G. M. (1936). *J. Infect. Diseases* **64**, 302.

Davison, E., Dack, G. M., and Cary, W. E. (1938). *J. Infect. Diseases* **62**, 219.

Dick, G. F., and Dick, G. H. (1924). *J. Am. Med. Assoc.* **82**, 265.

Dolman, C. E., and Wilson, R. J. (1938). *J. Immunol.* **35**, 13.

Dolman, C. E., Wilson, R. J., and Cockcroft, W. H. (1936). *Can. Public Health J.* **27**, 489.

Donnelly, C. B., Shideler, J. E., Black, L. A., and Lewis, K. H. (1963). *Bacteriol. Proc.* p. 15.

Donnelly, C. B., Black, L. A., and Lewis, K. H. (1964). *Appl. Microbiol.* **12**, 311.

Favorite, G. O., and Hammon, W. McD. (1941). *J. Bacteriol.* **41**, 305.

Fildes, P., and Richardson, G. M. (1937). *Brit. J. Exptl. Pathol.* **18**, 292.

Fildes, P. R., Richardson, G. M., Knight, B. C. J. G., and Gladstone, G. P. (1936). *Brit. J. Exptl. Pathol.* **17**, 481.

Frea, J. I., McCoy, E., and Strong, F. M. (1963). *J. Bacteriol.* **86**, 1308.

Genigeorgis, C., and Sadler, W. W. (1966a). *J. Food Sci.* **31**, 441.

Genigeorgis, C., and Sadler, W. W. (1966b). *J. Food Sci.* **31**, 605.

Gillespie, W. A., and Alder, V. G. (1952). *J. Pathol. Bacteriol.* **64**, 187.

Gladstone, G. P. (1937). *Brit. J. Exptl. Pathol.* **18**, 322.

Gladstone, G. P., Fildes, P., and Richardson, G. M. (1935). *Brit. J. Exptl. Pathol.* **16**, 335.

Goslings, W. R. O., and Buchli, K. (1958). *A.M.A. Arch. Internal. Med.* **102**, 691.

Hall, H. E., Angelotti, R., and Lewis, K. H. (1963). *Public Health Rept. (U.S.)* **78**, 1089.

Hall, H. E., Angelotti, R., and Lewis, K. H. (1965). *Health Lab. Sci.* **2**, 179.

Hammon, W. M. (1941). *Am. J. Public Health* **31**, 1191.

Hancock, R. (1960). *J. Gen. Microbiol.* **23**, 179.

Haynes, W. C., Kuehne, R. W., and Rhodes, L. J. (1954). *Appl. Microbiol.* **2**, 339.

Innes, A. G. (1960). *J. Appl. Bacteriol.* **23**, 108.

Johnson, H. M., Hall, H. E., and Simon, M. (1967). *Appl. Microbiol.* **15**, 815.

Jordan, E. O. (1930). *J. Am. Med. Assoc.* **94**, 1649.

Jordan, E. O., and Burrows, W. (1934). *Am. J. Hyg.* **20**, 604.

Knight, B. C. J. G. (1937a). *Biochem. J.* **31**, 731.

Knight, B. C. J. G. (1937b). *Biochem. J.* **31**, 966.

Knight, B. C. J. G., and McIlwain, H. (1938). *Biochem. J.* **32**, 1241.

Ludlam, G. B. (1949). *Monthly Bull. Min. Health* **8**, 15.

Macfarlane, D. A., Murrel, J. S., Shooter, R. A., and Curwen, M. P. (1960). *Brit. Med. J.* **2**, 900.

Martin, T. D. M., and Whitehead, J. E. M. (1949). *Brit. Med. J.* **4595**, 173.

Miles, A. A., Williams, R. E. O., and Clayton-Cooper, B. (1944). *J. Pathol. Bacteriol.* **56,** 513.

Minnet, F. C. (1938). *J. Hyg.* **38,** 623.

Morse, S. A., and Mah, R. A. (1967). *Appl. Microbiol.* **15,** 58.

Munch-Petersen, E. (1963). *J. Food Sci.* **28,** 692.

Oakley, C. L., and Fulthorpe, A. J. (1953). *J. Pathol. Bacteriol.* **65,** 49.

O'Kane, D. J. (1941). *J. Bacteriol.* **41,** 441.

Packalen, T., and Bergqvist, S. (1947). *Acta. Med. Scand.* **127,** 291.

Parker, J. T. (1924). *J. Exptl. Med.* **40,** 761.

Peterson, A. C., Black, J. J., and Gunderson, M. F. (1962a). *Appl. Microbiol.* **10,** 16.

Peterson, A. C., Black, J. J., and Gunderson, M. F. (1962b). *Appl. Microbiol.* **10,** 23.

Porter, J. R., and Pelczar, M. J., Jr. (1940). *Science* **91,** 576.

Raj, H., and Liston, J. (1961). *Bacteriol. Proc.* p. 68.

Read, R. B., Jr., and Bradshaw, J. G. (1966). *Appl. Microbiol.* **14,** 130.

Read, R. B., Jr., and Pritchard, W. L. (1963). *Can. J. Microbiol.* **9,** 879.

Read, R. B., Jr., Pritchard, W. L., Bradshaw, J., and Black, L. A. (1965a). *J. Dairy Sci.* **48,** 411.

Read, R. B., Jr., Bradshaw, J., Pritchard, W. L., and Black, L. A. (1965b). *J. Dairy Sci.* **48,** 420–424.

Report of the Public Health Laboratory Service—Food Poisoning in England and Wales, Food Poisoning of All Types, *Monthly Bull. Min. Health Public Health Lab. Serv.* **21,** 180, (1962); **22,** 200, (1963); **23,** 189, (1964).

Richardson, G. M. (1936). *Biochem. J.* **30,** 2184.

Ridley, M. (1959). *Brit. Med. J.* **5117,** 270.

Robinson, J., and Thatcher, F. S. (1965). *Bacteriol. Proc.* p. 72.

Rosendal, K., Bulow, P., and Jensen, O. (1964). *Nature* **204,** No. 4964, 1222.

Rountree, P. M., and Barbour, R. G. H. (1951). *J. Pathol. Bacteriol.* **63,** 313.

Schantz, E. J., Roessler, W. G., Wagman, J., Spero, L., Dunney, D. A., and Bergdoll, M. S. (1965). *Biochemistry* **4,** 1011.

Segalove, M., and Dack, G. M. (1941). *Food Res.* **6,** 127.

Sevag, M. G., and Green, M. N. (1944). *J. Biol. Chem.* **154,** 719.

Shooter, R. A., and Wyatt, H. V. (1955). *Brit. J. Exptl. Pathol.* **36,** 341.

Shooter, R. A., and Wyatt, H. V. (1956). *Brit. J. Exptl. Pathol.* **37,** 311.

Shooter, R. A., and Wyatt, H. V. (1957). *Brit. J. Exptl. Pathol.* **38,** 473.

Spero, L., Stefanye, D., Brecker, P. I., Jacoby, H. M., Dalidowicz, J. E., and Schantz, E. J. (1965). *Biochemistry* **4,** 1024–1030.

Straka, R. P., and Combs, F. M. (1952). *Food Res.* **17,** 448.

Strasters, K. C., and Winkler, K. C. (1963). *J. Gen. Microbiol.* **33,** 213.

Surgalla, M. J. (1947). *J. Infect. Diseases* **81,** 97.

Surgalla, M. J., Kadavy, J. L., Bergdoll, M. S., and Dack, G. M. (1951). *J. Infect. Diseases* **89,** 180.

Surgalla, M. J., Bergdoll, M. S., and Dack, G. M. (1953). *J. Lab. Clin. Med.* **41,** 782.

Surgalla, M. J., Bergdoll, M. S., and Dack, G. M. (1954). *J. Immunol.* **72,** 398.

Thatcher, F. S., and Matheson, B. H. (1955). *Can. J. Microbiol.* **1,** 382.

Thatcher, F. S., and Simon, W. (1956). *Can. J. Microbiol.* **2,** 703.

Troeller, J. A., and Frazier, W. C. (1963a). *Appl. Microbiol.* **11,** 11.

Troeller, J. A., and Frazier, W. C. (1963b). *Appl. Microbiol.* **11,** 163.

Vogelsang, T. M. (1951). *Acta Pathol. Microbiol. Scand.* **29,** 363.

Vogelsang, T. M. (1958). *Acta Pathol. Microbiol. Scand.* **43,** 196.

Wagman, J., Edwards, R. C., and Schantz, E. J. (1965). *Biochemistry* **4**, 1017.

Williams, R. E. O. (1946). *J. Pathol. Bacteriol.* **58**, 259.

Williams, R. E. O. (1963). *Bacteriol. Rev.* **27**, 56.

Williams, R. E. O., Jevans, M. P., Shooter, R. A., Hunter, C. J. W., Girling, J. A., Griffiths, J. D., and Taylor, G. W. (1959). *Brit. Med. J.* **2**, 658.

Wilson, G. S., and Miles, A. A. (eds.) (1964). "Topley and Wilson's Principles of Bacteriology and Immunity," 5th ed. p. 751. Williams and Wilkins, Baltimore.

Zebovitz, E., Evans, J. B., and Niven, C. F. (1955). *J. Bacteriol.* **70**, 686.

CHAPTER X | **ALIMENTARY MYCOTOXICOSES**

Gerald N. Wogan

I. INTRODUCTION

Physiologically active chemical contaminants may find their way into the human or animal food supply through a variety of routes. Some of these substances are known by experimental proof or by long experience of usage to be nondeleterious. Certain such materials, however, have been clearly established as causative agents of animal and human diseases. Toxins produced by microbial food contaminants are important examples of the latter group, and an extensive body of knowledge has developed concerning the bacterial toxins which are discussed elsewhere in this volume.

Within the past two decades, increasing attention has been focused on the real and potential significance to the safety of the food supply posed by toxic compounds produced by fungal contaminants of foods and food raw materials. These compounds, referred to by the generic term "myco-toxins," comprise a group of chemical compounds widely diverse in their nature and biological effects. The literature dealing with the development of antibiotics is replete with examples of compounds of microbial origin which ultimately found little or no clinical application because of toxicity to animals. In this sense, the mycotoxins might be regarded as antibiotics against animals rather than as microorganisms. It would not be possible, nor would it be germane in the present context, to review the literature on toxic antibiotics. Rather, this discussion will be limited to a considera-tion of those compounds that find their way into the human or animal food chain by virtue of the presence and growth of fungi that produce them in commodities used for food. It is appropriate at this point to consider briefly several general aspects of fungal spoilage of foods before entering into detailed discussions of the known mycotoxins and their effects.

A. Fungal Spoilage of Food Crops

Mold spores are ubiquitously distributed, and foodstuffs may be con-taminated with a wide variety of fungi. For example, Christensen (1965) has described a wide variety of fungi that invade the seeds of cereal grains and their products, and has grouped these organisms into categories (field fungi, storage fungi, and advanced decay fungi) according to the stage at which invasion and growth occur. If an equally complex fungal flora is assumed to be present in other commodities, it is apparent that the majority of food raw materials are liable, in varying degrees, to growth of contaminating fungi at some stage during their harvest, storage, transport, or processing. The possibility for spore germination and growth on a given product is governed by several factors including moisture content, relative humidity, and temperature. During the entire postharvest period, food crops are essentially in a state of storage, and mold growth on them is avoidable only by careful regulation of moisture content, temperature, and other environmental conditions. Commodities particularly prone to mold damage include grains, oilseeds, fruits, and vegetables.

Total losses of foods or raw materials due to mold damage are difficult to estimate accurately. However, the magnitude this problem can attain is illustrated by the conservative estimate (Johnson, 1948) that 2% of the annual world grain production is lost because of the activities of microorganisms. In tropical and subtropical climates, estimates of annual losses have been as great as 45% of the annual production. Mold damage constitutes a major factor in these losses, which have great impact on the total world food supply.

Many types of crop damage result from mold growth which diminishes or destroys the usefulness of the commodity for food purposes. Examples of serious losses of tropical foods attributed to fungal damage include discoloration of rice and cocoa butter, loss of seed viability, mustiness of coffee, and lipolysis of plant oils (Hiscocks, 1965). Until rather recently, interest in mold-induced damage of foodstuffs has centered principally on the economic losses incurred by virtue of deterioration of quality or other attributes of the affected commodity such as those mentioned above.

B. Mycotoxicoses

Although these deleterious effects have had serious economic consequences, fungal spoilage has created additional problems by contaminating the affected commodity with toxic metabolites ("mycotoxins") produced during their growth. Toxicity syndromes resulting from ingestion of contaminated foodstuffs have been referred to as "mycotoxicoses" (Forgacs and Carll, 1962) and have been encountered in many forms. All the recognized mycotoxicoses have posed direct hazards to human consumers, or have caused significant losses in domestic animals used for food. The net result in either case has been a reduction in the utilizable food supply, either by directly limiting human consumption or by decreasing available animal protein sources.

The following discussion of mycotoxicoses has been organized to provide a summary of currently available information on each recognized syndrome. Most emphasis has been placed on those toxicoses that have directly affected the human food supply or have caused serious losses of animals used for food. The aflatoxins are discussed in considerably more detail than other mycotoxins because of their recent discovery and the proportionately large amount of research devoted to this problem within a relatively short time. Furthermore, although various aspects have been reviewed, unified considerations of the general problem are not yet available. In most cases, no attempt has been made to provide an exhaustive survey of existing literature. In the main, information has been drawn from original publications, and reviews are cited where they are available.

Mycotoxicoses result from the ingestion of toxic substances produced by fungi. Thus, poisoning by certain species of mushrooms may be regarded as a classic example of this type of toxic syndrome. This form of the toxicity, however, has been thoroughly investigated and documented and therefore will not be discussed here. Probably the first mycotoxicosis to have been recognized was ergotism, a syndrome having its origin with the ingestion of rye and other grains infested with the mold *Claviceps purpurea*. In its chronic form, ergotism was known to the ancients as "St. Anthony's fire," and apparently occurred in frequent epidemics. In the mid-sixteenth century the toxic symptoms were first associated with ergot (scabrous grain), and extracts of the fungus were introduced into medicine as oxytocic substances early in the nineteenth century (Barger, 1931).

The substances responsible for the syndrome were isolated and chemically identified during the 1930's; they include a group of six alkaloids, all derivatives of lysergic acid (Rothlin and Bircher, 1952). Epidemics of ergotism are now rare in the United States, the last outbreak being in 1825. However, serious epidemics occurred in Russia in 1926–1927 and in England in 1928 (Barger, 1931), and occasional isolated episodes are still reported. Although the hazard to the human food supply has largely been eliminated, contamination of certain types of pasture grasses by the fungus continues to cause problems in the veterinary practice in some areas (Garner, 1961).

It has been commonplace to regard mold-damaged crops as acceptable for use in animal feeds, for which purpose they have generally been regarded as harmless. A substantial literature has developed, in fact, attesting to the lack of deleterious (and, on occasion, to the beneficial) effects of feeding moldy diets to a variety of domestic animal species. These findings are compatible with the currently available evidence which, though limited, suggests that probably a relatively small proportion of molds commonly found on foodstuffs are capable of producing toxic metabolites. They are, however, somewhat surprising in view of the biochemical capabilities of fungi to produce a wide variety of complex organic molecules as metabolic products (Miller, 1961).

Despite the extensive knowledge concerning ergotism, the potential significance of food-borne mycotoxins in problems of animal and human health has not been generally recognized. Occasional reports have appeared during the past three decades which have clearly associated toxicity syndromes in domestic animals (and, in one instance, in man) with the ingestion of mold-contaminated foodstuffs. The following discussion is intended to summarize the available knowledge concerning these mycotoxicoses, which have been reviewed in detail elsewhere (for example, Forgacs and Carll, 1932; Wogan, 1965a, 1966a).

II. AFLATOXICOSIS

This syndrome, first discovered as a veterinary entity in 1960, is the most recent mycotoxicosis to have been recognized. Its discovery and the rapid development of information concerning the potency of the active agents, as well as the nature of their biological effects, have been important factors in stimulating interest and research activity on the mycotoxin problem in general.

A. History of Field Outbreaks

The syndrome was first recognized in 1960 when serious outbreaks of unexplained mortality were encountered in turkey flocks in the south and east of England (Blount, 1961). Total losses in the initial outbreaks involved an estimated 100,000 animals, and the syndrome was referred to as "Turkey-X" disease. Within a short time, a similar disease was reported in ducklings and chickens (Asplin and Carnaghan, 1961), and also in swine (Harding et al., 1963; Loosmore and Harding, 1961) and young cattle (Loosmore and Markson, 1961).

Although the etiologic agent in these outbreaks was initially unknown, the syndrome was soon attributed to toxic substances in animal feeds giving rise to symptoms in cattle similar to those of ragwort poisoning. Systematic examinations of the various feeds involved revealed the common factor to be a peanut meal of Brazilian origin. Intensive investigations for the presence of known toxic agents in this meal failed to demonstrate any known poisonous agent. It was found, however, that an active principle could be extracted from toxic meals (Allcroft et al., 1961; Sargeant et al., 1961a) which produced death and typical pathologic signs in susceptible animal species such as the duckling.

Subsequently, many samples of feedstuffs and peanut meals were tested for toxicity, with the result that some samples from at least fourteen peanut-producing countries were found to be contaminated with the toxic agent (Allcroft and Carnaghan, 1963a; Newberne, 1965). These findings indicated that toxic meals were not restricted to a single source, but rather arose from widely scattered geographic areas. Subsequent studies revealed that commodities other than peanuts can become contaminated. Thus, significant levels of aflatoxins have been found in samples of oilseeds (cottonseed, soybeans), grains (maize, rice, wheat, millet, barley, sorghum), pulses (peans, beans, cowpeas) and other food staples (cassava, yams, etc.) from various parts of the world (Allcroft and Carnaghan, 1963a; Loosmore et al., 1964; Wogan, 1968a). Available information is fragmentary and does not permit an accurate estimation of levels of these contaminants entering the human food supply. It has been suggested, by examination of several types of agricultural commodities (Hiscocks, 1965), that feed-grade peanut meals may be more frequently contaminated than other crops. This has been attributed to such factors as high moisture content and methods of harvest and storage, which tend to favor germination and growth of contaminating molds.

B. Microbiological Aspects

Attempts to identify the active agents in samples of toxic peanut meals included various types of chemical and other analyses for the presence of known poisonous compounds. In the course of microscopic examinations for fragments of poisonous plants, it was noted that up to 20% of the peanut cotyledon fragments contained fungal hyphae, whereas few were present in a nontoxic meal (Austwick and Ayerst, 1963). This observation along with the sporadic but widespread distribution of the toxic agent in whole peanuts and meals, suggested that the toxin was of fungal origin. Investigation of this hypothesis resulted in the isolation, from a sample of toxic meal, of several pure fungal cultures which produced the toxic material when grown on sterilized nontoxic peanuts (Sargeant et al., 1961b). The fungus was identified as Aspergillus flavus, and the toxic substance was named "aflatoxin" to denote its origin. It has subsequently been found that the material originally isolated comprises several fractions which have been individually identified as described below.

Aspergillus flavus is taxonomically placed in the A. flavus-oryzae group, whose members are ubiquitously distributed and grow on virtually any substrate suitable for fungal growth. Aspergillus flavus has been reported as a frequent contaminant of a wide variety of food crops. It is important to note, however, that not all strains of this fungus are capable of producing aflatoxin. Since the original isolation of a toxin-producing strain from toxic peanut meal, other strains of this species have been selected and studied. The greater proportion of randomly selected A. flavus do not appear capable of aflatoxin production. In a selected group of isolates from a highly toxic consignment of peanuts, Austwick and Ayerst (1963) found that 9 of 59 isolates produced aflatoxins. These results, because of the selection procedure used, probably give a biased indication of the frequency of toxin-producing isolates. In other studies, Parrish *et al.* (1966) found 26 aflatoxin-producing strains among 93 strains of *A. flavus* examined.

The discovery of toxin-producing *A. flavus* strains has logically led to examinations of other *Aspergillus* species and other fungal genera to determine the distribution of aflatoxin-producing capabilities among the fungi. For example, Austwick and Ayerst (1963) found no aflatoxin-producing organisms among isolates of *A. tamarii, A. fumigatus, Botryodiplodia theobromae, Rhizopus arrhizus,* and *Phoma* spp. derived from toxic peanut meal. Diener *et al.* (1963), who isolated a toxin-producing *A. flavus* from domestic U. S. peanuts, reported no toxin production by seven other *Aspergillus* species and one *Penicillium* species isolated from the same source. Similarly, Parrish *et al.* (1966) reported no aflatoxin production in 49 strains of *A. flavus* or in nine *Aspergillus* species other than *A. flavus,* and also obtained negative results with ten strains of four *Penicillium* species. The original taxonomic classification of the toxin-producing mold as *A. flavus* was based on conventional criteria, principally involving morphologic characteristics. On these bases, the strain was poorly differentiated from *A. parasiticus.* Consequently, several reports (Austwick and Ayerst, 1963; Parrish *et al.,* 1966) have indicated that several strains of this species also produce aflatoxin. There has been one unconfirmed report (Hodges *et al.,* 1964) that the toxin is produced by *P. puberulum.* Thus, the metabolic capability for aflatoxin production appears to be rather narrowly restricted to members of the *A. flavus* group.

Various investigations have dealt with conditions that favor growth of *A. flavus* and production of aflatoxin under laboratory as well as under field conditions. Austwick and Ayerst (1963) showed that the conditions for growth of a toxin-producing *A. flavus* include an ambient temperature of 10° to 45°C (optimum 30°C) and relative humidity of 75% or greater. The equilibrium moisture content of peanuts at this humidity is about 9% on a wet-weight basis. This observation is of considerable practical consequence in considering control measures under field conditions. A series of investigations have been carried out in several African peanut-producing areas dealing with the influence of agricultural practices on the contamination of peanuts by aflatoxin. Bampton (1963) has reviewed the general problem, and subsequent reports have dealt with the various aspects of individual factors (McDonald and Harkness, 1963, 1964;

MacDonald and A'Brook, 1963; Burrell *et al.*, 1964; McDonald *et al.*, 1964; Schroeder and Ashworth, 1965).

The earlier findings had made it clear that, since *A. flavus* will grow only when the moisture content of the substrate exceeds 9%, aflatoxin production can occur only before postharvest drying reduces moisture content below this critical level. While the nuts are still in the ground, their moisture content is still at least 25%, but mold growth seems not to occur at this stage except when the shell is broken or damaged. Insect or other damage to shell and seed coat tend to increase the frequency of fungal attack. Probably the most critical stage is that immediately following harvest. The nuts are still at high moisture content and, unless both atmospheric conditions and harvesting practice are good, drying to the safe moisture level may not take place quickly enough to avoid mold development. In such cases, the risk of mold growth and aflatoxin production is great.

Variations occur in the extent of aflatoxin contamination from year to year in a given producing area and also in the aflatoxin content in individual kernels. Even in highly toxic batches of peanuts, the toxin generally is confined to a relatively small proportion of the individual kernels. This finding is presumably attributable to factors such as shell damage which tend to be distributed randomly. There appears to be no pattern of resistance to mold attack or aflatoxin production among the several varieties of peanuts that have been studied.

If the crop is harvested and dried under optimum conditions which prevent mold damage, there is little further opportunity for contamination with aflatoxin unless the kernels or meal are exposed to conditions that permit the moisture content to rise above the critical minimum level. Thus, conditions of storage and transport after harvest are important determinants of postharvest contamination.

Upon recognition of the fungal origin of aflatoxin and isolation of toxin-producing fungal strains, various methods have been devised for culture of the organism for the purpose of accumulating toxin (see Wogan and Mateles, 1968). These generally have consisted in culture of the mold on natural substrates such as sterilized peanuts (Sargeant *et al.*, 1961*b;* de Iongh *et al.*, 1962), wheat (Asao *et al.*, 1963; Armbrecht and Fitzhugh, 1964), rice (Hesseltine *et al.*, 1966), or others. In one series of experiments, Codner *et al.* (1963) cultured *A. parasiticus* and six strains of *A. flavus-oryzae* on peanuts and achieved yields of aflatoxin ranging from 24 to 265 mg/kg after growth at 30°C for 10 to 13 days. In a comparison of the efficiency of several substrates for support of aflatoxin production, Hesseltine *et al.* (1966) found that three toxin-producing *A.*

flavus strains grew well on rice, sorghum, peanuts, corn, wheat, and soybeans. Aflatoxin was produced on all substrates, although yields were poor on soybeans, maximal on rice, and intermediate on the other commodities. In all cases, increased aeration improved the toxin yield.

Early attempts to utilize synthetic mycologic media for aflatoxin production met with limited success. Thus, Sargeant *et al.* (1963) reported that Czapek's medium supported growth of the fungus, but gave poor yields of toxin. Nesbitt *et al.* (1962) found that addition of zinc to the medium slightly increased toxin production, which, however, was still significantly smaller than yields on natural substrates. Subsequently, Mateles and Adye (1965) systematically studied the influence of medium composition on toxin production and devised a synthetic medium which supported toxin yields of 60 to 80 mg/liter in submerged growth. These investigators demonstrated that glucose, fructose, and sucrose are preferred carbon sources, and casamino acids are the preferred nitrogen source. Zinc is also required, as suggested by previous investigations. Fermentations using this procedure have been used successfully in the production of quantities of aflatoxins for experimental purposes (see Wogan and Mateles, 1968).

C. Chemistry of the Aflatoxins

1. ISOLATION, CHARACTERIZATION, AND CHEMICAL PROPERTIES

The aflatoxins were first isolated from samples of highly toxic peanut meals (Sargeant *et al.*, 1961*b*, 1963) by conventional extraction and concentration procedures. The isolation procedures were greatly facilitated by the early discovery that the toxic substances were fluorescent in ultraviolet light. This property has been utilized as a convenient means of monitoring purification procedures, and it also forms the basis for chemical assays for aflatoxin in foods as described below.

The compounds are soluble in moderately polar solvents such as methanol, chloroform, and acetone, but they are virtually insoluble in nonpolar solvents (for example, hexane, petroleum ether) and in water. Initial isolation was accomplished by extraction with hot methanol (Allcroft *et al.* 1961; Sargeant *et al.*, 1961*a*), and a variety of extraction solvents have since been used for various purposes. These include, for example, 55% aqueous methanol (Campbell *et al.*, 1964; Nesheim, 1964; Trager *et al.*, 1964), 70% aqueous acetone (Pons and Goldblatt, 1964), and a hexane–water–acetone azeotrope (Goldblatt, 1965).

In the production and isolation of quantities of aflatoxins from mold cultures on solid substrates, a convenient extraction and concentration procedure involves total extraction of the culture with chloroform and subsequent precipitation of the aflatoxins in petroleum ether (Asao *et al.*, 1963, 1965). The aflatoxins produced in cultures on liquid media are almost quantitatively removed by partitioning into chloroform (Adye and Mateles, 1964).

Extracts produced by these procedures usually contain mixtures of fluorescent compounds, which are separable into their individual components by chromatographic techniques. Resolution on filter paper is incomplete (Sargeant *et al.*, 1961*b*), but it is greatly improved by thin-layer chromatographic procedures. Although several such systems have been developed, including the use of alumina as the support medium (Broadbent *et al.*, 1963), the conditions most widely used involve separation on silica gel plates developed with 3 to 5% methanol in chloroform (Asao *et al.*, 1965; de Iongh *et al.*, 1965; Nesheim, 1964), or with 5 to 15% acetone in chloroform (Eppley, 1966*b*).

When chromatograms of extracts containing aflatoxins are viewed under ultraviolet light, a complex array of fluorescent compounds is generally present. The known aflatoxins are four of these components. Two emit blue visible light and were therefore named aflatoxins B_1 and B_2, and two fluoresce yellow-green (aflatoxins G_1 and G_2). On silica gel plates developed in 97–3 chloroform–methanol (Asao *et al.*, 1965), aflatoxin B_1 migrates with an R_f of the order of 0.56; B_2, 0.53; G_1, 0.48; and G_2, 0.46; although absolute R_f values are poorly reproducible. The amounts and relative proportions of these four compounds present in culture extracts are variable, depending on such factors as mold strain, medium composition, and culture conditions. Typically, aflatoxins B_2 and G_2 are present in small relative amounts, whereas B_1 is usually present in largest yield; G_1 concentration is intermediate.

These four compounds were originally isolated by groups of investigators in England (Nesbitt *et al.*, 1962; Sargeant *et al.*, 1961*b*) and Holland (van der Zijden *et al.*, 1962). The molecular formula of aflatoxin B_1 was established as $C_{17}H_{12}O_6$ and that of aflatoxin G_1 as $C_{17}H_{12}O_7$; aflatoxins B_2 and G_2 were found to be the dihydro derivatives of the parent compounds, $C_{17}H_{14}O_6$ and $C_{17}H_{14}O_7$ (Hartley *et al.*, 1963). Pertinent physical and chemical properties of the compounds are summarized in Table I.

Structures based largely on interpretation of spectral data were proposed for aflatoxins B_1 and G_1 in 1963 (Asao *et al.*, 1963, 1965) and for B_2 (Chang *et al.*, 1963; van Dorp *et al.*, 1963) and G_2 shortly thereafter. These are shown in Fig. 1. The proposed structure of G_1 has been supported by X-ray crystallography (Cheung and Sim, 1964), and the abso-

TABLE I

Selected Physical and Chemical Data on Aflatoxins

Aflatoxin	Molecular formula	Molecular weight	Ultraviolet absorption (ϵ)		Fluorescence emission maximum (mμ)
			265 mμ	363 mμ	
B$_1$	C$_{17}$H$_{12}$O$_6$	312	13,400	21,800	425
B$_2$	C$_{17}$H$_{14}$O$_6$	314	9,200	14,700	425
G$_1$	C$_{17}$H$_{12}$O$_7$	328	10,000	16,100	450
G$_2$	C$_{17}$H$_{14}$O$_7$	330	11,200	19,300	450

lute configuration established by stereochemical means (Brechbühler *et al.*, 1967). Laboratory synthesis of aflatoxin B$_1$ has recently been accomplished (Büchi *et al.*, 1967).

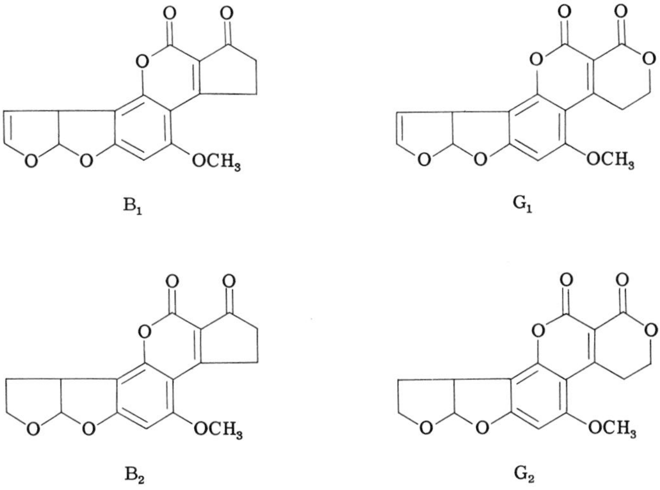

FIG. 1. Structures of the aflatoxins.

The spectral characteristics of the four aflatoxins have been determined by several investigators (Asao *et al.*, 1963, 1965; de Iongh *et al.*, 1962; Hartley *et al.*, 1963; van der Zijden *et al.*, 1962; van Dorp *et al.*, 1963) and are summarized in Table I. The ultraviolet absorption spectra are very similar, each showing maxima at 223, 265, and 363 mμ. Because of the close similarities in structural configuration, the infrared absorption spectra of the four compounds are also very similar. The fluorescence emission maximum for B$_1$ and B$_2$ has been reported to be 425 mμ, and that for G$_1$ and G$_2$ is 450 mμ (Hartley *et al.*, 1963).

The chemical reactivity and behavior of the aflatoxins have not yet been systematically studied. However, it has been shown (Asao et al., 1963, 1965; van Dorp et al., 1963) that catalytic hydrogenation of aflatoxin B_1 to completion results in the uptake of 3 moles of hydrogen with the production of the tetrahydrodeoxy derivative. Interruption of the hydrogenation procedure after the uptake of 1 mole of hydrogen results in the production of aflatoxin B_2 in quantitative yield (Chang et al., 1963; van Dorp et al., 1963).

The presence of the lactone configuration in the molecules makes the compounds labile to alkaline hydrolysis (de Iongh et al., 1962). This fact has an important practical consequence in the manufacture of peanut oil, which involves, among other procedures, an alkali refining process that destroys any aflatoxin that may have been present in oil from contaminated nuts. No aflatoxin has been detected in highly refined food-grade oil.

The stability of the aflatoxins to other treatments has not received systematic study. Although they appear to decompose partially when exposed to heat and/or light, particularly when in solution in strongly polar solvents, they persist in contaminated meals for long periods. Economically feasible procedures for decontamination of foodstuffs by heating or other means have not been reported. However, Fischbach and Campbell (1965) found that aflatoxin extracts rapidly lost their fluorescence and were converted to nontoxic products on exposure to 5% NaOCl. A highly toxic meal was also rendered nontoxic by exposure to a 10% chlorine gas atmosphere. As these treatments may adversely affect the nutritional quality of the meal, their usefulness as decontaminants is not yet established.

2. PHYSICOCHEMICAL ASSAYS FOR AFLATOXINS

Various chemical properties described above have been exploited in the development of chemical procedures for the detection and quantitative assay of aflatoxins in various foodstuffs. Many assay procedures have been devised (Broadbent et al., 1963; Genest and Smith, 1963; de Iongh et al., 1964a; Nesheim, 1964; Heusinkveld et al., 1965; Lee, 1965; Pons and Goldblatt, 1965; Cucullu et al., 1966; Eppley, 1966b). All these methods involve extraction, concentration, and quantitative estimation of concentration of aflatoxins in various food commodities. Various solvents have been applied for the extraction process, which is carried out on previously defatted material. These include aqueous methanol, chloroform, acetone, and chloroform–methanol mixtures. Similarly, a variety of procedures, usually column chromatography, have been devised for concentrating the aflatoxins in the original extracts.

However, all these methods include detection of the aflatoxins by their chromatographic properties on thin-layer chromatograms and quantitative estimation of concentration based on visual estimation of their fluorescence intensity in ultraviolet light.

The efficacy of several of these methods in the detection and quantitative estimation of aflatoxins in peanuts and peanut products has recently been investigated comparatively (Trager *et al.*, 1964). Although variations in precision and accuracy exist among the various methods, there is general agreement that the lower limit of their sensitivity is of the order of 5 μg/kg (5 ppb). These assay procedures are used in the routine examination of peanuts and other commodities to avoid contamination of the food supply. It is not yet clear whether any of the above methods can be applied to all foodstuffs with equally accurate results.

D. Biological Effects of the Aflatoxins

The discovery of these compounds as contaminants of animal feeds and recognition of their potency have stimulated a great deal of research concerned with their effects in various biological systems. The toxic properties of the aflatoxins manifest themselves differently, depending on the test system, the dose, and the duration of exposure. Thus, they are lethal to animals and animal cells in culture when administered acutely in large doses, and they cause histologic changes when sublethal doses are given subacutely. Long-term exposure for extended periods results in chronic toxicity, including tumor induction in several species.

1. AFLATOXICOSIS OF DOMESTIC ANIMALS

The events that led to the discovery of the aflatoxins consisted in a series of outbreaks of unexplained toxicity in various domestic animal species as described earlier. These outbreaks involved poisoning of turkey poults, ducklings, pigs, and calves. The field observations led to experimental studies in domestic species, carried out by feeding peanut meal contaminated with known concentrations of aflatoxins (Allcroft, 1965).

a. Susceptibility. Susceptibility to aflatoxins varies among different species, and in all cases the young are more sensitive than mature animals. Ducklings are particularly susceptible, a finding which was utilized in the isolation and purification of the compounds (Asplin and Carnaghan, 1961), and also in the development of a biological assay procedure as described below. Turkey poults are less sensitive, and chickens are comparatively resistant. Among larger animals, pigs (3 to 12 weeks old), pregnant sows, and calves (1 to 6 months old) are very susceptible, while older cattle are less sensitive to aflatoxin poisoning. Sheep are remarkably resistant.

Ducklings fed rations containing approximately 0.3 ppm of aflatoxin for 6 weeks showed 30% mortality, while the same level caused no mortality in turkey poults. Survivors of this level showed marked growth suppression. In chickens, a toxin content of 1.8 ppm caused only slight weight suppression when fed for a 10-week period (Asplin and Carnaghan, 1961).

Growing pigs were not adversely affected until the aflatoxin level reached 0.41 to 0.69 ppm, at which concentration growth suppression occurred. Diminished growth rate and eventually death occurred in calves fed rations containing 2.2 ppm for 16 weeks or longer, whereas heifers and older cattle were able to tolerate rations containing 2.4 ppm of aflatoxin for 13 months without clinical signs of toxicity (Allcroft, 1965). Lactating cows fed 2.7 ppm of aflatoxin for 20 to 53 days showed significant reduction in milk yield (Allcroft and Lewis, 1963). It is important to note that the latter animals excrete a toxic metabolite of aflatoxin in milk, a point which will be discussed in more detail subsequently.

The comparative resistance of sheep is illustrated by the findings that a ration containing 2.4 ppm of aflatoxin B_1 failed to cause demonstrable signs of toxicity when fed for more than 3 years (Allcroft, 1965).

b. Pathology. In all domestic and experimental animal species, the primary lesions caused by the aflatoxins occur in the liver. Animals dying from acute or subacute poisoning exhibit a variety of hepatic lesions. The major types of histologic change which occur in domestic animal species after feeding aflatoxin-contaminated rations are summarized in Table II.

Acute poisoning by large doses of aflatoxins is generally accompanied by hepatic necrosis and hemorrhage, which appears most clearly within a short time following acute administration. In chronic poisoning, fibrosis is well-marked in cattle and pigs, but not in other species. The veno-occlusive lesion has not been reported in animals other than cattle, and seems to be distinctive in this species. The information in Table II indicates that bile ductule hyperplasia is the histopathologic lesion observed with most consistency, and it is this lesion which has been used as a criterion of activity in the duckling bioassay for aflatoxin.

2. Toxicity of Aflatoxins in Experimental Animals

The observations in domestic animals have been extensively confirmed and supplemented by studies in many laboratory animal species. Such investigations were begun with aflatoxin-contaminated peanut

TABLE II

COMPARATIVE PATHOLOGY IN ANIMALS FED AFLATOXIN-CONTAMINATED FEED

Liver lesions	Species							
	Calves	Cattle	Swine	Sheep	Duckling	Adult duck	Turkey poult	Chick
Acute necrosis and hemorrhage	−	−	+	−	+	−	+	−
Chronic fibrosis	+	+	+	0	−	+	−	−
Regeneration nodules	−	+	+	0	±	+	+	−
Bile duct hyperplasia	+	+	+	0	+	+	+	±
Veno-occlusive disease	+	+	−	0	−	−	−	−
Enlarged hepatic cells	+	+	+	0	+	+	+	−

meals, and, as improvements were made in production and isolation methods, the individual compounds became available in pure form. These have been used in both acute and subacute experiments.

a. Acute Toxicity in Animals. Aflatoxin B_1 has been most extensively studied, as regards its lethal potency; its *in vivo* lethality to various experimental animals is summarized in Table III. The LD_{50} of 1-day-old ducklings is of the order of 0.5 mg/kg. This value is considerably smaller than those for the rat and hamster. However, the dog, rabbit, guinea pig, and pig have LD_{50} values of the same order of magnitude as the duckling (Table III), and the same observation is true for the rainbow trout (Ashley *et al.,* 1965).

The LD_{50} values presented were calculated from mortality over 7-day periods. In most species, death usually occurred within the first 72 hours following administration of the compound, and necropsy at this stage generally revealed liver damage as the most prominent pathologic sign. Hemorrhage in the intestinal tract and peritoneal cavity as well as ascites were occasionally seen in some species. Butler (1964) has found the principal histologic changes in rat liver to be the development of a periportal zone of necrosis over 3 to 4 days following dosing with marked biliary proliferation. The latter lesion persists after 1 month.

In all species studied, sensitivity decreases with age, as illustrated by the rat data in Table III. The female rat is less susceptible to the acute effects even at 21 days of age (weaning), and this sex difference appears to become more marked as sexual maturity is approached.

TABLE III
COMPARATIVE LETHALITY OF AFLATOXIN B_1 (SINGLE DOSE)

Species	Strain	Age	Sex	Route	LD_{50} (mg/kg)
Duckling	Khaki Campbell	1 day	M	Oral	0.37
	White Pekin	1 day	M	Oral	0.56
Trout	Rainbow	100 gm	M–F	Oral	ca. 0.5
Dog	Mongrel	Adult	M–F	Oral	ca. 1.0
		Adult	M–F	IP^a	ca. 0.5
Rabbit	Dutch Belted	Weanling	M–F	IP	ca. 0.5
Pig	Poland-China	Weanling	M–F	Oral	ca. 0.5
Guinea pig		Adult	M	IP	ca. 1.0
Rat	Fischer	1 day	M–F	Oral	1.0
		21 days	M	Oral	5.5
		21 days	F	Oral	7.4
	MRC	100 gm	M	Oral	7.2
		100 gm	M	IP	6.0
		150 gm	F	Oral	17.9
Hamster	Golden	30 days	M	Oral	10.2
Mouse	CF-1	Adult	M	IP	62.0

[a] Intraperitoneal.

The relative lethal potencies of the four aflatoxins in the day-old duckling have been examined by Carnaghan et al. (1963). The oral 7-day LD_{50} values reported, in milligrams per kilogram of body weight, for each compound were: aflatoxin B_1, 0.36; B_2, 1.80; G_1, 0.79; and G_2, 3.45. These values illustrate relationships of structural configuration to acute lethality. Aflatoxin B_1 is most potent, followed by G_1, B_2, and G_2 in order of decreasing potency. The presence of the additional oxygen in the G compounds results in activity decreased by a factor of about 2, while the unsaturated compounds are approximately 4.5 times as potent as the dihydro derivatives.

Toxic effects of the aflatoxins in monkeys have been reported by Tulpule et al. (1964) and by Madhaven et al. (1965). In these experiments, young (1.5 to 2.0 kg) rhesus monkeys were fed either 1.0 mg/day of aflatoxin or 0.5 mg/day for 18 days, then 1.0 mg/day thereafter. All animals developed anorexia and died in 14 to 28 days. The principal histopathologic findings included liver lesions similar to those seen in ducklings (portal inflammation and fatty change) and suggestive of biliary cirrhosis. The animals received a total dose of 10 to 15 mg/kg of body weight of a preparation containing 60% aflatoxin B_1 and 40% G_1. Svoboda et al. (1966) reported that monkeys injected with a single dose of aflatoxin B_1 (0.45 to 2.6 mg/kg) developed liver lesions resembling those of viral hepatitis in man. Lesions appeared within 48 hours after dosing.

   b. *Acute Toxicity in Cell Cultures and Embryos.* The toxic effects of aflatoxin B_1 have also been investigated in several *in vitro* cell culture systems and in avian embryos. These experimental systems are very sensitive in terms of the quantity of compound required to cause detectable effects. The results of several such studies are shown in Table IV.

   The early studies of Juhasz and Greczi (1964) demonstrated that methanolic extracts of aflatoxin-contaminated peanut meals were toxic to calf kidney cells. Legator and Withrow (1964) have shown that crude aflatoxin mixtures as well as purified aflatoxin B_1 suppressed mitosis in heteroploid and diploid human embryonic lung cells. Inhibition occurred 4 hours after exposure of the cells to the toxin and was maximal in 8 to 12 hours. A concentration of 0.01 μg/ml (ppm) of the toxin could be detected by this method, and a concentration of 0.03 ppm of the toxin resulted in 51% reduction in mitosis. In subsequent studies, Legator *et al.* (1965) investigated the effects of the compound on cell growth, morphology, and DNA synthesis in heteroploid human embryonic lung cells. Cell growth was inhibited after 48 hours of exposure to as little as 0.05 ppm of a mixture of aflatoxin B_1 and G_1, and was completely suppressed by 5.0 ppm. After exposure to aflatoxin B_1 at a concentration of 1.0 ppm for 6 to 8 hours, there was a 92% increase in numbers of giant cells present as compared to control cultures. Giant cell formation was attributed to a suppression of DNA synthesis which was studied by tritiated thymidine uptake. Inhibition of thymidine incorporation was evident within 4 hours after exposure to 1 ppm of aflatoxin B_1.

   Gabliks *et al.* (1965) have reported that concentrations of aflatoxin B_1 in the range of 1 to 5 ppm of medium cause destruction of human liver (Chang) cells, HeLa cells, and primary cell cultures from whole duck and chick embryos. Smaller concentrations resulted in demonstrable inhibition of protein synthesis in these cell lines. Daniels (1965) studied the effects of a mixed aflatoxin B_1 (36%) and G_1 (62%) preparation on rat fibroblast cultures. In this system, 0.06 ppm caused 50% inhibition of growth (cell numbers), and a detectable suppression of growth was obtained by a concentration of 0.02 ppm.

   Cytotoxicity of aflatoxin B_1 at levels of 10 ppm in cultures of human embryonic liver cells was reported by Zuckerman *et al.* (1967b). This toxic response involved the loss of cytoplasmic RNA and marked changes in nuclear morphology. Using the same cell type, Zuckerman *et al.* (1967a) reported that the LD_{50} values for aflatoxins B_1, G_1 and G_2 were 1 ppm, 5 ppm, and 16 ppm, respectively. On the basis of effects on

TABLE IV

TOXICITY OF AFLATOXIN B_1 IN CELL CULTURES

Cell type	Aflatoxin concentration (μg/ml)	Effects	Reference
Calf kidney	?	Nuclear and cytoplasmic destruction in 48 hours	Juhasz and Greczi (1964)
Human embryonic lung	0.03	Reduction of mitotic rate (51% in 8 to 12 hours). Giant cell formation. Decrease in ^{3}H-thymidine incorporation into DNA	Legator and Withrow (1964) Legator et al. (1965)
Rat fibroblasts	0.06	Growth inhibition (50%)	Daniels (1965)
Chang (human) liver	1.0	Cell destruction (TD_{50}) in 48 hours	Gabliks et al. (1965)
HeLa	5.0	Cell destruction (TD_{50}) in 48 hours	Gabliks et al. (1965)
Primary duck embryo	1.0	Cell destruction (TD_{50}) in 48 hours	Gabliks et al. (1965)
Primary chick embryo	5.0	Cell destruction (TD_{50}) in 48 hours	Gabliks et al. (1965)

uptake of radioactive precursors, they further demonstrated that aflatoxin B_1 inhibited the synthesis of RNA and DNA in these cells.

Platt *et al.* (1962) observed that aflatoxin mixtures were toxic to chick embryos, and this observation has been confirmed and extended with aflatoxin B_1 (Table V). The data of Verrett *et al.* (1964) indicate the

TABLE V

TOXICITY OF AFLATOXIN IN AVIAN EMBRYOS

	LD_{50} (μg/egg)	Route	Reference
Chick (5 days)	0.3	Yolk	Platt *et al.* (1962)
Chick (preincubation)	0.048	Yolk	Verrett *et al.* (1964)
	0.025	Air cell	
Chick (10 days)	2.0–5.0	C-A cavity[a]	Gabliks *et al.* (1965)
Duck (15 days)	0.5–1.0	C-A cavity	

[a] Chorio-allantoic cavity

sensitivity of the chick embryo, in which the LD_{50} is 0.048 μg/egg when administration is made via the yolk, and 0.025 μg/egg when the compound is applied in the air cell. In both cases, the fertilized eggs were treated prior to incubation. The data of Gabliks *et al.* (1965) indicate somewhat reduced sensitivity in older embryos receiving the toxin via the chorio-allantoic cavity and suggest that the duckling embryo is more susceptible than the chick.

3. BIOASSAYS FOR AFLATOXINS

Several of the biological effects of aflatoxins described in the preceding sections have been adapted in the development of quantitative bioassay procedures. These assays have been useful in monitoring the isolation and purification of the compounds as well as in the examination of extracts of foodstuffs for the presence of the toxins.

One assay is based on the sensitivity of the young duckling to the hepatotoxic effects of the compounds, as noted earlier. Sargeant *et al.* (1961*b*) developed the first form of the assay in which test material (peanut meal) was fed to these animals for 5 days, and the extent of liver damage assessed after 8 days. With the aid of a modification of this assay, in which extracts of toxic meals were force-fed to ducklings, Sargeant *et al.* (1961*a*) were able to isolate the active substance from toxic peanut meal. Other modifications of the duckling assay (Armbrecht and Fitzhugh, 1964) have shown it to be useful as a confirmatory supplement to chemical assay procedures.

When purified aflatoxins are force-fed to ducklings for 5 consecutive days and the livers are examined at 8 days, the extent of biliary hyperplasia can be evaluated semiquantitatively and used as an index of toxicity. It has been shown (Newberne *et al.*, 1964*a;* Wogan, 1965*b*) that as little as 2 μg of B_1 administered over 5 days reproducibly causes bile ductule lesions. Aflatoxin G_1 and B_2 cause similar lesions when administered at larger doses. As in the case of acute lethality, B_1 is most toxic, followed by G_1 and B_2 in that order. Application of these data to the 5-day feeding assay for aflatoxins in natural products reveals that the duckling assay as described is capable of detecting as little as 30 to 50 ppb of aflatoxin under conditions where feed intake is normal.

The susceptibility of avian embryos and of cell cultures to the toxicity of aflatoxins provides potential assay systems with great sensitivity, and several such assays have been proposed. In the experiments of Legator and Withrow (1964), it was found that mitotic rate of human embryonic lung cells was suppressed to the extent of 51% by 0.03 ppm of aflatoxin B_1. Daniels (1965) found a linear dose-response relationship over the range of 0.025 to 0.25 μg of aflatoxin B_1 per milliliter of medium, using the inhibition growth of rat fibroblasts as the index of toxicity. The concentration of aflatoxin required to cause 50% inhibition of growth was 0.062 μg/ml. The sensitivity of this response as an assay system is more than 100 times as great as that of the duckling in terms of the quantity of test material required to cause detectable effects.

Recently, Townsley and Lee (1967) have reported that aflatoxins inhibit cell cleavage in the egg of the marine borer, *Bankia setacea*. A bioassay based on this response reportedly has a sensitivity of 0.05 ppm of aflatoxin concentration. Also, de Waart (1967) found that the waterflea *(Daphnia)* and the brine shrimp *(Artemia)* are sensitive to aflatoxins B_1 and G_1 at levels of 15 ppm.

Although these *in vitro* assays have the advantage of great sensitivity, they also suffer the disadvantage of low specificity, requiring that test material be of relatively high purity to avoid possible interference from non-aflatoxin toxicants in extracts of test materials. Since they also require relatively sophisticated apparatus and interpretation, they have not as yet found extensive application for routine examination of foodstuffs.

4. CARCINOGENIC PROPERTIES OF THE AFLATOXINS

The potency of the aflatoxins as acute poisons is clearly demonstrated by the effects described in the preceding sections. It has become equally well established that the compounds are also potent carcinogens in several animal species.

The carcinogenic properties of aflatoxin-contaminated peanut meals were first described by Lancaster *et al.* (1961). These investigators reported that the inclusion of a toxic peanut meal into a purified rat diet at a level of 20% resulted in hepatoma development in nine of eleven survivors after 6 months of feeding. This was the first indication that aflatoxin-contaminated peanut meals were not only toxic but also carcinogenic. These findings have since been confirmed and extended by a number of investigators. Results of experiments involving the use of peanut meals known or suspected to have contained aflatoxins are summarized in Table VI.

TABLE VI

SUMMARY OF RAT HEPATOMA INCIDENCE ASSOCIATED WITH
PEANUT MEAL IN DIET

Authors	Peanut meal in diet (%)	Aflatoxin content (ppm)	Feeding time (weeks)	Tumor incidence
Schoental (1961)	15	?	52	2/5
Le Breton *et al.* (1962)	16	?	72	26/48
Lancaster *et al.* (1961)	20	?	24	9/11
Newberne *et al.* (1964b)	34	?	42–53	47/94
Salmon and Newberne (1963)	33.3	0.1 –3.5	48–73	64/73
Butler and Barnes (1964)	40	2.8 –4.0	35–38	5/6
		2.8 –4.0	16	4/6
		1.4 –1.6	12	5/7[a]
		1.4 –1.6	26	9/10[b]
		0.35–0.40	81	2/7

[a] 38 to 73 weeks after withdrawal of toxic diet.
[b] 40 to 58 weeks after withdrawal of toxic diet.

The experiments of Barnes and Butler (1964), Butler and Barnes (1964), and Newberne *et al.* (1964b) involved the use of peanut meals in which the presence of aflatoxins was established by chemical assay and offered strong presumptive evidence that the aflatoxins were the carcinogenic agents. These observations have made possible the interpretation of previously unexplained tumor incidence in two series of experiments. Schoental (1961), in investigating the properties of a toxic guinea pig diet, fed the diet to a small group of rats for one year. Hepatomas were found in two of five survivors, which led to the conclusion that a toxic factor, unidentified at that time, was responsible. Le Breton *et al.* (1962) had reported, over a period of years, the incidence of "spontaneous hepatoma" in Wistar rats. These were found to be associated with the use of a diet containing 16% of peanut meal. Controlled experiments comparing this with purified casein-sucrose diets resulted in clear association of carcinogenesis with peanut meals. In retrospect, it seems probable that both of these observations can be attributed to aflatoxin contamination of the diets.

In an extensive series of experiments in which Salmon and Newberne (1963) fed to rats diets containing 33.3% peanut meals of United States origin, very high incidence (64/73) of hepatoma was found after 48 to 72 weeks. Chemical analyses performed subsequently demonstrated that these and similar meals contained varying concentrations of aflatoxins.

Although precise dose-response conditions have not yet been established, some information is available regarding the dose-response relationships between tumor incidence in rats and aflatoxin content of contaminated peanut meals. The results of studies on several such meals have been described by Newberne (1965), who reported good correlation between liver tumor incidence and dietary aflatoxin content over the range of 0.07 to 1.8 ppm of aflatoxin. The highest level resulted in more than 90% tumor incidence when fed over a period of 370 days. The lowest level of aflatoxin detected (0.005 ppm) failed to induce liver tumors within a similar time period (384 days).

Carnaghan (1965) has reported induction of hepatomas in ducks fed aflatoxin-containing peanut meal. In a group of animals fed a ration containing 5% peanut meal for 14 months, eight of eleven survivors had developed hepatic carcinomas. The aflatoxin content of the peanut meal was approximately 7 ppm, giving a concentration of 0.03 ppm in the hepatoma-inducing ration. Thus, the duck appears to be considerably more susceptible to the hepatocarcinogenic action than the rat.

The carcinogenic potency of purified aflatoxin preparations has been established by recent experiments, which have also indicated that continuous feeding is not required for hepatoma induction in rats. The results of two such investigations are summarized in Table VII. In the studies of Barnes and Butler (1964), rats were fed 1.75 ppm of aflatoxin (containing 80% aflatoxins with G_1 present in higher quantities than B_1) in the diet for 89 days and then returned to an aflatoxin-free diet. All of three treated animals ultimately developed liver cancer after more than 300 days following withdrawal.

In somewhat similar experiments (Wogan, 1965b) an unfractionated mixture of partially purified aflatoxins (approximately 30% B_1, 20% G_1) was administered to rats by stomach tube. Each animal was treated

TABLE VII
RAT HEPATOMA INCIDENCE FOLLOWING ADMINSTRATION
OF PURIFIED AFLATOXINS

Aflatoxin	Feeding time	Tumor incidence
1.75 ppm in diet[a]	89 days	1/3 (316 days later)
		2/2 (485 days later)
150 μg/rat/day[b]	30 days	3/3 (5 months later)
		2/2 (10 months later)
75 μg/rat/day	30 days	4/5 (10 months later)
37.5 μg/rat/day	30 days	5/5 precancerous lesions (10 months later)
15 μg/rat/day	30 days	4/5 precancerous lesions (10 months later)

[a] Barnes and Butler (1964).
[b] Wogan (1965b).

daily for 30 days, then held without further treatment for 10 months. Animals that received the highest dose (150 μg/rat/day) had fully developed liver tumors 5 months following withdrawal of treatment. Even those receiving the lowest dose (15 μg/rat/day) showed significant incidence of precancerous lesions at the same time interval.

Recently, we have conducted a dose-response study of the carcinogenic potency of highly purified aflatoxin B_1 in rats (Wogan and Newberne, 1967). Male animals fed a diet containing 0.015 ppm of aflatoxin B_1 developed liver carcinomas at an incidence of 12/12 survivors after 68 weeks of feeding. A high incidence (18/22 animals) of liver tumors was induced in a shorter time (35 to 40 weeks) when the dietary level was 1.0 ppm. Administration of ten consecutive daily doses (40 μg/dose) at the beginning of the experiment induced, without further treatment, hepatocellular carcinomas in 4/24 animals 35 to 82 weeks after treatment.

Carnaghan (1967) reported that liver tumors were induced in approximately 50% of a group of female rats given a single oral dose of aflatoxin B_1 at a level of 7.6 mg/kg. The average time for tumor formation was about 24 months.

On the basis of these data, it has been estimated that the effective dose of aflatoxin B_1 for liver tumor induction in rats is of the order of 10 μg/animal/day (Butler, 1965). The potency of the compound in this regard is readily apparent when compared with similar estimates for other liver carcinogens. Effective doses of dimethylnitrosamine are about

750 μg/animal/day, and of butter yellow 9000 μg/animal/day. Aflatoxin B_1 is, on this basis, among the most potent chemical carcinogens known. Studies by Ashley *et al.* (1964, 1965) and by Sinnhuber *et al.* (1965) have suggested that the rainbow trout may be considerably more sensitive than the rat to the hepatocarcinogenic effects of aflatoxins. These investigators have shown that rainbow trout develop liver tumors at significant incidence rates when fed purified diets containing only 0.5 to 2.0 μg of aflatoxin B_1 per kilogram (0.5 to 2.0 ppb). The apparent sensitivity of this species has led to the recognition (Ashley *et al.*, 1965) of the potential role of the aflatoxins as etiologic agents in the so-called "trout hepatoma syndrome" (Halver, 1964; Hueper and Payne, 1961).

In a different test system, Dickens and Jones (1964) studied the effects of multiple subcutaneous injections of aflatoxins in rats. A mixed preparation (about 38% B_1 and 56% G_1) of the compounds suspended in peanut oil was administered to groups of rats twice weekly. One group received 50 μg at each injection, and the treatment was continued for 50 weeks; a second group received 500 μg at each injection for a period of 8 weeks, after which treatment was discontinued. In the former group, 6/6 animals developed sarcomas or fibrosarcomas at the injection site within a 60-week period. At the higher dose level, 5/5 animals developed tumors within a 30-week period. These observations indicate that the compounds are also carcinogenic for the subcutaneous tissues of the rat. Additional carcinogenic responses have recently been suggested by the findings of low incidences of gastric carcinomas and colonic carcinomas in rats fed aflatoxin-containing diets (Butler and Barnes, 1966; Wogan and Newberne, 1967).

E. Metabolism of Aflatoxins

The metabolic fate of aflatoxins in animals is of considerable interest from several points of view. A knowledge of their metabolism provides important information relevant to distribution and sites of action of the compounds in animal tissues. In terms of protection of the food supply, it is obviously important to determine whether the compounds are likely to appear in animal tissues or products used for human food derived from animals which may have been fed aflatoxin-contaminated rations.

Although available information on this point is not yet definitive, results of several studies of different types have been reported (see Wogan, 1968*b*). Investigations based on excretion of fluorescent aflatoxin metabolities have revealed several pathways of metabolism. Allcroft and Carnaghan (1963*b*) reported that cows fed aflatoxin-containing rations excreted in their milk a factor that was toxic to ducklings; the lesions caused in ducklings were similar to those caused by aflatoxins. The toxic material was not found in bulk milk supplies or in tissues (blood, liver) of similarly treated animals. Platonow (1965) also reported no detectable toxicity in muscle or liver from chickens fed rations containing 3.1 ppm of aflatoxin.

The transmission of toxicity into milk by lactating cattle was also reported by de Iongh *et al.* (1964*b*), who found that the toxicity was associated with a fluorescent compound apparently derived from aflatoxin B_1, but with markedly different chromatographic properties. No aflatoxin B_1 per se was found in the milk. These investigators found that a similar substance, which they call the "milk toxin," was produced by lactating rats dosed with aflatoxin B_1. Van der Linde *et al.* (1965) reported that cattle excrete approximately 1% of the ingested dose of aflatoxin in milk as determined by duckling toxicity and chemical assay.

Butler and Clifford (1965) showed the presence of a substance with properties similar to the "milk toxin" in the livers of rats treated with purified aflatoxin B_1. The substance appeared at maximum concentration at 0.5 to 1.0 hour, and only traces remained 24 hours after dosing.

Sheep are relatively resistant to the toxic actions of aflatoxins (Lewis *et al.*, 1967), and the excretion of the so-called "milk toxin" has been studied extensively in this species. Allcroft *et al.* (1966) showed that "milk toxin" is detectable in liver, kidneys, and urine of sheep dosed with purified aflatoxin preparations. Because the metabolite is found in tissues and excreta as well as in milk, these workers suggested giving the compound the name "aflatoxin M" to denote its original derivation from milk, but permitting more general applicability. This suggestion has been generally accepted, and the substance is now referred to as aflatoxin M. Furthermore, the "milk toxin" isolated under these conditions has been found to contain two related compounds, aflatoxins M_1 and M_2, which will be described subsequently. Subsequent studies in sheep by Nabney *et al.* (1967) revealed that sheep dosed with aflatoxin mixtures excreted aflatoxin M_1 as the principal metabolite in milk, urine, and feces. Aflatoxin G_1 was also excreted in significant amounts in urine.

The chemical nature of aflatoxin M has recently been elucidated by Holzapfel *et al.* (1966). These investigators isolated aflatoxin M from the urine of sheep dosed with a mixed aflatoxin preparation. They found, however, that this preparation could be resolved chromatographically into two components, which they designated aflatoxins M_1 and M_2. On the basis of spectral and other data, they concluded that aflatoxins M_1 and M_2 were monohydroxylated derivatives of aflatoxins B_1 and B_2, respectively, with structures as shown in Fig. 2. It is now thought that these compounds are identical with "milk toxin" or "aflatoxin M" isolated from milk and excreta of animals dosed with aflatoxins B_1 and B_2. Con-

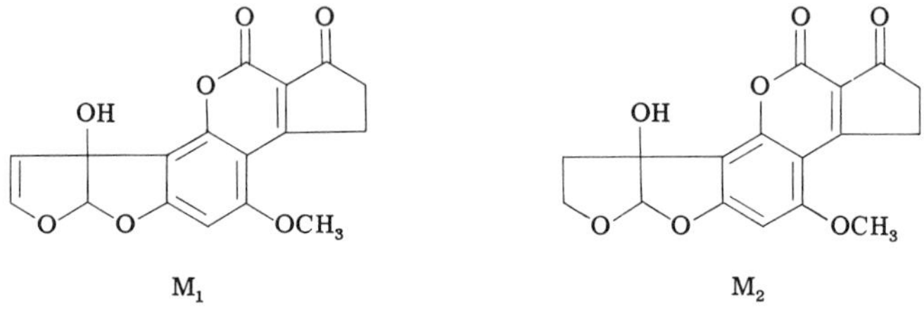

FIG. 2. Structures of aflatoxins M_1 and M_2.

clusive evidence of the identity of these substances with "milk toxin" isolated from cow's milk was provided by the work of Masri *et al.* (1967), in which the isolation and structural elucidation of aflatoxin M_1 from cow's milk was accomplished.

The toxicity of aflatoxins M_1 and M_2 in ducklings was studied by Holzapfel *et al.* (1966), and by Purchase (1967). The LD_{50} values for aflatoxins M_1 and M_2 were 16 and 61.4 μg/animal, respectively, in ducklings weighing 40 to 50 gm. Under comparable conditions, the value for aflatoxin B_1 was 12 μg/animal. It thus appears that the metabolite M_1 retains the toxic properties of the parent compound, B_1.

The kinetics of tissue distribution and excretion of aflatoxin B_1 in rats has been investigated with the use of ^{14}C-labeled compound (Shank and Wogan, 1965; Wogan, 1966*b;* Wogan *et al.,* 1967). After intraperitoneal injection, the compound appears rapidly in liver, and reaches a maximum concentration in that tissue within 30 minutes after injection. It was also found that 70 to 80% of the injected dose was excreted within 24 hours. A major excretory route was through biliary excretion into feces, which accounted for nearly 60% of the administered dose. A further 20% was excreted in urine. By use of selectively labeled compounds, it was also found that a substantial proportion of aflatoxin B_1 undergoes O-demethylation, presumably in the liver.

At all times after dosing, the liver contained 5 to 15 times as much aflatoxin as any other tissue, and at the end of 24 hours this organ contained as much of the compound as the remainder of the carcass. This relatively great affinity of liver for the compound is thought to be related to its specificity as a hepatotoxin.

These results, although still inconclusive, suggest that aflatoxin residues in animal tissues and products do not represent a serious food contamination problem. The exception to this generalization is the possible contamination of milk if dairy cattle are fed contaminated rations. This possibility is readily amenable to prevention by avoidance of aflatoxin-contaminated feed ingredients.

F. Biochemical Effects of Aflatoxins

The toxic and other manifestations of aflatoxins in biological systems must ultimately be attributed to biochemical alterations caused by the compounds. Considerable research activity has been concerned with attempts to define the biochemical locus and mode of action of the toxins. Presently available information does not provide sufficient evidence for definition of a complete sequence of biochemical events leading to gross manifestations of toxicity and carcinogenicity. However, biochemical changes occurring immediately after exposure of animals and cell cultures to aflatoxins have revealed a general pattern of responses which, because of their early and consistent appearance and their central position in certain cellular metabolic pathways, are thought to play an important role in the toxicity of the compound. The reactions involved concern alterations caused by aflatoxins in nucleic acid and protein metabolism.

Interaction of aflatoxins with DNA is envisioned as the initial and critical event in this sequence. This interaction is thought to interfere with DNA transcription, thus resulting in impaired synthesis of DNA and of DNA-dependent RNA synthesis. Ultimately, protein synthesis is also inhibited under some circumstances. Most of these responses have been observed in one or more animal systems sensitive to aflatoxins, and the evidence supporting this hypothesis regarding their mode of action is briefly summarized as follows (see Wogan, 1968b).

Interaction of aflatoxin B_1 with DNA has been observed *in vitro* by several groups of investigators. Sporn *et al.* (1966) demonstrated *in vitro* binding of aflatoxin B_1 to calf thymus DNA and also to certain types of RNA. Similar conclusions were reached by Clifford and Rees (1966) and by Black and Jirgensons (1967).

Inhibition of DNA synthesis caused by aflatoxins has been demonstrated in regenerating rat liver (Frayssinet *et al.*, 1964; De Recondo *et al.*, 1965, 1966). On the basis of the characteristics of the responses, these authors concluded that aflatoxin B_1 acted directly upon the DNA molecule and inhibited its ability to act as a primer for DNA synthesis. Inhibition of DNA polymerase has also been observed in *E. coli* grown in the presence of aflatoxin (Wragg *et al.*, 1967).

Alterations in RNA metabolism of rat liver caused by aflatoxin treatment have been reported by a number of investigators. Lafarge *et al.* (1965), Clifford and Rees (1966), and Sporn *et al.* (1966) have reported that aflatoxin administration to rats results in dramatic and rapid inhibition of precursor incorporation into rat liver nuclear RNA. Under similar conditions, the activity of RNA polymerase is also inhibited (Gelboin *et al.*, 1966). These effects are accompanied by changes in nucleolar morphology, administration of the toxin causing a segregation of nucleolar components (Bernhard *et al.*, 1965; Svoboda *et al.*, 1966; Unuma *et al.*, 1967).

Although protein synthesis is inhibited in liver slices exposed to aflatoxins *in vitro* (Smith, 1963), total liver protein synthesis as measured *in vivo* in rats is largely unaffected by aflatoxins (Shank and Wogan, 1966). In contrast, the synthesis of certain specific liver enzymes is completely blocked by aflatoxin administration. Specifically, the toxin inhibits the increase in tryptophan pyrrolase induced by hydrocortisone (Wogan and Friedman, 1965; Friedman and Wogan, 1966). By virtue of the characteristics of the inhibition, it was concluded that the effects on protein synthesis in this system are secondary to blockade of RNA synthesis.

The aflatoxins also have interesting effects in tissues of plant origin. Black and Altschul (1965) have reported the finding that gibberellic acid-induced increases in lipase and α-amylase activity of the germinating cottonseed are inhibited by aflatoxin. This effect is similar qualitatively to the inhibition of enzyme induction in rat liver described above, although it is not yet clear whether the mechanism of inhibition is related in the two systems. Schoental and White (1965) have shown that aflatoxins in concentrations of 25 μg/ml inhibit the germination of the seeds of cress (*Lepidium sativum* L.), and smaller concentrations apparently interfere with chlorophyll synthesis.

Considerable further experimental data will be required in order to relate directly the observed biochemical effects of aflatoxins to cellular events manifest in toxicity or carcinogenicity. It would appear, on the basis of present evidence, that biochemical changes induced by single doses of aflatoxins are associated with the acute toxicity of the compounds, Association of these specific changes with tumor induction is less certain. The need for further experimentation concerning the biochemical changes associated with carcinoma induction by aflatoxins is clearly indicated. In such studies, the aflatoxins may provide important model compounds for gaining additional insight into the mechanisms involved in the carcinogenic process.

III. ALIMENTARY TOXIC ALEUKIA

Alimentary toxic aleukia (ATA) is a form of panmyelotoxicosis of man which has occurred in certain districts of the U.S.S.R. It appears to be the only form of mycotoxicosis other than ergotism recognized to have seriously affected human populations directly. It is of interest in the present context for, although the conditions leading to its development as well as the causative agents have been established, control measures seem not to have been completely successful in eradication of the syndrome.

Incidence of the disease has been restricted to the U.S.S.R., and there are no published reports of its occurrence elsewhere. This geographic localization is generally attributed to a combination of climate and agricultural practices in the districts where outbreaks occur. Although most attention has been paid to the human disease, concurrent outbreaks in domestic animals have been reported in the same areas.

Because of the frequency and severity of some episodes of the toxicosis, it has been intensively investigated, and a large Russian literature exists concerning various aspects of the problem. This literature has been thoroughly reviewed by Mayer (1953a, b), and much of the present discussion is derived from this source. Other reviews include those of Gajdusek (1953) and Forgacs and Carll (1962). The interested reader is referred to these sources for more detailed information.

A. History of Field Outbreaks

Alimentary toxic aleukia is an often fatal disease whose essential features are basically identical with those of diseases described as aplastic anemia, hemorrhagic aleukia, or panmyelopathy. The chain of pathological alterations in the hematopoietic systems is initiated by the ingestion of cereal grains, particularly proso millet, when these are left unharvested in the fields over the winter.

Although earlier outbreaks of the syndrome are suspected, concerted attention was first drawn to the toxicosis by the appearance of a peculiar disease of the blood and blood-forming tissues in the spring of 1932 in certain districts of Western Siberia. The death rate in the initial outbreaks was high, but with the recognition of mold-damage grain as the causative agent, control measures have reduced the fatality of the disease. In the decades following the first major episode, the geographic area in which the toxicosis occurred widened, and scattered annual outbreaks have been reported in districts within the area 40° to 140° East and 50° to

60° North. Reports indicate that the spread of the syndrome became especially rapid during the World War II years of 1941–1945.

ATA is observed almost exclusively in rural areas where locally grown grain forms a major part of the diets and where, for reasons of expediency or custom, cereal grain crops are allowed to overwinter in the field without harvest until spring. The toxicosis occurs in individuals consuming toxic grain irrespective of age or sex. The severity of the poisoning is increased by long-standing malnutrition or undernutrition, and also is greater in age groups under 5 or over 50 years. The disease has a seasonal character, the first outbreaks generally appearing in the spring months. As the syndrome follows a prolonged clinical course, incidence of diagnosed cases progresses into the summer months, reaching a peak in June and virtually disappearing by November.

Climatic conditions which favor the development of toxicity in grains are a mild winter with heavy snow, followed by slowly progressing spring thawing. These conditions are associated with growth and toxin production by fungi contaminating the grain. Millet is an extensive vector of ATA because it is widely grown in the affected areas, but various outbreaks have also been associated with wheat, rye, oats, and buckwheat. Although the association between ingestion of certain batches of grain and incidence of ATA was soon apparent, the deleterious nature of the grains was variously attributed to infectious agents, vitamin deficiencies, toxic plants or seeds, and other factors. The observation was made that toxic grain samples were mold-damaged, and a large number of mold strains were isolated and proposed as the causative agents. However, in 1945, Sarkisov (see Mayer, 1953b) isolated from a sample of toxic millet a strain of *Fusarium sporotrichioides,* cultures of which caused ATA symptoms in experimental animals. The view has been expressed that *F. sporotrichioides* is the sole causative fungus for ATA (Mayer, 1953b), but subsequent studies have implicated other fungi as well (Joffe, 1965).

B. Microbiological Aspects

Among the fungi isolated from various lots of toxic grains, organisms in the following genera and species were reported with significant frequency: *Hymenopsis, Phoma* spp., *Macrosporium, Aspergillus, Cladosporium, Stachybotrys alternans,* and *Fusarium* species including *F. sporotrichioides, F. roseum,* and *F. graminareum* (Mayer, 1953b).

Joffe (1965) has summarized a series of studies carried out in the Orenburg district of the U.S.S.R. over the period 1943–1950. Samples of grain collected in the homes of diseased persons were subjected to toxicologic study (by bioassay), and contaminating fungi were isolated and identified (Joffe, 1960). The results of tests on more than 950 samples are summarized in Table VIII. These data indicate that a considerable

TABLE VIII

GENERA OF FUNGI ASSOCIATED WITH TOXIN PRODUCTION
IN OVERWINTERED GRAIN[a]

	Isolates			Species		
Genus	Total number	Percent highly toxic	Percent mildly toxic	Number isolated	Percent highly toxic	Percent mildly toxic
Fusarium	501	22.4	13.3	25	60	28
Cladosporium	480	5.4	8.5	15	60	20
Alternaria	506	2.8	5.3	6	30	0
Penicillium	830	1.6	3.8	36	22	33
Mucor	335	3.0	7.2	18	33	22

[a] Modified from Joffe (1965)

number of fungal genera and species are capable of toxin production as determined by the various biological assays used. Of the various organisms found, three were dominant in terms of their frequency in occurrence and in preponderance of toxin-producing isolates. These organisms were *Fusarium poae, Fusarium sporotrichioides,* and *Cladosporium epiphyllum*. Among the various bioassay systems used in screening of the toxin-producing molds, it was found that the cat responded to the toxins in a manner very similar to that of man and represented the only species in which the entire clinical picture of ATA could be experimentally reproduced.

Joffe (1962), in laboratory experiments, studied the biological properties of the toxic fungi isolated from overwintered cereal grains, and found that toxin-producing strains of *Fusarium* and *Cladosporium* species (but not nontoxic strains) grew and sporulated at temperatures below 0°C. Toxin production, which occurred in cultures grown at $-2°$ to $-10°C$, was maximal in heavily sporulated cultures.

Climatic factors conducive to toxin production in wheat, millet, and barley allowed to remain in the field over the winter have been studied in well-controlled field experiments in the Orenburg district (Joffe, 1963). As mentioned earlier, toxin production was most pronounced in the spring, following winters in which snowfall was unusually heavy. Alternate freezing and thawing also enhanced toxin production, which occurred only on grain that had come into contact with the soil. These climatic factors provide ideal conditions for growth and heavy sporulation which were essential factors in toxin production.

C. Chemical Characteristics of Toxic Compounds

Attempts to isolate the active agents from toxic millet revealed the compounds to be soluble in lipid solvents such as petroleum ether, chloroform, acetone, ether, and ethanol. The optimum solvent which has been adapted for assay purposes is diethyl ether. In the preparation of extracts for bioassays, a standardized procedure has been developed consisting in two successive Soxhlet extractions of 50 gm of test material for 6 hours and for 3 days (Mayer, 1953*a*).

Several observations indicate that the toxic compounds are relatively stable. They persist in toxic grain stored for periods up to six years and are apparently not destroyed by baking. Toxicity of grain is not reduced by heating (125°C, 30 minutes; 110°C, 18 hours), acids, or alkali treatment (Mayer, 1953*b*). Joffe (1962) has shown that toxicity of *F. poae* and *F. sporotrichioides* extracts are not reduced by heating for 30 minutes at 100°C. The toxins are apparently not excreted in milk, since infants suckling on diseased mothers showed no symptoms of toxicity (Mayer, 1953*a*).

Structures proposed for toxic compounds produced by *Fusarium* and *Cladosporium* species are shown in Fig. 3 (see Joffe, 1965).

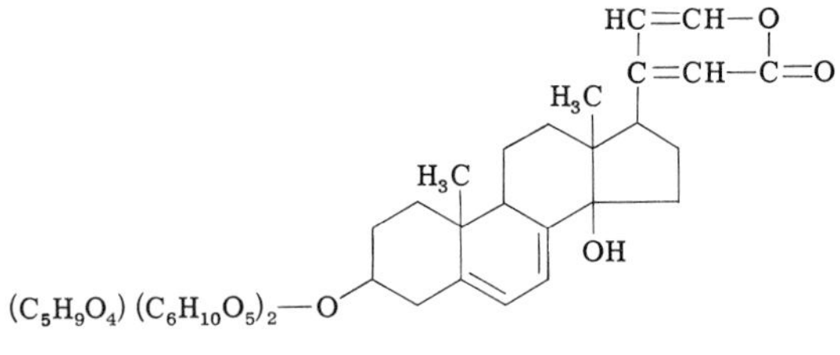

Fusariogenin

$CH_3(CH_2)_9CH{=}CH(CH_2)_{13}C(=O)SH$

Epicladosporic acid

$CH_3(CH_2)_9CH{=}CH(CH_2)_9C(=O)SH$

Fagicladosporic acid

FIG. 3. Formulas of toxins associated with ATA.

D. Clinical Features of ATA

The toxic syndrome in man progresses through four stages, described on the basis of the appearance of symptoms and pathological findings. The first stage consists in a local inflammatory response which begins within hours following ingestion of toxic grain. There is a burning sensation of the buccal and pharyngeal tissues which progresses along the gastrointestinal tract. Within 1 to 3 days, acute gastroenteritis develops with diarrhea, nausea, and vomiting, which persists up to 9 days and then ends, even on continued ingestion of toxic grain.

Disappearance of gastrointestinal symptoms is followed by an asymptomatic period of 2 weeks to 2 months during which progressive bone marrow destruction occurs. In the latter portion of this second stage, hematologic findings include leukopenia, agranulocytosis, anemia, and decreased platelet count. Finally, the marrow changes reach a degree where petechial hemorrhages appear on the skin, which marks the transition into the third stage.

In the third stage, which lasts for 5 to 20 days, the symptoms reflect total atrophy of the bone marrow, hemorrhagic diathesis, necrotic angina, sepsis, and total agranulocytosis. Hematologic changes initiated earlier are intensified, and affected individuals show moderate fever (38° to 40°C). Widespread petechial hemorrhages become larger and confluent and are followed by development of necrotic ulcers of the skin and adjacent musculature. Necrotic lesions in the mouth, pharyngal cavity, and esophagus become severe. In more than 30% of cases in this stage, death is caused by acute stenosis of the glottis. Pulmonary complications, notably bronchopneumonia, abscesses, or hemorrhages are frequent.

If toxic grain is removed from the diet and appropriate therapeutic measures are applied to reverse the extensive hematopoietic damage and to control intercurrent infections, recovery from the toxicosis can be achieved within 2 months. Relapses have been infrequently reported.

Mortality rates from ATA are variable, depending on the toxin concentration in grain, the duration of ingestion, and the nutritional status. In the 1932 outbreak, mortality was 62%, whereas in subsequent outbreaks, rates were in the range of 2% to 80% in the 1945 peak year (Mayer, 1953a).

E. Experimental ATA in Animals

Early attempts to reproduce ATA in animals were not successful. Various types of toxicity were produced in cattle, sheep, chickens, horses, swine, rabbits, and mice by feeding toxic grain or extracts, but in none of these was the clinical syndrome of man reproduced. The entire ATA syndrome was experimentally produced in cats in 1945 by feeding cultures of a toxin-producing *Fusarium* strain on millet. Other species which reportedly develop the ATA syndrome include the guinea pig, monkey, and dog.

A commonly used bioassay for toxicity of grains is the dermal toxicity response in rabbits (Mayer, 1953*b*). Ether extracts of test grain are applied to shaved areas of the skin of rabbits, and the extent of toxicity present is evaluated by the severity of dermal inflammatory response, if present. This response varies in intensity from slight erythema of brief duration to necrotic lesions several days later. Although useful as a screening tool for general toxicity, the response is not specific for toxins causing ATA, and its use probably has given rise to the findings of large numbers of toxin-producing fungal strains not related to the ATA syndrome.

IV. TOXICITY OF MOLD-DAMAGED RICE

Fungal damage of rice provides a further example of the contamination of an important human foodstuff by mycotoxins. This commodity, in common with others, is prone to mold spoilage if storage conditions provide moisture content and temperature suitable for spore germination and growth. Stored rice can contain an abundant mycoflora, as illustrated by one systematic survey on rice of Taiwanese origin in Japan (see Kinosita and Shikata, 1965), which revealed the presence of some 200 genera and species including more than 50 in the *Penicillium* genus. The magnitude that the problem of fungal spoilage of rice can attain is illustrated by the estimate that more than 100,000 tons of rice imported into Japan during the period of 1947–1954 were unfit for human consumption because of fungal damage (Uraguchi *et al.,* 1961*c*). Heavily molded rice is commonly referred to as "yellowed" or "yellowsis" rice, owing to the presence of pigments produced by the contaminating fungi. Although there have been no published reports of large-scale outbreaks of toxicity in humans, allusions have been made to isolated instances of human disease associated with the ingestion of "yellowed" rice (Uraguchi *et al.,* 1961*c*).

A. Toxin-Producing Fungi Associated with Rice

An extensive literature is available concerning the fungi found on stored rice. Kinosita and Shikata (1965) have summarized Japanese and other investigations on certain aspects of this subject. Although the fungal flora of rice can be extensive, species presently known to be capable of producing toxic metabolites seem to be restricted to members of the *Fusarium, Rhizopus, Aspergillus,* and *Penicillium* genera.

In the *Fusarium* group, the growth of *F. roseum* Link on barley and rice has been associated with toxicity syndromes in swine and in man (Kinosita and Shikata, 1965), and also *F. nivale* has given rise to toxicity in horses. Among the *Rhizopus* species, Kinosita and Shikata have found that culture extracts of a strain of *R. oryzae* prove lethal to mice. Of 32 species of *Aspergillus* found on rice, 12 are known to be capable of producing a variety of toxic compounds. The known toxic metabolites of this group include such compounds as kojic acid, terreic acid, aspergillic acid, citrinin, fumagillin, nidulin, and patulin (Miller, 1961). None of these compounds has specifically been implicated in toxicity syndromes in man.

The penicillia found on rice included 47 species, among which 11 are known to be capable of production of toxic compounds. These include, among others, citreoviridin, citrinin, notatin, patulin, rugulosin, and frequentic acid (Miller, 1961). All these compounds have toxic properties in experimental animals. However, only patulin has been implicated as the causative agent in a field outbreak of toxicity, in which cattle were killed by ingestion of a ration heavily infested by *P. urticae,* Bainer (Uraguchi *et al.,* 1961c).

B. Toxicity Caused by *Penicillium islandicum* Sopp.

Among the various fungi associated with toxicity of moldy rice, the most systematic and detailed investigations have been devoted to the organism *P. islandicum* Sopp. and its metabolites. These investigations were stimulated by the finding that rice infested by this fungus caused acute and chronic liver damage when fed to mice or rats. This finding was of considerable interest in view of the high incidence of liver diseases, including primary hepatic carcinoma, in Asiatic populations in which rice forms a major portion of the diet (Uraguchi *et al.,* 1961c). The results of a number of pathological, chemical, and toxicological investigations concerning this problem have recently been summarized by Miyake and Saito (1965).

Following preliminary observations of hepatotoxic effects in mice of "yellowed" rice heavily contaminated by *P. islandicum,* Kobayashi *et al.* (1959) carried out a series of feeding experiments utilizing 14-day rice cultures of the mold (strain Ea, isolated from Egyptian rice). In rats fed diets containing the rice culture at a level of 5% of the diet for up to 620 days, fibrosis, bile ductule hyperplasia, and nodular hyperplasia of hepatic cells were found in most survivors. A small incidence of hepatomas was also reported. More extensive studies were carried out by Miyake *et al.* (1960), who used rice cultures of *P. islandicum* (strain Ud) isolated from Californian rice. Moldy rice was incorporated into diets at levels of 1%, 10%, 30%, and 100% and fed to mice throughout their life span. At a level of 100%, most animals died within 3 to 8 days with acute liver atrophy. Animals surviving lower dietary levels showed cirrhosis in the later stages of the experiment – 220 to 600 days. Similar results were obtained with rats, in which cirrhosis was induced at a small incidence at the 1% dietary level. In addition, adenoma-like nodules developed in rats after long-term feeding, showing cytological aberrations and abnormal cellular arrangement (Enomoto, 1959).

In an experiment to determine whether cultures of this fungus were capable of hepatoma induction, cultures were fed to rats alone and in combination with a known hepatocarcinogen, *p*-dimethylaminoazobenzene (DAB) (Miyake and Saito, 1965). In animals fed only the mold culture at a level of 5% of the diet, 5/30 developed liver tumors; when fed in combination with 0.06% DAB, 14/30 showed hepatomas compared with 6/30 receiving DAB alone. These results reveal carcinogenic properties of cultures of *P. islandicum* as well as their promoting effect when fed in combination with DAB.

Definitive association of hepatotoxicity with products of fungal growth was provided by the investigations of Uraguchi *et al.* (1961a), which demonstrated similar properties of fungal mats grown on synthetic medium. When fed to mice and rats, fungal mats of *P. islandicum* gave rise to toxic liver injuries ranging from acute atrophy to chronic cirrhosis and other changes closely related to carcinoma induction. These findings, which were dose-related, indicated that the toxic agents arose directly from the mold.

Efforts to isolate the toxic agents involved the integration of toxicologic evaluations (Uraguchi *et al.*, 1961c) with chemical extraction and purification procedures (Uraguchi *et al.*, 1961b). As a result of these approaches, two toxic substances were isolated from cultures of *P. islandicum* – luteoskyrin and a chlorine-containing peptide. The proposed structures of these compounds are shown in Fig. 4 and some of their biological properties are summarized in Tables IX and X.

Luteoskyrin is a lipid-soluble pigment which is relatively slow-acting. Acute liver injuries of mice and rats surviving more than 24 hours after luteoskyrin administration include centrolobular necrosis and diffuse fatty metamorphosis. Electron microscopic findings revealed mitochondrial swelling and cystic dilatation of the endoplasmic reticulum (Miyake and Saito, 1965). Chronic administration causes periportal fibrosis and a small incidence of hepatoma formation. In contrast, the chlorine-contain-

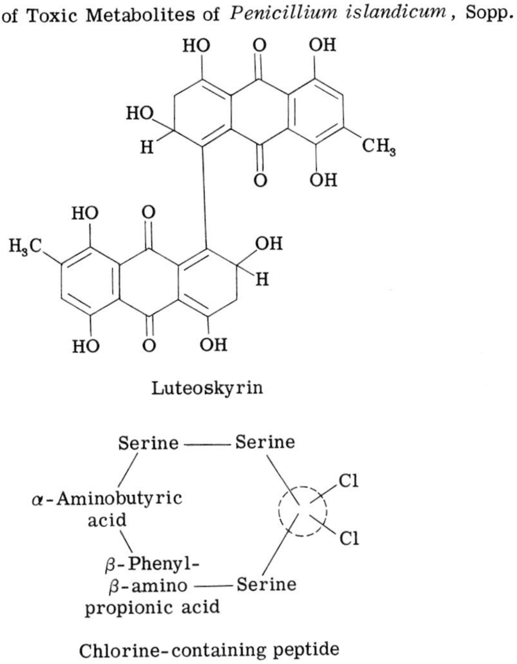

Structures of Toxic Metabolites of *Penicillium islandicum*, Sopp.

Luteoskyrin

Chlorine-containing peptide

FIG. 4. Structures of toxic metabolites of *Penicillum islandicum* Sopp.

TABLE IX

TOXICITY OF LUTEOSKYRIN TO MICE AND CELL CULTURES[a]

LD_{50} in Mice
Intravenous: 6.65 mg/kg
Subcutaneous: 147.0 mg/kg
Oral: 221.0 mg/kg

Liver Injury
Acute lesion: Centrolobular necrosis and fatty change
Chronic lesion: Periportal fibrosis (slight) and hepatoma formation

Tissue Culture (Chang Liver Cells)
MED[b] (growth inhibition): 0.1 μg/ml
LD[c] (cell destruction):   1.0 μg/ml

[a] Adapted from Miyake and Saito (1965)
[b] Minimum effective dose.
[c] Lethal dose.

TABLE X

TOXICITY OF CHLORINE-CONTAINING PEPTIDE TO MICE AND CELL CULTURES[a]

LD_{50} in Mice
Intravenous: 338 μg/kg
Subcutaneous: 475 μg/kg
Oral: 6550 μg/kg

Liver Injury
Acute lesion: Vacuolar degeneration, cell destruction, periportal hemorrhage
Chronic lesion: Liver cirrhosis, monolobular

Tissue Culture (Chang liver cells)
MED^b (growth inhibition): $10^{0.5}$ μg/ml
LD^c (cell destruction): $10^{1.5}$ μg/ml

[a] Adapted from Miyake and Saito (1965).
[b] Minimum effective dose.
[c] Lethal dose.

ing peptide is water-soluble, acts more rapidly, and has a somewhat higher potency as an acute poison. These observations clearly establish an association of these compounds with the hepatotoxic action of fungal cultures containing them (Uraguchi et al., 1961c). However, definitive proof that they are hepatocarcinogenic agents awaits further experimental verification.

V. ANIMAL TOXICOSES OF FUNGAL ORIGIN

Sporadic episodes of poorly defined animal diseases of unknown etiology are frequently encountered in veterinary medicine. Since outbreaks of this kind tend to be localized, solutions to the problems are generally achieved by empirical means, and the etiologic factors are rarely identified. However, when diseases appear in widespread proportions and result in serious animal losses, efforts are stimulated to establish and control the causative factors. The facial eczema syndrome is one example of such a situation, and other veterinary toxicoses have also been attributed to mold-produced toxins. They are distinguishable by the animal species generally involved in field outbreaks as well as by the substrates or fungi associated with the toxicity. The present state of knowledge concerning these syndromes is highly variable; in some instances toxicity is clearly attributable to a specific fungus, whereas in others only apparent association has been made between moldy feed and toxicity symptoms.

The facial eczema syndrome has received extensive study in New Zealand and Australia, where it has been of serious economic consequence. The etiology of the disease has been established, and sporidesmin, a metabolite of *Pithomyces chartarum*, has been identified as the causative agent. In addition to this syndrome, five other veterinary toxicoses are included in the present discussion. These diseases include stachybotryotoxicosis, moldy corn poisoning, toxicity associated with *Aspergilli*, hemorrhagic disease of poultry, and a genital hypertrophy syndrome of swine, all associated with moldy feeds. Most of these syndromes have been described in detail by Forgacs and Carll (1962) and by Forgacs (1965), and only their essential features will be summarized.

A. Facial Eczema in Ruminants

The facial eczema syndrome of sheep and cattle is one of a group of photosensitization diseases that occur in most of the livestock-producing areas of the world. Although they are generally sporadic, of low incidence, and transient in nature, in some areas they may be of more serious economic consequence than diseases of an infectious nature. This has been particularly true of facial eczema, for it is responsible for serious annual losses of sheep in New Zealand and Australia. Toxic syndromes of similar characteristics but as yet unproved etiologic agents have been encountered in the southern United States (Gibbons, 1958; Glenn et al., 1964; Kidder et al., 1961).

Facial eczema is a disease of hepatogenous origin, in which the photosensitization phenomenon is secondary to severe liver damage. Obliterative cholangitis with accompanying bile ductule hyperplasia, fibrosis, and parenchymal cell damage impairs the biliary excretion of phyloerythrin, a ruminant digestion product of chlorophyll. Accumulation of this compound as well as bilirubin results in photosensitization and icterus. The liver damage has been found to be caused by sporidesmin, a toxic metabolite of *Pithomyces chartarum*, a fungus which grows profusely on ryegrass pastures under certain climatic conditions.

The present discussion will summarize pertinent information concerning the disease, which has been reviewed in detail by Filmer (1958), Forgacs and Carll (1962), and Dodd (1965).

1. HISTORICAL AND MICROBIOLOGICAL ASPECTS OF FACIAL ECZEMA

Facial eczema was apparently recognized in New Zealand as early as 1900 and has been subjected to considerable investigation (Filmer, 1958). More recent outbreaks have been reported in Australia (Janes,

1959), and a very similar disease of cattle has been noted in the United States (Gibbons, 1958; Glenn *et al.,* 1964; Kidder *et al.,* 1961).

Several features of the disease formed the basis for research into nature of the syndrome. In the New Zealand outbreaks, the disease appears primarily during the summer and early autumn. Sheep and cattle of all ages are affected, but swine are not susceptible. The first cases always appear 1 to 2 weeks following a rain after a relatively dry period, and invariably occur only in animals grazing on rapidly growing ryegrass pastures. The incidence of icterus appearing in sheep is variable, being absent in some flocks and affecting every animal in others.

The earliest clinical manifestations of the toxicosis, which develops rapidly, include lacrymation and nasal discharge. These are followed within a brief period by edema of skin areas unprotected by the fleece (ears, eyelids, face, lips, etc.). The edematous tissues undergo serum exudation and encrustation and eventually become necrotic and slough. The period of dermal lesions may last for several weeks, whereas icterus tends to appear in the terminal phases, when liver damage is severe.

The disease was recognized as a mycotoxicosis after attempts to implicate infectious agents and toxic plants as etiologic agents gave negative findings. In 1959, a clear association was made between outbreaks of the syndrome and the presence of the fungus *Sporidesmium bakeri* on highly toxic pastures (Percival, 1959). In outbreaks of a similar disease in cattle grazing on Bermuda grass pastures, Kidder *et al.* (1961) have identified the predominant fungus as *Periconia minutissima.* It is not known whether this organism or toxins produced by it are the same as those responsible for the syndrome in sheep.

Sporidesmium bakeri (later reclassified as *Pithomyces chartarum*) is a saprophytic fungus which proliferates on dead plant material (Thornton and Percival, 1959). The syndrome is confined to ryegrass pastures because of the fibrous nature and slow decomposition of the leaves of this grass. When the plant dies during dry summer weather, the dead leaves remain at the base of the plant, and, after periods of rainfall, an abundant culture medium is present. The fungus sporulates heavily, and spore density can be so great on a pasture that the muzzles of grazing animals become covered with a visible black rim of spores (Dodd, 1965).

The toxin of *S. bakeri* is present in both the conidia and the mycelium of the fungus (Percival, 1959). Production of toxin by the fungus depends on strain, medium, and incubation period. Although toxin production did not occur when the mold was grown on Raulin-Thom or Czapek-Dox media, cultures on potato-carrot extract medium gave good toxin yields (Thornton and Percival, 1959).

2. CHEMICAL NATURE OF SPORIDESMIN

The development of chemical procedures for extraction, purification, and isolation of the toxic substance responsible for facial eczema was closely interrelated with the development of biological and chemical assays for monitoring isolation and purification stages. These various aspects are therefore conveniently discussed concurrently. The chemistry of sporidesmin and related compounds has recently been reviewed (Taylor, 1967).

An empirical chemical assay devised for the detection of facial eczema toxicity in pastures, described as the so-called "beaker-test" (Perrin, 1959), was found to correlate well with toxicity and the presence of mold spores (Percival and Thornton, 1958). In its revised form (Sandos et al., 1959), the test consists in extraction of pasture grass with acetone followed by chromatography on alumina. The chromatographic eluate is evaporated to dryness, and a positive test is indicated by the presence of a white residue. It is not clear whether the residual matter is identical with the toxin, but results of this assay correlate well with biological assays of the extracts.

Isolation of the toxic compound was also greatly facilitated by the discovery of Evans et al. (1957) that ryegrass which produced facial eczema in sheep caused liver lesions of a similar nature in guinea pigs. The subsequent adaptation of this response as a bioassay method (Perrin, 1957) enabled White (1958) to achieve concentration of the active agent from toxic grass. Evidence was also presented that the concentration of toxin in relatively toxic grass was of the order of 10 ppm. Subsequently, Synge and White (1959) isolated the toxin in pure form and applied the name "sporidesmin" to indicate its fungal origin. The proposed structure for sporidesmin (Fridrichsons and Mathieson, 1962; Hodges et al., 1963) is shown in Fig. 5.

In highly purified form, sporidesmin is soluble in ether, acetone, benzene, methanol, chloroform, and carbon disulfide; it is virtually insoluble

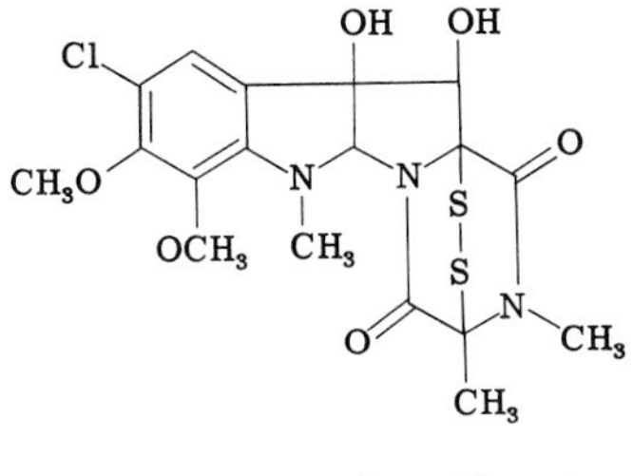

FIG. 5. Structure of sporidesmin.

in water or petroleum ether (White, 1958). Although its stability apparently has not been systematically studied, some information is available from observations made during the isolation procedure. The compound is readily decomposed (rendered nontoxic) by dilute alkali, and also by exposure to highly activated alumina or charcoal. It apparently decomposes readily under field conditions, since considerable care had to be used to avoid loss in the initial attempts at isolation from grass samples. The conditions governing this process are not known, but decomposition of toxin is undoubtedly among the factors responsible for the inconstant and varied nature of the field syndrome.

3. BIOLOGICAL AND BIOCHEMICAL EFFECTS OF SPORIDESMIN

The clinical and pathological aspects in field outbreaks of facial eczema in sheep have been described by Cunningham et al. (1942), by McFarlane et al. (1959), and by Done et al. (1960). These investigators concluded that the disease was of hepatogenous origin, and that the essential liver lesion was acute cholangitus which progresses to obliteration of bile ducts by fibrous tissue. A recent and extensive series of experiments on sheep revealed that the clinical manifestations resulting from administration of sporidesmin were indistinguishable from those of field cases of the syndrome (Mortimer and Taylor, 1962). In these experiments, a dose of 1.0 gm of sporidesmin per kilogram of body weight given orally or intravenously uniformly produced severe liver lesions and caused 75% mortality with 24 days after administration. A single dose of 3.0 mg/kg caused 100% mortality within 10 days. Single or multiple doses of 0.3 mg/kg resulted in mild liver lesions without clinical signs, whereas 0.5 mg/kg caused photosensitization as well as liver lesions and also resulted in small (8%) mortality.

Changes in serum constituents of these animals were studied by Done et al. (1962), who reported that doses of sporidesmin of 0.5 mg/kg or larger caused increased serum glutamic-oxalacetic transaminase levels, as well as increases of serum cholesterol, lipid phosphorus, and bilirubin. The changes were consistently evident 10 days after sporidesmin administration and were dose-dependent. Mortimer (1962) found that these dose levels resulted in increased leukocyte counts and also in alterations of serum proteins associated with progressive liver injury. Liver function, as determined by conventional clinical tests, was also impaired. The histopathological alterations following administration of 1.0 mg of sporidesmin per kilogram were also investigated by Mortimer (1963). He found early (2 to 4 days) acute, reversible parenchymal cell degeneration with cholangitis, repair of which resulted in partial stenosis of bile ducts. From the tenth day, biliary obstruction and photosensitization were manifest.

Various observations have indicated wide species variation in susceptibility to sporidesmin poisoning. Field studies have shown that swine and horses are not affected by the same pastures which cause severe facial eczema in sheep and cattle. Among the experimental animal groups, guinea pigs are highly susceptible, while rats, mice, and chickens are insensitive (Perrin, 1957). Rabbits show approximately the same sensitivity as guinea pigs (Dodd, 1960), and the portal entry of the toxin into the liver as well as its excretion in the bile has been studied in this species (Worker, 1960).

Sporidesmin has been reported to cause cytopathological changes in HeLa cell cultures at a concentration of 3.0 mμg in the culture medium (Done et al., 1961). The same investigators found that conjunctival instillation of as little as 5 μg of the compound in the rabbit eye resulted in corneal opacity. These observations indicate great potency of the compound in susceptible systems.

A number of recent investigations have been concerned with attempts to define the biochemical alterations responsible for the observed clinical and histopathological effects of sporidesmin. Peters (1963) and Peters and Smith (1964) have conducted extensive studies on the early effects of the compound on the hepatic lipids of sheep. An oral dose of 1.0 mg/kg resulted in an early and significant rise in liver triglycerides, with no changes in concentration of other lipid fractions. This effect, which is similar to that of other hepatotoxins, was interpreted as a malfunction of hepatic triglyceride-secreting mechanism.

When incubated in vitro with homogenates of sheep or guinea pig liver, sporidesmin at concentrations of 75 μM or 150 μM inhibits respiration in the presence of substrates which require nicotinamide coenzymes for their oxidation (Wright and Forrester, 1965). Succinate oxidation was far less sensitive. Similar results were obtained when liver mitochondria were incubated with the compound, and slight swelling of mitochondria was also reported.

In studies on the rat, a species relatively resistant to acute poisoning, Slater et al. (1964) found that a dose of 6 mg of sporidesmin per kilogram failed to alter liver nucleotides, nucleic acids, or activity of several enzymes. Serum bilirubin and alkaline phosphatase were also unaffected. On the other hand, the compound was shown to increase the permeability of capillaries, giving rise to ascites and pleural effusions.

Thus, it is not yet possible to describe the primary locus of action of sporidesmin, and an explanation of the great variation in species susceptibility awaits further study. Similarly, it is difficult to estimate accurately the magnitude of total losses of domestic animals attributable to sporidesmin or similar mycotoxins. However, photosensitization diseases occur with some frequency in virtually every major livestock-raising area of the world, and toxic fungal metabolites may play a role in more of the unexplained incidence than has been recognized.

B. Stachybotryotoxicosis

This toxicosis, caused by toxic metabolites of *Stachybotrys atra,* was first recognized in Russia in 1931 and has been extensively studied by Russian investigators. Some years after recognition of the disease entity, the fungal origin of the causative agent was discovered when highly toxic straw was found to be heavily contaminated by this fungus, and the syndrome was experimentally reproduced by mold cultures.

Field cases have involved large numbers of horses, although occasional outbreaks have also been reported in cattle. The disease is seasonal in character, appearing when animals are fed hay and straw during winter months. The toxicosis may run a typical clinical course in three stages, or it can assume an atypical form. In the typical form, the first signs of toxicity are reflected in stomatitis, resulting in local response to the toxic agents. After a period of 8 to 12 days, the second stage develops, in which systemic effects appear. This stage is characterized by progressive leukopenia, agranulocytosis, and thrombocytopenia, and lasts for 15 to 20 days. With continued intake of toxic straw, the toxicosis proceeds into a third stage, accompanied by fever and intensification of blood changes. Blood-clotting mechanisms are almost completely impaired. The similarities between these symptoms and those of ATA in humans described in a preceding section are noteworthy.

In its atypical form, stachybotryotoxicosis is characterized by sudden onset (72 hours), absence of blood dyscrasia, and signs of acute shock. Death ensues in all animals in this less common form of the syndrome. Pathologic findings in the disease are the same in both forms, consisting mainly in widespread necrosis in most tissues as well as generalized hemorrhages.

Although field cases of stachybotryotoxicosis occur principally in horses, Forgacs *et al.* (1958) reproduced the syndrome experimentally in the calf, sheep, and pig by feeding straw cultures of *S. atra.* Among the experimental animals, cultures of the mold have been shown to be toxic to mice, guinea pigs, rabbits, and dogs. It has also been noted (Forgacs and Carll, 1962) that men exposed to toxic straw frequently develop signs of dermal toxicity on exposed skin, and, in a few cases, moderate to severe leukopenia has been reported. In such cases, exposure probably occurs through dust or aerosol suspensions of the toxic substances.

Stachybotrys atra is widely distributed and grows freely on cellulosic materials over a wide range of temperatures and moisture contents. The toxic substance, which is extractable from toxic straw by diethyl ether, is stable to heat, light, X-rays, and acids, but is readily destroyed by alkalis.

C. Moldy Corn Poisoning

In 1952, a disease of unknown etiology occurred in swine in the southeastern United States. In a series of separate outbreaks, mortality ranged from 5 to 55%, and a large number of animals were lost. According to Sippel *et al.* (1953), earlier, less extensive outbreaks of a similar nature had been reported, and a further severe occurrence appeared in 1955 (Forgacs, 1965). In many instances, the disease was associated with the practice of allowing swine to forage in fields where corn was lying in contact with the ground, or with feeding of heavily molded corn in storage.

Sippel *et al.* (1953) reproduced the syndrome by force-feeding corn collected from fields where the toxicosis had occurred. Subsequently, Burnside *et al.* (1957) conducted a mycologic examination of toxic corn and isolated, among several other fungi, two toxin-producing organisms. These molds, one a strain of *Aspergillus flavus* and the other a *Penicillium rubrum,* were cultured on sterile corn and fed to swine. The *A. flavus* cultures were lethal when force-fed to young pigs, causing death in 3 to 5 days (Forgacs and Carll, 1962). Cultures of the *P. rubrum* strain caused death within 8 to 36 hours, with symptoms characteristic of the field syndrome. These observations, as well as results of similar studies in mice, suggest that the *P. rubrum* strain was the more toxic of the two fungi used. These fungi are generally regarded to be the primary, but not necessarily the sole, sources of toxicity in the field syndrome.

Although moldy corn is the primary vector of the syndrome, there are also reports of toxicity in swine caused by feeding moldy peanuts, oats, and stale bread. There is also evidence that rations compounded from moldy ingredients have caused a similar toxicity in dogs (Bailey and Groth, 1959).

The symptomatology of the syndrome is to a large extent dependent on the degree of toxicity of the corn as well as the rate and duration of its ingestion. Animals ingesting lethal amounts of toxic corn usually die within 2 to 5 days, whereas those consuming smaller quantities over a longer period develop a subacute toxicity syndrome. In the acute disease, the chief pathologic findings comprise widespread and massive hemorrhages, whereas chronic toxicity results in liver damage and its sequelae (icterus, ascites, etc.) as well as hemorrhages in many tissues.

The nature and identity of the toxic agents produced by the fungi involved in this syndrome are uncertain. The possibility that the *A. flavus* strain may have produced aflatoxins cannot be discounted. This suggestion is supported by the findings of Newberne *et al.* (1966) that many of the features of the so-called "hepatitis X" syndrome in dogs mentioned earlier (Bailey and Groth, 1959) are reproduced by aflatoxins and also that certain samples of toxic dog food contain aflatoxins. As regards the *P. rubrum,* Wilson and Wilson (1962*a, b*) report the extraction and partial purification of an agent hepatotoxic to mice, guinea pigs, rabbits, and dogs from a strain of the fungus originally isolated from a moldy corn poisoning episode. The compounds involved have not yet been isolated in pure form.

D. Toxic *Aspergilli* and Bovine Hyperkeratosis

A toxicity syndrome of cattle accompanied by hyperkeratinization of epithelial tissues (commonly referred to as bovine hyperkeratosis or "X-disease") has occurred with irregular frequency in cattle-raising areas of the United States, and some outbreaks have resulted in serious economic losses (Garner, 1961). Many outbreaks of the disease have been associated with contamination of the feed by chlorinated naphthalenes (Bell, 1954), a group of compounds used for a variety of industrial purposes. Feed contamination apparently resulted from lubricating oil used on processing machinery.

However, outbreaks of the disease have been reported in which chlorinated naphthalenes were ruled out as causative agents. Similarities of certain features of these outbreaks with recognized veterinary mycotoxicoses led to an examination of toxic feeds for the presence of toxin-producing molds. In a series of such investigations summarized by Forgacs and Carll (1962), toxic strains of *Aspergillus chevaleri* and *A. clavatus* were isolated from feedstuffs shown by various workers to produce the hyperkeratosis syndrome under field conditions. Cultures of these organisms as well as *A. fumigatus* resulted in acute and chronic toxicity symptoms when force-fed to calves. In the acute poisoning, animals died within 5 to 20 days, depending on the daily intake of fungal cultures. In chronic cases, where feeding continued for 30 days or longer, various pathologic changes were produced, including hyperkeratinization.

Although the field syndrome is undoubtedly of multiple etiology, these results indicate the possibility that mycotoxins may play a role in its origin.

E. Mycotoxicoses in Poultry

In a manner somewhat analogous to the bovine hyperkeratosis syndrome, an apparently new disease entity was encountered in poultry flocks in the United States about 1950 (Forgacs and Carll, 1962). Prominent among the symptoms in the malady were widespread petechial and larger hemorrhages in many tissues, and the descriptive name "poultry hemorrhagic syndrome" has been applied to it. Hemorrhages are accompanied by bone-marrow suppression and erosion of the gastrointestinal mucosa. This syndrome, which mainly affects broiler flocks during the rapid growth period, causes up to 50% mortality, and results in significant losses of animals.

Although various theories have been proposed as to the character of the etiologic agents of the syndrome, their precise nature has not yet been determined. Forgacs and Carll (1962) have summarized a series of investigations which indicate that toxicoses of a similar nature can be induced by several fungi.

In a survey of the fungal flora of feed and litter in areas where the syndrome was endemic, these workers isolated a number of toxic molds. Cultures were fed to chickens under laboratory conditions in order to determine the characteristics of their toxicity. Chickens that consumed grain infected with *Aspergillus clavatus, A. flavus, Penicillium purpeurogenum,* or a species of *Alternaria* died within 3 to 27 days. Pathologic findings in most instances included hemorrhages in skeletal and cardiac musculature as well as gastrointestinal lesions.

These results, together with the findings of Forgacs et al. (1963) that the antifungal agent 8-hydroxyquinoline appears to reduce the incidence of the syndrome in poultry flocks, suggest that fungi may be involved in the etiology of the disease.

F. Genital Hypertrophy in Swine

Several reports have associated feeding of moldy grain with a syndrome involving vulvar hypertrophy, vaginal prolapse, and mammary hypertrophy in swine. Stob *et al.* (1962) investigated the mycoflora of corn involved in several outbreaks of the syndrome and isolated a strain of *Gibberella zeae* (the perfect form of several *Fusarium* species). Cultures of this organism on corn reproduced the syndrome in immature pigs and mice, and ethanol extracts of the cultures caused similar effects in mice. A fluorescent compound which produced the effects was extracted and purified from fungal cultures.

Subsequently, Christensen *et al.* (1965) studied grain involved in other outbreaks of the syndrome, and demonstrated the potency of twelve *Fusarium* isolates in causing estrogenic symptoms. Extraction of cultures and purification of the active agent revealed that the utero-trophic compound (F-2) in these cultures was identical with that isolated from *Gibberella zeae* by Stob and co-workers (1962).

VI. OTHER MYCOTOXINS

Several recent reports have dealt with isolation of toxin-producing fungi or mycotoxins from fungi commonly associated with foodstuffs. For the most part, these organisms or compounds have not been associated with outbreaks of mycotoxicoses, but rather in some instances have been discovered in the course of intentional searches for previously un-recognized fungal toxins.

For example, Perone and co-workers (1963) have isolated and char-acterized two toxic metabolites of *Sclerotinia sclerotiorum,* a fungus causing so-called "pink-rot" of celery. These compounds, 8-methoxy-psoralen and 4,5′,8-trimethylpsoralen, on contact with the skin of man or animals produce a severe blistering dermatitis after exposure to ultra-violet or sunlight. These compounds are of interest as potential food contaminants because of the presence in their structures of the furocou-marin moiety similar in some respects to the aflatoxins. No evidence seems to be available concerning the carcinogenic or other toxic properties of the psoralens.

Wilson and Wilson (1964) have extracted and partially purified a sub-stance causing prolonged tremors in mice from cultures of *Aspergillus flavus,* a fungus widely distributed on foodstuffs. More recently, van der Merwe *et al.* (1965) isolated and chemically characterized "ochratoxin A" from cultures of *Aspergillus ochraceus* isolated from South African cereal and legume crops. This compound, whose structure is shown in Fig. 6, is a potent hepatotoxin in ducklings, in which species the LD_{50} is of the order of 0.4 mg/kg, a value similar to that for aflatoxin B_1.

The significance of these recently described toxic metabolites of molds as food contaminants is not yet established. Their discoveries, however, reflect the results of experiments designed to investigate the toxin-

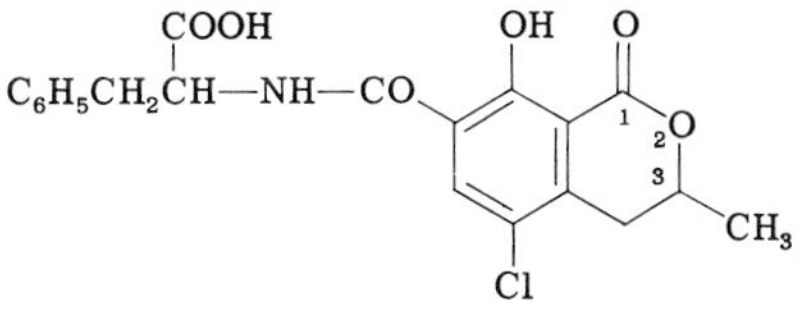

FIG. 6. Structure of ochratoxin A.

producing capacity of fungi commonly found on foodstuffs. The purposes of such investigations contrast with those of most previous studies in which mycotoxins were discovered retrospectively following episodes of human or animal toxicoses. As these and similar investigations proceed, it is virtually certain that additional mycotoxins will be discovered.

VII. PUBLIC HEALTH AND OTHER IMPLICATIONS OF MYCOTOXINS

The mycotoxin problem has considerable significance, on a worldwide basis, in terms of public health, agriculture, and economics. The veterinary public health aspects of the problem are clearly indicated by the various mycotoxicoses that have been described in domestic animals. In fact, most of the research to date on mycotoxicoses has been focused on this aspect as illustrated by information in preceding sections.

The degree to which various mycotoxins represent a health hazard to man has not been fully explored. In the case of the alimentary toxic aleukia syndrome, major epidemics of human toxicity have clearly been caused by toxic fungal metabolites. None of the other mycotoxins described are known with certainty to have played an etiologic role in human diseases. However, recent developments of knowledge in this field have led to considerations of the possible role of such compounds in diseases of unknown etiology.

Important among these is the possible association of consumption of diets contaminated by aflatoxins or other mycotoxins with incidence of liver disease, including primary liver carcinoma (Kraybill and Shimkin, 1964; Carnaghan and Crawford, 1964; Wogan, 1968a). Primary hepatic carcinoma occurs with a distinct geographic localization in certain tropical and subtropical areas. According to Berman (1951), this disease is observed with a frequency of about 1% of autopsies in Orientals, or about 14% of all carcinomas in tropical areas of the Orient or in Central Africa. Comparatively, liver cancer is found in 0.3% of autopsies or less than 2.5% of all carcinomas in the United States and Europe. Reported incidence within zones of relatively high occurrence is variable. For example, in a Japanese survey, hepatomas were found in 1.5% of all autopsies or about 7.6% of all carcinomas, according to Takeda and Aizawa (1956). In Thailand, Bhamarapravati and Nimsomburna (1963) reported a liver carcinoma incidence of 1.9% in autopsies performed over a two-year period. The highest reported incidence of hepatoma occurs in the male African Bantu, where this malignancy represents 68% of all carcinomas (Oettle, 1964). Although many racial, genetic, and environmental factors have been considered in the etiology of the disease, its causes remain obscure.

Increasing attention is being paid to the possibility that aflatoxins or similar food contaminants may be involved in the etiology of this and other diseases, and some speculation has been advanced on association of moldy diet components with high rate of liver carcinoma. This speculation is supported by high incidence rates in areas where climatic conditions and agricultural practice would appear to favor mold damage of major diet ingredients. In many such areas, food habits and customs also include the use of foods fermented under uncontrolled conditions by poorly defined microbial inocula or regular use of diet components prone to mold damage by virtue of storage conditions used.

An accurate assessment of the importance of mycotoxins in this regard will be possible only by systematic surveys for their presence in diets of affected populations. Further research efforts into the significance of mycotoxins as food contaminants must necessarily involve many disciplines and approaches of wide scope. It will be necessary to identify the fungi responsible for toxin production on various foodstuffs and to define the environmental conditions influencing toxin production. The toxic compounds must be isolated in pure form and chemically identified, and suitable chemical and biological assay methods should be developed for their detection in foods. In will also be necessary to characterize their biological effects by animal experiments.

By virtue of the possible public health significance of mycotoxins, the problem of mold damage of foodstuffs has extensive agricultural and economic implications in addition to the well-recognized detrimental changes in physical, nutritional, organoleptic, and other attributes arising from fungal growth. This consideration has had particular importance in connection with the development of protein-rich foods based in part on peanuts, cottonseed, soya, and other plant sources for use in alleviation of world protein malnutrition (Milner, 1965). Recognition of the possible contamination of protein sources such as peanut flour by aflatoxins has necessitated precautionary measures to assure uncontaminated products. It has been found, for example, that mechanical or electronic sorting based on such parameters as size and discoloration effectively eliminates aflatoxin-containing kernels from peanuts destined for human consumption.

Control of the mycotoxin problem on a worldwide basis involves several considerations in connection with agricultural practice and technology. Most effective control could probably be achieved by preventive measures which would minimize or eliminate growth of fungal contaminants on food crops. Improvements in harvesting methods and introduction of postharvest rapid drying processes are of considerable importance in this regard. Similarly, storage and transport under conditions of controlled moisture content and temperature as well as the use of fumigants and fungicides offer possibilities for effective control of the problem. Therefore, a wide field for exploration has evolved as a result of the recognition of the importance of mycotoxins in foodstuffs as components of the increasingly complex chemical environment.

REFERENCES

Adye, J., and Mateles, R. I. (1964). *Biochim. Biophys. Acta* **86**, 418.

Allcroft, R. (1965). *In* "Mycotoxins in Foodstuffs" (G. N. Wogan, editor), pp. 153–162. MIT Press, Cambridge.

Allcroft, R., and Carnaghan, R. B. A. (1963a). *Chem. Ind. (London)*, p. 50.

Allcroft, R., and Carnaghan, R. B. A. (1963b). *Vet. Record* **75**, 259.

Allcroft, R., and Lewis, G. (1963). *Vet. Record* **75**, 487.

Allcroft, R., Carnaghan, R. B. A., Sargeant, K., and O'Kelly, J. (1961). *Vet. Record* **73**, 428.

Allcroft, R., Robers, H., Lewis, G., Nabney, J., and Best, P. E. (1966). *Nature* **209**, 154.

Armbrecht, B. H., and Fitzhugh, O. G. (1964). *Toxicol. Appl. Pharmacol.* **6**, 421.

Asao, T., Büchi, G., Abdel-Kader, M. M., Chang, S. B., Wick, E. L., and Wogan, G. N. (1963). *J. Am. Chem. Soc.* **85**, 1706.

Asao, T., Büchi, G., Abdel-Kader, M. M., Chang, S. B., Wick, E. L., and Wogan, G. N. (1965). *J. Am. Chem. Soc.* **87**, 882.

Ashley, L. M., Halver, J. E., and Wogan, G. N. (1964). *Federation Proc.* **23**, 105.

Ashley, L. M., Halver, J. E., Gardner, W. K. Jr., and Wogan, G. N. (1965). *Federation Proc.* **24**, 627.

Asplin, F. D., and Carnaghan, R. B. A. (1961). *Vet. Record* **73**, 1215.

Austwick, P. K. C., and Ayerst, G. (1963). *Chem. Ind. (London)*, p. 55.

Bailey, W. S., and Groth, A. H., Jr. (1959). *J. Am. Vet. Med. Assoc.* **134**, 514.

Bampton, S. S. (1963). *Trop. Sci.* **5**, 74.

Barger, G. (1931). "Ergot and Ergotism." Gurney and Jackson, London

Barnes, J. M., and Butler, W. H. (1964). *Nature* **202**, 1016.

Bell, W. B. (1954). *J. Am. Vet. Med. Assoc.* **124**, 289.

Berman, C. (1951). "Primary Carinoma of the Liver." H. K. Lewis and Co., London.

Bernhard, W., Frayssinet, C., Lafarge, C., and Le Breton, E. (1965). *Compt. Rend.* **261**, 1785.

Bhamarapravati, N., and Nimsomburna, P. (1963). *Proc. 16th Assembly Japan. Med. Congr.* **3**, 376.

Black, H. S., and Altschul, A. M. (1965). *Biochem. Biophys. Res. Commun.* **19**, 661.

Black, H. S., and Jirgensons, B. (1967). *Plant Physiol.* **42**, 731.

Blount, W. P. (1961). *Turkeys* **9**, 52, 55–58, 61, 77.

Brechbühler, S., Büchi, G., and Milne, G. (1967). *J. Org. Chem.* **32**, 2641.

Broadbent, J. H., Cornelius, J. A., and Shone, G. (1963). *Analyst* **88**, 214.

Büchi, G., Foulkes, D. M., Kurono, M., Mitchell, G. F., and Schneider, R. S. (1967). *J. Am. Chem. Soc.* **89**, 6745.

Burnside, J. E., Sippel, W. L., Forgacs, J., Carll, W. T., Atwood, M. B., and Doll, E. R. (1957). *Am. J. Vet. Res.* **18**, 817.

Burrell, N. J., Grundey, J. K., and Harkness, C. (1964). *Trop. Sci.* **6**, 74.

Butler, W. H. (1964). *Brit. J. Cancer* **18**, 756.

Butler, W. H. (1965). *In* "Mycotoxins in Foodstuffs" (G. N. Wogan, ed.), pp. 175–186. MIT Press, Cambridge.

Butler, W. H., and Barnes, J. M. (1964). *Brit. J. Cancer* **17**, 699.

Butler, W. H., and Barnes, J. M. (1966). *Nature* **209**, 90.

Butler, W. H., and Clifford, J. I. (1965). *Nature* **206**, 1045.

Campbell, A. D., Dorsey, E., and Eppley, R. M. (1964). *J. Assoc. Offic. Agr. Chemists* **47**, 1002.

Carnaghan, R. B. A. (1965). *Nature* **208**, 308.

Carnaghan, R. B. A. (1967). *Brit. J. Cancer* **21**, 811.

Carnaghan, R. B. A., and Crawford, M. (1964). *Brit. Vet. J.* **120**, 201.

Carnaghan, R. B. A., Hartley, R. D., and O'Kelly, J. (1963). *Nature* **200**, 1101.

Chang, S. B., Abdel-Kader, M. M., Wick, E. L., and Wogan, G. N. (1963). *Science* **142**, 1191.

Cheung, K. K., and Sim, G. A. (1964). *Nature* **201**, 1185.

Christensen, C. M. (1965). *In* "Mycotoxins in Foodstuffs" (G. N. Wogan, ed.), pp. 9–14. MIT Press, Cambridge.

Christensen, C. M., Nelson, G. H., and Mirocha, C. J. (1965). *Appl. Microbiol.* **13**, 653.

Clifford, J. I., and Rees, K. R. (1966). *Nature* **209**, 312.

Codner, R. C., Sargeant, K., and Yeo, R. (1963). *Biotechnol. Bioeng.* **5**, 185.

Cucullu, A. F., Lee, L. S., Mayne, R. Y., and Goldblatt, L. A. (1966). *J. Am. Oil Chemists' Soc.* **43**, 89.

Cunningham, I. J., Hopkirk, C. S. M., and Filmer, J. F. (1942). *New Zealand J. Sci. Technol.* **A24**, 185.

Daniels, M. R. (1965). *Brit. J. Exptl. Pathol.* **46**, 183.

de Iongh, H., Beerthuis, R. K., Vles, R. O., Barrett, C. B., and Ord, W. O. (1962). *Biochim. Biophys. Acta* **65**, 548.

de Iongh, H., Van Pelt, J. G., Ord, W. O., and Barrett, C. B. (1964a). *Vet. Record* **76**, 901.

de Iongh, H., Vles, R. O., and Van Pelt, J. G. (1964b). *Nature* **202**, 466.

de Iongh, H., Vles, R. O., and de Vogel, P. (1965). *In* "Mycotoxins in Foodstuffs" (G. N. Wogan, ed.), pp. 235–246. MIT Press, Cambridge.

De Recondo, A. M., Frayssinet, C., Lafarge, C., and Le Breton, E. (1965). *Compt. Rend.* **261.** 1409.

De Recondo, A. M., Frayssinet, C., Lafarge, C., and Le Breton, E. (1966). *Biochim. Biophys. Acta* **119,** 322.

de Waart, J. (1967). *Antonie Van Leeuwenhoek J. Microbiol. Serol.* **33,** 230.

Dickens, F., and Jones, H. E. H. (1964). *Brit. J. Cancer* **17,** 691.

Diener, U. L., Davis, N. D., Salmon, W. D., and Prickett, C. O. (1963). *Science* **142,** 1491.

Dodd, D. C. (1960). *New Zealand J. Agr. Res.* **3,** 491.

Dodd, D. C. (1965). *In* "Mycotoxins in Foodstuffs" (G. N. Wogan, ed.), pp. 105–110. MIT Press, Cambridge.

Done, J., Mortimer, P. H., and Taylor, A. (1960). *Res. Vet. Sci.* **1,** 76.

Done, J., Mortimer, P. H., Taylor, A., and Russell, D. W. (1961). *J. Gen. Microbiol.* **26,** 207.

Done, J., Mortimer, P. H., and Taylor, A. (1962). *Res. Vet. Sci.* **3,** 161.

Enomoto, M. (1959). *Acta Pathol. Japon.* **9,** 189.

Eppley, R. M. (1966a). *J. Assoc. Offic. Anal. Chemists* **49,** 473.

Eppley, R. M. (1966b). *J. Assoc. Offic. Anal. Chemists* **49,** 1218.

Evans, J. V., McFarland, D., Reid, C. S. W., and Perrin, D. D. (1957). *New Zealand J. Sci. Technol.* **A38,** 491.

Filmer, J. F. (1958). *New Zealand J. Agr.* **97,** 202.

Fischbach, H., and Campbell, A. D. (1965). *J. Assoc. Offic. Agr. Chemists* **48,** 28.

Forgacs, J. (1965). *In* "Mycotoxins in Foodstuffs" (G. N. Wogan, ed.), pp. 87–104. MIT Press, Cambridge.

Forgacs, J., and Carll, W. T. (1962). *Advan. Vet. Sci.* **7,** 273.

Forgacs, J., Carll, W. T., Herring, A. S., and Hinshaw, W. R. (1958). *Trans. N.Y. Acad. Sci.* **20,** 787.

Forgacs, J., Koch, H., and White-Stevens, R. H. (1963). *Avian Diseases* **7,** 56.

Frayssinet, C., Lafarge, C., de Recondo, A. M., and Le Breton, E. (1964). *Compt. Rend.* **259,** 2143.

Fridrichsons, J., and Mathieson, A. (1962). *Tetrahedron Letters* **26,** 1265.

Friedman, M. A., and Wogan, G. N. (1966). *Federation Proc.* **25,** 662.

Gabliks, J. Z., Schaeffer, W., Friedman, L., and Wogan, G. N. (1965). *J. Bacteriol.* **90,** 720.

Gajdusek, D. C. (1953). *Med. Sci. Publ.* **2,** p. 107. Army Medical Service Graduate School, Walter Reed Army Medical Center, Washington, D.C., pp. 107–111.

Garner, R. J. (1961). "Veterinary Toxicology," p. 303. Williams and Wilkins, Baltimore.

Gelboin, H. V., Wortham, J. S., Wilson, R. G., Friedman, M. A., and Wogan, G. N. (1966). *Science* **154,** 1205.

Genest, C., and Smith, D. M. (1963). *J. Assoc. Offic. Agr. Chemists* **46,** 817.

Gibbons, W. J. (1958). *Vet. Med.* **52,** 297.

Glenn, B. L., Monlux, A. W., and Panciera, R. J. (1964). *Pathol. Vet.* **1,** 469.

Goldblatt, L. A. (1965). *In* "Mycotoxins in Foodstuffs" (G. N. Wogan, ed.), pp. 261–264. MIT Press, Cambridge.

Halver, J. E. (1964). *In* "Mycotoxins in Foodstuffs" (G. N. Wogan, ed.), pp. 209–230. MIT Press, Cambridge.

Harding, J. D. J., Done, J. T., Lewis, G., and Allcroft, R. (1963). *Res. Vet. Sci.* **4,** 217–229.

Hartley, R. D., Nesbitt, B. F., and O'Kelly, J. (1963). *Nature* **198,** 1056.

Hesseltine, C. W., Shotwell, O. L., Ellis, J. J., and Stubblefield, R. D. (1966). *Bacteriol. Rev.* **30,** 795.

Heusinkveld, M. R., Shera, C. C., and Baur, F. J. (1965). *J. Assoc. Offic. Agr. Chemists* **48**, 448.

Hiscocks, E. S. (1965). *In* "Mycotoxins in Foodstuffs" (G. N. Wogan, ed.), pp. 15–26. MIT Press, Cambridge.

Hodges, F., Allen, J. R., Zust, H. R., Nelson, A. A., Armbrecht, B. H., and Campbell, A. D. (1964). *Science* **145**, 1439.

Hodges, R., Ronaldson, J. W., Taylor, A., and White, E. P. (1963). *Chem. Ind. (London)* p. 42.

Holzapfel, C. W., Steyn, P. S., and Purchase, I. F. H. (1966). *Tetrahedron Letters* **25**, 2799.

Hueper, W. C., and Payne, W. W. (1961). *J. Natl. Cancer Inst.* **27**, 1123.

Janes, B. S. (1959). *Nature* **184**, 1327.

Joffe, A. Z. (1960). *Bull Res. Council Israel* **8D**, 81.

Joffe, A. Z. (1962). *Mycopathol. Mycol. Appl.* **16**, 201.

Joffe, A. Z. (1963). *Plant and Soil* **18**, 31.

Joffe, A. Z. (1965). *In* "Mycotoxins in Foodstuffs" (G. N. Wogan, ed.), pp. 77–85. MIT Press, Cambridge.

Johnson, A. J. (1948). *Food Agr. Organ. U.N., FAO Agr. Studies* No. 2, 88.

Juhasz, S., and Greczi, E. (1964). *Nature* **203**, 861.

Kidder, R. W., Beardsley, D. W., and Erwin, T. C. (1961). *Univ. Florida Gr. Exptl. Sta. Bull.* **630**, 1.

Kinosita, R., and Shikata, T. (1965). *In* "Mycotoxins in Foodstuffs" (G. N. Wogan, ed.), pp. 111–132. MIT Press, Cambridge.

Kobayashi, Y., Uraguchi, K., Sakai, F., Tatsuno, T., Tsukioka, M., Noguchi, Y., Tsunoda, H., Miyake, M., Saito, M., Enomoto, M., Shikata, T., and Ishiko, T. (1959). *Proc. Japan. Acad.* **35**, 501.

Kraybill, H. F., and Shimkin, M. B. (1964). *Advan. Cancer. Res.* **8**, 191.

Lafarge, C., Frayssinet, C., and De Recondo, A. M. (1965). *Bull. Soc. Chim. Biol.* **47**, 1724.

Lancaster, M. C., Jenkins, F. P., and Philp, J. McL. (1961). *Nature* **192**, 1095.

Le Breton, E., Frayssinet, C., and Boy, J. (1962). *Compt. Rend.* **255**, 784.

Lee, W. V. (1965). *Analyst* **90**, 305.

Legator, M. S., and Withrow, A. (1964). *J. Assoc. Offic. Agr. Chemists* **47**, 1007.

Legator, M. S., Zuffante, S. M., and Harp, A. R. (1965). *Nature* **208**, 345.

Lewis, G., Markson, L. M., and Allcroft, R. (1967). *Vet. Record* **80**, 312.

Loosmore, R. M., and Harding, J. D. J. (1961). *Vet. Record* **73**, 1362.

Loosmore, R. M., and Markson, L. M. (1961). *Vet. Record* **73**, 813.

Loosmore, R. M., Allcroft, R., Tutton, E. A., and Carnaghan, R. B. A. (1964). *Vet. Record* **76**, 64.

Madhavan, T. V., Tulpule, P. G., and Gopalan, C. (1965). *Arch. Pathol.* **79**, 466.

Masri, M. S., Lundin, R. E., Page, J. R., and Garcia, V. C. (1967). *Nature* **215**, 753.

Mateles, R. I., and Adye, J. C. (1965). *Appl. Microbiol.* **13**, 208.

Mayer, C. F. (1953*a*). *Military Surgeon* **113**, 173.

Mayer, C. F. (1953*b*). *Military Surgeon* **113**, 295.

McDonald, D., and A'Brook, J. (1963). *Trop. Sci.* **5**, 208.

McDonald, D., and Harkness, C. (1963). *Trop. Sci.* **5**, 143.

McDonald, D., and Harkness, C. (1964). *Trop. Sci.* **6**, 12.

McDonald, D., Harkness, C., and Stonebridge, W. C. (1964). *Trop. Sci.* **6**, 131.

McFarlane, D., Evans, J. V., and Reid, C. S. W. (1959). *New Zealand J. Agr. Res.* **2**, 194.

Miller, M. W. (1961). "The Pfizer Handbook of Microbial Metabolites." McGraw-Hill, New York.

Milner, M. (1965). *In* "Mycotoxins in Foodstuffs" (G. N. Wogan, ed.), pp. 69–73. MIT Press, Cambridge.

Miyake, M., and Saito, M. (1965). *In* "Mycotoxins in Foodstuffs" (G. N. Wogan, ed.), pp. 133–146. MIT Press, Cambridge.

Miyake, M., Saito, M., Enomoto, M., Shikata, T., Ishiko, T., Uraguchi, K., Sakai, F., Tatsuno, T., Tsukioka, M., and Sakai, Y. (1960). *Acta Pathol. Japon.* **10,** 75.

Mortimer, P. H. (1962). *Res. Vet. Sci.* **3,** 269.

Mortimer, P. H. (1963). *Res. Vet. Sci.* **4,** 166.

Mortimer, P. H., and Taylor, A. (1962). *Res. Vet. Sci.* **3,** 147.

Nabney, J., Burbage, M. B., Allcroft, R., and Lewis, G. (1967). *Food Cosmet. Toxicol.* **5,** 11.

Nesbitt, B. F., O'Kelly, J., Sargeant, K., and Sheridan, A. (1962). *Nature* **195,** 1062.

Nesheim, S. (1964). *J. Assoc. Offic. Agr. Chemists* **47,** 1010.

Newberne, P. M. (1965). *In* "Mycotoxins in Foodstuffs" (G. N. Wogan, ed.), pp. 187–208. MIT Press, Cambridge.

Newberne, P. M., Wogan, G. N., Carlton, W. W., and Abdel-Kader, M. M. (1964a). *Toxicol. Appl. Pharmacol.* **6,** 542.

Newberne, P. M., Carlton, W. W., and Wogan, G. N. (1964b). *Pathol. Vet.* **1,** 105.

Newberne, P. M., Russo, R., and Wogan, G. N. (1966). *Pathol. Vet.* **3,** 331.

Oettle, A. G. (1964). *J. Natl. Cancer Inst.* **33,** 383.

Parrish, F. W., Wiley, B. J., Simmons, E. G., and Long, L. Jr. (1966). *Appl. Microbiol.* **14,** 139.

Percival, J. C. (1959). *New Zealand J. Agr. Res.* **2,** 1041.

Percival, J. C., and Thornton, R. H. (1958). *Nature* **182,** 1095.

Perone, V. B., Scheel, L. D., and Meitus, R. J. (1963). *Am. Chem. Soc., Div. Biol. Chem. Abstr.* **22,** 14A.

Perrin, D. D. (1957). *New Zealand J. Sci. Technol.* **A38,** 669.

Perrin, D. D. (1959). *New Zealand J. Agr. Res.* **2,** 266.

Peters, J. A. (1963). *Nature* **200,** 286.

Peters, J. A., and Smith, L. M. (1964). *Biochem. J.* **92,** 379.

Platonow, N. (1965). *Vet. Record* **77,** 1028.

Platt, B. S., Stewart, R. J. C., and Gupta, S. R. (1962). *Nutr. Soc. Proc.* **21,** 30–31.

Pons, W. A., and Goldblatt, L. A. (1964). *J. Am. Oil Chemists' Soc.* **41,** 59.

Pons, W. A., and Goldblatt, L. A. (1965). *J. Am. Oil Chemists' Soc.* **42,** 471.

Purchase, I. F. H. (1967). *Food Cosmet. Toxicol.* **5,** 339.

Rothlin, E., and Bircher, R. (1952). *Progr. Allergy* **3,** 434.

Salmon, W. D., and Newberne, P. M. (1963). *Cancer Res.* **23,** 571.

Sandos, J., Clare, N. T., and White, E. P. (1959). *New Zealand J. Agr. Res.* **2,** 623.

Sargeant, K., O'Kelly, J., Carnaghan, R. B. A., and Allcroft, R. (1961a). *Vet. Record* **73,** 1219.

Sargeant, K., Sheridan, A., O'Kelly, J., and Carnaghan, R. B. A. (1961b). *Nature* **192,** 1096.

Sargeant, K., Carnaghan, R. B. A., and Allcroft, R. (1963). *Chem. Ind. (London),* p. 53.

Schoental, R. (1961). *Brit. J. Cancer* **15,** 812.

Schoental, R., and White, A. F. (1965). *Nature* **205,** 57.

Schroeder, H. W., and Ashworth, L. J., Jr. (1965). *Phytopathology* **55,** 464.

Shank, R. C., and Wogan, G. N. (1964). *Federation Proc.* **23,** 200.

Shank, R. C., and Wogan, G. N. (1965). *Federation Proc.* **24,** 627.

Shank, R. C., and Wogan, G. N. (1966). *Toxicol. Appl. Pharmacol.* **9,** 468.

Sinnhuber, R. O., Wales, J. H., Engebrecht, R. H., Amend, D. F., Kray, W. D., Ayres, J. C., and Ashton, W. E. (1965). *Federation Proc.* **24,** 627.

Sippel, W. L., Burnside, J. E., and Atwood, M. B. (1953). *Proc. 90th Ann. Meeting Am. Vet. Med. Assoc., Toronto, Canada,* p. 174.

Slater, T. F., Strauli, U. D., and Sawyer, B. (1964). *Res. Vet. Sci.* **5,** 450.

Smith, R. H. (1963). *Biochem. J.* **88,** 50P.

Sporn, M. B., Dingman, C. W., Phelps, H. L., and Wogan, G. N. (1966). *Science* **151,** 1539.

Stob, M., Baldwin, R. S., Tuite, J., Andrews, F. N., and Gillette, K. G. (1962). *Nature* **196,** 1318.

Svoboda, D., Grady, H., and Higginson, J. (1966). *Am. J. Pathol.* **49,** 1023.

Synge, R. L. M., and White, E. P. (1959). *Chem. Ind. (London),* p. 1546.

Takeda, J., and Aizawa, M. (1956). *Trans. Soc. Pathol. Japon.* **45,** 1.

Taylor, A. (1967). *In* "Biochemistry of Some Foodborne Microbial Toxins" (R. I. Mateles and G. N. Wogan, eds.), pp. 69–107. MIT Press, Cambridge.

Thornton, R. H., and Percival, J. C. (1959). *Nature* **183,** 63.

Townsley, P. M., and Lee, E. G. H. (1967). *J. Assoc. Offic. Anal. Chemists* **50,** 361.

Trager, W. T., Stoloff, L., and Campbell, A. D. (1964). *J. Assoc. Offic. Agr. Chemists* **47,** 993.

Tulpule, P. G., Madhavan, T. V., and Gopalan, C. (1964). *Lancet* **i,** 962.

Unuma, T., Morris, H. P., and Busch, H. (1967). *Cancer Res.* **27,** 2221.

Uraguchi, K., Sakai, F., Tsukioka, M., Noguchi, Y., Tatsuno, T., Saito, M., Enomoto, M., Ishiko, T., Shikata, T., and Miyake, M. (1961*a*). *Japan. J. Exptl. Med.* **31,** 435.

Uraguchi, K., Tatsuno, T., Sakai, F., Tsukioka, M., Sakai, Y., Yonemitsu, O., Ito, H., Miyake, M., Saito, M., Enomoto, M., Shikata, T., and Ishiko, T. (1961*b*). *Japan J. Exptl. Med.* **31,** 19.

Uraguchi, K., Tatsuno, T., Tsukioka, M., Sakai, Y., Sakai, F., Kobayashi, Y., Saito, M., Enomoto, M., and Miyake, M. (1961*c*). *Japan J. Exptl. Med.* **31,** 1.

van der Linde, J. A., Frens, A. M., and van Esch, G. J. (1965). *In* "Mycotoxins in Foodstuffs" (G. N. Wogan, ed.), pp. 247–249. MIT Press, Cambridge.

van der Merwe, K. J., Steyn, P. S., Fourie, L., Scott, De B., and Theron, J. J. (1965). *Nature* **205,** 1112.

van der Zijden, A. S. M., Koelensmid, W. A. A., Boldingh, J., Barrett, C. B., Ord, W. O., and Philp, J. (1962). *Nature* **195,** 1060.

van Dorp, D. A., van der Zijden, A. S. M., Beerthuis, R. K., Sparreboom, S., Ord, W. O., de Jong, K., and Keuning, R. (1963). *Rec. Trav. Chim. Pays-Bas* **82,** 587.

Verrett, M. J., Marliac, J. P., and McLaughlin, J., Jr. (1964). *J. Assoc. Offic. Agr. Chemists* **47,** 1003.

White, E. P. (1958). *New Zealand J. Agr. Res.* **1,** 433.

Wilson, B. J., and Wilson, C. H. (1962*a*). *J. Bacteriol.* **84,** 283.

Wilson, B. J., and Wilson, C. H. (1962*b*). *J. Bacteriol.* **83,** 693.

Wilson, B. J., and Wilson, C. H. (1964). *Science* **144,** 177.

Wogan, G. N. (ed.) (1965*a*). "Mycotoxins in Foodstuffs." MIT Press, Cambridge.

Wogan, G. N. (1965*b*). *In* "Mycotoxins in Foodstuffs" (G. N. Wogan, ed.), pp. 163–173. MIT Press, Cambridge.

Wogan, G. N. (1966a). *Advan. Chem. Ser.* **57,** 195.

Wogan, G. N. (1966b). *Bacteriol. Rev.* **30,** 460.

Wogan, G. N. (1968a). *Federation Proc.* **27,** 932.

Wogan, G. N. (1968b). *In* "Aflatoxin—Scientific Background, Control and Implications" (L. A. Goldblatt, ed.). Academic Press, New York.

Wogan, G. N., and Friedman, M. A. (1965). *Federation Proc.* **24,** 627.

Wogan, G. M., and Mateles, R. I. (1968). Mycotoxins. *In* "Progress in Industrial Microbiology" (D. J. D. Hockenhull, ed.), Vol. 7. Heywood, London.

Wogan, G. N., and Newberne, P. M. (1967). *Cancer Res.* **27,** 2370.

Wogan, G. N., Edwards, G. S., and Shank, R. C. (1967). *Cancer Res.* **27,** 1729.

Worker, N. A. (1960). *Nature* **185,** 909.

Wragg, J. B., Ross, V. C., and Legator, M. S. (1967). *Proc. Soc. Exptl. Biol. Med.* **125,** 1052.

Wright, D. E., and Forrester, I. T. (1965). *Can. J. Biochem.* **43,** 881.

Zuckerman, A. J., Rees, K. R., Inman, D., and Petts, V. (1967a). *Nature* **214,** 814.

Zuckerman, A. J., Tsiquaye, K. N., and Fulton, F. (1967b). *Brit. J. Exptl. Pathol.* **48,** 20.

PART III **MISCELLANEOUS**

CHAPTER XI | LABORATORY METHODS

John H. Silliker and Richard A. Greenberg

I. GENERAL CONSIDERATIONS

A. Introduction

There is little doubt that the discrepancy between the real and the reported incidence of food-borne illness is considerable. Approximately 21,000 cases of human salmonellosis are reported yearly in the United States, but there is reason to believe that the true incidence may be as high as 2,000,000 (Ager *et al.*, 1964). The reporting of other forms of food-borne illness is probably equally inadequate. In many instances, particularly in small, self-limited outbreaks, food is never suspected. In other cases, the attending physician satisfies himself and his patients with the unscientific diagnosis of "ptomaine poisoning," and no attempt is made to ascertain the responsible food, let alone the causative organism.

Although inadequate clinical and epidemiological investigations contribute heavily to the discrepancy between true and reported incidences of food poisoning, incompetent analysis of suspect foods by the microbiologist is likewise a contributing factor. The Food Protection Committee of the Food and Nutrition Board (1964) reported that no specific etiological agent has been determined in about half of the outbreaks recorded since 1957. In view of the present state of our knowledge of food poisoning, this record is reprehensible. As Niven (1962) has pointed out: "Among the thousands of bacterial species and the hundreds of metabolites produced by them, relatively few bacteria are known to be capable of producing substances that are toxic to man and animals. Even among these few microorganisms, most of them must invade and infect the host in order to produce illness as a result of toxin production . . . Fortuitously, only 2 bacterial species are recognized which are capable of producing extracellular metabolites when growing in foods, and which are toxic when introduced via the oral route; namely, *Staphylococcus aureus* and *Clostridium botulinum.*"

A perusal of the table of contents of this volume indicates that the field is similarly limited when one considers food-borne infections. The conclusion is inescapable that the etiological agents of food-borne infections and intoxications comprise a rather small and specific group of microorganisms, albeit the scope of biological hazards in foods is broadening to include such new agents as halophilic vibrios, viruses, mycotoxins, and parasites. Although, as indicated in the report of the Food Protection Committee (see above), agents not detectable by conventional laboratory procedures are probably responsible for many food-borne illnesses, we feel that the most frequent cause of failure lies in inadequate laboratory support. Progress toward better recognition and control of food-borne illnesses demands competent laboratory analysis of suspect foods.

When the microbiologist receives samples purported to represent foods that have caused human illness, the diagnostic role which he must play is analogous to that of the medical microbiologist who analyzes clinical specimens. The medical microbiologist must use methods which permit him to study the total microflora of the specimen. He must ultimately judge the significance of his laboratory results in terms of his knowledge of the normal and abnormal microflora of the specimen which he has examined. In some cases, quantitative aspects must be considered;

in other instances, the qualitative detection of a particular pathogen may be diagnostic. He must take into account the previous history of the specimen he analyzes and weigh the significance of his findings accordingly. The confirmation of initial impressions may require controlled animal inoculation experiments before a definitive diagnosis can be rendered. Finally, the results of laboratory tests must be interpreted in terms of the clinical, and sometimes epidemiological, facts known with respect to the patient from whom the specimen was obtained.

B. Nature of the Food Involved

It is of great importance for the food microbiologist to have some familiarity with the nature of the food examined. If it is a processed food, then a knowledge of the manufacturing procedures will aid in determining the significance of the microorganisms found in the sample. Even severely heat-processed shelf-stable canned foods may not be sterile. In a broader sense, food processing procedures change the physical and chemical properties of the raw material in manners which ultimately result in a finished product showing a more or less characteristic microflora. The number of microorganisms *normally* found in a particular type of food is a reflection of the severity of the processing methods and/or the presence of microbial inhibitors. Shelf-stable foods owe their stability to the inability of surviving organisms to multiply and thus cause spoilage. In such products one frequently observes a steady decline in microbial numbers. But it should be emphasized that the mere presence of microorganisms in a shelf-stable food product is not necessarily indicative of abnormality. The microflora of individual categories of shelf-stable foods follows patterns which are more or less unique to the particular food product. Thus, in severely heat-processed low-acid vegetable products one may encounter small numbers of thermophilic spores which tend to die during normal storage. Mesophilic spores may be encountered in such diverse shelf-stable products as canned cured meats, mayonnaise, and peanut butter. In all these instances the total numbers of contaminants are small. But this is not always the case. With dried fermented sausage products, meats with sufficiently high salt and acid contents as to render them virtually shelf-stable, large numbers of bacteria are encountered, and this flora consists almost exclusively of lactic acid-producing organisms such as lactobacilli, streptococci, and leuconostocs.

In his examination of suspect shelf-stable foods, the microbiologist must be keenly aware that he may encounter viable microorganisms. He must correlate his findings with knowledge of the normal flora of the product and available information with respect to the food poisoning episode.

In contrast, perishable foods are never sterile, and their microflora at any given time is a composite reflection of processing procedures, post-processing handling, and storage conditions. It is axiomatic that perishable foods will undergo spoilage as a natural consequence of not being eaten. This will occur even though the foods are prepared from the highest-grade raw materials, processed under proper conditions, and subsequently stored in an ideal manner. Frequently it is concluded that a particular food has been responsible for illness simply because it contains a large microbial population — this despite our knowledge that a relatively small number of microbial species are capable of causing food poisoning.

Initially, perishable foods will generally contain a heterogeneous flora, but with storage a particular group, or species, of microorganisms usually multiplies. The proliferating organisms are responsible for characteristic spoilage patterns in the product. The analyst studying perishable foods must become aware of the expected spoilage flora. For example, the normal spoilage of fresh meats involves the growth of pseudomonads and closely related gram-negative aerobes. Yeasts and micrococci constitute the normal flora of stored non-vacuum-packaged cured meats, whereas lactic acid bacteria reach large numbers when the same products are packaged under anaerobic conditions. All too frequently the specific nature of the spoilage flora of a perishable food has led the uninitiated analyst to claim the discovery of a new food poisoning agent and, perhaps, to overlook the true etiological agent of a food poisoning episode simply because the latter was quantitatively overwhelmed by the normal spoilage flora of the product. Again, it is essential that the analyst know the normal flora of the suspect food and interpret his analytical findings accordingly. The problem is more difficult with perishable foods, since spoilage microorganisms may be present in far greater numbers than the sought-after food poisoning bacteria.

C. History of the Sample

In many instances the history of the food sample received poses serious problems to the analyst. Ideally, the sample submitted for analysis should be one which has been refrigerated as soon as possible after being served and thereafter promptly delivered to the laboratory for analysis. Unfortunately, this is seldom the case, as in most instances the food is not suspected of being the cause of illness for some period of time, often days, after its consumption. In the interim, it may have been held under refrigeration, but frequently this is not the case. At times the only sample available may have been retrieved from the depths of a garbage receptacle several days following the food poisoning episode. The suspect food may have been incorporated into a "leftover" dish, such as a meat salad or stew. Under these conditions the inciting agent of the episode may have been destroyed by cooking or by some constituent of the leftover dish. Even though mistreatment of the sample makes interpretation of results exceedingly difficult, the specimen should not be rejected for analysis, as careful study may lead to an explanation of the episode.

D. Handling and Treatment of Food by Consumer

Important to any interpretation of analytical results is information relative to the conditions under which the food was prepared, stored, and served by the consumer. A knowledge of the cooking procedures employed may indicate whether the organisms subsequently detected in the sample might have been present before preparation of the food or represent postcooking contamination. Further, if the food was stored by the consumer subsequent to cooking, such storage might have resulted in the observed contamination, and the holding conditions might have been such as to allow the development of bacteria at a time subsequent to that at which the food was served. At times it may even be necessary to reproduce the cooking procedures used in order to interpret the significance of microorganisms isolated from the suspect sample.

E. Sampling

As with any analytical procedure, the results obtained are no more accurate than the sample received. In many instances, selective sampling of the product may indicate its mode of contamination. For example, ham is frequently implicated in food poisoning outbreaks. The analyst may receive samples consisting of the ham as it was actually served to the afflicted individuals. A careful sampling of the interior of the ham, the cut surface, and the two varieties of sliced product may lead to an explanation for the episode. If, as in most cases of food poisoning from ham, *Staphylococcus* intoxication was involved, then many possibilities exist. Among them are the following: (1) Large numbers of staphylococci on the cut surface as well as in the uncooked meat slices would point toward contamination during slicing. (2) If appreciably higher *Staphylococcus* counts were found in the uncooked slices as compared with the surface of the ham, this would suggest mistreatment of the contaminated slices prior to cooking. (3) If only the cooked slices showed viable staphylococci, this would suggest contamination subsequent to cooking. Still other combinations of results are possible, and each points to a plausible sequence of events leading ultimately to the food poisoning episode.

All too frequently, however, the analyst elects to "composite" the submitted samples so as to get an "average" evaluation of the product. Such "averages" seldom lead to definitive conclusions, whereas a selective approach to sampling may clearly suggest the sequence of events which ultimately were responsible for the food poisoning episode.

II.   GENERAL LABORATORY PROCEDURES

A.   Examination of Packaging Material

If the product received was commercially packaged, as in a can or plastic film, then an attempt should be made to obtain the packaging material for examination. If a canned product is involved, the can should be carefully examined by a person competent to render an opinion relative to its condition. The determination of can defects is beyond the scope of this discussion. However, specialists in this area can determine whether a particular can is sound, and such studies should be made on any can involved in a food poisoning outbreak.

If the product was packaged in a plastic film or some other material, determination of the soundness of the packaging film and, if indicated, its permeability to moisture and gas should be made.

B.   Organoleptic Examination

Laboratory analysis of the sample should start with a thorough organoleptic examination. Note should be made of the temperature of the sample at the time of receipt and its history during laboratory storage up until the time of this examination. The organoleptic study should involve determination of the color, odor, and texture of the product. Are these normal, or is there evidence of deterioration? Is there overt evidence of spoilage? If the product is perishable, then are the evidences of spoilage consistent with the normal spoilage pattern of the food product? If there is spoilage, is there localization, or is it distributed throughout the body of the sample?

The organoleptic examination of the product may indicate the nature of the sampling to be done for microbiological analyses. For this reason, aseptic precautions must be taken during the examination. For example, if the sample consists of a chunk of meat, both interior and surface samples must be obtained and separate determinations made. If the specimen shows localized areas of discoloration, abnormal texture, or odor, then portions of the sample from these areas should be removed and examined separately. The sampling of the product for microbiological analyses demands extreme care, and careful organoleptic examination will often allow wise judgments to be made relative to appropriate sampling procedures to be used.

C. Chemical Tests

Although we are not concerned here with a discussion of the detection of chemical poisons, the organoleptic examination of the sample may indicate certain chemical tests, the results of which would aid in the ultimate interpretation of microbiological results. The nature of these tests will be dictated by the food involved and the observations made in the course of organoleptic examinations. For example, if a cured meat product shows localized green or brown discoloration, it may be instructive to carry out nitrite, nitrate, and salt determinations in an effort to determine whether the product had been properly cured. If an acid product, such as mayonnaise, shows evidence of fermentation, pH and total acidity determinations may point to a reasonable explanation for the observed instability. If a canned product obviously contains gas, then an analysis of this gas may be of great aid in establishing the cause of spoilage. In the majority of cases no chemical tests will be necessary, but the analyst should always consider the possibility that chemical as well as microbiological analyses may be instructive.

D. Sample Preparation

The sample for microbiological studies should be removed with the aid of appropriate instruments, using aseptic techniques. Commonly, an 11-gm sample is used for analysis, but the amount of material examined must be dictated by the sample received. Sometimes, far less than 11 gm of food is available for examination, and appropriate adjustments must be made. In other instances it may be appropriate to remove much larger amounts of the specimen. In any event, that portion of the sample which is to be analyzed must be weighed accurately and to it an appropriate amount of sterile dilution fluid added. The sample should then be blended so as to prepare a uniform emulsion or solution. If the specimen received is completely soluble in water, mechanical blending is not necessary. On the other hand, if the specimen contains considerable amounts of fat, the addition of a wetting agent may facilitate dispersion and emulsification. Generally, the amount of diluting fluid added to the sample will be sufficient to prepare a 1–10 dilution—the addition of 99 ml of diluting fluid to 11 gm of sample. However, with certain types of foods, such as gums and stabilizers, the primary dilution must be higher, owing to the high viscosity at low dilutions. With other materials, preliminary dilution is unnecessary—for example, milk or water. These details notwithstanding, the first step to be taken in the microbiological analysis of a suspect food is the aseptic removal of a representative portion of the sample and the preparation of this in such a form that it can be subjected to quantitative microbiological studies.

E.   Direct Microscopic Examination

A stained smear of the blended sample should always be prepared and examined. Although a gram stain may be used, we have found that a simple methylene blue preparation is generally more satisfactory with concentrated food specimens. If a fatty food is being studied, it may be beneficial to defat the prepared slide with xylene prior to staining. However, this is seldom necessary.

The stained smear should be examined under an oil immersion lens. Under these conditions, the microscopic factor will be quite large. If 0.01 ml of a 1–10 dilution of the sample is spread over 1 sq cm of the slide, it requires approximately 5,000,000 microorganisms per gram of sample if one is to observe an average of 1 microorganism per field. Thus, it is advisable to examine a large number of fields on samples which appear to be negative on cursory examination.

The results of microscopic and culture studies must be carefully correlated. If the microscopic examination of the sample shows evidence of large numbers of microorganisms, whereas culture studies are negative, this is evidence that the organisms seen microscopically were no longer viable at the time the sample was received for analysis. This might suggest that the organisms were present in a raw material and were destroyed during processing or, alternatively, that culinary treatment had rendered the food free of living microorganisms. The latter situation would be expected in an outbreak of *Staphylococcus* food poisoning involving mistreatment of a contaminated food to a sufficient degree that enterotoxin was formed, this followed by adequate cooking of the food to destroy living staphylococci. In such a circumstance, the direct microscopic examination of the food sample should reveal the presence of staphylococci, even though culture studies would yield negative results. Failure to conduct the microscopic examination of the sample would have the effect of obliterating the one link which could connect the specimen with epidemiological findings in connection with the outbreak.

Completely negative findings on the microscopic test would certainly suggest that at no time was the food grossly contaminated with microorganisms. In a processed food this constitutes a measure of the general microbiological quality of the raw materials.

F. Total Count

The analysis of a suspect sample should include a determination of the "total count." We are, however, disinclined to recommend a "standard" procedure for this purpose. The reader is referred to the recent review by Angelotti (1964) concerning the significance of "total counts" in the bacteriological examination of foods. Obviously, where standard methods have been evolved – for example, milk, water, and shellfish – these should be applied. But the same methods are not necessarily applicable to all types of foods nor to all situations. If the objective of the analyst is to enumerate the total flora of the sample, then he must exercise considerable discretion. For example, many of the lactic acid bacteria which develop on cured meat products are incapable of growing on tryptone glucose yeast extract agar, the medium so frequently used in the determination of total plate counts, whereas the commercially available APT agar was specifically developed for the determination of these organisms. We have found with certain highly acid foods, such as French dressing and tomato paste, that the flora of lactic acid bacteria actually develops better on the highly acid LBS agar than on APT, and here only after extended incubation (7 to 10 days) at room temperature. Certainly some judgment must be exercised in the selection of incubation temperature, for with products that are merchandised under refrigeration it seems imprudent to conduct "total count" analyses at 35° to 37°C.

Thus, we would include the determination of "total count" as a general laboratory procedure to be followed in the analysis of a sample suspected of having produced food poisoning. But wide latitude should be given to the analyst in his selection of the method to be used. This must be dictated by a knowledge of the food product under consideration. So also must the ultimate interpretation of results. All too often it is concluded that a particular food was responsible for an outbreak of food poisoning simply because it showed a high total count. As previously mentioned, bacterial food poisoning is highly specific, and there is, to our knowledge, no good evidence that the mere presence of large numbers of microorganisms in a food product will lead to illness if the food is consumed. The total count should be conducted to give a measure of the general microbiological quality of the sample, and the results should be correlated with the epidemiological findings. Again, the interpretation of these results demands knowledge of the microbiological problems of the product under consideration. Under no circumstances would we consider it appropriate to conclude that a particular food had been responsible for poisoning on the sole basis that analysis showed a high total count.

G. Coliforms

Although the coliforms do not produce food poisoning, their position as indicators of fecal contamination makes it advisable that any suspect food be analyzed for the presence of this group.

The coliforms can be determined by direct plating procedures, the selective violet red bile and deoxycholate lactose media being most commonly employed. In both media, the coliforms form red colonies which distinguish them from the colorless colonies produced by non-lactose fermenters. However, atypical colonies may be encountered, making confirmatory tests necessary. For example, in the experience of the authors, fecal streptococci occurring in fermented egg albumen may give rise to small red colonies on deoxycholate agar. Ross and Thatcher (1957) emphasized that care must be taken to record only typical red colonies surrounded by a precipitate of bile when using violet red bile agar. On occasion, food particles may absorb the dyes in these media and produce artifacts that are difficult to distinguish from colonies. Despite these shortcomings, the two media have proved valuable for the quantitative analysis of foods for coliform bacteria. As with most analytical methods, the experience of the analyst in applying the procedure is of considerable importance.

A shortcoming of the direct plating procedures is the fact that they are not applicable to foods with low coliform densities. Generally, no greater than 1 ml of a 1–10 dilution can be used, meaning that the analytical procedure can detect no fewer than 10 organisms per gram. Further, it is generally recognized that the greatest accuracy in direct plating procedures is achieved when the counted plate contains between 30 and 300 colonies.

With foods containing small numbers of coliforms, an MPN (most probable number) procedure must be applied. For this purpose, lauryl sulfate tryptose broth and brilliant green bile lactose broth have been commonly used. For details concerning the application of the MPN technique, the reader is referred to Standard Methods for the Examination of Water and Waste Water (1960). With the MPN procedure, it is possible to determine very small numbers of coliforms; for example, with a five-tube MPN procedure and an inoculum of 10 gm per tube, as few as 0.018 coliform per gram can be detected.

The coliforms are a group of bacteria which, by definition, are gram-negative, nonsporeforming rods that ferment lactose with the production of acid and gas. Within the group are a number of biotypes which are distinguished from one another on the basis of the IMViC reaction (indole, methyl red, Voges-Proskauer, and citrate tests). *Escherichia coli* type I, showing a + + − − pattern, is generally acknowledged to be of fecal origin, whereas *Aerobacter aerogenes* type I (− − + +) is usually of nonfecal origin. In studying the IMViC reaction of naturally occurring coliforms, one is apt to encounter isolates showing a spectrum of patterns intermediate between these extremes, and this leaves a broad area for interpretation of their significance. The EC confirmation test of Hajna and Perry (1943) has been widely used for the detection of type I *E. coli* in water. The test has been found to have definite limitations in its application to food analysis (Tennant *et al.,* 1959; Kelly, 1960; Raj and Liston, 1961).

Despite the acknowledged limitations of analytical procedures for the determination of coliforms, analysis of a suspect food for these organisms is often of considerable value. If the food involved is normally processed at temperatures sufficiently high to kill coliforms, then their presence is evidence of postprocessing contamination or gross under-processing. Further, if the history of the food indicates that it had been cooked by the consumer, then coliform organisms suggest postcooking contamination. If the coliforms present are those generally considered to be of fecal origin, this then suggests the possibility that enteric pathogens may be present. Quantitative aspects of the coliform contamination may also be important, since if the level is high it may indicate not only contamination of the food but also its mistreatment in terms of exposure to incubation temperatures for extended times.

As with other analytical procedures which may be carried out on a suspect food, the interpretation of results demands a knowledge of the normal microbiological flora of the product. For example, grains frequently contain *A. aerogenes,* but this coliform contamination is definitely not of fecal origin and is without public health significance. Similarly, it has been our experience that a variety of fresh vegetables, including celery, lettuce, and carrots, commonly are contaminated with coliforms. We have recently conducted a survey involving the analysis of a large number of coliform-contaminated celery samples which were obtained from retail markets. Each of these was negative for salmonellae in 25 gm. We are led to conclude that such coliform contamination is without sanitary significance. Thus, the presence of coliforms in an uncooked product containing cereal or a prepared dish containing uncooked vegetables cannot be interpreted a priori as evidence of fecal contamination. In such foods the coliform determination is obviously of limited value in assessing sanitary quality.

Procedures for sample collection and laboratory diagnosis of specific enteropathogenic *E. coli* are mentioned in the chapter on miscellaneous organisms.

H. Enterococci

Although foods containing large numbers of enterococci have occasionally been implicated in food poisoning outbreaks, these organisms are of dubious food poisoning significance (Deibel and Silliker, 1963; Shattock, 1962; Greenberg, 1965). However, determination of enterococci as an index of fecal contamination has proved useful in the analysis of water. In this respect, the presence of enterococci is considered to have the same significance as coliforms, but it is felt by some that the relatively high resistance of these bacteria to unfavorable conditions makes them more suitable than coliforms as indicators of fecal contamination. This subject has been discussed in some detail by Buttiaux and Mossel (1961). In the opinion of the authors, the resistance of the enterococci to heat, combined with their ability to grow at temperatures as low as 10°C and their resistance to salt, all mitigate against their use as indicators of fecal contamination of food products. Deibel (1964) has pointed out that in processed meat products, for example, the occurrence of enterococci does not necessarily suggest direct fecal contamination. More recently, Greenberg (1965) reported the regular occurrence of small numbers of fecal streptococci in both bacon and sausage known to have been produced under good sanitary conditions. Combined with this is the known capability of these organisms to grow in such products, even when they are stored under adequate refrigeration conditions. Thus, it is not surprising that such foods may ultimately support exceedingly large populations of enterococci, and it is inappropriate to attach sanitary significance to their presence.

Despite the foregoing, there is continued interest in the detection and enumeration of enterococci in food products. The subject has recently been reviewed by Hall (1964). It is our opinion that sanitary significance should be attributed to the presence of these bacteria in food products only if adequate study has shown that such a relationship exists in the particular product under consideration. Thus, we would be inclined to agree that the presence of enterococci in water purported to be potable is indicative of fecal contamination, whereas the same conclusions would seem unwarranted in products such as cured meats and even pasteurized milk.

I. Yeasts and Molds

These organisms are generally not recognized as etiological agents of food poisoning, albeit the recently recognized mycotoxins may ultimately fall in this category. Nevertheless, the inclusion of a yeast and mold analysis may yield useful information with regard to the history of a sample. The normal spoilage of many foods involves the development of yeasts and/or molds, and the presence of large numbers of these organisms in a suspect food is in some cases, at least, a measure of the storage history of the product. A number of selective media have been used for determination of these microorganisms; however, we have found acidified potato dextrose agar highly satisfactory for routine purposes.

Methods for cultivation of aflatoxin-producing fungi and for detection of toxins are described in the chapter on alimentary mycotoxicoses.

III. ANALYSIS OF FOODS FOR SPECIFIC FOOD POISONING ORGANISMS

A. *Salmonella*

1. INTRODUCTION

Probably no microbiological procedure has presented more difficulties than the analysis of food products for salmonellae. In his early approach to this task, the food microbiologist attempted, unsuccessfully, to apply the methods of the clinical laboratory in his search for *Salmonella* in foods. During the last decade, significant advances have been made in the evolution of procedures specifically designed for the analysis of foods. However, errors of both omission and commission are still common; salmonellae are frequently overlooked, and foods are commonly falsely condemned as contaminated with these organisms. Fortunately, *Salmonella* methodology is being pursued actively in many laboratories throughout the world, and it is to be hoped that the next decade may see the evolution of a "standard" methodology for the analysis of food products.

2. ENRICHMENT PROCEDURES

a. Introduction. When salmonellae occur in food products, they are generally present in small numbers relative to the competing microflora. As a consequence, it is necessary to employ enrichment procedures. For this purpose, tetrathionate (Kauffman, 1935) and selenite F (Leifson, 1936) broths have been used. While these media have been of great utility in the analysis of clinical materials, they have definite shortcomings when used for the direct enrichment of foods. At least two problems are involved: (1) The sample itself may impair the selectivity of the medium (Schneider, 1946; Hurley and Ayres, 1953; Silliker and Taylor, 1958; Sugiyama *et al.*, 1960; North, 1961). Silliker and Taylor (1958) demonstrated that water-soluble food components are partially responsible for the effect. It is of interest to note that feces does not have a detrimental effect on the selectivity of selenite broth; indeed, it appears to enhance the selectivity of the medium (Silliker *et al.*, 1964a,b). (2) The salmonellae found in food products are less capable of initiating growth in these enrichment media. Apparently, exposure to heat, drying, freezing, acids, etc., tends to render the salmonellae in food products physiologically dormant. These two factors—namely, the effect of the food sample on the selectivity of the medium and the physiological dormancy of the organism itself—make necessary the utilization of entirely different analytical procedures than are applied to clinical materials.

Silliker and Taylor (1958) demonstrated that the adverse effects of water-soluble components in the medium could be circumvented by centrifugation of food suspensions in water, with subsequent enrichment of the sediment containing the contaminating organisms. However, this procedure is exceedingly cumbersome and does not overcome the problem of physiological dormancy encountered among the food-type salmonellae.

b. Pre-enrichment. Sugiyama *et al.* (1960) observed increased *Salmonella* recovery when egg samples were pre-enriched in a noninhibitory medium prior to employment of a selective broth. The sensitivity of this procedure was greater than that observed by direct culture of sediments from centrifuged aqueous albumen samples. Subsequently, North (1961) reported that pre-enrichment of egg samples in lactose broth, followed by culture in selenite broth, effected significantly increased *Salmonella* recovery. These findings were confirmed by Taylor and Silliker (1961) and Taylor (1961).

Currently, most laboratories employ a pre-enrichment step in the analysis of food samples for salmonellae. This is followed by subculture in selenite and/or tetrathionate broths. The pre-enrichment procedure appears to perform two functions: (1) It rejuvenates physiologically dormant cells which are unable to initiate growth in selective media. (2) It increases the numbers of salmonellae that are ultimately introduced into the selective media. Only to the extent that the food sample is diluted through pre-enrichment does this procedure overcome the adverse effects which the sample itself may have on the efficiency of the selective medium.

Whereas pre-enrichment in a noninhibitory broth may greatly increase the chances of *Salmonella* recovery, it must be recognized that overgrowth of salmonellae may occur as a result of this procedure. North (1961) recommended a 24-hour pre-enrichment, whereas Montford and Thatcher (1961) suggested a 6-hour incubation of the lactose cultures prior to subculturing in selective media. Silliker *et al.* (1964b), working with dried egg products, found that the time of lactose subculture was critical. Prolonged incubation in the noninhibitory medium may lead to overgrowth by non-salmonellae, particularly coliforms. On the other hand, if the *Salmonella* level in the specimen is low, then early subculture to enrichment broths may "dilute out" the salmonellae and cause a false negative result. While there seems little doubt that pre-enrichment of food samples in a noninhibitory medium promotes recovery of debilitated salmonellae, considerable discretion must be exercised when employing this methodology. If the sample contains a large coliform population, then early subculture from the nonselective medium to an enrichment broth is indicated.

c. Selective Media. A number of modifications in tetrathionate and selenite broths have been recommended. Galton *et al.* (1954) found that the addition of a wetting agent, sodium heptadecyl, to tetrathionate broth aided in dispersion and emulsification of foods containing fat. Galton *et al.* (1955) noted that this agent appears to enhance *Salmonella* growth within a wide range of dilutions and was helpful in the rehydration of dried dog meals. Galton (1961) suggested the addition of brilliant green (1 ml of a 1–1000 solution per 100 ml of tetrathionate enrichment broth) for the isolation of salmonellae from food products.

Modifications in selenite broth have included the addition of cystine (North and Bartram, 1953), brilliant green (Osborne and Stokes, 1955), and sulfapyridine (Osborne and Stokes, 1955).

The addition of cystine to selenite broth is an almost universally accepted modification. We have found that the addition of tergitol and brilliant green to tetrathionate broth improved the selectivity of this medium. However, sulfa drugs in the enrichment broths have not proved similarly efficacious.

Although it has been common for individual investigators to favor one of the two enrichment media for analysis of food samples, the present trend is toward use of both media. Hobbs (1962) reported results obtained on foodstuffs, including frozen whole egg, dried whole egg, frozen egg albumen, frozen boned meat, and coconut. Her data clearly indicate increased recovery when both selenite and tetrathionate broths were used. Further, in foods with low counts and a mixed flora, such as frozen albumen, dried whole egg, and coconut, increased efficiency of *Salmonella* isolation was obtained through the use of nutrient broth as a primary culture medium.

In one of our laboratories (JHS), 2905 food samples have been analyzed for *Salmonella* during the past year and a half. The technique has involved pre-enrichment of 25-gm samples in 250 ml of lactose broth, followed by subculture into both cystine selenite and tetrathionate broths containing tergitol and brilliant green. Salmonellae have been isolated from 222 (7.6%) of these samples. With reference to the positive samples only, salmonellae were detected after enrichment in both selenite and tetrathionate broths with 188 of the samples (84.7%). Selenite only yielded positives from 7 samples (3.15%); similarly, tetrathionate broth alone was positive with 27 samples (12.15%). We do not suggest that these data constitute proof that tetrathionate broth is universally a better medium than selenite broth for the enrichment of food samples. Rather, we feel that the statistics confirm Hobbs' (1962) findings that increased recovery is obtained when both media are used. Certainly, in the analysis of a suspect food sample for salmonellae, both tetrathionate and selenite broths should be inoculated with the sample, and, in addition, pre-enrichment of the specimen in a noninhibitory medium followed by subculture in selenite and tetrathionate broths is indicated.

While we have recommended the use of both selective media, with and without pre-enrichment, for the analysis of suspect food samples, such extensive analysis of a single sample is not indicated in connection with *Salmonella* control programs on specific products. Here we feel that the investigation of increased numbers of samples employing a single selective medium may be productive of more "total positives" and therefore may effect better "total control." Whether pre-enrichment is practiced must be dictated by the nature of the food material. As previously indicated, increased numbers of positive samples are generally detected after pre-enrichment of dried foods; but with moist foods, which generally carry a larger competing flora, pre-enrichment is contraindicated. Again, if a single selective medium is to be used, the choice should be dictated on the basis of experience with the particular food in question.

Although most laboratories practice subculture of enrichment media after 24-hour incubation, there is some evidence that more prolonged incubation may result in further positives. Nottingham and Urselmann (1961), studying *Salmonella* infection in calves and other animals, found that plating after 24 and 72 hours increased the chances of detecting small numbers of salmonellae. Hobbs (1962) obtained 4 to 28% additional positives when enrichment media were subcultured after 3 days of incubation in instances where initial plating at 24 hours yielded negative results. However, Montford and Thatcher (1961), working with egg specimens, found that continued incubation of selenite and tetrathionate for up to 3 days did not in a single instance provide an increase in the number of *Salmonella* but instead tended to promote a heavy growth of coliforms, *Proteus, Pseudomonas,* and other organisms, and fewer recognizable *Salmonella* colonies.

Incubation of enrichment cultures at 37°C is almost universally practiced. However, Harvey and Thomson (1953) found that 43°C incubation was preferable. Hobbs (1962) was unable to demonstrate any advantage to the higher incubation temperature. Further work relative to the influence of incubation temperature is definitely indicated.

3. PLATING MEDIA

Numerous differential and selective media have been developed for the detection of salmonellae. To a large degree the choice between these remains in the province of the analyst himself. For this reason, no attempt will be made to discuss the merits of the innumerable substrates available. Among those widely used are eosin–methylene blue agar, MacConkey's agar, Leifson's deoxycholate agar, bismuth sulfite agar, deoxycholate citrate agar, SS agar, and brilliant green agar. With a few exceptions, the salmonellae grow well on all these media. The salmonellae are differentiated on the basis of their failure to ferment lactose, and the media are so constituted that lactose and non-lactose-fermenting colonies are easily distinguished. However, since a variety of non-salmonellae are also lactose-negative or fail rapidly to ferment lactose, none of these media are completely differential.

In the United States, brilliant green agar has become the medium of choice for the isolation of salmonellae from food products. In other parts of the world other differential media have gained favor. In general, it is recommended that enrichment cultures be streaked on more than one agar medium. As Hobbs (1962) has indicated, this provides different conditions for subculture and further benefits the individual investigator, who may have greater preference for one medium than another. In our laboratories we routinely streak enrichment cultures on brilliant green and SS agars. Results indicate that a significant proportion of *Salmonella*-containing samples would be overlooked if either one of the two media were used exclusively.

4. IDENTIFICATION

Non-lactose-fermenting colonies developing on the differential agar media are suspect salmonellae. However, individual colonies must be picked and subcultured. Pure cultures of the isolates must be subjected to further testing in order to confirm or deny their identity as members of the genus *Salmonella*. In the selection of colonies for further study there is a broad area for subjective judgment on the part of the analyst. The questions of which colonies to select and how many always loom large. There are, of course, no easy answers. Certainly, the chances of detecting salmonellae are increased in direct relationship to the number of colonies selected for study, but ultimately practical considerations must intercede. It is recommended that at least three non-lactose-fermenting colonies be selected from each of the differential media showing them, if this many suspect colonies develop. Experience teaches that certain types of foods present more difficult problems than others in that in some non-lactose-fermenting organisms other than salmonellae comprise a part of their normal flora, for example, pseudomonads in fresh meats and eggs, *Proteus* in rendered animal by-products. In these situations, *Salmonella* colonies may be greatly outnumbered by non-lactose-fermenting colonies other than salmonellae. Only by picking and examining a multiplicity of colonies can the analyst hope to detect *Salmonella* colonies.

It is common practice to subculture suspect colonies in one of a variety of media which are so formulated as to effect further biochemical differentiation. Among the media more commonly used for this purpose are triple sugar iron agar, Kligler's iron agar, Russell's double sugar agar, and dulcitol lactose iron agar. These media have considerable utility, as they screen out some of the non-salmonellae on the basis of biochemical reactions.

In most laboratories the next step in analysis involves further biochemical screening. A wide variety of procedures are available, and no attempt will be made to delineate them here. The reader is referred to

Edwards and Ewing (1962) in which intensive treatment is given to the biochemical differentiation of enteric bacteria. We have found the dulcitol, urea, lysine, and KCN tests most useful. The salmonellae almost uniformly ferment dulcitol and decarboxylate lysine, but fail to hydrolyze urea or to grow in KCN broth.

While biochemical tests are useful screening procedures, it should be emphasized that the ultimate classification of an organism as a member of the genus *Salmonella* is based on antigenic analysis. Many analysts place far too much emphasis on the biochemical aspects of *Salmonella* identification, postponing the inevitable necessity of conducting serological tests in order to reach a final decision. It is our opinion that immunological studies should be carried out at the earliest possible stage in the analysis.

While both O and H agglutinations should be done, the H antigens are far more specific than the O (Edwards and Ewing, 1962). Despite this, all too frequently the analyst satisfies himself with the conclusion that an organism agglutinating in polyvalent O antiserum may be properly designated a *Salmonella*. There is an ubiquitous distribution of non-salmonellae which agglutinate promptly in commercially available polyvalent O antisera. The same may be said, to a lesser extent, for non-salmonellae which agglutinate in group-specific O antisera. This problem is not encountered with H agglutinations. Edwards and Ewing (1962) have stated: "While it is true that there are many H antigens common to the *Salmonella* and Arizona groups, this community of H antigens does not extend to other groups. If an Arizona strain should be reported erroneously as a member of the *Salmonella* group, no harm has been done since both are pathogenic forms and persons infected with either should be similarly treated."

Recently, a procedure designed to permit the earliest possible serological testing was developed (Silliker *et al.,* 1965). The technique entails subculture of non-lactose-fermenting colonies growing on differential agar media directly into infusion broth. These cultures are incubated for 4 to 6 hours, at which time they are sufficiently turbid to permit their use as antigens in H agglutination tests. To the broth culture is added an equal volume of formalinized saline, and the antigen so prepared is added to an equal volume of polyvalent H antiserum prepared by mixing the individual sera in the Spicer-Edwards "kit." To date, over 1000 different *Salmonella* cultures have been tested by this procedure, and only one false positive reaction has occurred, this being with an Arizona-type paracolon isolated from dried meat scrap. Less than 1% of the salmonellae encountered have failed to agglutinate in the polyvalent H antisera, and the majority of these have proved to be nonmotile *Salmonella*. These results indicate the value of serological screening of non-lactose-fermenting colonies. Both false positive and false negative reactions with the rapid polyvalent H agglutination procedure are readily detected when confirmatory O agglutinations and biochemical testing are done.

The final identification (serotyping) of *Salmonella* isolates remains within the province of specialized testing laboratories. However, the errors will be few if it is concluded that a non-lactose-fermenting organism agglutinating in polyvalent H antiserum, fermenting dulcitol, decarboxylating lysine, or failing to hydrolyze urea or grow in KCN broth is a member of the genus *Salmonella*. The rare isolate which fails to agglutinate in H antisera but which shows biochemical conformity with the genus must still be considered a possible *Salmonella,* particularly if it agglutinates in group-specific O antisera. In such instances, complete antigenic analysis is certainly necessary before a definitive conclusion can be drawn.

5.   FLUORESCENT ANTIBODY TECHNIQUES

Recent work (Haglund *et al.,* 1964; Silliker *et al.,* 1966) indicates that fluorescent antibody techniques hold considerable promise as a means of detecting salmonellae in food products. These procedures, of necessity, combine culture techniques with the fluorescent antibody procedure. A considerable saving in time accrues to this procedure; however, an obvious disadvantage is the fact that the organism itself is not isolated. Silliker *et al.* (1965) have found that the fluorescent antibody technique is far more sensitive than conventional culture techniques for the detection of *Salmonella* in liquid egg products. This may likewise prove true in other food products, particularly ones with a large non-*Salmonella* gram-negative flora.

The analysis of food products for salmonellae is diagramed in Fig. 1. It will be noted that the procedure provides for pre-enrichment of the specimen or direct inoculation of selective medium. If the fluorescent antibody technique is to be employed, the antigen is prepared by subculturing the enrichment broths into BHI broth which is then incubated for a suitable length of time. To the latter is added formalinized saline, and a slide prepared from this antigen is examined for *Salmonella* by the indirect fluorescent antibody technique (Silliker *et al.,* 1966).

It will be noted that Fig. 1 provides for polyvalent H screening of non-lactose-fermenting colonies developing on differential media. Alternatively, such colonies may be inoculated directly into a medium, such as TSI agar, from which differential biochemical and serological testing procedures can be initiated. It is emphasized, once again, that the ultimate classification of an isolate as a *Salmonella* must rest upon serological tests and that such procedures must include H agglutinations. When complete serotyping of a *Salmonella* isolate is mandatory, this work must of necessity be conducted by a qualified typing laboratory.

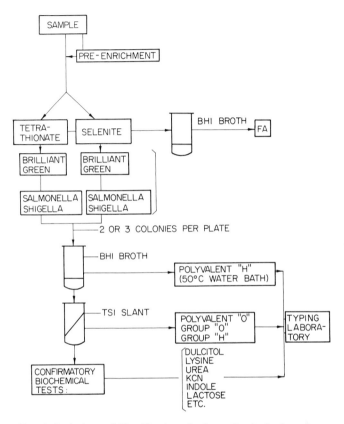

FIG. 1. Isolation and identification of salmonellae in food products.

B. Staphylococci

A food causing *Staphylococcus* food poisoning must, at some time in its history, have supported sufficient growth of enterotoxin-producing staphylococci to allow the accumulation of toxin. This does not happen until extensive staphylococcal growth has occurred, and foods causing *Staphylococcus* food poisoning will uniformly contain millions of staphylococci per gram. However, if the food has been heated subsequent to the formation of enterotoxin, then all viable staphylococci may be destroyed, and presumptive diagnosis must be based on microscopic observations and clinical findings. However, the finding of large numbers of staphylococci in a suspect food, by either cultural or microscopic procedures, does not indicate a priori a *Staphylococcus* etiology. This must rest on the demonstration of enterotoxin in the food product. If large numbers of coagulase-positive staphylococci are found in the food, this constitutes strong presumptive evidence, since virtually all enterotoxin-producing organisms are coagulase-positive (Evans and Niven, 1950). But not all coagulase-positive staphylococci are enterotoxin producers. Further, if presumptive conclusions are based on microscopic demonstration of *Staphylococcus*-like cocci, it should be realized that coagulase-negative organisms will show the same morphology as will also the closely related nontoxigenic organisms in the genus *Micrococcus*.

Innumerable media have been developed for the selective isolation of staphylococci from foods and clinical materials. Methods for the isolation and enumeration of staphylococci in foods have been reviewed by Crisley (1964). More recently, Crisley *et al.* (1965) have reported a statistical study on the performance of the media commonly used for the isolation of staphylococci from foods. Methods for detection of staphylococci and for phage typing are also mentioned in the chapter on staphylococcal intoxications. In view of these excellent references, no attempt will be made to evaluate the merits of the available media for *Staphylococcus* isolation.

Methods for the detection of enterotoxin are discussed elsewhere in this book. Ideally, the suspect food should be analyzed for the presence of enterotoxin, and, if this is found, then a diagnosis of *Staphylococcus* food poisoning is of course appropriate. Alternatively, if large numbers of staphylococci are present and it can be demonstrated that these are capable of producing enterotoxin, then *Staphylococcus* food poisoning is certainly indicated.

Unfortunately, relatively few laboratories are currently capable of conducting analyses for enterotoxin. If, however, a suspect sample is shown to contain large numbers of coagulase-positive staphylococci (for example, in the millions per gram), then this fact, combined with compatible clinical findings, permits a diagnosis of *Staphylococcus* food poisoning with a reasonable degree of certainty.

All too frequently it is concluded that a suspect food has caused *Staphylococcus* food poisoning simply on the basis of qualitative demonstration of coagulase-positive staphylococci. Quantitative studies are an absolute necessity, since virtually any food product handled by human hands or exposed to the air may contain small numbers of coagulase-positive staphylococci. Only by demonstrating the presence of large numbers of these organisms is it permissible to assume a *Staphylococcus*

enterotoxin etiology, in the absence of actual demonstration of enterotoxin in the food product.

Further, it should be emphasized that organisms in the genus *Micrococcus* are widely distributed in nature and are commonly found as a part
of the normal flora in many food products. These organisms will grow on
many of the media commonly used for the selective isolation of staphylococci and will form colonies which are frequently indistinguishable
from those produced by staphylococci. Further, microscopic studies do
not permit one to distinguish between micrococci and staphylococci,
since both genera consist of bacteria which are gram-positive cocci,
forming irregular grape-like clusters.

The seventh edition of Bergey's Manual (Breed *et al.*, 1957) clearly
distinguishes the two genera on the basis that the micrococci are aerobic
organisms and the staphylococci are facultatively anaerobic. If colonies
being subcultured for coagulase tests are introduced into a semisolid
medium, such as thioglycolate broth, then a separation between micrococci and staphylococci is easily made. Characteristically, the staphylococci will grow luxuriantly throughout the medium; whereas micrococci
will exhibit only scanty growth confined to the surface of the medium.
The latter organisms—namely, micrococci—can be eliminated from
further consideration, since they are never associated with food poisoning. The former organisms must be subjected to coagulase testing so
as to distinguish between *Staphylococcus aureus,* which is coagulase-
positive and may produce enterotoxin, and *Staphylococcus epidermidis*
which is coagulase-negative and does not produce enterotoxin. Although
it is reported that *Staphylococcus aureus* can be further distinguished
from *Staphylococcus epidermidis* on the basis of the ability of the former
to ferment mannitol (Evans and Niven, 1950), we have not found this to
be a uniform characteristic. Indeed, we have encountered coagulase-
positive staphylococci which proved incapable of the anaerobic fermentation of mannitol, and coagulase-negative staphylococci which promptly
fermented the sugar. Thus, as in clinical bacteriology, the coagulase test
remains as the most reliable tool for distinguishing between pathogenic
and nonpathogenic staphylococci. In the food laboratory, valuable epidemiological information can be obtained as the result of routinely
segregating the facultative staphylococci from the aerobic micrococci.

C. Clostridia

1. INTRODUCTION

Two members of the genus *Clostridium, C. perfringens* and *C. botu-*

linum, must be considered in the laboratory analysis of any sample associated with a food poisoning outbreak. Both organisms occur widely in nature. They are inert in the spore stage and must grow extensively in the food product if its consumption is to elicit symptoms. The presence of small numbers of *C. perfringens* in foods is to be expected. Thus, the presence of this organism should not be interpreted as automatically signifying the cause of an outbreak. On the other hand, *C. botulinum* spores are seldom found in foods, and their detection must be viewed with alarm. Any suspect food should be analyzed presumptively for these organisms by determining the level of putrefactive anaerobic spore-forming bacteria present. Processed foods contain typically very low levels of clostridia. The presence of 100 or more per gram is indicative of some failure in sanitation during manufacture or in temperature control during storage.

The nature of the food sample must be considered when carrying out a putrefactive anaerobe analysis, in order that the proper culture medium may be utilized. Wheaton and Pratt (1961), in a classic study, demonstrated that clostridial spores severely injured by thermal processes required highly nutritious "home-made" media for germination and growth on subculture. The need for such media does not exist with spores subjected to mild (pasteurizing) thermal processes. Greenberg *et al.* (1958) showed that in certain cases it was actually better to employ nutritionally poorer media to detect putrefactive anaerobes in spoiled products containing saprophytic contaminants in large numbers, to prevent overgrowth and false negative results. Thus, should the sample be a spoiled product, Difco's peptone colloid medium would be considered superior to freshly prepared liver infusion broth. On the other hand, should a container of, for example, green peas be suspected as having been underprocessed, liver infusion broth would be the medium of choice.

2. *Clostridium perfringens*

Although *C. perfringens* is not accepted universally as a cause of food poisoning, there is sufficient evidence to include it in any listing of harmful microorganisms. The organism is particularly ubiquitous. It is found in soil, water, milk, dust, sewage, and the intestinal tract of man and animal. It has been considered by some authorities so widespread in nature as to be impossible to exclude from foods. Hence, arguments exist that some if not all food poisoning outbreaks attributed to *C. perfringens* are the result of other agents, with *C. perfringens* considered responsible solely on the basis of circumstantial evidence. Most authorities feel that *C. perfringens* must be considered as a food poisoning agent and that *C.*

perfringens food poisoning is the result of lecithinase-producing strains growing in lecithin-rich foods.

Hobbs *et al.* (1953) reported that heat-resistant strains dominated in foods associated with outbreaks. About 90% of persons who had eaten suspect food had heat-resistant *C. perfringens* in their feces, in contrast to only 5% in the feces of normal persons. Smith (1964) argued that the question of heat resistance was probably simply a matter of whether contamination had occurred before or after cooking. Hobbs *et al.* (1953) attribute most cases of *C. perfringens* food poisoning to meat or poultry dishes that are allowed to cool slowly overnight.

There is general agreement that large numbers of organisms (at least in the millions per gram) are required before enteric symptoms will occur.

Foods neutral in pH, containing meat or poultry ingredients, previously cooked and kept for some hours at temperatures between 20° and 50°C, are prime suspects when symptoms suggest *C. perfringens* food poisoning. *Clostridium perfringens* food poisoning is discussed in detail in the chapter on *Clostridium perfringens* and *Bacillus cereus*.

Clostridium perfringens are nonmotile, nitrate-reducing, sulfite-reducing, nonhemolytic, gram-positive anaerobic sporeforming rods.

Some authorities recommend enrichment culturing of suspect food samples. Thatcher (1964) suggests a primary enrichment in freshly exhausted deep tubes of meat broth for not more than 5 hours at 45°C. Because extremely large numbers of *C. perfringens* are required to elicit enteric symptoms and, as previously pointed out, the organism is an expected contaminant, pre-enrichment is generally unnecessary.

Sulfite–polymycin–sulfadiazine agar (SPS agar) (Angelotti *et al.,* 1962) has been demonstrated to be effective for the quantitative recovery of *C. perfringens* from food samples and feces (Southworth and Strong, 1964). The basal medium contains 1.5% Bacto tryptone, 1.0% Bacto yeast extract, 1.5% agar, and 0.05% iron citrate. The medium is adjusted to pH 7.0 and heat-sterilized. Just before use, the following Seitz-filtered solutions are added per liter of basal medium: 5 ml of fresh 10% sodium sulfite ($NA_2SO_3 \cdot 7H_2O$), 10 ml of an 0.1% solution of polymyxin B sulfate, and 10 ml of a solution of sodium sulfadiazine containing 12 mg/ml.

After inoculation with sample material, SPS agar plates are incubated for 48 to 72 hours at 37°C in an anaerobic jar. Plastic film pouches (Bladel and Greenberg, 1965) can be substituted for petri plates and incubated directly without recourse to anaerobic jars. Colonies on SPS agar are black. Blood agar plates should also be prepared. Blood agar colonies are nonhemolytic, 2 to 5 mm in diameter, and smooth, with entire margins. Motility and nitrate-reducing capacity are tested by incubation of suspect colony stabs in motility-nitrate agar (Bacto-nitrate broth plus 0.3% agar). Lecithinase activity is tested by means of the Nagler reaction, using the egg yolk medium described by McClung and Toabe (1947).

Diagnostic procedures are discussed in more details in the chapter on *Clostridium perfringens* and *Bacillus cereus*.

3. *Clostridium botulinum*

Although the American public accepts philosophically the high mortality rates inherent in such afflictions as cancer and heart disease, occasional deaths from botulism evoke great publicity and public outcry. In 1963, commercially produced foods were implicated in two major outbreaks in the United States. The first of these, occurring in a suburb of Detroit, Michigan, resulted in the death of two women who ate sandwiches prepared from commercially packaged tuna fish. Later in the year, commercially packaged smoked fish killed eight persons in two southern states. The tuna canning industry suffered tremendous economic losses following the outbreak. The smoked fish industry was virtually destroyed.

These incidents illustrate forcibly, to industry and regulatory agencies as well, that the word "botulism" is charged with emotion and should not be used or taken lightly. Positive diagnosis by both clinician and laboratory technician must be the result of laboratory proof. Diagnosis of botulism based solely on symptomology is responsible for some "documented cases" which have found their way into the literature and have become a part of the historical, statistical record of botulism. The 1963 outbreaks have served to alert the medical profession to the existence of botulism. It is important that the technician bear in mind that detection of toxin from the incriminated foodstuff or in a blood sample of the patient is the minimum evidence necessary for laboratory diagnosis.

Historically, foods in which the botulinal organisms grow poorly are the usual sources of human botulism. In such substrates, insufficient growth to produce obvious spoilage results in the consumer's not being warned of possible hazard. However, modern packaging and distribution procedures have evolved so many innovations as to render practically all foods possible, if not probable, sources of botulism.

When foods of marine origin are involved, the technician should suspect type E, the most insidious member of the *Clostridium botulinum* spectrum. Type E organisms are nonproteolytic and are capable of growth and toxin production at temperatures well within the range of acceptable refrigeration. One must not lose sight of the possibility that type E organisms might occur in nonmarine foods or also that other types such as A and B have often been implicated in botulism outbreaks involving seafoods.

The single, most meaningful test involves the use of intraperitoneal injection of suspect material or subcultures into white mice. Depending on the strain involved, death followed by typical paralytic symptoms will occur within about 5 hours up to about 3 days. Simple challenge of mice is insufficient evidence of botulinal toxin. Many of the clostridia produce toxins lethal to mice. Specific tests on presumptively positive samples are thereby required.

4. BACTERIOLOGICAL PROCEDURE

Recommended procedures for the enrichment, quantitation, and identification of food poisoning clostridia are diagramed in Fig. 2. (See also the chapter on *C. botulinum* type E and the chapter on *C. perfringens* and *B. cereus*.)

a. Enrichment and Quantification. The sample is inoculated into a suitable medium for primary enrichment and quantification. Severely processed ("sterile") foods should be subcultured in freshly prepared complex media, such as pork pea infusion broth or liver infusion broth (Wheaton and Pratt, 1961). Unprocessed or lightly processed foods should be subcultured in less complex media such as differential reinforced clostridial medium DRCM (Gibbs and Freame, 1965) or modified peptone colloid broth (Greenberg *et al.,* 1966). Three- or five-tube most-probable-number determinations may be carried out with the appropriate medium in order to give a quantitative estimate of the clostridial population. Alternatively, the less cumbersome albeit less accurate decimal dilution technique may be employed. A portion of the original sample preparation is set aside for subsequent toxin evaluation. All tubes should be incubated at 32°C for 72 hours.

b. Pasteurization. Aliquots from tubes showing growth are diluted in phosphate buffer, physiological saline, peptone water, or other diluent and used for inoculation of modified Angelotti SPS agar (Greenberg *et al.,* 1966) contained in petri dishes or anaerobic pouches. The remaining contents of the positive enrichment tubes are then pasteurized for 20 minutes at 80°C before dilution and subculture. The inoculated plates are incubated for 48 to 72 hours at 32°C.

c. Isolation. Two or three well-isolated black colonies from each plate or pouch are "picked" in freshly prepared tubes of DRCM or modified peptone colloid broth and incubated for 24 to 48 hours at 32°C.

Clostridium perfringens is ultimately identified by using procedures previously described.

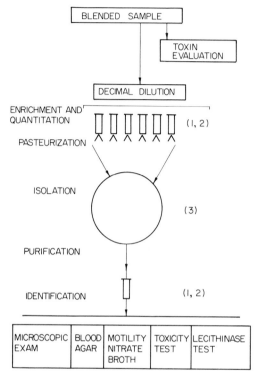

FIG. 2. Analysis for clostridia of food poisoning significance. (1) Severely processed ("sterile") foods are subcultured in freshly prepared complex media, such as pork pea infusion, beef infusion, or liver infusion. (2) Unprocessed or lightly processed foods are subcultured in less complex media, such as DRCM (Gibbs and Freame, 1965) or modified peptone colloid broth (Greenberg *et al.*, 1966). (3) Modified Angelotti SPS agar (Greenberg *et al.*, 1966). (4) DRCM or peptone colloid broth. *Clostridium perfringens:* Short gram-positive rods with square ends. Spores do not swell the cell. Nonhemolytic, nonmotile. Reduces nitrate, toxic to mice, nonprotectible with botulinal antitoxins. Food poisoning strains usually lecithinase-positive. *Clostridium botulinum:* Short to medium gram-positive rods with rounded ends. Spores swell the cell. Hemolytic (type E only slightly). Motile. Reduces nitrate. Usually lecithinase-negative (type E positive). Toxic to mice. Protectible with type specific antitoxins.

5. BOTULINAL TOXIN EVALUATION

The isolation procedure is tedious and, most important, time-consuming. It is of practical value, therefore, to attempt toxicity evaluations by using the original enrichment tubes and primary subcultures. However, the technician must be aware that mixed cultures often give equivocal toxicity results. If clear-cut toxicity patterns can be established at some early point, there is obvious value. However, ultimate reliance can be placed on results obtained with "isolation" subcultures.

We are interested in knowing two facts about the food sample suspected of having caused botulism. We must attempt to verify, first, the presence of botulinal toxin and, second, the presence of *C. botulinum* organisms (Fig. 3).

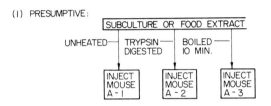

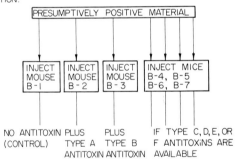

Fɪɢ. 3. Detection of botulinal toxin.

Aliquots of the prepared food sample and isolated subcultures should be set aside for heat inactivation and trypsin digestion. Heat inactivation is accomplished by boiling for 10 minutes. Trypsin digestion is considered to be of value in activating type E toxin and therefore should routinely be carried out in testing foods of marine origin. The procedure originally devised by Duff *et al.* (1956) has been adapted by several workers. Overnight incubation of the sample or subculture with 1% trypsin is both convenient and effective. Bonventure and Kempe (1959) demonstrated enhanced toxicity with young (12-hour) type A and B cultures. However, 96-hour cultures showed no trypsin activation and, indeed, suffered a slight loss in toxicity. Schmidt (1964) has demonstrated a fourfold activation of a proteolytic type B strain isolated from canned liver paste. It is possible, therefore, that including trypsinized cultures or samples routinely in the toxicity evaluation procedure can be of considerable value.

Harmon and Kautter (1967) recommend the addition of 0.1% trypsin to enrichment media for type E. Besides activating toxin and destroying bacteriocin-like substances, trypsin in the growth medium also appears to prevent formation of other antagonistic factors.

White mice, usually 12 to 25 gm, are injected intraperitoneally with 0.5 ml. of the untreated, heat-inactivated, or trypsinized material. Death of the control and/or trypsinized animal and survival of the mouse challenged with the heat-inactivated material is considered presumptive evidence of botulinal toxin. Symptoms of toxicity (hind-quarter paralysis and labored breathing) may develop as early as 3 hours or as late as 48 hours after injection, depending on the amount of toxin present in the challenge dose. Death usually occurs within 24 hours with type E toxin and within 3 days with types A and B. The test animal should be observed 96 hours before considering the test negative.

Confirmation of presumptively positive material is carried out by means of protection tests. Each of 4 mice is injected with 0.5 ml of subculture or prepared sample. One mouse receives no antitoxin and thus serves as an unprotected control. The other mice receive 0.5 ml of A and B antitoxins. Additional mice should be used if C, D, E, and F. antitoxins are available. Only the animal protected with the antitoxin corresponding to the botulinal toxin present in the challenge dose should survive. If all the animals die within 96 hours, the test should be repeated with diluted material to rule out the possibility of having overwhelmed the antotoxin dose administered to the protected animals.

The toxin content of a food sample can be calculated by establishing how much the sample can be diluted before a challenged animal survives. For example, a sample is diluted 1:10 in gelatin phosphate buffer. Death occurs after injection of 0.5 ml. Tenfold dilutions of the original 1 : 10 blend result in death through the 1 : 1000 but not the 1 : 10,000 dilution. Multiplying the 1 : 10 original dilution factor by 1000 and by 2 (because our challenging dose was 0.5 ml), we arrive at a toxicity value of approximately 20,000 MLD. More precise determinations can be made using larger numbers of mice and closer dilution intervals.

The mouse test for detection of botulinal toxins is sensitive, but quantitation of toxin by the mouse test is cumbersome and expensive. Attempts are being made to apply other techniques (hemagglutination, bentonite flocculation, gel diffusion) for detection and quantitation of botulinal toxins (Johnson *et al.,* 1966; Vermilyea *et al.,* 1967). A fluorescent antibody technique for detection of toxin in the individual bacterial cells has been described by Inukai and Riemann (1967). Fluorescent antibody technique based on somatic antigens has been used experimentally for identification of *C. botulinum* in cultures (Walker and Batty, 1964; Boothroyd and Georgala, 1964).

D. *Bacillus cereus*

While food poisoning incidents involving *Bacillus cereus* have been well-documented in Europe (particularly Norway), little evidence exists that the organism is of consequence as a cause of food poisoning in North America. Thus, determination of the presence or level of this organism is not included in a routine examination. *Bacillus cereus* food poisoning is discussed in the chapter on *Clostridium perfringens* and *Bacillus cereus.*

E. Other Food-borne Disease Agents

Techniques for the detection of *Vibrio parahaemolyticus,* viruses, fungal toxins, parasites, and poisonous animals and plants are mentioned elsewhere in this book.

REFERENCES

Agar, E. A., Nelson, A. E., Galton, M. M., and Boring, J. R. (1964). Recent Outbreaks of Egg-borne Salmonellosis and Implications for Their Prevention. Presented at the National Convention, American Public Health Association, New York.
Angelotti, R. (1964). *In* "Examination of Foods for Enteropathogenic and Indicator Bacteria" (Keith H. Lewis and Robert Angelotti, eds.), U.S. Department of Health, Education, and Welfare, Public Health Service.
Angelotti, R., Hall, H. E., Foter, M. J., and Lewis, K. H. (1962). *Anal. Microbiol.* **10,** 193.
Bladel, B. O., and Greenberg, R. A. (1965). *Appl. Microbiol.* **13,** 281.
Bonventure, P. F., and Kempe, L. L. (1959). *J. Bacteriol.* **78,** 892.
Boothroyd, M., and Georgala, D. L. (1964). *Nature* **202,** 515.
Breed, R. S., Murray, E. G. D., and Smith, N. R. (1957). "Bergey's Manual of Determinative Bacteriology," 7th ed. Williams and Wilkins, Baltimore.
Buttiaux, R., and Mossel, D. A. A. (1961). *J. Appl. Bacteriol.* **24,** 353.
Crisley, F. D. (1964). *In* "Examination of Foods for Enteropathogenic and Indicator Bacteria" (K. H. Lewis and R. Angelotti, eds.), U.S. Department of Health, Education, and Welfare, Public Health Service.
Crisley, F. D., Peeler, J. T., and Angelotti, R. (1965). *Appl. Microbiol.* **13,** 140.
Deibel, R. H. (1964). *Bacteriol. Rev.* **28,** 330.
Deibel, R. H., and Silliker, J. H. (1963). *J. Bacteriol.* **85,** 827.

Duff, J. T., Wright, C. G., and Yarinsky, A. (1956). *J. Bacteriol.* **72,** 455.

Edwards, P. R., and Ewing, W. H. (1962). "Identification of Enterobacteriaceae." Burgess, Minneapolis.

Evans, J. B., and Niven, C. F. (1950). *J. Bacteriol.* **59,** 548.

Food Protection Committee (1964). An Evaluation of Public Health Hazards from Microbiological Contamination of Foods. *Natl. Acad. Sci.–Natl. Res. Council Publ.* **1195.**

Galton, M. M. (1961). *Proc. 65th Am. Meeting U.S. Livestock Sanitary Assoc.,* pp. 434–440.

Galton, M. M., Scatterday, J. E., and Hardy, A. V. (1952). *J. Infect. Diseases* **91,** 1.

Galton, M. M., Lowery, W. D., and Hardy, A. V. (1954). *J. Infect. Diseases* **96,** 232.

Galton, M. M., Harless, M., and Hardy, A. V. (1955). *J. Am. Vet. Med. Assoc.* **126,** 57.

Gibbs, B. M., and Freame, B. (1965). *J. Appl. Bacteriol.* **28,** 95.

Greenberg, R. A. (1965). *In* "Activities Report XVII, Research and Development Associates," p. 62.

Greenberg, R. A., Silliker, J. H., and Basa, K. B. (1958). *Bacteriol. Proc.* p. 17.

Greenberg, R. A., Bladel, B. O., and Zingelmann, W. J. (1966). *Appl. Microbiol.* **14,** 223.

Haglund, J. R., Ayres, J. C., Paton, A. M., Kraft, A. A., and Quinn, L. Y. (1964). *Appl. Microbiol.* **12,** 447.

Hajna, A. A., and Perry, C. A. (1943). *Am. J. Public Health* **33,** 550.

Hall, H. E. (1964). *In* "Examination of Foods for Enteropathogenic and Indicator Organisms" (K. H. Lewis and R. Angelotti, eds.), p. 13. U.S. Department of Health, Education, and Welfare, Public Health Service.

Harmon, S. M., and Kautter, D. A. (1967). *Bacteriol. Proc.* p. 5.

Harvey, R. W. S., and Thompson, S. (1953). *Monthly Bull. Min. Health Lab.-Serv.* **12,** 149.

Hobbs, B. C. (1962). *In* "Chemical and Biological Hazards in Foods" (J. C. Ayres, A. A. Kraft, H. E. Snyder, H. W. Walker, eds.), p. 224. Iowa State University Press.

Hobbs, B. C., Smith, M. E., Oakley, C. L., Warrack, G. H., and Cruickshank, J. C. (1953). *J. Hyg.* **51,** 75.

Hurley, N. A., and Ayres, J. C. (1953). *Appl. Microbiol.* **1,** 302.

Inukai, Y., and Riemann, H. (1967).

Johnson, H. M., Brenner, K., Angelotti, R., and Hall, H. E. (1966). *J. Bacteriol.* **91,** 967.

Kauffman, F. (1935). *Z. Hyg.* **117,** 26.

Kelly, C. B. (1960). Robert A. Taft San. Engr. Cent. (Cincinnati), Technical Report F60-2.

Leifson, E. (1936). *Am. J. Hyg.* **4,** 423.

McClung, L. S., and Toabe, R. (1953). *J. Bacteriol.* **53,** 139.

Montford, J., and Thatcher, F. S. (1961). *J. Food Sci.* **26,** 510.

Niven, C. F. (1962). *In* "Chemical and Biological Hazards in Foods" (J. C. Ayres, A. A. Kraft, H. E. Snyder and H. W. Walker, eds.), p. 330. Iowa State University Press.

North, W. R. (1961). *Appl. Microbiol.* **9,** 188.

North, W. R., and Bartram, M. T. (1953). *Appl. Microbiol.* **1,** 130.

Nottingham, P. M., and Urselmann, A. J. (1961). *N. Zealand J. Agr. Res.* **4,** 449.

Raj, H., and Liston, J. (1961). *Appl. Microbiol.* **9,** 171.

Ross, A. D., and Thatcher, F. S. (1957). *Food Technol.* **12,** 369.

Schmidt, C. F. (1964). Personal communication.

Schneider, M. D. (1946). *Food Res.* **11,** 313.

Shattock, P. M. F. (1962). *In* "Chemical and Biological Hazards in Food" (J. C. Ayres, A. A. Kraft, eds.), p. 303. Iowa State University Press.

Silliker, J. H., and Taylor, W. I. (1958). *Appl. Microbiol.* **6,** 228.

Silliker, J. H., Deibel, R. H., and Fagan, P. T. (1964*a*). *Appl. Microbiol.* **12**, 100.

Silliker, J. H., Deibel, R. H., and Fagan, P. T. (1964*b*). *Appl. Microbiol.* **12**, 224.

Silliker, J. H., Fagan, P. T., Chiu, J. Y., and Williams, A. (1965). *Am. J. Clin. Pathol.* **43**, 548.

Silliker, J. H., Schmall, A., and Chiu, J. Y. (1966). *J. Food Sci.* **31**, 240.

Smith, L. D. S. (1964). Personal communication.

Southworth, J. M. L., and Strong, D. H. (1964). *J. Milk Food Technol.* **25**, 205.

"Standard Methods for the Examination of Water and Waste Water," 11th ed. American Public Health Association, New York.

Stokes, J. L., and Osborne, W. W. (1955). *Appl. Microbiol.* **3**, 217.

Sugiyama, H., Dack, G. M., and Lippitz, G. (1960). *Appl. Microbiol.* **8**, 205.

Taylor, W. I. (1961). *Appl. Microbiol.* **9**, 487.

Taylor, W. I., and Silliker, J. H. (1961). *Appl. Microbiol.* **9**, 484.

Tennant, A. D., Reid, J. E., and Rockwell, L. J. (1959). Canadian Department of National Health and Welfare. Manuscript reports of Labotatory of Hygiene No. 59-6.

Thatcher, F. S. (1964). *In* "Laboratory Procedures of the Food and Drug Directorate of Canada."

Vermilyea, B. L., Ayres, J. C., and Walker, H. W. (1967). *Bacteriol Proc.* p. 4.

Walker, P. D., and Batty, I. (1964). *J. Appl. Bacteriol.* **27**, 140.

Wheaton, E., and Pratt, G. B. (1961). *J. Food Sci.* **26**, 1.

CHAPTER XII | FOOD PROCESSING AND PRESERVATION EFFECTS

Hans Riemann

I. INTRODUCTION

The original purpose of food preservation was to prevent spoilage, but the preservation also involved preparation and caused considerable changes in the nature of the food. Most foods preserved by traditional methods (drying, curing, fermentation) are rather different from the fresh raw material, and some food products are hardly consumed in their fresh state any more (sardines).

The outstanding characteristic of the most recent development in food processing has been a more extensive transfer of food preparation from the home to the factory. As a result, convenience foods are produced which require a minimum of preparation in the homes and which resemble the usual "fresh" products as much as possible. A wide variety of such prepared "ready-to-cook/ready-to-eat" foods are being manufactured in fewer but larger places and distributed over larger areas than before. Two-thirds of the approximately 8000 food items now available in supermarkets in the United States were not available before World War II or are radically modified prewar products (Food Protection Committee, 1964). Considerable concern has been expressed with regard to the safety of some of the new foods such as frozen precooked foods, foods sold from vending machines, and mildly processed "moist" foods. The question has been asked whether some of the new processing and storage methods such as aseptic packaging and vacuum packaging may not fail to give protection against disease organisms and reduce the safety of such foods. The outbreaks of botulism caused by vacuum-packed smoked fish a few years ago illustrated that it is necessary to evaluate carefully any new technology. Another example—where the risk is more difficult to predict—is mentioned in the chapter on parasites. Here it is suspected that a new practice of packing uneviserated fish in ice may increase the risk of infection with a parasite.

A large proportion of food poisoning cases are caused not by commercially prepared foods but by foods prepared from raw materials in the home. A recent survey of food poisoning in Denmark from 1954 to 1963 showed that food processing plants were implicated in less than 25% of the outbreaks (Horne-Jensen, 1967). Faulty preparation technique applied in kitchens (institutional and private) or unexpected contamination of the raw food material was responsible for the majority of the outbreaks. Such observations stress two other important factors in food safety, namely, education of the consumer and control of contamination of raw food materials.

Education of the consumer is an important task, but it will not be discussed further. Control of contamination of raw materials is a very complex problem, since it involves the food producing industries as well as the processing industry. However, excellent results can be achieved, as is illustrated in the case of milk. Systematic efforts to improve the hygiene of milk production and to eliminate disease problems in milk cows have done much to reduce contamination of milk. The subsequent pasteurization in processing plants eliminates practically all milk-borne disease organisms which may be present. The result is that milk, although an excellent growth medium for microorganisms, is a very safe food which accounts for less than 4% of all cases of food poisoning in the United States. It seems likely that elaborate control systems will be introduced for other raw food materials in order to control salmonellosis and other food-borne diseases.

II. HEAT PROCESSING

A. General Considerations

1. MECHANISM OF HEAT INACTIVATION

When bacterial cells or spores are exposed to heat, the proportion surviving at any time can be plotted against the heating time to give a survival curve. This curve is often exponential, and a straight line is obtained when the logarithms of survivors are plotted against time in arithmetic units. This type of curve is typical for first-order chemical reactions, and it has therefore been assumed that death of microbial cells is due to a single lethal event occurring at random. The event could be denaturation of an essential protein.

However, the cell contains many molecules, probably at least 10^4 of each essential protein and it is difficult to imagine that destruction of just one molecule causes death unless this molecule is an essential part of the genetic mechanism (Wood, 1956). A number of studies indicate that death is not due to a single event.

The killing of microbial cells is an experimentally defined term and is not an all-or-none process. Viability depends on the medium, the diluent, and the incubation temperature used for recovery of heat-treated cells. It is a common experience that cells which have been exposed to heat are more fastidious than nonheated cells. This has been observed with vegetative cells (Hansen and Riemann, 1963) as well as with bacterial spores. Soluble starch or charcoal in the recovery medium increases viability of heat-treated *Clostridium botulinum* spores (Olsen and Scott,

1950), and recovery may be better at 24°C and 31°C than at 37°C (Williams and Reed, 1942).

Thus, heat-treatment may damage cells with the result that they behave as if they were dead in one medium but show viability under other conditions.

Heat damage of bacterial spores may result in delayed germination, so-called dormancy. The germination-stimulating effect of a mild heat shock was mentioned in the chapter on botulism. If the heat shock is severe, the spores may become dormant. It is not possible to draw a precise line between heat-treatments that activate spores and treatments that will result in dormancy. The effect of a heat-treatment depends on the resistance of the spores as well as on the environment in which the spores are heated and incubated. Dormancy has been known for a long time. Esty and Meyer observed that heated *C. botulinum* spores could remain dormant in laboratory media for a year at 36° to 37°C, and Dickson (1928) recorded 6 years' dormancy. Dormancy can be partially or completely eliminated in appropriate media by adding charcoal or soluble starch. The mechanism involved is unknown.

Recent studies of vegetative bacterial cells exposed to heat stress support the view that death is a gradual and partially reversible process. Heating causes a breakdown of ribosomes which may be followed by substrate-accelerated death (Strange and Shon, 1964) or by recovery if the cells (staphylococcus) are placed in a suitable medium (Sogin and Ordal, 1967). Heating also results in loss of the osmotic function of bacterial cells, and an exudate containing peptides, amino acids, and nucleic acids leaks out. Heat resistance can be increased when this exudate is added to dilute suspensions of cells (Prudhomme, 1966). The increased resistance of cells in dense suspensions (the so-called population effect) is probably due to the protective effect of cell exudate.

Progressive changes have also been observed in spores exposed to high temperatures (Hunnell, 1961).

These observations indicate that microbial cells when exposed to heat are killed by the joint action of a number of events. This also seems reasonable under the assumption that an economically built cell would have the same safety margin of all cell components (Mitchell, 1951).

Viability in a spore suspension or a vegetative culture may vary as a result of individual cell variation. There may also be orderly periodic changes in the culture as a whole. Bacterial cells which are in the exponential growth phase are generally less resistant than cells in the stationary phase or lag phase (Hansen and Riemann, 1963).

The considerations above suggest that some form of statistical distribution of the survival time for individual cells exists. Such a hypothesis is not compatible with a strict exponential survivor curve. However, a log-normal distribution of survival times in a population of cells may result in survivor curves that cannot be distinguished from exponential curves, considering the experimental error involved (Meynell and Meynell, 1965). Furthermore, deviations from an exponential survivor curve are frequently found, especially with vegetative cells. The survivor curves for spores generally approach the exponential course well enough to permit its use in calculation of heat processes for canned foods, as pointed out by Stumbo (1965) in a discussion of methods for the calculation of safe heat processes.

2. PARAMETERS OF HEAT INACTIVATION

An exponential survivor curve makes it possible to express the heat resistance as the number of minutes it takes to kill 90% of the spores or vegetative cells at a defined temperature; this is the so-called D value. The reference temperature is usually 250°F (121.1°C) in the case of spores, and resistant botulinum A and B spores have D_{250} values of 0.10 to 0.20 minute. The highest D_{250} value found for C. *botulinum* spores is 0.204 minutes (Schmidt, 1964), which agrees with the highest resistance found in experiments forty years ago by Esty and Meyer. This does not mean that all botulinum spores have a high resistance. Esty (1923) found a 25-fold difference in heat resistance between more than 100 strains of botulinum A and B and concluded that the majority of strains in nature are probably not very resistant. The survival time (D value) increases dramatically when the heating temperature decreases, usually by a factor of 10 for a temperature decrease of 15° to 18°F (8° to 10°C). In canning practice it has been customary to assume that 18°F corresponds to a 10-fold change in D value, and this is the so-called z value.

The D_{250} and the z values are used to calculate heat processes for canned foods. Because of the nature of thermal destruction of spores, no heat-treatment will destroy all spores in an infinite number of cans, but calculation makes it possible to design heat-treatments that will reduce the numbers of viable spores to a very low level. Suppose that 10^{12} (thousand billion) cans of food contain one C. *botulinum* spore each, with a resistance that corresponds to a D_{250} of 0.21 minute, which is a very high resistance. If these cans are heated so that every particle in the food is exposed to 250°F for 2.52 minutes (or to 232°F for 25.2 minutes), we can use the following equation (Stumbo, 1965) to calculate the number of spores that will survive:

$$2.52 = 0.21 \, (\log 10^{12} - \log b)$$

and we find that b, the number of surviving spores, equals 1; 2.52 minutes at 250°F or correspondingly longer times at lower temperatures will thus reduce 10^{12} resistant botulinum spores to 1 spore. Such treatments have become the minimum processes applied for low-acid foods (pH higher than 4.5) by the canning industry and represent the so-called 12-D concept. In practice, a value of 3 minutes or more at 250°F ($F_0 = 3.00$) is more often applied to provide a safety margin with regard to spoilage caused by more heat-resistant organisms. Experience from the last forty years has shown that application of the 12-D concept gives a high degree of safety against $C.$ $botulinum$ in canned foods.

Vegetative cells are far less heat-resistant than bacterial spores, and they also show more variation in heat resistance. Most studies have been carried out with the cells in aqueous suspensions. Very little is known about the heat destruction of vegetative cells in solid media, although the majority of foods are solid or contain solid particles where cells may be present as clumps or microcolonies.

For these reasons and also because vegetative cells are so heat-sensitive that even a mild heat-treatment kills most of them, no system has been developed for calculation of minimum safe heat processes (pasteurization of milk is an exception). In recent years, however, studies of the heat resistance of *Salmonella* have led to formulation of pasteurization processes of liquid egg products. This will be discussed later.

3. Effect of Environmental Factors on Heat Resistance

A number of factors in the heating menstruum and in the recovery medium influence the ability of microorganisms to survive a heat-treatment.

a. Effect of pH. The pH has a pronounced effect on heat resistance of vegetative cells. Generally, there is a rather narrow pH range where the resistance is highest. Outside this range there is a rapid decrease in resistance (Hansen and Riemann, 1963). The effect of acids does not depend solely on a decrease in pH. Lactic acid and acetic acid differ from hydrochloric acid in their effect on the heat resistance of *Salmonella* (Lategan and Vaughn, 1964). *Clostridium botulinum* spores seem to have maximum resistance at pH values between pH 6.3 and 6.9 (Esty and Meyer, 1922). Lang (1935) was unable to find correlation between heat resistance of botulinum spores and pH values within the range 5.2 to 6.8, but a marked reduction was found at pH 4.9.

Later studies (Xezones and Hutchings, 1965) showed that the decimal reduction time for *C. botulinum* 62A spores increased 3- to 5-fold as the pH of the medium in which the spores were heated increased from 4.5 to

6.7. The increase was somewhat larger at 230°F than at higher temperatures.

b. Solutes and Water Activity. (i) Vegetative cells. A decrease in the water activity (a_w) of the heating menstruum increases heat resistance. This has been demonstrated with, for example, *Salmonella*, which was heated in dry egg white or dry animal feeds containing different amounts of residual water (Banwart and Ayres, 1956; Riemann, 1967). A rapid decrease in heat resistance was observed when the water content was increased to above 10% or at a water activity (a_w) above 0.90.

Soluble carbohydrates such as sucrose generally increase the heat resistance. This is probably due to a reduction of the water activity, and often 50% sucrose or more is required to produce a significant increase in heat resistance. The water activity is also reduced when soluble salts are added to a medium. Addition of sodium chloride could therefore be expected to increase heat resistance. However, species of non-salt-tolerant organisms—as, for example, *Salmonella*—seem to be damaged when a relatively large concentration of salt is added to an aqueous suspension, and this effect eliminates the protective effect of the decreased water activity. The effect of smaller concentrations of salts seems to vary with the type of salt, its concentration, and the nature of the heating medium (watery solution or foods). Different microorganisms also respond differently to salt. For these reasons, it is not possible to give general rules concerning the effect of salt.

The heat resistance is generally highest in complex organic media such as foods. In some cases, this may be due to a low water activity—for example, if drying occurs during heating—but protective substances such as proteins, peptides, and amino acids also play a role.

ii. Spores. There is no general agreement in the early literature with regard to the effect of NaCl on heat resistance of *C. botulinum* spores. Weiss (1921) found that 3% NaCl decreased heat resistance, while Esty and Meyer (1922) observed increased heat resistance in the presence of 1 to 2% NaCl. Yesair and Cameron (1942) noticed a decrease of heat resistance when 3.5% NaCl was added to phosphate buffer, which may be due to a drop in pH caused by salt addition. Stumbo *et al.* (1945), working with a nontoxic anaerobic sporeformer, found that NaCl and other curing ingredients in concentrations normally used in canned cured meat had negligible effect on heat resistance in commercial food processing. There is little doubt, however, that NaCl and other curing ingredients have a considerable inhibitory effect on heated spores; they cause dormancy, and this makes it possible to preserve canned cured meats by heat-treatments that are much lower than the classical 12-*D* treatments used for other low-acid canned foods. Recent studies (Perigo *et al.*, 1967) have shown that the inhibitory effect of nitrite on vegetative cells of clostridia is increased significantly and becomes far less pH-dependent if nitrite is autoclaved in the medium. It is believed that nitrite, on heating, reacts with some component of the medium, producing an unknown inhibitory substance. Sucrose in concentration of 12.5% or higher has been shown to increase the heat resistance of *C. botulinum* (Sugiyama, 1951).

In more recent studies of the effect of water activity on heat resistance of bacterial spores (Murrell and Scott, 1966), the highest resistance of spores was found at $a_w = 0.4$ to 0.6. The increase in resistance caused by reduced a_w was highest for spores that have a low resistance. Thus the thermal resistance of *C. botulinum* E spores was increased tenfold when the water activity was reduced from 1 to 0.9.

   c. Preservatives and Antibiotics. (i) Vegetative cells. Bactericidal compounds are generally more effective at higher temperatures, and they decrease the heat resistance of bacteria. This has been demonstrated with sulfur dioxide, essential oils, and nitrite. Such compounds may also change the heat resistance pattern. It has thus been found that 100 ppm of nitrite reduced the heat resistance of fecal streptococci 100- to 500-fold at 68° to 70°C but had little effect at 65°C (Greenberg and Silliker, 1961).

   ii. Spores. Chemical preservatives, except acids and curing salts, have not been applied to control *C. botulinum* spores in canned foods. The use of antibiotics for this purpose was suggested by experiments that showed that several antibiotics have an inhibitory action on mildly heat-treated botulinum spores. The antibiotics seem to interfere with the outgrowth of spores rather than reduce their heat resistance. Some antibiotics may also decrease heat resistance, since spores heated in pea puree containing 14 ppm of subtilin had a 37% shorter survival time (LeBlanc *et al.*, 1953).

   Similar results were found when botulinum spores were heated in the presence of nisin (O'Brien *et al.*, 1956). These effects could also be due to a carryover of small amounts of antibiotics absorbed on the spores. Even 0.03 ppm of subtilin in the recovery medium markedly reduces survivor counts of heated clostridial spores (O'Brien and Titus, 1955).

   d. Fat Protection. Most heat resistance studies have been carried out with the cells suspended in an aqueous medium. Few experiments have been made with cells suspended in oil or fat, but it has been demonstrated that such cells may have a significantly increased resistance. It is believed that this is because cells or spores in oil are in a dry state. Although it is true that oils may contain small amounts of water, the solubility of water is temperature-dependent, and drying may take place during slow heating (Hansen, 1967). Striking results have been obtained with *C. botulinum* as described below.

i. Vegetative cells. Fat protection of vegetative cells has been demonstrated (Jensen, 1954). It has often been assumed that fat protection has little practical significance, since bacterial cells tend to remain in the fat phase, where they are unable to grow. It has been observed, however, that cells that have been trapped in the fat phase can migrate into the water phase of a food and cause spoilage. Transfer from the fat to the water phase may be of significance in luncheon meat and similar canned meat emulsions if these products are exposed to agitation at a temperature where the fat phase is liquid.

ii. Spores. The most striking demonstration of fat protection has been reported by Lang (1935), who showed that *C. botulinum* spores which were present in the oil phase of canned fish packed in oil survived heat-treatments that killed botulinum spores in aqueous media, and that the spores could germinate and grow out if they later passed from the oil phase to the aqueous phase. This migration into the aqueous phase could be aided by physical shaking. Recent studies (Molin and Snygg, 1967) suggest that various oils may have specific protective effects – besides the absence of water.

e. Other Factors. A number of other factors such as nutrients and degree of anaerobiosis have been found to influence recovery of heated cells and spores (Hansen and Riemann, 1963; Hersom and Hulland, 1965), but only little information relates to disease agents in foods.

B. Heat Resistance of Various Disease Agents

1. *Salmonella*

Numerous studies of the heat resistance of *Salmonella* have been published, and some of the recent results are presented in Table I. The various serotypes and strains of *Salmonella* have almost the same resistance to heat (Ng, 1966) with the exception of *Salmonella senftenberg* 775W. This strain is ten to twenty times as resistant in liquid media or in moist foods as are other strains. It does not seem to be more resistant in dry materials at a water activity below 0.90 (Riemann, 1967). The z value of the thermal death time curve for most *Salmonella* is about 4.0° to 5.0°C. Thus, if it takes 40 minutes to kill a certain number of *Salmonella* at

TABLE I

Heat Resistance Of *Salmonella*

Medium	Temperature	Time	Destruction	References
Chicken meat	60°C	5 minutes	Complete kill of 3×10^8 cells of *S. typhimurium*.	Bayre *et al.* (1965)
	65°C	10–15 minutes	Complete kill of 3×10^8 cells of	Bayre *et al.* (1965)
Egg yolk	60°C	0.5 minutes	90% destruction of *S. senftenberg* 775W	Licciardello *et al.* (1965)
	60°C	15 minutes	90% destruction of *S. senftenberg*	Licciardello *et al.* (1965)
Liquid egg, pH 5.5	60°C	7 to 9 minutes	Complete kill of 10^5 cells of most *Salmonella* types	Anellis *et al.* (1954)
	60°C	84 minutes	Complete kill of 10^5 cells of *S. senftenberg* 775W	Anellis *et al.* (1954)
Liquid egg, pH 8	60°C	2.0–3.5 minutes	Complete kill of 10^5 cells of most *Salmonella* strains	Anellis *et al.* (1954)
	60°C	11 minutes	Complete kill of 10^5 cells of *S. senftenberg* 775W	Anellis *et al.* (1954)
Liquid egg	64.4°C	2½ minutes	10^5-fold destruction of *S. senftenberg*	Shrimpton *et al.* (1962)
Shell eggs	59°–60°C	Few minutes	Complete kill of 10^9 *S. typhimurium* present at the center of the egg	Clarenburg and Burger (1950)
Saline	60°C	3.7–7.5 minutes	90% destruction of *S. senftenberg* 775W	Thomas *et al.* (1966)
Green pea soup	60°C	5.2–10.0 minutes	90% destruction of *S. senftenberg* 775W	Thomas *et al.* (1966)
Chicken meat	75–85°C	40 minutes	Approximately 10^{10}-fold destruction of *S. typhimurium*	Thomas *et al.* (1966)
Liquid egg, pH 5.5	55°C	10 minutes	90% destruction of *S. typhimurium*	Lategan and Vaughn (1964)

Stabilized egg white, pH 7	60–62°C	3 ½–4 minutes	10^8-fold destruction of most *Salmonella*	USDA (1964)
Egg shell	65.6°C	1–3 minutes	Most *Salmonella* destroyed during washing	Bierer and Barrett (1965)
Dried egg white	49–54°C	1–2 weeks	Most *Salmonella* destroyed	Slanetz et al. (1963)
Dry animal feed	82°C	1 hour	Most naturally occurring *Salmonella* killed	Rasmussen et al. (1964)

56°C, only 4 minutes will be required at 60° to 61°C. Somewhat higher z values, 5.6° to 6.4°C, have been found for $S.$ *senftenberg* 775W.

2. *Staphylococcus aureus*

Heat-resistant strains of *Staphylococcus aureus* have the same degree of resistance as $S.$ *senftenberg* 775W. There is, however, considerable difference between strains, as illustrated in Table II. The reported z values for the heat destruction of *Staphylococcus aureus* vary from 4.7°C to 7.3°C.

3. MISCELLANEOUS NONSPORING ORGANISMS

The heat resistance of other nonsporing organisms varies considerably, as shown in Table III.

4. SPOREFORMING ORGANISMS

The heat resistance of $C.$ *botulinum* spores was reviewed recently by Perkins (1964). Spores of A and B are reduced by a factor of at least 10^{12} by exposure to 250°F (121.1°C) for 2½ minutes. The z value of the thermal death time curve is approximately 18°F (10°C), so that a heat-treatment at 111.1°C for 25 minutes corresponds to 121.1°C for 2½ minutes.

Clostridium botulinum E spores are less resistant. Reported D values in buffer and media vary from 0.6 to 3.3 minutes at 80°C, and the z value is generally 7.4° to 10.7°C but may be as low as 5.6°C (Schmidt, 1964; Stumbo, 1965). Recent studies have indicated D values of 1.6 to 4.3 minutes in whitefish chubs heated at 80°C with a z value from 13.1° to 13.6°F (7.3° to 7.6°C) (Crisley *et al.,* 1967). The heat resistance of type E spores has been studied less, compared to types A and B, but there are indications that E spores may sometimes have an unexpected high resistance.

The heat resistance of $C.$ *perfringens* spores may approach the resistance of $C.$ *botulinum* type A spores. Spores of nonhemolytic strains have shown D values of 4 to 17 minutes when heated in water at 100°C. The less-resistant spores of β-hemolytic strains had D values of less than 1 minute at 100°C and 3 to 5 minutes at 90°C (see Chapter 4, *Clostridium perfringens* and *Bacillus cereus* Infections, Sections II, A and IV). No information seems to be available for food poisoning strains of $B.$ *cereus.*

5. PARASITES

Trichina and several other parasites are killed by exposure to 137°F

TABLE II

HEAT RESISTANCE OF *Staphylococcus Aureus*

Medium	Temperature	Time	Destruction	References
Saline	60°C	2.02–2.25 minutes	90% destruction	Thomas *et al.* (1966)
Milk	60°C	3.16–3.29 minutes	90% destruction	Thomas *et al.* (1966)
Beef soup	60°C	3.20–2.48 minutes	90% destruction	Thomas *et al.* (1966)
Green pea soup	60°C	6.67–6.87 minutes	90% destruction	Thomas *et al.* (1966)
Precooked fish	49°C or higher	5 minutes	10^8-fold destruction	Haughtley and Liston (1965)
Turkey roll	60–66°C	Minutes	10^8-fold destruction	Wilkinson *et al.* (1965)
Cured ham	58°C	Minutes	Complete kill	Lechowich *et al.* (1956)
Cured ham	39°C–45°C	2 hours	50% destruction	Silliker *et al.* (1962)
	38°C–59°C	4½ hours	More than 10^3-fold destruction	Silliker *et al.* (1962)
Laboratory medium	59°C	10 minutes	From 10^1- to 10^5-fold destruction, depending on the strain	Silliker *et al.* (1962)
	59°C	60 minutes	10^5-fold destruction or more	Silliker *et al.* (1962)
Cream filling	55°C	8 minutes	10^1–10^2-fold destruction	Husseman and Tanner (1947)
	65°C	4 minutes	10^2–10^3-fold destruction	Husseman and Tanner (1947)
Turkey stuffing	60°C	Few minutes	10^3–10^4-fold destruction	Castellani and Niven (1955)
Custard	60°C	7.7 minutes	90% destruction	Angelotti *et al.* (1961)
Chicken a la king	60°C	5.2 minutes	90% destruction	Angelotti *et al.* (1961)

TABLE III

HEAT RESISTANCE OF VARIOUS NONSPORING ORGANISMS

Medium	Temperature	Time	Organism	Destruction	Reference
Milk	65.6°C	35 minutes	Fecal streptococci	90%	Statens Forsøgsmejeri (1960)
Cured ham	65.6°C	10–100 minutes	Fecal streptococci	90%	Riemann (1957)
	56–58°C	10 minutes	*Pasteurella tularensis*	Complete	Chapter VI
	63–80°C	5 minutes	*Listeria monocytogenes*	Incomplete	Chapter VI
	61.7°C	35 minutes	*Listeria monocytogenes*	Less than 10^5-fold	Chapter VI
	100°C	15 seconds	*Listeria monocytogenes*	Incomplete	Chapter VI
	55°C	1 hour	*Brucella abortus*	Complete	Chapter VI
	62–63°C	3 minutes	*Brucellus abortus*	Complete	Chapter VI
Milk	60°C	30 minutes	*Mycobacterium tuberculosis*	Incomplete	Chapter VI
	62.8°C	30 minutes	*Mycobacterium-tuberculosis*	Complete	Chapter VI
	71.7°C	15 seconds	*Mycobacterium-tuberculosis*	Complete	Chapter VI

(58.3°C), and all known parasites seem to be effectively destroyed at 100°C.

6. VIRUSES

Most viruses seem to be inactivated by ½ to 1 hour of heating at 65°C or a few seconds of heating at 72°C. Some forms may require boiling for complete destruction (see Chapter 2, Viral Infections, Section V, B).

7. TOXINS

Most fungal toxins including the aflatoxins are resistant to heat and are not destroyed by boiling of foods (see Chapter 10, Mycotoxicoses, Section II, C, 1).

Straphylococcal enterotoxins are very heat-resistant (Denny *et al.,* 1966; Read and Bradshaw, 1966; Read *et al.,* 1965). Autoclaving food does not ensure destruction. Botulinal toxins are heat-sensitive and are readily destroyed by boiling. Many plant and animal toxins are heat-resistant, but several are water-soluble, and their concentration in foods may be effectively reduced by boiling in water (see chapters on *Staphylococcus aureus, C. botulinum,* and poisonous plants and animals).

C. Heat Processing of Foods

Heating is the most important method of eliminating viruses, microorganisms, toxins, and parasites from food. If all foods were boiled (heated to 100°C) shortly before consumption, food-borne diseases would never occur with the exception of those caused by staphylococcus enterotoxin and toxins of plant, animal, or fungal origin. However, preparation by heating is often done a considerable time before consumption. This means that pathogenic or toxigenic organisms that have survived heat processing or have contaminated the food after processing may cause food poisoning.

The degree of heat-treatment used in commercial food processing depends on the type of the food (composition), the storage temperature, and the microorganisms which most commonly cause problems. Low-acid canned foods (pH above 4.5) which are distributed at ambient temperature are generally given a heat-treatment that is equivalent to 3 minutes at 250°F (121.1°C) to make sure that no botulinum spores will survive in the cans. In some cases more severe heat processes are used to eliminate spores which are more resistant than *C. botulinum.*

The standards for heat processing of low-acid canned foods have been adopted voluntarily by the canning industry. Only a few states have made such standards official. Acid canned foods (pH less than 4.5) are inhibitory to all known food poisoning organisms except fungi, and they are heat-processed only to ensure shelf-life. However, growth of spoilage organisms may cause a rise in pH, and this may result in a health hazard. Some types of low-acid canned foods, such as cured meats, are partially or completely inhibitory to *C. botulinum* and other sporeformers. They are given much milder heat-treatments than are other low-acid canned foods. The treatments generally correspond to 0.3 to 0.6 minutes of heating at 121.1°C and result in about 10^4-fold reduction in the number of resistant clostridial spores which can grow out in the product. The main inhibitory factors in these products are the curing salts.

Products distributed frozen or refrigerated are generally not heat-processed with the purpose of eliminating bacterial spores. An exception is smoked fish, which must now be heated to 82°C for 30 minutes to safeguard against *C. botulinum* E spores. Some frozen or refrigerated products are not heat-processed at all (fresh meat and poultry). Some are processed at temperatures from 60° to 100°C to ensure shelf-life (blanching of vegetables) or as a step in their preparation (precooked frozen dinners, several types of sausages and cured lunch meats). Others are heated with the purpose of eliminating *Salmonella* (frozen egg). The heat processes which are used kill most nonsporing disease organisms but have little or no effect on spores with exception of type E botulinum spores. Most of the heat processes are based on commercial practice, but in some cases official regulations have been introduced. Pasteurization of egg is compulsory in England (64.4°C for 2½ minutes) and in some states in the United States (60°C for 4 minutes or more). Some states have specifications for precooked frozen foods—for example, heating to at least 80°C followed by refrigeration to 5° to 10°C in less than 3 hours. The United States Department of Agriculture requires that precooked poultry in inspected plants be heated to 77°C, and processed pork products must be heated to 57°C to destroy trichina.

The safety of refrigerated and frozen products rests upon the maintenance of a low temperature, since the heat processes do not destroy bacterial spores effectively, and the products may have become contaminated after processing. The heat processes are important, however, since *Salmonella* and other vegetative cells and also parasites and probably a number of viruses which may be present in the raw food material are eliminated. There is some concern, however, that mild heat-treatment, by eliminating competing vegetative spoilage organisms might favor growth of sporeformers in the case that the food is kept at too high a temperature. Public health hazards, except for those caused by mishandling at the consumer level, could be almost completely eliminated by heat pasteurization after packaging and insurance against growth of sporeformers by a low pH and/or a low water activity in the product. For technical and culinary reasons this is possible only with some products.

Dried foods and heavily salted foods are sometimes heat-processed. Dried fruits are frequently pasteurized, and dried eggs are treated with heat before drying to eliminate *Salmonella*. Some dried foods seem susceptible to *Salmonella* contamination—for example, from dusty raw materials—which easily spreads throughout the plant. Such dust may become an excellent growth medium in localized "wet" spots. If a food is pasteurized after drying with the purpose of destroying *Salmonella* or other organisms, it must be kept in mind that the heat resistance is increased at reduced water activities.

III. RADIATION

A. General Considerations

The development of nuclear technology and the increasing availability of radioactive isotopes have increased the interest in the use of radiation for food preservation. The types of radiation which are useful for this purpose are the gamma rays produced by radioisotopes and electron beams generated by electron accelerators. The quantity of radiation is expressed in rads. One rad is equivalent to the absorption of 100 ergs per gram of food. One megarad (Mrad) is equivalent to one million rads.

The killing of microorganisms by irradiation is caused by two effects, a direct and indirect. The direct is due to "hits" and is independent of the environment and also relatively independent of temperature. The indirect effect is caused by reactive compounds created in the medium and in the bacterial cell as a result of irradiation. The indirect effect, therefore, depends on the composition of the medium. It also depends on temperature, since temperature affects the reactivity of compounds produced by irradiation. Vegetative cells are killed to a larger extent by indirect effects than are bacterial spores.

The survival curve for irradiated spores is generally not a straight line but has a characteristic initial lag or "shoulder" corresponding to initial radiation doses of 0.4 to 1 Mrad which cause little reduction in the number of viable spores (Wheaton and Pratt, 1962; Kempe, 1964; Roberts and Ingram, 1965). The lag has been explained to mean that a number of direct hits, about 15, are required to kill a spore. Because of the presence of the initial lag, doses are often expressed in terms of 99% or higher levels of destruction rather than by D values. The initial lag is independent of spore concentration. Spore destruction usually proceeds exponentially after the lag, but data obtained when large numbers of botulinum spores were exposed to radiation indicate that the destruction curve is in fact sigmoid. A small fraction of spores, $10^{-1}\%$ to $10^{-6}\%$, showed a higher resistance than that predicted from the exponential section of the destruction curve (Wheaton and Pratt, 1962).

The minimal radiation for food sterilization is defined as the dose which has the same effect as heat processing of canned low-acid food — that is, a dose that reduces viable *C. botulinum* spores by a factor of 10^{12}. This dose can be calculated as $L + 12D$, where L is the lag and D is the dose required for destruction of 90% of the spores.

The $L + 12D$ for a highly resistant type B strain suspended in phosphate buffer varied from 3.5 Mrads (million rads) at 20°C to 5.28 Mrads at −196°C (El Bisi, 1967). The resistance in food is higher.

B.   Effect of Environment

1.   EFFECT OF pH

The radioresistance of spores was not affected by pH changes within a range of pH 3.2 to 8.4 (Williams-Walls, 1960).

2.   TEMPERATURE

a. Nonsporeformers.   A slightly increased temperature may make vegetative cells more sensitive to irradiation. This has been demonstrated with *Salmonella* irradiated at 43°C (Licciardello *et al.*, 1965). Freezing has the opposite effect, and the explanation is probably that immobilization of water tends to arrest indirect chemical effect. The protective effect of freezing has been demonstrated with *Staphylococcus aureus* (Bellamy and Lawton, 1954).

b. Spores.   Freezing may also increase the radioresistance of bacterial spores, and the effect of freezing depends on the medium in which the spores are irradiated. A 3- to 10-fold increase in resistance has been found for *C. botulinum* A spores irradiated in phosphate buffer, while the increase for spores irradiated in a pork pea medium was only 2- to 3-fold. The pork pea medium apparently contains compounds that can scavenge radicals formed by irradiation (Denny *et al.*, 1959; Grecz *et al.*, 1967).

A reduction in temperature from 20°C to −196°C increased the initial lag of destruction of *C. Botulinum* B spores from 0.316 Mrad to 0.476 Mrad (El Bisi, 1967).

In other studies, no significant difference in survival was found when botulinum spores were irradiated in food or buffer at temperatures ranging from +80°C to −80°C (Ingram and Thornley, 1961; Kempe et al., 1956).

Heating of C. botulinum spores has little or no effect on the radioresistance. It has been demonstrated that heating before or during irradiation may actually increase the radiation resistance of botulinum spores in beef (Anderson et al., 1967). Irradiation at levels of 1 Mrad or more makes spores more sensitive to heat. The heat-treatment required to destroy 5×10^6 type B spores dropped from 1 minute at 250°F to 0.15 minute at 250°F when the spores had been irradiated with 1.2 to 1.5 Mrads; radiation sterilization without heat processing required 3.9 Mrads (Kempe et al., 1957a,b).

3. CURING SALTS AND OTHER FOOD ADDITIVES

Two megarads completely destroyed 10^6 A and B botulinum spores in beef containing 4% NaCl; 10^6 type A spores were also destroyed by this dose when the meat contained only 3% NaCl. Two megarads caused complete destruction of 10^6 spores when 0.1% mustard oil had been added to the meat. Sodium citrate and Versene also reduced the radiation resistance of botulinum A and B spores. When 1000 ppm of sodium nitrate together with 2.5% sodium chloride was added to ground beef inoculated with 10^8 type A spores, no spoilage or toxin formation could be detected after irradiation with 2 Mrads.

Sodium nitrite (200 ppm) had a similar potentiating effect on radiation sterilization of botulinum spores (Anderson et al., 1967; Anderson, 1964; Krabbenhoft et al., 1964). Other observations suggest that irradiation may reduce the inhibitory properties of curing ingredients, probably by destroying nitrite, and thus make a cured product less stable if it is contaminated *after* irradiation. If spores are irradiated in the product, damage of the spores seems to be more extensive than reduction of nitrite level (Kempe, 1964; Kempe and Graikoski, 1964a; Riemann, 1957).

4. PROTECTIVE FACTORS

A low water activity may increase radioresistance of vegetative cells, since indirect radiation effect is limited in a dehydrated environment. Reduction of the oxygen pressure increases radiation resistance. Complete removal of oxygen has been found to increase the resistance of Escherichia coli threefold (Hollaender et al., 1951).

Gluthathione, sodium hydrosulfite, and catalase have been found to protect botulinum spores against indirect radiation effects (Kempe,

1964). The effect of protective compounds can be observed when spores from the same spore suspension are irradiated in different foods. Thus the radioresistance of *C. botulinum* spores was found to be ten times as high in raw or cooked beef as in green beans (Kempe, 1964; Wheaton *et al.*, 1961), and a 10^5-fold reduction in numbers of viable botulinum A or B spores required 1.3, 1.9, and 2.5 Mrads in phosphate buffer, peas, and pork, respectively (Denny *et al.*, 1959). The radiation resistance of *C. botulinum* B spores has been found to be 55 to 58% higher in meat than in phosphate buffer and showed little variation with temperature (El Bisi, 1967). The radiation resistance of *C. botulinum* spores seems to be identical in raw and cooked beef (Kempe, 1964).

5. DORMANCY

Very little is known about the changes which may take place in surviving spores when irradiated food is stored. It seems likely that lethal processes caused by the indirect radiation effect continue, but this may depend on the storage temperature. Thus it has been found that the percentage of the botulinum spores that survive increases with time if the irradiated product is kept frozen (Wheaton *et al.*, 1961). It has recently been found that botulinum A spores irradiated in distilled water continued to die during storage, but spores irradiated in the dry state showed increasing recovery during storage (Upadhyay and Grecz, 1967).

C. Radiation Resistance of Disease Agents

1. NONSPOREFORMERS

Vegetative cells are much less radioresistant than bacterial spores. Reported *D* values for *Salmonella* and *Staphylococcus aureus* vary from 0.02 to 0.04 Mrad (Lineweaver, 1966; Niven, 1958; Stabyj *et al.*, 1965). Doses of 0.36 to 0.5 Mrad killed 10^7 *Salmonella* cells in frozen eggs; 0.65 Mrad was required to destroy 10^5 cells in frozen horsemeat, and 0.45 Mrad destroyed 10^3 in dried coconut (Lew *et al.*, 1963; Comer *et al.*, 1963).

2. SPOREFORMERS

Spores of *C. botulinum* A and B are among the most radioresistant microorganisms. Spores of some putrefactive clostridia, PA3679 and S_2 are more heat-resistant than botulinum spores but seem to have a slightly lower radioresistance (Anellis and Koch, 1962; Roberts and Ingram, 1965), although higher resistance has occasionally been found for spores of S_2. The radioresistance of *C. botulinum* spores varies considerably.

Overall, type A spores seem more resistant than type B, but the variation between strains of the same type can be as large as that found for heat resistance (Wheaton *et al.*, 1961). Determination of *D* values for 100 type A and B strains in phosphate buffer gave the following results (Anellis and Koch, 1962):

Most-resistant strains	0.32–0.34 Mrad
Intermediate strains	0.27–0.25 Mrad
Most-sensitive strains	0.13 Mrad

A 10^4-fold destruction of botulinum spores in aqueous suspensions was found to require 1.1 to 1.2 Mrads for type A, 1.0 to 1.1 Mrads for type B, 0.7 to 0.9 Mrad for type E, and 1.2 Mrads for type F (Roberts and Ingram, 1965; Watts, 1967). The D_{12} dose for type A spores in meat has been calculated to be 4.7 to 5.4 Mrad (Ingram and Thornley, 1961). In the United States, 4.8 Mrads has become accepted as the D_{12} dose.

Relatively few studies have been published concerning the radiation resistance of *C. perfringens*. Decimal reduction doses of 0.21 to 0.25 Mrad have been reported for spores in buffer. Heat shocking of spores before irradiation increased the required dose to 0.43 to 0.49; heat shocking after irradiation increased the dose to 0.34 Mrad (Matsuyama *et al.*, 1964). These results suggest that *C. perfringens* spores have a resistance which falls between the resistance of botulinum A and B and that of botulinum E.

3. PARASITES

Parasites are rather sensitive to radiation, compared to microorganisms. Although 1 Mrad may be required to kill trichina larvae, 10,000 rads (0.01 Mrads) is sufficient to sexually sterilize the female larvae. Such action would break the trichina cycle (Brandley *et al.*, 1966).

4. VIRUSES

Viruses are more radiation-resistant than bacterial spores. A number of viruses require 2.5 to 4 Mrads for 10^5- to 10^6-fold reduction of infectivity. However, the dose applied for destruction of *C. botulinum* will probably be sufficient to inactivate viruses which may be naturally present in foods (see Chapter II, Viral Infections, Section V, B).

5. TOXINS

The radiation resistance of botulinum toxins (see Chapter VII, botulinum, Section VII, B) and staphylococcal enterotoxins (see Chapter IX; Read and Bradshaw, 1967) is so high that a dose which effectively destroys bacterial spores cannot be expected to inactivate preformed toxins in foods. This is probably the case also with other toxins.

D. Radiation Preservation of Foods

Commercial irradiation of foods is still at its beginning but may be expected to play an important role in the future. There is no doubt that irradiation of foods will come under more strict regulation than most other food preservation techniques. This is motivated by the health aspects of the use of high-energy radiation and by the fact that radiation preservation is a new technique, the effect and consequences of which are still difficult to evaluate.

The possibility that radiation might induce radioactivity in foods or give rise to formation of toxic substances has been intensively studied. It has been suggested that food irradiated with doses up to 5 Mrads and energy levels below 10 MeV are as wholesome as conventionally preserved foods (Raica, 1965).

The interest in irradiation of food is focused on three applications: (1) sterilization of foods; (2) substerilization to prolong shelf-life or to eliminate certain pathogenic organisms such as *Salmonella;* (3) insect deinfestation of grain, etc. Only the two first aspects will be considered here.

1. STERILIZATION

Principles of microbial safety of irradiated foods have been discussed by El-Bisi (1965). A minimum dose of 4.5 to 4.8 Mrads has been suggested for low-acid foods (pH above 4.5) stored at ambient temperature. The argument behind this dose requirement is that irradiation should give the same degree of safety as conventional heat sterilization. Bacon was for a time cleared for radiation sterilization in the United States, with a dose requirement of 4.5 Mrads (Brandley *et al.,* 1966). However, bacon conventionally contains rather high concentrations of curing salts, and 2.7 to 2.9 Mrads has been found sufficient to give a 10^{12}-fold reduction in the number of *C. botulinum* spores that are able to produce toxin in the product (Anellis *et al.,* 1965). A somewhat lower standard dose, 2.5 Mrads, has been suggested for cured meats (Dean and Howie, 1964). Even this dose may seem excessive compared to the effect of conventional heat processing of canned bacon (heating to 70° to 75°C internal temperature). Studies with cured ham indicate that 2.0 to 2.5 Mrads is sufficient to prevent toxin production in cans inoculated with 10^6 A or B spores (Anellis *et al.,* 1967). Even lower doses seem to be equivalent to the heat processing conventionally applied for canned ham (Riemann,

1957). The interest in using lower doses is due to the experience that even doses below 2.5 Mrads give significant organoleptic changes. The dose requirement could possibly be reduced by methods other than addition of curing salts. One way would be to apply heat-treatment after irradiation. In the case of solid-pack foods which have slow heat penetration, directed radiation could be used to sterilize the center, and the surface layers could be sterilized by heat (Hansen, 1965). None of these methods seems to have been explored commercially.

A fully sterilizing radiation dose will, in contrast to heat processing, not inactivate preformed botulinum toxin. This may require special precautions.

2. SUBSTERILIZATION

The application of substerilization involves certain considerations with regard to safety. Radiation doses of 0.25 to 0.75 Mrad intended to eliminate *Salmonella* or to extend the shelf-life of a food have little destructive effect on *C. botulinum* spores. This is due to the presence of the "shoulder" on the survivor curve for spores, which was discussed earlier. Substerilization may therefore be selective for *C. botulinum,* and extended shelf-life could give this organism a chance to grow and produce toxin. The probability that such a situation would occur depends on the nature of the food and the storage conditions. There would seem to be little risk involved in substerilization of dried foods such as coconut and dried eggs, since the water activity is too low for growth of *C. botulinum.* The same applies to foods that are inhibitory to botulinum because of high salt concentration or low pH (fresh fruit). However, extension of shelf-life of certain fruits may increase the risk of growth of toxin-producing fungi.

The risk involved in the application of substerilization doses would also be low in the case of refrigerated and frozen foods if there were no chance of temperature mishandling and misjudgments of the organoleptic condition (spoilage) of the food. Such misjudgment could be caused by a change in spoilage pattern. Substerilized foods are, in a microbiological sense, new products. The radiation treatment not only reduces the total number of vegetative spoilage bacteria but also causes changes in the composition of the flora, since different microbial species have different radioresistance. Refrigerated meat and chicken which are normally spoiled by organisms such as *Pseudomonas* and *Achromobacter* will gradually spoil because of the growth of *Microbacterium thermosphactum* and *Streptococcus faecium* after substerilization with 0.1 to 0.25 Mrad (Wolin *et al.,* 1957; Thornley, 1957). Studies of marine products have also demonstrated that changes in the spoilage flora may occur and as a result a higher total number of bacteria may be formed before spoilage becomes detectable (Shiflett *et al.*, 1966; Spinelli *et al.*, 1964). Very little is known about the possible effect of a changed spoilage flora on the consumer or on the ability of *C. botulinum* to grow and form toxin (see Section X on microbial competition).

Aspects of substerilization of marine foods have been discussed by Goldblith (1966). When fish were inoculated with 10^4 to 10^6 spores of *C. botulinum* E and irradiated with 0.35 to 0.8 Mrad, toxin was produced in 7 to 13 weeks at 4.5°C but not in 24 weeks at 1.7°C. It was concluded that substerilization can be regarded as a safe process, since the natural spore load is generally low, less than 0.2 type E spores per gram, and since the product was no longer marketable at a time when toxin was produced from a much larger spore inoculum. Furthermore, normal cooking of fish reduced 21 man-lethal doses of toxin to less than one.

Substerilization is still in the development stage, since detailed studies have been carried out with only a few food products.

IV. REFRIGERATION–CHILLING

A. General Considerations

Food poisoning microorganisms can multiply within a temperature range from about 2°C to 55°C. Growth in the lower temperature range becomes increasingly slow, and the temperature at which growth stops completely depends on the composition of the food. A low pH or water activity reduces the ability of microorganisms to grow at low temperatures.

Disease agents may remain viable for a long time at temperatures that are too low for growth, and growth may be resumed if the food is exposed to a higher temperature.

B. Effect on Various Disease Agents

1. Nonsporeformers

The growth of *Salmonella* is arrested at temperatures between 5° and 10°C. Slow growth has been observed at 5.7°C in chicken à la king, pH 6.2, but no growth occurred at 10°C in ham salad, pH 5.6, or in custard containing 30% sucrose, pH 6.8. The upper temperature limit for growth was found to be 44° to 47°C (Angelotti *et al.*, 1961). *Staphylococcus aureus* has temperature requirements similar to those of *Salmonella*. Both organisms may survive for long periods in refrigerated foods. A number of other nonsporing organisms, such as *Brucella, Shigella,* and *Vibrio,* are also able to survive for extended periods in foods. The enteropathogenic *V. parahaemolyticus* has a growth range from 15° to 40°C.

2. SPOREFORMERS

The temperature range for growth of *C. botulinum* A and B is from 10° to 12°C to 48° to 50°C (Ohye and Scott, 1953). The range for *C. botulinum* E is about 8°C lower than that of types A and B. Type F seems to be able to grow at the same low temperature as type E.

The minimum growth temperature for *C. perfringens* is 6.5°C, but growth becomes restricted at temperatures below 15° to 20°C. Optimum growth occurs at 43° to 45°C. *Bacillus cereus* multiplies at temperatures from 10° to 45°C, and *B. subtilis* at 12° to 55°C.

3. PARASITES, VIRUSES, AND TOXINS

These disease agents do not multiply in foods but may remain viable for a long time, especially at refrigeration temperatures.

C. Refrigeration of Foods

Refrigeration is one of the most important methods for preservation of foods. It is widely applied for raw materials as well as for finished products, and in industry as well as in private households. Correct use of refrigeration could prevent most outbreaks of food poisoning. The frequent abuse of the requirements for cold storage is probably due to a lack of understanding that most foods are poor heat conductors and that the rate of growth of microorganisms may increase tremendously at temperatures slightly above the minimum temperature for growth. Thus, the rate of growth of *C. botulinum* A or B may increase tenfold when the temperature increases from 15°C to 25°C, and the lag period, the period before multiplication starts, may decrease tenfold.

The critical effect of temperature has been observed in studies with inoculated packs. Thus, type A produced toxin in 5 days in ham (4.8% brine) at 30°C, but not in 30 days at 25°C (Pivnick and Barnett, 1967). Similarly, no A toxin was formed in bologna at temperatures below 18°C (Steinke and Foster, 1951).

Most foods cool slowly in air, and the rate of cooling depends on the size, shape, and composition of the container. Generally, the rate varies with the square of the dimension, which means that the cooling rate is decreased by 75% if the smallest dimension of the container is doubled. It is most difficult to maintain proper cooling conditions when large volumes of foods are handled and prepared — for example, in institutional kitchens or at large group picnics. There seems to be a need for introduction of more efficient methods, similar to those used for rapid cooling of carcasses in the meat industry.

V. REFRIGERATION–FREEZING

A. Effect on Disease Agents

1. BACTERIA

Freezing not only arrests the multiplication of microorganisms but also causes a rapid initial decrease in the number of viable vegetative organisms. This is followed by a slower decrease during frozen storage. Although more than 90% of vegetative cells may be destroyed by freezing, this preservation method cannot be relied upon as a technique to eliminate food poisoning microorganisms. When the food is thawed, growth is resumed, and if drip is formed during thawing or if water condenses on the food, the spread of organisms may be more rapid than that on similar nonfrozen products. Freezing has little or no influence on bacterial spores.

2. PARASITES

Freezing kills many parasites and is used to eliminate trichina in pork.

3. VIRUSES AND TOXINS

Viruses and toxins are generally very stable in frozen condition.

B. Freezing of Foods

Freezing is the best method for long-time preservation of the "fresh" character of foods. It is therefore used extensively for prepared convenience foods such as frozen dinners. The manufacturing of such products is complex and may cause microbiological problems. The temperature treatments of the dinners before freezing must be carefully controlled to avoid survival and growth of contaminating bacteria, many of which would later be preserved in a living state by the freezing. Frozen foods have a good public health safety record in spite of these risks. It is possible that competing spoilage organisms in these foods often suppress the growth of food poisoning organisms in the case of temperature mishandling (storage of thawed food at too high a temperature).

VI. ACIDIFICATION

A. General Considerations

The effect of acids in destruction or inhibition of microorganisms may be due to the hydrogen ion concentration and/or to toxicity of the undissociated acid or the anion. Some organic acids such as formic and acetic acids have a toxic effect, and they are more active at a given pH than mineral acids that act mainly through their effect on pH.

B. Effect on Disease Agents

1. NONSPORING BACTERIA

Staphylococcus aureus is completely inhibited in meat at pH values 4.8 to 5.0, and this organism is most sensitive to low pH under anaerobic conditions (Lechowich *et al.*, 1956). Slow growth has been observed at pH 4.5 but not at pH 4.3 in a slurry prepared from frozen pot pie (Dack and Lippitz, 1962). *Salmonella* is inhibited at pH values below 4.5 or above 8 (Banwart and Ayres, 1957). Both *Staphylococcus aureus* and *Salmonella* are destroyed in mayonnaise and similar products where the pH has been adjusted to 3.2 to 4 by addition of acetic acid. Rapid destruction takes place in lemon juice and lime juice (pH 2 to 3), and a slower destruction, 1 to 4 weeks, occurs in tomato juice at pH 4.3 to 4.4 (Ayres, 1966).

Many other nonsporing bacteria are destroyed in acid foods.

2. SPOREFORMING BACTERIA

No growth of *C. botulinum* A, B, or E has been observed at pH values below 4.6. Germination in a nutrient environment is also inhibited at low pH. Fatty acids have an inhibitory effect at neutral pH if they are rancid. Influence of fatty acids is found not only for clostridial but also for *Bacillus* spores (Ramsey and Kempe, 1963; Casolari, 1963). This may be of importance in some special food items.

Clostridium perfringens seems to be more sensitive to acids than *C. botulinum,* and growth becomes variable at pH values 5.7 to 5.8. Most *Bacillus* spores are inhibited at pH values below 4.5 and above 8.5 (Banwart and Ayres, 1957; Bohrer, 1963). Bacterial spores seem to die slowly in inhibitory acid media (Bonventre and Kempe, 1959; Ingram and Robinson, 1951a, b; Townsend *et al.*, 1954).

3. PARASITES, VIRUSES, AND TOXINS

Very little information seems to be available concerning the des-

tructive effect of acids on parasites. But food acidification might be unreliable as a practical method for the control of parasites.

This is true also with viruses. Although some viruses are readily destroyed at an acid pH, other types can survive for several hours at pH 2.0
in hydrochloric acid. (See the chapter on viral infections.) Food-borne
toxins are generally quite stable at low pH, but some of them, such as
botulinum toxin and several fungal toxins, become unstable at alkaline
pH values (see Chaper 7, Section VII, C, and Chapter 10, Section II, C).

C. Acidification of Foods

No food poisoning microorganisms except toxin-producing fungi are
able to grow at pH values below 4.5. A large group of foods including
most fruits and several fermented products is therefore relatively safe
and seldom causes food poisoning.

Acidification of food by added acid is not used widely. One problem
is that uniform distribution is difficult to achieve in some foods, and
the addition of strong acids required to lower the pH in heavily buffered
foods such as meat cause undesirable organoleptic changes. This problem can be overcome by allowing acid formation in the product either by
fermentation or by adding a compound such as glucono-δ-lactone which
slowly hydrolyzes and forms acid (Sair, 1965).

Many foods, especially meats, are preserved by a combination of low
pH and reduced water activity, each at a subinhibitory level. Such combination treatments will be discussed later.

It is important to keep the pH stable when acid is the main preservation
factor (see Section X, B, 1).

VII. LIMITATION OF MOISTURE

A. General Considerations

Not only drying but also addition of sugar or salt can be considered as
preservation by moisture limitation, since the result in both cases depends on a reduction of available moisture to a level where microbial
growth is prevented. The effects are not completely identical, however,
salts may have specific toxic effects, and survival of organisms in concentrated salt solutions is generally much shorter than in dried foods.

Different strains of the same microbial species are generally homogeneous in their requirement for moisture to support growth. This has
been found for *Salmonella, Staphylococcus,* and *C. botulinum* (Ohye

et al., 1967). The effect of moisture limitation on the growth of micro-organisms has been reviewed by Scott (1957) and Christian (1963).

B. Effect on Disease Agents

1. NONSPORING BACTERIA

a. Salmonella. Salmonella has a rather low salt tolerance. Growth occurs in meat with 3.2% brine* and scantily in meat with 5.3% brine (Dack, 1963). The inhibitory level of sodium chloride in nutrient broth and ham with gelatin is between 5.8% and 7.5% brine, depending on pH. Growth stopped at brine concentrations 3.7 to 5.8% when pH was 5.4, but 5.8 to 6.7% brine was required at pH 5.8, and more than 6.7% brine at pH 6.35 (Blanche Koelensmid and Van Rhee, 1964).

Motile *Salmonella* have been found to multiply in an a_w (water activity) range of 0.999 to 0.945 under aerobic and anaerobic conditions (Ohye *et al.,* 1967). Nonmotile *Salmonella* were slightly more sensitive (Christian and Scott, 1953). The ability of microorganisms to grow at a low water activity is improved in rich media. Proline and its analogs but not other amino acids have been found to stimulate the growth of *Salmonella* at low a_w (Christian and Waltho, 1966).

Salmonella has a considerable ability to survive at a low water activity. Limited destruction (67 to 99%) takes place during freeze-drying where the cells are exposed to elevated temperatures during the final drying period (Silverman and Goldblith, 1965; Sinskey *et al.,* 1964). Increased temperature has also been found to accelerate death in dried egg white, the effect being dependent on the amount of residual water in the product. Six to ten days at 50°C or 2 to 3 days at 70°C was required for a 10^5- to 10^6-fold reduction in numbers of viable *Salmonella* when the water content was 1.5 to 6%. Three to four days at 50°C or 1 day at 70°C caused the same reduction when the water content was 12% (Banwart and Ayres, 1956).

At lower temperatures, *Salmonella* can survive for long periods of time, sometimes for years, in food and in other media if the water content is less than 10 to 12%.

Salmonella may also survive for a long time in cured products. Two months storage of naturally contaminated salami (23% brine) at room temperature did not cause complete destruction (Marazza and Crespi, 1963). Ninety percent destruction of *Salmonella* in bacon with 9.1 to 10.7% brine took place in 4 weeks at 5°C, and 98.5% destruction was observed at 25°C (Bardsley and Taylor, 1960).

---

*Brine concentration is defined as $\dfrac{\text{percent NaCl}}{\text{Percent water} + \text{percent NaCl}} \times 100$.

b. Staphylococcus aureus. *Staphylococcus aureus* can grow in laboratory media saturated with NaCl (Parfentjen and Catelli, 1964), but the optimum salt concentration for growth is about 3.4% brine (Nunheimer and Fabian, 1940). Growth and enterotoxin B production has also been observed in media with pH 6.9 and 10% NaCl or pH 5.1 and 4% NaCl (Genigeorgis and Sadler, 1966). Enterotoxin B production has been detected at 10° to 30°C in inoculated ham with pH higher than 5.3 and 9.2% brine. At 10°C toxin could be detected after 2 to 8 weeks of anaerobic incubation (Genigeorgis *et al.*, 1967).

The salt tolerance of *Staphylococcus aureus* is decreased if the cells have been exposed to a heat stress. Heated cells may fail to grow on salt containing staphylococcal media (Busta and Jezeski, 1963).

Staphylococcal strains seem to be homogeneous in their a_w requirements. Aerobic growth takes place in media with water activity from 0.99 to 0.86 (Scott, 1953; Ohye *et al.*, 1967), but the rate of growth is reduced considerably at a_w values below 0.94. Staphylococci may be slightly more tolerant to low a_w in foods.

Staphylococcus aureus may survive for an indefinite time in a dry environment. Cells dried on textiles have shown 100% survival after 20 days and 10% survival after 40 days of storage at 21°C and 42 to 50% relative humidity (Rountree, 1963). Destruction of *Staphylococcus aureus* during freeze-drying is limited. Fourteen to fifty percent survivors have been formed in freeze-dried beef and egg (Silverman and Goldblith, 1965; Sinskey *et al.*, 1964). Progressive death may occur in cured foods stored at low temperature. Ninety to ninety-nine percent destruction has been reported in bacon during 6 weeks of storage at 5°C (Bardsley and Taylor, 1960).

c. Other Nonsporeformers and Fungi. Many nonsporing disease agents such as *Listeria* and *Erysipelothrix* may survive for indefinite periods of time in dried or salted foods. *Vibrio parahaemolyticus* can grow in the presence of 7 to 10% NaCl in laboratory media.

Fungi capable of producing mycotoxins are able to grow at lower moisture levels than bacteria. Water activity values as low as 0.70 may be required for complete inhibition of fungi.

2. SPOREFORMING BACTERIA

a. Clostridium botulinum. The effect of moisture on *C. botulinum* has been extensively studied. The following limiting a_w values have been found for outgrowth of spores at optimum temperature in a laboratory medium (Ohye and Christian, 1967): *Clostridium botulinum* type A, 0.95; type B, 0.94; type E, 0.97.

Outgrowth at pH 6 was delayed at decreased a_w. Types A and B did not grow out at $a_w = 0.96$ at pH 6.0 and 20°C. Type E did not grow out at $a_w = 0.98$ at 10°C.

Other studies (Emodi and Lechowich, 1967) have indicated that incubation temperature has no consistent effect on the a_w required for outgrowth of type E. The type of food or medium probably plays a role.

Germination may sometimes seem to take place at a_w values lower than those indicated above. It is a common experience that the numbers of viable spores in dry materials decreases during storage (Wagenaar and Dack, 1954, 1960; Ulrich and Halvorson, 1949; Schmidt, 1964). The spores probably lose their ability to form colonies without undergoing true germinal changes.

The growth of vegetative cells is also very dependent on water activity (Ohye et al., 1967). Clostridium botulinum E showed maximum growth at $a_w = 0.995$ in a temperature range from 15° to 30°C. Reduction of a_w to 0.98 reduced the growth rate at 30°C by 35%, but toxin was formed even at the lowest a_w which permitted growth. The minimum a_w supporting growth of type E increased appreciably when the temperature was decreased by 15°C in a range above refrigeration temperature.

Vegetative growth of types A and B is inhibited at $a_w = 0.93$ to 0.94. Salt brine (10%) has a water activity of 0.935, and early studies indicated that 10.7 to 12% NaCl is required for complete inhibition of C. botulinum A and B in a medium containing meat particles (Tanner and Evans, 1933a). However, some strains seem to be less tolerant. Type A was inhibited in liquid laboratory media by 6.8 to 6.9% brine and type B by 5.9 to 6.0% in studies reported by Pedersen (1957).

i. Effect of salt in meat products.   The salt tolerance may be lower in some foods than in laboratory media. This is true for cured meats, where the effect of salt is supported by other curing salts, especially nitrite and pH (Scott, 1955). The inhibitory action of nitrate and nitrite is discussed later.

Salt in lower concentrations (5 to 7% brine) than that required for complete inhibition in laboratory media inhibits or delays germination and outgrowth of clostridial spores which have been heat-processed in cured meats (Stumbo et al., 1945; Silliker et al., 1958).

It has been observed that salt in meat affects putrefaction and toxin production by C. botulinum to different degrees. Spoilage of meat inoculated with type A or type B spores was inhibited by 6 to 7% brine, but detectable amounts of toxin were found in some of the unspoiled samples. Nine percent brine was required to inhibit toxin production completely (Greenberg et al., 1959).

A number of studies of growth and toxin production in vacuum-packed cured meats have been published. In products with high salt concentration or low pH, as cervelat sausage or summer sausage, no toxin production takes place (Hall, 1941), but in items with a lower salt concentration or a higher pH, such as bologna, toxin production may occur and is unaffected by vacuum packaging (Christiansen and Foster, 1965). Toxin production in such meat items depends not only on the salt concentration and the pH and nitrite content but also on the type of product; thus, more toxin was produced by *C. botulinum* A in ham than in bologna sausage (Pivnick and Bird, 1965). Type A and B produced toxin in bologna containing 3.8 to 5.1% brine in 1 week at 30°C but not in 8 weeks at 25°C or lower. In ham containing 3.4 to 4.5% brine, toxin production took place in 2 to 15 days at 20° to 30°C. At 30°C toxin was produced in the presence of 4.8% brine. When the amount of nitrite added to ham containing 4.1% brine was increased from 25 ppm to 150 ppm, the time required for toxin production at 25°C was increased from 4 days to 25 days. Type A and B produced toxin in pork tongue with 2.2 to 4.4% brine in 2 to 4 days at 30°C (Pivnick *et al.*, 1967; Pivnick and Barnett, 1965).

Similar results have been published by Spencer (1967). In cured meats with pH 6.3 to 6.5 toxin was formed in the presence of 3.3% brine. Type A also formed toxin in the presence of 4.5% brine, but no toxin could be detected in cured meat with 4.7% brine. The nitrite concentration in these meat products varied from 4.6 to 15.7 ppm. Toxin could not be detected in cured meat incubated up to 14 days when the incubation temperature was lowered to 20°C.

When 10^5 to 10^6 vegetative cells of type A or B were inoculated into a meat slurry, pH 6.8, growth took place in 1 to 7 days at 30°C in the presence of 7.5% brine.

ii.   Effect of other curing salts (nitrate and nitrite) in meats. Nitrate has a very limited inhibitory effect on *C. botulinum;* 2.2 to 4.7% is required to inhibit growth (Tanner and Evans, 1933*b*). Nitrite is much more effective, and the importance of nitrite as a microbial inhibitor has been demonstrated in a number of publications (Abrahamsson, 1964; Stumbo *et al.*, 1945; Silliker *et al.*, 1958; Pivnick *et al.*, 1967; Riemann, 1963; Tarr, 1940; Castellani and Niven, 1955). Originally only nitrate (saltpeter) was added in meat curing, and nitrite was formed by bacterial reduction of nitrate. Later, direct addition of nitrite became common practice, since very little nitrate is reduced to nitrite in the rapid curing processes which are now widely used. In some prolonged curing processes, such as curing of Wiltshire bacon, addition of nitrate, as well as nitrite, is used. The nitrate forms a reserve for nitrite, which is a rather unstable compound. The amount of nitrite added is usually about 0.5% of the amount of sodium chloride, and the legal limit of nitrite in meat is 200 ppm in most countries.

The mechanism by which nitrite acts on microorganisms is little known except that it is the undissociated HNO_2 which is the active compound; therefore, the effect increases with decreasing pH, but the maximum can be expected at pH values around 5.6, since nitrous acid is very unstable at lower pH values (Ingram, 1959). A number of other factors have been shown to influence the action of nitrite (Castellani and Niven, 1955), but they do not seem to have been explored in detail.

Since nitrite, or rather the undissociated HNO_2, is very reactive, up to 90% or more disappears during the processing of meat. The reactions responsible for its disappearance may lead to formation of other inhibitory compounds, since it has been found — as mentioned earlier — that the addition of small amounts of nitrite results in pronounced inhibition of clostridia provided the nitrite is autoclaved in the medium.

iii. Effect of salt in other foods.   The brine concentration required to inhibit *C. botulinum* is similar to what has been found for cured meats. The highest brine concentration which permitted production of type A toxin at 30°C in surface-ripened cheese was 8%; type B toxin production did not take place at brine concentrations higher than 6%. However, type A toxin production was inhibited by 6.5 to 4.9% brine in another type of cheese. This was probably due to a drop in pH caused by the accumulation of 1 to 2% free fatty acids or water-soluble antibiotics produced during the cheese ripening (Wagenaar and Dack, 1958; Grecz *et al.*, 1959a,b).

b. Clostridium perfringens.   This organism has been studied less extensively than *C. botulinum,* but published data indicate that it has a salt tolerance similar to that of *C. botulinum* type A (Hobbs, 1965; Oxhøj, 1943). *Bacillus cereus* and *B. mesentericus* have a higher salt tolerance but have not been reported as a cause of food poisoning in cured foods.

c. Survival of Spores.   Bacterial spores survive for a long time in food with low water activity. *Clostridium botulinum* spores have been found to survive for several months in bread (Ingram and Hanford, 1957; Kadavy and Dack, 1951), and the number of viable spores remained practically unchanged during a month of storage in fish and meat curing brines (Pedersen, 1957; Beerens, 1957). Likewise, spores of *C. perfringens* remained unchanged in numbers during 3 weeks of storage at 6° to 18°C in meat curing brine, but the number of vegetative cells was reduced 10^3- to 10^4-fold in 20 hours. Fecal strains of *C. perfringens* were found to be more sensitive than strains from soil (Beerens, 1957). *Clostridium perfringens* spores may survive for 1 to 2 months in cured meats (Gough and Alford, 1965).

3. PARASITES

Some parasites such as *Trichina* are killed by prolonged curing, and such curing techniques are used to control trichinosis.

4. VIRUSES AND TOXINS

These two types of disease agents may actually be preserved by drying or curing. Sodium chloride in concentrations up to 12% protects some viruses, and viruses dried in organic material may survive for years. Toxins are also preserved by drying.

C. Curing and Drying of Foods

Addition of 9 to 10% sodium chloride (brine concentration) inhibits growth of all food poisoning microorganisms with the exception of *Staphylococcus aureus* and fungi. Even these two types of microorganisms have reduced growth rate, and there is no example of production of fungal toxins in cured foods. The amount of salt which must be added depends on the water content of the food. Thus 6.9% salt must be added to fresh lean meat containing 70% water to get a brine concentration of 9%, but 4% salt is sufficient in meat product that contains only 40% water, as, for example, meat that is high in fat content or partially dried during smoking.

Some care should be taken in evaluation of water activity or brine concentration in heterogeneous foods. It has been reported that *C. botulinum* type E produced toxin in bacon with 5.46% brine (Warnecke *et al.*, 1967) and that type A produced toxin in the presence of 16% brine or more (Anellis *et al.*, 1965). Both concentrations are considerably higher than what has been found to be inhibitory in other media, and it may be possible that the water activity remains different for some time in different parts of a heterogeneous food product. Local differences in water activity in a food may also be caused by water migration if the food is exposed to changing temperatures.

Salt is often used together with other factors, each being applied at a subinhibitory level. Such factors are pH, sodium nitrite, and application of heat. Hot smoking of cured meats will reduce the possibility of staphylococcal growth, since this process will cause heat damage of the cells as well as a decrease in pH. Heat processing of canned cured meat similarly makes spores more sensitive to salt and sodium nitrite, and the addition of vinegar to pickled fish reduces the concentration of salt required.

Although the interactions between pH, sodium chloride, heat processing, storage temperature, and nitrite have been demonstrated, no quantitative data exist which permit the prediction of the required level of any one of these factors when the other factors are known.

Information about the effect of degree contamination, with, for example, *C. botulinum* spores, on the concentration of required inhibitory compounds is limited. Studies (Riemann, 1967) indicate that the required level of pH decreases and the level of salt increases when more spores of *C. botulinum* are present. More important is the fact that even a few spores can germinate and grow out at salt concentrations only slightly below the inhibitory level. This suggests that in curing of food it is even more important to keep the curing process in control than to control contamination.

Although drying or addition of salt can successfully inhibit food poisoning organisms, the killing effect is limited. Survival of contaminating agents must therefore be expected, and great care must be taken to control the temperature and time when such foods are reconstituted with water before use.

VIII. CHEMICAL PRESERVATIVES AND OTHER INHIBITORS

A number of chemicals are used in food to control microbial growth, but practically all are applied with the purpose of controlling spoilage rather than growth of food poisoning organisms. However, it should be a requirement that a preservative must be at least as effective against food poisoning organisms which can grow in the food as against spoilage organisms (Ingram *et al.,* 1964). If this were not the case, selective growth of food poisoning organisms could result from the addition of the preservative.

In the case of "symbiosis," however, the situation may be somewhat different. A preservative may be inactive against *C. botulinum* itself but inhibit organisms which promote the growth of *C. botulinum* — for example, inhibition of fungi which may raise the pH in acid products. This seems to be the case with preservatives such as caproic, propionic, and sorbic acids (Hansen and Appleman, 1955). Several aspects of antimicrobial agents in foods have recently been discussed at a symposium (Molin, 1964).

Various metal salts have been found to influence outgrowth of *C.*

botulinum in food products. It has been observed that dissolved tin from cans inhibits *C. botulinum*. Growth from spores of *C. botulinum* A and B in canned beet root was prevented by 150 ppm of an added tin citrate complex, and 30 to 60 ppm was sufficient in carrot (Scott and Steward, 1944). When 0.25 to 0.5 gm of powdered metals was mixed into 100 gm of pork luncheon meat inoculated with type A spores, growth developed faster with metals of greater electrochemical activity. Iron was more active, however, and aluminum less active than predicted. Toxin production was erratic but was in no instance found in cans with added magnesium, nickel, cobalt, or tin. No growth took place in cans with added nickel or cobalt (Kempe and Graikoski, 1964*b*).

Various inhibitors may occur naturally in foods. It has already been mentioned that small amounts of inhibitors such as rancid fatty acids and other unknown compounds can inhibit germination of botulinum spores in laboratory media. The inhibition can in many cases be overcome by addition of 0.1% soluble starch or by treatment with charcoal. Compounds have also been found in solutions of glucose autoclaved together with phosphate, which influence growth of *C. botulinum* in a synthetic medium. Some of these compounds were inhibitory and their effect could be removed by treatment with charcoal (Bowers and Williams, 1962). It is doubtful that such inhibitors play an important role in complex food media, but it has been noticed that flavonoids related to compounds found in asparagus are inhibitory to types A and B spores (Andersen and Berry, 1947). A few types of spices may also exert an inhibitory effect.

Antibiotics have probably received more publicity than all other chemical preservatives. Several antibiotics have shown promising effect, and proper use of antibiotics in foods might decrease the incidence of food poisoning (Ingram and Barnes, 1955). It has been found that 0.4 ppm of subtilin prevents growth from an inoculum of *C. botulinum* type A spores in laboratory media and foods (Andersen, 1952), but potential commercial use of this antibiotic was abandoned after it was discovered that the effect was only temporary because of instability of subtilin. Furthermore, *C. botulinum* can become resistant to 100 ppm of subtilin by successive transfers in a subtilin-containing medium (Campbell and Wrenarski, 1959). Nisin, which is another polypeptide antibiotic produced by *S. lactis*, was first used to prevent growth of butyric acid-forming clostridia in cheese. It also affects *C. botulinum*. Nisin is being used in canned foods in England to control thermoresistant sporeformers, but the foods must be given an adequate heating to eliminate the risk of botulism. Interest has recently been focused on another antibiotic, tylosin, which inhibits growth of *C. botulinum* types A and B for extended periods of time. Many spoilage organisms seem to be more resistant to tylosin (Greenberg and Silliker, 1964). A water-soluble antimicrobial compound which inhibits *C. botulinum* in a concentration of 0.1% has been found in surface-ripened cheese, and a similar compound is formed by *B. linens* in a cheese medium (Grecz *et al.*, 1961; Grecz, 1964).

So far, no purposely added antibiotic has been applied commercially to control botulism or other types of food poisoning, since it is very difficult to prove that a given antibiotic would be effective under all circumstances. One problem is that continued use of an antibiotic may create an environment where resistant strains will emerge. It has indeed been shown that use of chlorotetracycline as a preservative resulted in selection of resistant *Salmonella* strains (Thatcher and Lait, 1961), and *C. botulinum* A may become resistant to subtilin. The selective pressure of antibiotics will also apply to numerous spoilage organisms; this and other ecological aspects of preserving food with antibiotics have not yet been fully explored. The situation created by the use of antibiotics will probably vary considerably with the type of antibiotic and the type of food for which it is used.

It is clearly not advisable to use the same antibiotic as a feed supplement for food animals, which frequently harbor pathogens, and as an additive intended to preserve food. However, tetracyclines are being used as feed additives for chickens and as a preservative for poultry meat. The feeding practice may be responsible for the increase in tetracycline-resistant *Salmonella* isolated from man and domesticated animals from 0% in 1948 to 27 to 29% in 1962 (McWhorter *et al.*, 1963).

IX. SMOKING

There seems to be little information about the effect of smoking on food poisoning organisms. However, botulism caused by smoked fish has recently prompted a series of studies.

The heat applied in a hot smoking process will generally destroy parasites as well as many nonsporing food poisoning organisms. Hot smoking of bacon may reduce total bacterial count by a factor of 10^5 (Gibbons *et al.*, 1954) and hot smoking (82°C for 30 minutes) of fish has been found to cause a 10^3- to 10^4-fold reduction in total bacterial count (Pace *et al.*, 1967). Wood smoke also has chemical effects due to its content of phenolic compounds, acetic acids, etc. The pH in the superficial layers of fish may drop from 6.7 to 5.9 during smoking (Shewan, 1956). The chemical effects will be limited to a surface layer, however. Smoking has a selective effect on the microflora in foods. Lactic acid bacteria are generally less affected than staphylococci and micrococci. The bacterial flora of smoked meats will, therefore, tend to change from catalase-positive to catalase-negative (Hanford and Gibbs, 1964).

Clostridium botulinum A, B, and E inoculated into smoked salmon (2.0 to 4.1 ppm of phenols) produced toxin in the presence of 1.96% brine, but no toxin was formed in fish with 2.9% brine (Spencer, 1967). Small amounts of E toxin were produced at 20°C and 10°C in vacuum-packed kippers with 3.2% brine (Cann *et al.*, 1967).

Other studies with *C. botulinum* types E and B have indicated that neither rancidity products nor phenol had any significant inhibitory effect on the spores, but formaldehyde in concentrations normally found in the tissues of smoked salmon (35 to 40 μg/gm) had a strong inhibitory effect (Nielsen and Pedersen, 1967).

Clostridium botulinum type E was isolated from 2 of 10 samples of eel that had been smoked for 2 hours at 55°C and for $\frac{1}{2}$ hour at 60°C, but no toxin was formed when smoked eel was inoculated with 10^3 type E spores and incubated in plastic bags for 10 days at 10°C (Abrahamsson, 1967). The salt concentration in the eel was 2% (about 4 to 4.5% brine), and the pH was 4.5.

Pace *et al.* (1967) found that smoking of fish at 82°C for 30 minutes caused a 90% decrease in *C. botulinum* E count.

Liquid smoke preparations are being used increasingly in the food industry. These products contain a number of the important components of natural smoke (Halenbeck, 1964), and they have some bacteriostatic action. Carcinogenic compounds are removed by the production process for liquid smoke.

X. MICROBIAL COMPETITION AND SYNERGISM

A. General Considerations

The preservation of "moist" foods by freezing, refrigeration, and mild processing (curing, heating) often results in products which contain rather high numbers of various saprophytic organisms. This has created interest in "natural" antibiosis.

The saprophytes will generally greatly outnumber any pathogen which may be present, and if they grow they often exert a repressive effect on the pathogens. The situation is somewhat similar to that of classical fermented foods, although fermentation is generally not intended in the case of most frozen and refrigerated foods. The "fermentation" which takes place during prolonged storage or temperature mishandling of such foods is most often classified as spoilage and is regarded as undesirable. The general attitude is that the total bacterial numbers in these foods should be kept as low as possible. This attitude is supported by the fact that very little is known about the effect a continued consumption of fairly large numbers of mixed saprophytic organisms may have on human beings. Furthermore, it is generally believed that a high total bacterial count may indicate an increased risk that the food is also contaminated with pathogens.

However, the fact remains that many moist, mildly processed foods contain many saprophytic organisms. These will develop into larger numbers even when the initial count is low if a refrigerated food is stored for a long time or at an elevated temperature. This situation may be more common for foods packed under vacuum and/or preserved by irradiation (substerilization), since such treatments may change the composition of the microflora and therefore also change the spoilage pattern with the result that foods remain acceptable with higher total numbers of bacteria. One way to bring this situation under better control would be to deliberately inoculate the foods with microorganisms considered harmless or even beneficial for humans. Such organisms must be able to grow in the food so that they will make up the main bacterial flora in the case of prolonged storage or temperature mishandling. This procedure is parallel to the use of lactic acid starter cultures in dairy products and has been suggested as a method to prevent growth of *C. botulinum* in defrosting frozen foods. One of the few examples of development of starter cultures in recent years is the application of *Pediococcus cerevisiae* in the manufacture of summer sausage and similar meat products. The purpose in this case is the classical: The starter culture is added not only to control the microflora but also to produce a desirable flavor. The product is therefore deliberately treated in such a way that maximum bacterial growth occurs.

B. Effect on Various Disease Agents

1. Clostridium botulinum

The effect of other organisms on *C. botulinum* depends on the type of organism, the type of food, and the storage temperature.

A synergistic effect of lactic acid bacteria on *C. botulinum* has been observed. The effect may be due to a lowering of the oxidation-reduction potential or to production of growth factors (Benjamin *et al.,* 1956).

Yeast has been observed to improve growth of *C. botulinum* in acid food, apparently through oxidation of acid with a resulting increase in pH (Meyer and Gunnison, 1929). Cases have also been reported where growth of *Penicillium* and *Mycoderma* in fruit products increased pH from 3.3–3.8 to 4.9 and 5.4, with the result that toxin was formed by *C. botulinum* (Tanner *et al.*, 1940). It has been observed that strains of type C will form toxin only in mixed cultures with strains such as *C. sporosphaeroides* or *B. polymyxa* (Bulatova *et al.*, 1967). The function of the symbionts in this case seems to be a rapid decrease of the oxidation-reduction potential.

Growth of *C. botulinum* in easily fermentable foods such as vegetables and milk is often prevented by naturally occurring acid-forming bacteria. It has been noted that contaminants may interfere with growth and toxin formation when frozen vegetables are thawed. Rapid growth of contaminants may decrease the pH to 5 or less. Even in frozen foods, such as chicken á la king, where most of the contaminants have been eliminated by heat processing, acid production may occur to an extent that inhibits *C. botulinum* (Saleh and Ordal, 1955*a*). When lactic acid organisms were inoculated in chicken á la king, a pH drop to 3.5 to 4.4 took place during defrosting at elevated temperature, and no botulinum toxin was formed (Saleh and Ordal, 1955*b*).

A decrease in pH caused by the contaminants seems to be a main factor in "natural" inhibition of *C. botulinum*. Bacteria which spoil cream-style corn without a decrease in pH did not prevent toxin production by type A or B (Malin and Greenberg, 1964). However, other types of inhibition of growth or toxin production have also been observed. Filtrates of *C. sporogenes* cultures and to some extent filtrates of *Streptococcus faecalis* and *E. coli* cultures may have an inhibitory effect on *C. botulinum* spore germination, and proteolytic enzymes from *C. sporogenes* may destroy botulinum toxin. *Bacillus cereus* enzymes have a similar effect (Skulberg, 1964). Growth of *C. botulinum* is inhibited in certain types of surface-ripened cheese not only by free fatty acids which accumulate during maturation but also by antibiotic compounds formed by *Brevibacterium linens* (Grecz *et al.*, 1959*a,b*, 1961; Grecz, 1964). Another antibiotic that has an inhibitory effect on clostridia is nisin, which is produced by *Streptococcus lactis* in milk products.

It has been observed that *Pseudomonas fragi* and *Ps. fluorescens* facilitate production of type E toxin in fish tissue incubated under aerobic conditions at 7.2°C. *Streptococcus lactis* and *Lactobacillus viridescens* seemed to inhibit or delay toxin production (Valenzuela *et al.*, 1967). It is not known whether the inhibition is due to antibiotics.

Clostridium botulinum strains themselves produce antibacterial agents, boticins. The boticins are active against other botulinum strains, and there are cross reactions with *C. sporogenes* strains (Beerens and Tahon, 1967). Some nontoxic *C. botulinum* E strains produce boticins active against toxic strains (Kautter, 1964). Also *C. perfringens* produce bacteriocins (welchicins), some of which are active against a broad spectrum of sporeformers (Sasarman and Antohi, 1967).

It is not known what role such compounds play in foods, but it has been observed that isolation of *C. botulinum* E is difficult when boticin-producing strains are present (Harmon and Kautter, 1967). The effect can be overcome by adding trypsin to the culture medium.

2. *Staphylococcus aureus* AND *Salmonella*

The natural flora in chicken, turkey, and beef pie may have a strong inhibitory effect on the growth of *Staphylococcus aureus, Salmonella,* and *E. coli* in slurries made from the frozen pies. Evident inhibition has been observed even when the inoculated pathogens outnumbered the natural flora by a factor of 10^2 and seemed to be correlated with an increased acidity to pH 4 (Dack and Lippitz, 1962). *Staphylococcus aureus* was also outgrown by saprophytes in raw crab meat (Stabyj *et al.,* 1965).

Strong repression of *Staphylococcus aureus* has been observed in defrosted chicken pie, but not in macaroni-and-cheese dinner. However, in the latter, spoilage developed at the same rate as staphylococci (Peterson *et al.,* 1962*a*). Similarly, a mixed flora of psychrophilic bacteria repressed *Staphylococcus aureus* in laboratory media and produced organoleptic changes. The mixed population dominated over *Staphylococcus aureus* at all tested salt concentrations when the incubation temperature was 10°C. At 30° to 37°C the staphylococci were able to dominate in the presence of 5.5% NaCl, at 20° to 37°C in the presence of 2.5% NaCl (Peterson *et al.,* 1962*b*, 1964). Almost 50% of bacterial cultures isolated from various foods have been found to affect the growth of *Staphylococcus aureus,* and more than half of the strains had an inhibitory effect. Most of the inhibitory strains belonged to the genera *Streptococcus, Leuconostoc,* and *Lactobacillus* (Graves and Frazier, 1963). Mixed cultures containing *Achromobacter, E. coli, Pseudomonas, B. cereus, Serratia,* and *Aerobacter* repressed *Staphylococcus aureus. Serratia* and *Pseudomonas* seem to inhibit by "out-competing" the staphylococci. Other species inhibited by producing dialyzable, heat-stable antibiotics. Inhibition was not due to changes in pH (optimum pH for inhibition was 6.2 to 7.4), oxidation-reduction potential, or accumulation of peroxide or fatty acids.

Oxygen was not a factor in inhibition, but the type of medium had an influence. Some gram-negative species were found to stimulate staphylococcal growth at 15°C (DiGiacinto and Frazier, 1966; Seminiano and Frazier, 1966; Troller and Frazier, 1963a,b).

Other studies with food microorganisms have confirmed the fact that inhibition of *Staphylococcus* is more common than stimulation. Of 44 organisms studied, only *B. cereus* was observed to stimulate staphylococcal growth and enterotoxin formation (McCoy and Faber, 1966).

A micrococcus related to *M. aurantiacus* has been shown to delay spoilage in salami-type sausages and to repress *Staphylococcus aureus* (Nünivaara, 1957). This strain, together with *Streptococcus faecium,* is used as starter cultures for salami sausages in some European countries. Lactic acid bacteria may inhibit or stimulate growth of *Staphylococcus,* depending on the conditions (Oberhofer and Frazier, 1961; Kao and Frazier, 1966). Extracts of thermophilic lactobacilli have been found to contain an inhibitory principle (Dahiya and Speck, 1967).

3. VIRUSES

No effect of the saprophytic bacteria in foods has been observed.

XI. PACKAGING IN MATERIALS OF LOW PERMEABILITY

The microbiological aspects of packaging have been discussed by Ingram (1962).

Packaging of foods which are not sterile and which can support microbial growth involve several changes. The surface equilibrium relative humidity (water activity) of the product is no longer determined by the humidity of the surrounding atmosphere but by the water activity of the product. The gas phase, especially its content of oxygen and carbon dioxide, is determined by the respiratory activity of the product itself and its microflora. The latter is influenced by the water activity, the composition of the product, the storage temperature, and also the gases in the package. The composition of the microflora will therefore frequently change in coordination with changes in the gas phase. Generally oxygen will disappear and CO_2 will accumulate. As a result, the microflora will change from an aerobic to a facultative or obligate anaerobic, the extent and rate of change depending on initial contamination, product composition, and storage temperature. The less a product is preserved by other means (curing, smoking, drying) and the lower the storage temperature, the longer shelf-life will be when it is vacuum-packed in plastic films of low permeability.

Vacuum-packed smoked fish have caused outbreaks of botulism. In these cases a rather mild preservation (smoking process) permitted *C. botulinum* E to survive and to grow out in the package where spoilage or competing organisms were repressed. These outbreaks have caused suspicion of vacuum packaging. It has, however, been demonstrated that vacuum packaging is not a requirement for *C. botulinum* to grow and produce toxin on fish (Thatcher *et al.*, 1962; Johanssen, 1961). It is clear that clostridia in the interior of a fish or a piece of meat would hardly be influenced by vacuum packaging, and it has been demonstrated that *C. botulinum* spores inoculated between two slices of bologna produce toxin independent of vacuum packaging (Christiansen and Foster, 1965). On the other hand, clostridia on the surface might have their lag phase reduced by vacuum packaging. *Clostridium perfringens* in fresh horse meat showed increasing lag phase when the potential increased from $Eh = -45$ mV to $+ 230$ mV. However, when growth had commenced, it proceeded with a rate that was independent of the potential (Barnes and Ingram, 1956). The organisms themselves apparently are able to control the oxidation-reduction potential in the immediate vicinity of the cells once cell metabolism and growth has been initiated. Similarly, clostridial growth on the surface of unwrapped fish may be facilitated by facultative saprophytes that reduce the oxidation-reduction potential. An important question is whether organoleptic spoilage will proceed faster than toxin production and thus warn the consumer.

Not only smoked fish but also wet fish are being vacuum-packed. It seems likely that wet fish, with the exception of herring, will spoil before detectable amounts of E toxin are formed. Even so it is clear that all vacuum-packed fish should be stored at temperatures below 4°C (Cann *et al.*, 1967).

The microflora which develops in "moist" vacuum-packed foods frequently contains high numbers of lactobacilli which are relatively favored by a gas phase which is low in O_2 and high in CO_2. However, the flora is very dependent on storage temperature and water activity. In vacuum-packed sliced bacon stored at 20° to 30°C, catalase-positive organisms dominate when the salt content is high (9 to 11% brine), while group D streptococci and lactobacilli dominate in low-salt bacon (5 to 7% brine). Lactic acid bacteria seldom produce gross organoleptic changes in, for example, meats, and acid production is limited unless carbohydrates are added. Pivnick and Barnett (1967) observed that botulinum toxin was produced in inoculated meat packed in a variety of plastic pouches but spoilage was delayed in vacuum packs.

Staphylococcus aureus is often repressed in naturally contaminated vacuum-packed bacon, apparently by lactobacilli (Eddy and Ingram, 1962; Ingram, 1960, 1962; Tonge *et al.,* 1964). The same seems to be the case in vacuum-packed bologna (Christiansen and Foster, 1965). It has also been observed that enterotoxin A production in meat is mainly confined to surfaces exposed to air (Casman *et al.,* 1963).

XII. SAFETY OF FOOD PROCESSING

The degree of safety which constitutes wholesomeness is not defined. There seems to be a tacit agreement that any quantity of food should be rendered completely free of toxigenic or pathogenic microorganisms. It is clear that this situation does not exist, but no one has tried to define safety limits, although certain empirical standards have been adopted. A brief discussion of the present situation may be useful.

In discussing food processing in relation to the safety of food, a distinction must be made between foods in which food poisoning bacteria can multiply and foods in which they cannot. For example, no attempts are made to kill *C. botulinum* spores in tomato juice, since the pH of the juice inhibits growth of this organism. The situation is different with foods that offer good growth conditions for *C. botulinum* and other food poisoning organisms. For such foods, precautions must be taken to prevent growth. This is done either by refrigeration, freezing, or regulation of the water activity, or by application of heat or irradiation after the food has been placed in a hermetically sealed container.

The processing conditions required to protect against *C. botulinum* have been studied most intensively in the case of heat-processed canned foods and irradiation-processed foods. It is a general assumption that foods with pH higher than 4.5 must be given a treatment that will reduce viable *C. botulinum* spores by a factor of 10^{12}. This requirement is called the 12-*D* requirement, since it corresponds to 12 decimal reductions in the numbers of viable spores.

Experience from the past 40 years shows that a 12-*D* process gives a high degree of safety against *C. botulinum* in canned, low-acid foods (Perkins, 1964). However, the 12-*D* concept rests on very vague grounds, and it is interesting that a much milder heat-treatment corresponding to approximately 4-*D* values traditionally is used for one category of low-acid canned food, namely canned cured meat. Canned cured meats have maintained a good public health record in spite of the mild heat process, and it is believed that this is due to a low incidence of botulinum spores in meat, but the proof is lacking. Canned cured meat is not the only exception. Newer methods of heat processing such as high-temperature short-time canning and aseptic canning may have a higher survival probability for *C. botulinum* than conventional methods of retorting (Charm, 1966). These facts suggest that there is a need for a more accurate evaluation of safety margins in canned foods.

A number of food items are given a mild heat-treatment and then packed and distributed under refrigeration. Examples are smoked fish and various cured lunch meat and sausages. *Clostridium botulinum* A and B cannot develop in these foods if they are properly cooled, but type E can grow if the water activity and pH are high enough. No attempts were made to establish *D* requirements for this type of product until outbreaks of E botulism occurred a few years ago.

The outbreaks caused by smoked fish led to re-evaluation of processing techniques for this product. A standard heat process of 82°C for 30 minutes has been adopted to destroy *C. botulinum* E spores. This corresponds to more than 12-*D* values for type E spores heated in aqueous suspension, but spores may survive better on the surface of fish, and it is not yet possible to predict precisely the killing effect of the smoking process. More studies seem to be required to evaluate the smoking process in terms of *D* values and to establish *D* requirements that give a sufficient protection against botulinum. Other vacuum-packed pasteurized products are given a heat process which corresponds to only one or two *D* values for *C. botulinum* E spores. The safety of these products rests on proper cooling, low pH, and water activity and probably infrequent occurrence of botulinum spores.

The safety of home-cooked meals which are served cold also depends on proper cooking and refrigeration. Guidelines are given in the chapter on *C. perfringens* and *B cereus*. In the United States, the recommended 6 hours of cooking at 350°C of turkey may seem excessive in terms of the eating quality of the meat.

The 12-*D* requirement (for botulinum A and B spores) is applied in sterilization of foods by ionizing radiation. The logic behind this requirement is that products should possess the same degree of safety whatever the method of preservation. It has been suggested that the requirement should be lowered in the case of cured meats to correspond to the milder heat process traditionally used for these products.

There is considerable interest in application of low irradiation doses (substerilization, radio pasteurization) to prolong the shelf-life of low-acid foods, especially marine foods, which are distributed under refrigeration. The doses which have been suggested have very little destructive effect on botulinum spores, including type E, and the safety of these products will rest on proper refrigeration.

The 12-D concept for C. *botulinum* A and B has not been applied to other preservation methods such as curing, fermentation, and drying, although the safety of these methods rests on exactly the same principle as the safety of heat-canned or irradiated foods—namely, that one botulinum spore will initiate growth and give rise to toxin production. The safety margin of these processes is, therefore, often unknown. When testing has been carried out with inoculated packs, usually not more than 10^6 botulinum spores have been inoculated. This means that the test is based on a 6-D requirement. It is true that many products in practice are preserved by drying, curing, or fermentation to such an extent that any number of C. *botulinum* spores will be inhibited. However, there is a rapid change in processing technology with a tendency toward more "mild" foods, and future processes may become more marginal with regard to inhibition of botulinum spores. A very strict control in processing plants is required when marginal processes are applied. It has been found that a decrease in salt concentration of 0.5% in the critical range may cause a 10^5-fold increase in the numbers of spores which can germinate and grow out (Riemann, 1967). Present curing techniques which are applied—in meat curing, for example—seem to result in too great a variance in salt content to permit the use of marginal curing.

No general agreement with regard to D requirements seems to exist for organisms other than C *botulinum*. The literature describing heat processes designed to destroy *Salmonella* and *Staphylococcus* in foods usually suggests processes that vary from 10^5-fold reduction (5-D values) to 10^9-fold reduction (9-D values). In the case of liquid egg pasteurization, some states in the United States have regulations that call for a treatment at 140°F (60°C) for $3\frac{1}{2}$ to 4 minutes. This causes a 10^1- to 10^2-fold reduction in numbers of viable S. *senftenberg* 775W cells and a 10^5- to 10^6-fold reduction of other *Salmonella* cells. British regulations required 148°F (64.4°C for $2\frac{1}{2}$ minutes), which results in a 10^5-fold reduction of S. *senftenberg* W and probably more than 10^{12}-fold reduction of other *Salmonella*.

Ideally, preservation process requirements should be based on (1) the number of organisms which may be present in the food and (2) an accepted, low risk. At present, however, no degree of risk is generally accepted, and the information about the numbers of food poisoning organisms naturally occurring in foods is limited. The best approach for the food processor when he formulates new products or new processes is therefore to apply techniques that give a degree of safety equivalent to that of similar products which have proved successful.

REFERENCES

Abrahamsson, K. (1964). Report No. 150. Swedish Institute for Food Preservation Research, Gothenburg, Sweden.

Abrahamsson, K. (1967). *In* "Botulism 1966" (M. Ingram and T. A. Roberts, eds.), pp. 73–75. Chapmen and Hall, London.

Andersen, A. A. (1952). *J. Bacteriol.* **64,** 145.

Andersen, A. A., and Berry, J. A. (1947). *Science* **103,** 644.

Anderson, A. W. (1964). *U.S. Govt. Research Rept.* **39,** (17)S-6-7 (CA 61, 16698h).

Anderson, A. W., Corlett, D. A., and Krabbenhoft, K. L. (1967). *In* "Botulism 1966" (M. Ingram and T. A. Roberts, eds.), pp. 76–88. Chapman and Hall, London.

Anellis, A., and Koch, R. B. (1962). *Appl. Microbiol.* **10,** 326.

Anellis, A., Lubas, J., and Raymon, M. (1954). *Food Res.* **19,** 377.

Anellis, A., Grecz, N., Huber, D. A., Berkowitz, D., Schneider, M. D., and Simon, M. (1965). *Appl. Microbiol.* **13,** 37.

Anellis, A., Berkowitz, D., Jarboc, C., and El-Bisi, H. (1967). *Appl. Microbiol.* **15,** 166.

Angelotti, R., Foter, M. J., and Lewis, K. H. (1961). *Am. J. Public Health* **51,** 76.

Ayres, J. C. (1966). *In* "The Destruction of Salmonellae" (H. Lineweaver, ed.), pp. 22–26. U.S. Department of Agriculture.

Banwart, G. J., and Ayres, J. C. (1956). *Food Technol.* **10,** 68.

Banwart, G. J., and Ayres, J. C. (1957). *Food Technol.* **11,** 244.

Bardsley, A. J., and Taylor, A. Mc. (1960). *Brit. Food Manuf. Res. Assoc. Res. Rept.* **99.**

Barnes, E., and Ingram, M. (1956). *J. Appl. Bacteriol.* **19,** 117.

Bayre, H. G., Garibaldi, J. A., and Lineweaver, H. (1965). *Poultry Sci.* **44,** No. 5, 1281.

Beerens, H. (1957). *Proc. 2nd Intern. Symp. Food Microbiol.* pp. 235–247.

Beerens, H., and Tahon, M. M. (1967). *In* "Botulism 1966" (M. Ingram and T. A. Roberts, (eds.), pp. 424–428. Chapman and Hall, London.

Bellamy, W. D., and Lawton, E. J. (1954). *Nucleonics* **12,** No. 4, 54.

Benjamin, M. J. W., Wheather, D. M., and Shepherd, P. A. (1956). *J. Appl. Bacteriol.* **19,** 159.

Bierer, B. W., and Barrett, B. D. (1965). *J. Am. Vet. Med. Assoc.* **146,** 735.

Blanche Koelensmid, L. W., and Van Rhee, R. (1964). *Ann. Inst. Pasteur* **15,** 85.

Bohrer, C. W. (1963). *In* "Microbiological Quality of Food" (L. W. Slanetz, C. O. Chichester, A. R. Gaufins, and Z. J. Ordal, eds.), p. 198. Academic Press, New York.

Bonventre, P. F., and Kempe, L. L. (1959). *Appl. Microbiol.* **7,** 374.

Bowers, L. E., and Williams, O. B. (1962). *Antonie van Leeuwenhoek Microbiol. Serol.* **28,** 435.

Brandley, P. J., Mizaki, G., and Taylor, K. E. (1966). "Meat Hygiene." Lea and Febiger, Philadelphia.

Bulatova, T. I., Matveev, K. I., and Samsonova, V. S. (1967). *In* "Botulism 1966" (M. Ingram and T. A. Roberts, eds.), pp. 391–399. Chapman and Hall, London.

Busta, F., and Jezeski, J. J. (1963). *Appl. Microbiol.* **11,** 405.

Campbell, L. L., and Wrenarski, W. (1959). *Appl. Microbiol.* **7,** 285.

Cann, D. C., Wilson, Barbara B., Hobbs, G., and Shewan, J. M. (1967). *In* "Botulism 1966" (M. Ingram and T. A. Roberts, eds.), pp. 202–205. Chapman and Hall, London.

Casman, E. P., McCoy, D. W., and Brandley, P. J. (1963). *Appl. Microbiol.* **11**, 498.

Casolari, A. (1963). *Ind. Conserve (Parma)* **38**, 292.

Castellani, A. G., and Niven, C. F. (1955). *Appl. Microbiol.* **3**, 154.

Charm, S. E. (1966). *Food Technol.* **20**, 97.

Christian, J. H. B. (1963). *Recent Advan. Food Sci.* **3**, 248.

Christian, J. H. B., and Scott, W. J. (1953). *Australian J. Biol. Sci.* **6**, 565.

Christian, J. H. B., and Waltho, J. (1966). *J. Gen. Microbiol.* **43**, 345.

Christiansen, L. M., and Foster, E. M. (1965). *Appl. Microbiol.* **13**, 1023.

Clarenburg, A., and Burger, H. C. (1950). *Food Res.* **15**, 340.

Comer, A. G., Anderson, G. W., and Garrard, E. H. (1963). *Can. J. Microbiol.* **9**, 321.

Crisley, F. D., Peller, J. T., Angelotti, R., and Hall, H. E. (1967). *Bacteriol. Proc.* p. 5.

Dack, G. M. (1963). *In* "Diseases Transmitted from Animals to Man" (T. G. Hull, ed.), pp. 211–213. Thomas, Springfield, Illinois.

Dack, G. M., and Lippitz, G. (1962). *Appl. Microbiol.* **10**, 472.

Dahiya, R. S., and Speck, M. L. (1967). *Bacteriol. Proc.* **A65**. p. 11.

Dean, E., and Howie, E. L. (1964). *Bull. Parenteral Assoc.* **18**, 12.

Denny, C. B., Bohrer, C. W., Perkins, W. E., and Townsend, C. T. (1959). *Food Res.* **24**, 44.

Denny, C. B., Tain, P. L., and Bohrer, C. W. (1966). *J. Food Sci.* **31**, 762.

Dickson, E. C. (1928). *Proc. Soc. Exptl. Biol. Med.* **25**, 426.

DiGiacinto, J. V., and Frazier, W. C. (1966). *Appl. Microbiol.* **14**, 124.

Eddy, B. P., and Ingram, M. (1962). *J. Appl. Bacteriol.* **25**, 237.

El-Bisi, H. M. (1965). *In* "Radiation Preservation of Foods." *Natl. Acad. Sci. Natl. Res. Council Publ.* **1273**. pp. 223–232.

El-Bisi, H. M. (1967). *In* "Botulism 1966" (M. Ingram and T. A. Roberts, eds.), pp. 89–107. Chapman and Hall, London.

Emodi, E., and Lechowich, R. V. (1967). *Bacteriol. Proc.* p. 5.

Esty, J. R. (1923). *Am. J. Public Health* **13**, 108.

Esty, J. R., and Meyer, K. F. (1922). *J. Infect. Diseases* **31**, 650. p. 4.

Food Protection Committee (1964). "An Evaluation of Public Health Hazards from Microbiological Contamination of Foods." *Natl. Acad. Sci.–Natl. Res. Council Publ.* **1195**.

Genigeorgis, C., and Sadler, W. W. (1966). *J. Bacteriol.* **92**, 1383.

Genigeorgis, C., Riemann, H., and Sadler, W. W. (1967). *Proc. 18th World Vet. Congr.* p. 178.

Gibbons, N. E., Rose, D., and Hopkins, J. W. (1954). *Food Technol.* **8**, 155.

Goldblith, S. A. (1966). *Food Technol.* **20**, 191.

Gough, B. J., and Alford, J. A. (1965). *J. Food Sci.* **30**, 1025.

Graves, R. R., and Frazier, W. C. (1963). *Appl. Microbiol.* **11**, 513.

Grecz, N. (1964). *In* "Microbial Inhibitors in Food" (N. Molin, ed.), pp. 307–320. Almqvist and Wiksell, Stockholm.

Grecz, N., Wagenaar, R. O., and Dack, G. M. (1959a). *Appl. Microbiol.* **7**, 33.

Grecz, N., Wagenaar, R. O., and Dack, G. M. (1959b). *Appl. Microbiol.* **7**, 228.

Grecz, N., Dack, G. M., and Hedrick, L. R. (1961). *J. Food Sci.* **26**, 72.

Grecz, N., Upadhyay, T., and Tang, T. C. (1967). *Can. J. Microbiol.* **13**, 287.

Greenberg, R., and Silliker, J. H. (1961). *J. Food Sci.* **26**, 622.

Greenberg, R. A., and Silliker, J. H. (1964). *In* "Microbial Inhibitors in Food" (N. Molin, ed.), pp. 97–103. Almqvist and Wiksell, Stockholm.

Greenburg, R. A., and Silliker, J. H., and Fatta, L. D. (1959). *Food Technol.* **13,** 509.

Halenbeck, C. M. (1964). *Food Processing* **25,** 137.

Hall, I. C. (1941). *J. Colorado-Wyoming Acad. Sci.* **3,** 14.

Hanford, P. M., and Gibbs, B. M. (1964). *In* "Microbial Inhibitors in Food" (N. Molin, ed.), pp. 333–347. Almqvist and Wiksell, Stockholm.

Hansen, J. D., and Appleman, M. D. (1955). *Food Res.* **20,** 92.

Hansen, N. H. (1965). Personal communication, Danish Meat Research Institute, Roskilde, Denmark.

Hansen, N. H. (1967). Personal communication, Danish Meat Research Institute, Roskilde, Denmark.

Hansen, N. H., and Riemann, H. (1963). *J. Appl. Bacteriol.* **26,** 314.

Harmon, S. M., and Kautter, D. A. (1967). *Bacteriol. Proc.* p. 5.

Haughtley, G., and Liston, J. (1965). *Food Technol.* **19,** 874.

Hersom, A. C., and Hulland, E. D. (1965). "Canned Foods, An Introduction to Their Microbiology." Chemical Publishing, New York.

Hobbs, Betty C. (1965). *J. Appl. Bacteriol.* **28,** 74.

Hollaender, A., Stapleton, G. E., and Martin, F. L. (1951). *Nature* **167,** 103.

Horne-Jensen, B. (1967). *Medl. Danske Dyrlaegeforen.* **50,** 394.

Hunnell, J. (1961). *In* "Spores II" (H. O. Halvorson, ed.), pp. 101–113. Burgess, Minneapolis, Minnesota.

Husseman, D. L., and Tanner, F. W. (1947). *Am. J. Public Health* **37,** 1407.

Ingram, M. (1959). *Chem. Ind. (London)* 552.

Ingram, M. (1960). *J. Appl. Bacteriol.* **23,** 206.

Ingram, M. (1962). *J. Appl. Bacteriol.* **25,** 259.

Ingram, M. (1963). *In* "Microbiological Quality of Foods" (L. W. Slanetz, C. O. Chichester, A. R. Gaufin, and Z. J. Ordal, eds.), pp. 229–236. Academic Press, New York.

Ingram, M., and Barnes, Ella (1955). *J. Appl. Bacteriol.* **18,** 549.

Ingram, M., and Hanford, P. M. (1957). *J. Appl. Bacteriol.* **20,** 442.

Ingram, M., and Robinson, R. H. M. (1951*a*). *Proc. Soc. Appl. Bacteriol.* **14,** 62.

Ingram, M., and Robinson, R. H. M. (1951*b*). *Proc. Soc. Appl. Bacteriol.* **14,** 73.

Ingram, M., and Thornley, M. J. (1961). *J. Appl. Bacteriol.* **24,** 94.

Ingram, M., Buttiaux, R., and Mossel, D. A. A. (1964). *In* "Microbial Inhibitors in Food" (N. Molin, ed.), pp. 381–393. Almqvist and Wiksell, Sweden.

Jensen, L. B. (1954). "Microbiology of Meats." The Garrard Press, Campaign, Illinois.

Johanssen, A. (1961). SIK Rappert No. 100, Svenska Institutet for Konserveringsforskning, Goteborg, Sweden.

Kadavy, J. L., and Dack, G. M. (1951). *Food Res.* **16,** 328.

Kao, C. T., and Frazier, W. C. (1966). *Appl. Microbiol.* **14,** 251.

Kautter, D. A. (1964). *J. Food Sci.* **29,** 843.

Kempe, L. L. (1964). *In* "Botulism 1964" (K. Lewis and K. Cassel, Jr., eds.), pp. 205–221. U.S. Department of Health, Education, and Welfare, Public Health Service.

Kempe, L. L., and Graikoski, J. T. (1964*a*). *Food Technol.* **18,** 1078.

Kempe, L. L., and Graikoski, J. T. (1964*b*). *Food Technol.* **18,** 102.

Kempe, L. L., Bonventre, P. F., Graikoski, J. T., and Williams, N. J. (1956). U.S. Atomic Energy Commission Rept. No. TID-7512.

Kempe, L. L., Graikoski, J. T., and Bonventre, P. F. (1957*a*). *Appl. Microbiol.* **5,** 292.

Kempe, L. L., Graikoski, J. T., and Bonventre, P. F. (1957b). *Appl. Microbiol.* **6**, 261.

Krabbenhoft, K. L., Corlett, D. A., Anderson, A. W., and Elliker, P. R. (1964). *Appl. Microbiol.* **12**, 424.

Lang, O. W. (1935). *Univ. Calif. Publ. Public Health* **2**, No. 1, 1–182.

Lategan, P. M., and Vaughn, R. H. (1964). *J. Food Sci.* **29**, 339.

LeBlanc, F. R., Devlin, K. A., and Stumbo, C. R. (1953). *Food Technol.* **7**, 181.

Lechowich, R. V., Evans, J. B., and Niven, C. F. (1956). *Appl. Microbiol.* **4**, 360.

Lew, F. J., Freeman, B. M., and Hobbs, B. C. (1963). *J. Hyg.* **61**, 515.

Licciardello, J. J., Nickerson, J. T. R., and Goldblith, S. A. (1965). *Am. J. Public Health* **55**, 1622.

Lineweaver, H. (1966). *In* "The Destruction of Salmonellae" (H. Lineweaver, ed.), pp. 47–50. U.S. Department of Agriculture.

Malin, B., and Greenberg, R. A. (1964). *In* "Microbial Inhibitors in Food" (N. Molin, ed.), Almqvist and Wiksell, Stockholm.

Marazza, V., and Crespi, A. (1963). *Atti. Soc. Ital. Sci. Vet.* **17**, 537 (Chem. Abstr. **62**, 4526a).

Matsuyama, A., Thornley, M. J., and Ingram, J. (1964). *J. Appl. Bacteriol.* **27**, 125.

McCoy, D. W., and Faber, J. E. (1966). *Appl. Microbiol.* **14**, 372.

McWhorter, A. C., Murrell, M. C., and Edwards, P. R. (1963). *Appl. Microbiol.* **11**, 368.

Meyer, K. F., and Gunnison, J. B. (1929). *J. Infect. Diseases* **45**, 135.

Meynell, G. G., and Meynell, E. (1965). "Theory and Practice in Experimental Bacteriology," pp. 78–81. Cambridge University Press.

Mitchell, P. (1951). *In* "Bacterial Physiology" (G. H. Werkman and P. W. Wilson, eds.), pp. 127–177. Academic Press, New York.

Molin, N. (ed.) (1964). "Microbial Inhibitors in Food." Almqvist and Wiksell, Stockholm.

Molin, N., and Snygg, B. G. (1967). *Appl. Microbiol.* **15**, 1422.

Murrell, W. G., and Scott, W. J. (1966). *J. Gen. Microbiol.* **43**, 411.

Ng, H. (1966). *In* "The Destruction of Salmonellae" (H. Lineweaver, ed.), pp. 39–41. U.S. Department of Agriculture.

Nielsen, S. F., and Pedersen, H. O. (1967). *In* "Botulism 1966" (M. Ingram and T. A. Roberts, eds.), pp. 66–72. Chapman and Hall, London.

Niven, C. F. (1958). *Ann. Rev. Microbiol.* **12**, 507.

Nunheimer, T. D., and Fabian, F. W. (1940). *Am. J. Public Health* **30**, 1040.

Nünivaara, F. (1957). *Proc. 2nd Intern. Symp. Food Microbiol.* pp. 187–191.

Oberhofer, T. R., and Frazier, W. C. (1961). *J. Milk Food Technol.* **24**, 172.

O'Brien, R. T., and Titus, D. S. (1955). *J. Bacteriol.* **70**, 487.

O'Brien, R. T., Titus, D. S., Devlin, K. A., Sub, C. R., and Lewis, J. C. (1956). *Food Technol.* **10**, 352.

Ohye, D. F., and Christian, J. B. H. (1967). *In* "Botulism 1966" (M. Ingram and T. A. Roberts, eds.), pp. 217–221. Chapman and Hall, London.

Ohye, D. F., and Scott, W. J. (1953). *Australian J. Biol. Sci.* **6**, 178.

Ohye, D. F., Christian, J. H. B., and Scott, W. J. (1967). *In* "Botulism 1966" (M. Ingram and T. A. Roberts, eds.), pp. 136–143. Chapman and Hall, London.

Olsen, A. M., and Scott, W. J. (1950). *Australian J. Sci. Res.* **B3**, 219.

Oxhøj, P. (1943). *Chem. Zentr.* **11**, 34.

Pace, P. J., Krumbiegel, E. R., Wiesniewski, H. J., and Angelotti, R. (1967). *In* "Botulism 1966" (M. Ingram and T. A. Roberts, eds.), pp. 40–48. Chapman and Hall, London.

Parfentjen, I. A., and Catelli, R. R. (1964). *J. Bacteriol.* **88**, 1.

Pedersen, H. O. (1957). *Proc. 2nd Intern. Symp. Food Microbiol.* pp. 289–295.

Perigo, J. A., Whiting, E., and Bashford, T. E. (1967). *J. Food Technol.* **2**, 377.

Perkins, W. E. (1964). *In* "Botulism 1964" (K. H. Lewis and K. Cassel, eds.), pp. 187–204. U.S. Dept. of Health, Education, and Welfare, Public Health Service.

Peterson, A. C., Black, J. J., and Gunderson, M. F. (1962*a*). *Appl. Microbiol.* **10**, 16.

Peterson, A. C., Black, J. J., and Gunderson, M. F. (1962*b*). *Appl. Microbiol.* **10**, 23.

Peterson, A. C., Black, J. J., and Gunderson, M. F. (1964). *Appl. Microbiol.* **12**, 70.

Pivnick, H., and Barnett, H. (1965). *Food Technol.* **19**, 1164.

Pivnick, H., and Barnett, H. (1967). *In* "Botulism in 1966" (M. Ingram and T. A. Roberts, eds.), pp. 208–216. Chapman and Hall, London.

Pivnick, H., and Bird, H. (1965). *Food Technol.* **19**, 1156.

Pivnick, H., Rubin, L. J., Barnett, H. W., Nordin, H. R., Ferguson, P. A., and Perric, C. H. (1967). *Food Technol.* **21**, 204.

Prudhomme, R. (1966). M.S. Thesis, University of California, Davis.

Raica, N. (1965). Research Develop. Assoc. Activ. Rept. 17-2, p. 111. U.S. Army Medical Research and Nutrition Laboratory.

Ramsey, C. B., and Kempe, J. D. (1963). *J. Food Sci.* **28**, 562.

Rasmussen, O. G., Hansen, R., Jacobs, N. J., and Wilder, O. H. M. (1964). *Poultry Sci.* **43**, 1151.

Read, R. B., Jr., and Bradshaw, J. G. (1966). *Appl. Microbiol.* **14**, 130.

Read, R. B., Jr., and Bradshaw, J. G. (1967). *Appl. Microbiol.* **15**, 603.

Read, R. B., Jr., Bradshaw, J. G., and Dickeron, R. W., Jr. (1965). *J. Dairy Sci.* **48**, 770.

Riemann, H. (1957). *J. Appl. Bacteriol.* **20**, 404.

Riemann, H. (1963). *Food Technol.* **17**, 39.

Riemann, H. (1967). *In* "Botulism 1966" (M. Ingram and T. A. Roberts, eds.), pp. 148–157. Chapman and Hall, London.

Roberts, T. A., and Ingram, M. (1965). *J. Food Sci.* **30**, 879.

Rountree, Phyllis M. (1963). *J. Hyg.* **61**, 265.

Sair, L. (1965). *Meat* **31**, 44.

Saleh, M. A., and Ordal, Z. J. (1955*a*). *Food Res.* **20**, 332.

Saleh, M. A., and Ordal, Z. J. (1955*b*). *Food Res.* **20**, 340.

Sasarman, A., and Antohi, M. (1967). Proceedings, Workshop on Anaerobic Bacteria. Institut de Microbiologie et d'Hygiene de L'Université de Montreal, Canada.

Schmidt, C. F. (1964). *In* "Botulism 1964" (K. H. Lewis and K. Cassel, eds.), pp. 69–79. U.S. Department of Health, Education, and Welfare, Public Health Service.

Scott, W. J. (1953). *Australian J. Biol. Sci.* **6**, 549.

Scott, W. J. (1955). *Ann. Inst. Pasteur* **7**, 68.

Scott, W. J. (1957). *Advan. Food Res.* **7**, 83.

Scott, W. J., and Steward, D. F. (1944). *J. Council Sci. Indus. Res.* **17**, 16.

Seminiano, E. N., and Frazier, W. C. (1966). *J. Milk Food Technol.* **29**, 1961.

Shewan, J. M. (1956). *Mod. Refrig.* **59**, 423.

Shiflett, M. A., Lee, J. S., and Simhuber, R. O. (1966). *Appl. Microbiol.* **14**, 411.

Shrimpton, D. H., Monsey, J. B., Hobbs, Betty C., and Smith, Muriel E. (1962). *J. Hyg.* **60**, 153.

Silliker, J. H., Greenberg, R. A., and Schack, W. R. (1958). *Food Technol.* **12**, 551.

Silliker, J. H., Jansen, C. E., Voegeli, M. M., and Chmura, N. W. (1962). *J. Food Sci.* **27**, 50.

Silverman, G. T., and Goldblith, S. A. (1965). *Advan. Appl. Microbiol.* **7**, 305.

Sinskey, T. J., McIntosh, A. H., Pablo, I. S., Silverman, G. T., and Goldblith, S. A. (1964). *Health Lab. Sci.* **1**, 297.

540 HANS RIEMANN

Skulberg, A. (1964). Studies on the Formation of Toxin by *Clostridium botulinum.* Monograph. A/S Kaare Grytting, Orkanger, Norway.
Slanetz, L. W., Chichester, C. O., Gaufin, A. R., and Ordal, A. J. (eds.) (1963). "Microbiological Quality of Foods." Academic Press, New York.
Sogin, S. T., and Ordal, Z. J. (1967). *J. Bacteriol.* **94,** 1082.
Spencer, R. (1967). *In* "Botulism 1966" (M. Ingram and T. A. Roberts, eds.), pp. 123–129. Chapman and Hall, London.
Spinelli, J., Edlund, M., and Miyauchi, D. (1964). *Food Technol.* **18,** 933.
Stabyj, B. M., Dollar, A. M., and Liston, J. (1965). *J. Food Sci.* **30,** 344.
Statens Forsgøsmejeri. (1960). *Aarsberetning, 1960.*
Steinke, P. K. W., and Foster, E. M. (1951). *Food Res.* **15,** 477.
Strange, R. E., and Shon, M. (1964). *J. Gen. Microbiol.* **34,** 99.
Stumbo, C. R. (1965). "Thermobacteriology in Food Processing." Academic Press, New York.
Stumbo, C. R., Gross, C. E., and Vinton, C. (1945). *Food Res.* **10,** 293.
Sugiyama, H. (1951). *J. Bacteriol.* **62,** 81.
Tanner, F. W., and Evans, F. L. (1933a). *Zentr. Bakteriol. Parásitenk. Orig. Abt. II,* **88,** 44.
Tanner, F. W., and Evans, F. L. (1933b). *Zentr. Bakteriol. Parasitenk. Abt. II,* **89,** 48.
Tanner, F. W., Beamer, P. R., and Rickher, C. J. (1940). *Food Res.* **5,** 323–333.
Tarr, H. L. A. (1940). *Nature* **147,** 417.
Thatcher, F. S., and Lait, A. (1961). *Appl. Microbiol.* **9,** 39.
Thatcher, F. S., Robinson, J., and Endman, I. (1962). *J. Appl. Bacteriol.* **25,** 120.
Thomas, Constance C., White, J. C., and Longrée, Karla (1966). *Appl. Microbiol.* **14,** 815–819.
Thornly, M. J. (1957). *J. Appl. Bacteriol.* **20,** 286.
Tonge, R. J., Baird-Parker, A. C., and Covett, J. J. (1964). *J. Appl. Bacteriol.* **27,** 252.
Townsend, C. E., Yee, L., and Mercer, W. E. (1954). *Food Res.* **19,** 536.
Troller, J. A., and Frazier, W. C. (1963a). *Appl. Microbiol.* **11,** 11.
Troller, J. A., and Frazier, W. C. (1963b). *Appl. Microbiol.* **11,** 163.
Ulrich, J. A., and Halvorson, H. D. (1949). The Possibilities of the Growth of *Clostridium botulinum* in Canned Bread. University of Minnesota Hormel Institute Annual Report 1948049, 45–51.
Upadhyay, T., and Grecz, N. (1967). *Bacteriol. Proc.* p. 5.
U.S. Department of Agriculture Western Regional Research Laboratory (1964). WRRL Process for Pasteurizing Liquid Egg White.
Valenzuela, S., Niekerson, J. T. R., Campbell, C., and Goldblith, S. A. (1967). *In* "Botulism 1966" (M. Ingram and T. A. Roberts, eds.), pp. 224–235. Chapman and Hall, London.
Wagenaar, R. O., and Dack, G. M. (1954). *Food Res.* **19,** 521.
Wagenaar, R. O., and Dack, G. M. (1958). *J. Dairy Sci.* **41,** 1182, 1191, 1196.
Wagenaar, R. O., and Dack, G. M. (1960). *Food Res.* **25,** 646.
Warnecke, M. O., Carpenter, J. A., and Saffle, R. L. (1967). *Food Technol.* **21,** 433.
Watts, Nancy W. (1967). *In* "Botulism 1966" (M. Ingram and T. A. Roberts, eds.), pp. 158–168. Chapman and Hall, London.
Weiss, H. (1921). *J. Infect. Diseases* **28,** 70.
Wheaton, E., and Pratt, G. B. (1962). *J. Food Sci.* **27,** 327.
Wheaton, E., Pratt, G. B., and Jackson, J. M. (1961). *J. Food Sci.* **26,** 345.
Wilkinson, R. J., Mallman, W. L., Dawson, L. E., Irmiter, T. F., and Davidson, J. A. (1965). *Poultry Sci.* **44,** 131.

Williams, O. B., and Reed, J. M. (1942). *J. Infect. Diseases* **71,** 225.
Williams-Walls, N. J. (1960). *Dissert. Abstr.* **20,** 1975.
Wolin, E. F., Evans, J. B., and Niven, C. F., Jr. (1957). *Food Res.* **22,** 682.
Wood, Th. H. (1956). *Advan. Biol. Med. Phys.* **4,** 119.
Xezones, H., and Hutchings, I. J. (1965). *Food Technol.* **19,** 1003.
Yesair, J., and Cameron, E. J. (1942). *Canner* **94,** No. 13, 89.

CHAPTER XIII | **POISONOUS PLANTS AND ANIMALS**

Harold George Scott

I. INTRODUCTION

Since his earliest days on Earth man has tried to differentiate between those plants and animals which are safe to eat and those which are poisonous. Sometimes the decision has been simple and the lesson clear (certain organisms are always edible, others always poisonous). More often the decision has been complex and the lesson uncertain (an organism usually safe to eat is occasionally poisonous).

Perplexity prevails to modern times, and humans in all parts of the world are regularly injured or killed by ingestion of poisonous plants or animals. For example, in 1964 some 200 Japanese died as a result of eating fugu fish, and between 1952 and 1957 at least 40,000 persons suffered fish poisoning in the western Pacific. In 1965 four Montanans suffered nonfatal poisoning from ingestion of a combination of beer and inky-cap mushrooms; seven New Jerseyites (two of whom died) suffered severe poisoning from eating "wild mushrooms" gathered in their back yards; and seventeen Italians died after eating mushrooms purchased from "unauthorized vendors" in Rome (Reynolds and Lowe, 1965).

Persons forced by circumstance to select and eat nonpoisonous wild plants and animals in order to survive (Scott, 1963) have varying degrees of success, depending on perception, training, and (more often) sheer luck. Considerable perception was shown in April 1965 when a U. S. Army Master Sergeant, stranded by his downed aircraft near Watson Lake, Yukon Territory (Canada), ate "boiled grass" for seven days. In good condition when rescued he said, regarding his selection of the grass as food, "It keeps the moose alive." It is to be noted that he wisely boiled the grass to destroy heat-labile toxins and discarded the soup to eliminate water-soluble ones.

Results of training were dramatized by the 37-day survival of two University of Guam oceanographic researchers who drifted from Guam to Calayan Island (Philippines) on a 23-foot raft in April–May 1965. They fed on dolphin, fish, and shark, gathering rainwater in a can.

This chapter considers the problem of poisonous plants and animals, certain allergic phenomena related to poisoning, and certain special naturally occurring toxins. It does not consider nutritional deficiencies, venomous organisms, infectious organisms, drug overdoses, or certain plant and animal poisonings considered elsewhere in the text (enterotoxigenic staphylococci, botulism, mycotoxicoses).

With the vast numbers of inherently poisonous species, and the complex phenomena of secondary poisoning, it is considered judicious to confine this discussion to types of human poisoning which have actually occurred, and not to attempt to consider all situations in which man might conceivably become poisoned (Keegan and Toshioka, 1958; Keegan and MacFarlane, 1963).

A. History

The history of man's relationship to poisonous plants and animals is embroidered with superstition, mysticism, anecdotal humor, and tragedy. Even today, many cultures utilize plant and animal toxins in religious rites, as medicinals, or for pragmatic purposes such as hunting.

Indians of the southwestern United States and northern Mexico use peyote (from *Lophophora williamsi* or *Lophophora lewini*) during religious rites. An ongoing dispute occurs over this matter between the freedom of religious rites of the Indian and the narcotic control laws. The ancient Egyptians obtained the poison prussic acid from peaches (*Prunus persica*).

Toxic plants or plant extracts smeared on spear and arrowhead for hunting and warfare include curare (*Strychnos toxifera,* Central and South America, tropical Asia), strophanthin (*Strophanthus* spp., Africa), and ouabain (*Acocanthera* spp., Africa).

Some fantasy-producing toxins such as henbane (from *Hyoscyamus niger*), podophyllin (from *Podophyllum peltatum,* may apple), and solanines (from *Datura stramonium,* jimson weed and *Atropa belladonna,* deadly nightshade) are cultural inheritances from the ancient priesthoods. The narcotic effect of marijuana (from *Cannabis sativa,* Indian hemp) has been known for more than 2000 years.

Some ancient "mystic toxins" have been adapted for use as modern medicinals. Curare is used today as a muscle relaxant. Other modern drugs devised from ancient toxins include strychnine, opium, caffeine, strophanthin, cocaine, atropine, digitalis, ergotamine, and nicotine. In modern times, the isolation of anticoagulant drugs from many different plants and of rodenticide from molded sweetclover hay have been major advances (Link, 1959). Red squill *(Urginea maritima)* continues to be widely used as a rodenticide.

Many other plant toxins have been, and are being, used as insecticides (Jacobson, 1958) and rodenticides (Schery, 1952).

The Bible cautions against poisonous animals. "These you may eat of all that live in water: you may eat anything that has fins and scales. But you may not eat anything that has no fins and scales: it is unclean for you" (Deuteronomy 14:9–10). This same advice is found in modern survival manuals.

In 400 B.C., Xenophon reported poisoning of Greek soldiers from eating honey made by bees from *Rhododendron pontica.*

The writings of Meander of Colophon show that he was familiar with henbane, aconite, conium, and poisonous fungi as early as 184 B.C.

In 1818, several members of the Thomas Lincoln family, including Nancy Hanks Lincoln (mother of President Abraham Lincoln), died of "milk sickness" produced by drinking milk from cows which had fed on white snakeroot *(Eupatorium urticaefolium)* (Fig. 1).

A widely quoted anecdote of the Civil War involves the soldier who said he had eaten green persimmons *(Diospyros virginiana)* so as to shrink his stomach to fit his rations. "The Addams Family" television series made constant anecdotal references to toxic plants, especially wolfbane and henbane, and many toxic organisms are intimately inter-twined into the folklore of Halloween.

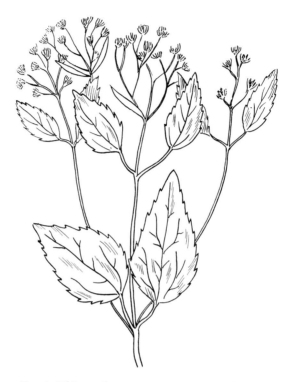

FIG. 1. White snakeroot *(Eupatorium urticaefolium)*.

In 1966, translation of a cookbook from French into English resulted in recommendation of poisonous laurel leaves (the *Laurus nobilis* of Europe) instead of bay leaves in one of the recipes. An immediate cor-rection was issued and, so far as is known, no poisonings resulted.

Intentional poisoning is not detailed herein. However, history is filled with references to suicides, murders, and "accidental" deaths of im-portant world figures in which the cause of death was a poisonous organ-ism. Socrates drank hemlock (from *Conium maculatum*) in 399 B.C., and Czar Alexis of Russia is considered to have died of *Russula* mush-room poisoning in 1676 A.D. Other historic figures who died as a result of mushroom poisoning include Pope Clement VII (1534 A.D.), Emperor Jovian of Rome (364 A.D.), Emperor Charles VI of the Holy Roman Empire (1740 A.D.), and possibly Emperor Claudius I of Rome (54 A.D.).

B. Medical Aspects

Prompt correct diagnosis is essential for effective treatment of poisoning victims. However, since poisoning syndromes can closely resemble those of metabolic or communicable diseases disassociated with ingestion of toxic substances, and since different toxic agents produce different signs and symptoms, diagnosis is often difficult and slow.

Differential diagnosis of poisoning is often a complex problem involving analysis of structural, functional, and biochemical changes in the victim, and efficient investigation of the epidemiologic circumstances leading to the poisoning. Medical procedure is outlined in numerous texts including Dreisbach (1966). Fundamental to both medical and investigative action is rapid precise identification of the poisonous material ingested. This allows employment of specific clinical treatment and serves as a base line for epidemiologic investigation. Food shock (adverse reaction on first or second contact with a new food) may be mistaken for and perhaps is a form of poisoning (von Oettingen, 1958).

Tolerance and *resistance* of individuals to plant and animal poisoning can be confusing factors.

Treatment for ingestion of poisonous plants or animals depends on the nature of the poison and the amount ingested. First aid is directed toward prevention of further absorption, removal of ingested materials, neutralization by antidotes, and symptomatic treatment to support failing body functions.

C. Epidemiologic Investigations

Many past epidemiologic investigations of poisoning from ingestion of animals and plants have been incomplete owing to a lack of comprehension of the investigators on the possibilities involved. Those investigations which have proved most valuable have completed at least the following: (1) Established actuality of the occurrence. (2) Surveyed known cases and common situations (time, place, person). (3) Gathered physical evidence. (4) Verified diagnosis. (5) Formed tentative hypothesis. (6) Conducted detailed investigation. (7) Analyzed data. (8) Tested hypothesis. (9) Formulated a conclusion. (10) Published a report.

D. Reporting

Determination of human morbidity and mortality associated with poisonous plants and animals is limited by poor reporting and by the problem of differentiating between illness produced by poisoning and that produced by the realization that a possible injurious substance has been ingested. For example, orthodox Jews have been known to become violently ill on discovering that they have unknowingly eaten pork.

In considering incidence it is convenient to differentiate "poisonous" organisms from those that are "venomous," "parasitic," or "allergenic." These terms are considered to have the following meanings in this chapter.

Poisonous: Capable of producing illness and/or death on being ingested for example, castor beans, polar bear liver).

Venomous: Capable of producing illness and/or death by injection or external contact (for example, poison ivy, rattlesnake).

Parasitic: Capable of producing illness and/or death via establishment in or on a host (for example, malaria protozoa, schistosome flukes).

Allergenic: Capable of producing illness and/or death in unusually sensitive hosts (for example, mayfly asthma, egg sensitivity).

Too few poisoning incidents are recorded in detail in scientific journals. Such publications form a base line for our knowledge, and their value is beyond measure.

E.   Inherent versus Secondary Toxicity

Some organisms develop toxins as the result of normal metabolic processes (for example, the *Amanita* mushrooms) and are therefore inherently poisonous. Other organisms contain toxins only secondarily as a result of involvement with an external agent (for example, ciguatera toxin originating in algae and passing up the food chain to large carnivorous fish such as the red snapper). Secondary toxicity may affect many species and is, therefore, the source of poisoning due to ingestion of a plant or animal which is generally regarded as safe to eat.

Sometimes poisoning attributed to a plant or animal is the result of industrial processes which add toxins to food products. For example, in November 1965, Japanese health workers tracked down and destroyed 230 tons of poisoned persimmons which had been shipped to stores all over the country. Poisoning of the fruit resulted from "depuckering" of the unripe persimmons with inexpensive methyl alcohol instead of with the usual safe (but more expensive) shochu wine. A similar situation occurs when plants or animals become toxic through pesticidal contamination.

Plant or animal poisoning may not obviously be associated with the food being consumed. For example, spearing steaks with oleander *(Nerium oleander)* sticks for open-fire roasting has resulted in fatal oleander poisoning.

II. POISONOUS PLANTS

Analysis of all plants involved in human poisoning would vastly exceed the space limitations of this chapter. Therefore, discussion is limited to representative species occurring in the United States. Kingsbury (1964) discusses poisonous plants in detail and presents 1785 references dated from 1701 to 1963.

A. "Rules of Thumb"

Although it is impossible to pinpoint all poisonous plants by a simple set of rules, the following generalities are commonly used in survival training: (1) Most poisonous plants have a bitter taste and/or milky juice. (2) Most plant poisons occur in the seed or nuts, some in the fruit and root, few in the leaves. (3) Wild fungi should not be eaten. (4) Laminated bulbs — for example, lily — are generally poisonous. (5) Some, *but by no means all,* poisons are heat-labile and water-soluble. (6) The "soup" remaining after boiling wild plants should not be consumed. (7) Children should be taught early not to put any wild plant berries, foliage, or any other parts, into their mouths. (8) Adults should show enough wisdom not to eat strange plants on dares or bets or to show courage. (9) Unusual parts of familiar food plants should not be eaten. (10) No part of the Earth is without poisonous plants.

That consuming unusual parts of familiar food plants may be dangerous is illustrated by the common apple *(Malus sylvestris).* Apples as usually eaten are nontoxic, but consumption of large numbers of apple seeds (which are cyanogenic) can be fatal (Reynard and Norton, 1942).

The danger from ingesting toxic plants can vary with quantity eaten, season, degree of mastication, geography, plant variations, and human variations. Also, different animals react differently to the various poisons. Thus, what birds and other mammals eat may not be harmless to man.

As the various poison plants are itemized it begins to seem that all must be dangerous. The reader must not lose perspective with regard to man's dependence on plants for his very existence (Little, 1953; Tucker and Kimball, 1960; Youngken and Karas, 1964).

B. Incidence

Itemizing 1700 ingestions of poisonous plants with 72 deaths in the United States between 1940 and 1960, O'Leary (1964) emphasizes that these "represent but a fraction of the ingestions which may have occurred during that period," and "until creation (1953–1956) of the system of voluntary, nationwide poison control centers . . ., little accurate information was available on the nature and extent of poisoning in humans resulting from the ingestion of plants . . . ." One source estimates 12,000 "plant poisonings" per year in children in the United States (May, 1965; Press, 1964).

Of poison clearing-house ingestions recorded by O'Leary (1964), 454 were of berries, 288 of mushrooms, 81 of seeds and beans, 60 of nuts, and fewer of flowers, leaves, stems, and roots.

Of the 288 mushroom ingestions, 266 were by children under 9 years of age, and 116 were by children under 2 years of age. Adults involved were 23 to 79 years of age.

C. Secondarily Toxic Plants

Some toxic minerals are absorbed by plants growing on soils containing them. Such secondarily toxic plants can poison animals which eat them. Selenium is a good example of this phenomenon.

Soil selenium content may reach 190 ppm, but most soils contain less than 5. Acid soils keep selenium insoluble and unavailable to plants. Alkaline soils put about half of the selenium in solution, but interference by other elements such as sulfur makes only part of this available to plants. Thus, even the most seleniferous soils offer less than 90 ppm to plants, and most soils offer less than 2 ppm. As little as 5 ppm of selenium in animal foods can cause chronic poisoning effects, and more than 200 ppm can cause acute poisoning.

Selenium content in most plants is the same as that available to them from the soil. However, some plants accumulate the element. Obligate accumulators require selenium for proper growth and thus are indicators of selenium soils. Facultative accumulators gather selenium when growing on seleniferous soils, but grow just as well in the absence of selenium.

Obligate selenium accumulators include poison vetches (*Astragulus* spp.), prince plume *(Stanleya* spp.), goldenweeds *(Oonopsis* spp.), and woody asters (*Xylorrhiza* spp.), some of which can accumulate as much as 15,000 ppm of selenium.

Facultative selenium accumulators include at least ten genera of wild and ornamental plants *(Aster, Atriplex, Castilleja, Comandra, Grayia, Grindelia, Gutierrezia, Machaeranthera, Penstemon,* and *Sideranthus)* and six crops (corn, wheat, oats, barley, grass, and hay).

Seleniferous soils are most common in Montana, South Dakota, North Dakota, Nebraska, Kansas, Oklahoma, Texas, Wyoming, Colorado, New Mexico, Idaho, Utah, Arizona, Nevada, California, Hawaii, and Western Canada.

Selenium poisoning is common in domestic animals, but poisoning in man has not been demonstrated even though some selenium-tainted foods undoubtedly are consumed. Studies of the situation continue, with selenium content of hair being used as a monitoring device (Olson *et al.,* 1954).

D. Major Inherently Toxic Plants

Algae are the source of toxin in certain poisonous animals, the toxin being passed up the food chain to fishes and mollusks. These will be discussed under the animals with which they are associated.

Only members of the Division Cyanophyta (blue-green algae) have been found directly toxic in the freshwater environment. Human epidemics of intestinal disorders have occurred from drinking water undergoing "blooms" of blue-green algae (especially *Anabaena* spp., *Aphanizomenon Flos-aquae,* and *Microcystis* spp., nicknamed Annie, Fannie, and Mike by workers in the field) (Senior, 1960; Schwimmer and Schwimmer, 1955).

Some of the larger seaweeds, especially brown algae (Division Phaeophyta), contain toxic principles (Habekost *et al.,* 1955).

Besides algae, some representative inherently toxic plant genera frequently reported as causing human poisoning in the United States are *Amanita, Helvella, Psilocybe, Dryopteris, Equisetum, Taxus, Ricinus, Phytolacca, Solanum, Atropa, Cicuta, Conium, Aesculus, Robinia, Ilex, Berberis, Datura,* and *Caesalpina.* At least 200 other inherently toxic plant genera (more than 700 species) are reported more or less commonly as causing human poisoning in the United States.

1. *Amanita*

About thirty species of *Amanita* mushrooms occur in the United States. Some are edible, some deadly. Since precise determination of species is virtually impossible with available keys, no *Amanita* should be ingested (Buck, 1961).

Amanita phalloides, death angel (Fig. 2), is reportedly the most dangerous poisonous plant in the United States. At least 22 human deaths due to its ingestion occurred in 1940–1960. Its white to brownish (olive to green in Europe) color, toxicity, and anthropomorphism give rise to the common name. The cap is 2 to 5 inches across, the stem 3 to 8 inches long. It occurs throughout the United States except, possibly, in the four Pacific Coast States, growing singly or in groups in open woods (less commonly in pastures or lawns) between June and September. It contains at least four toxic fractions: (1) Amanitahemolysin, heat-labile, detoxified by digestion, not responsible for major poisoning symptoms; (2) Phalloidine, heat-stable, polypeptide, produces rapid degeneration

FIG. 2. Death angel *(Amanita phalloides).*

of kidney, liver, and heart musculature; (3 and 4) alpha and beta amatines, heat-stable polypeptides with indole rings, slow acting, hypoglycemiogenic, responsible for major poisoning symptoms. After ingestion of *Amanita phalloides,* an asymptomatic period of 6 to 15 (rarely 40) hours is followed by seizures (severe abdominal pain, vomiting and diarrhea with blood and mucus, extreme thirst, anuria, screams of pain) alternating with remissions. Rapid loss of strength leads to prostration which is interrupted by the seizures. Death occurs in 50 to 90% of all cases, usually 2 to 8 days after ingestion. A laboratory test for *Amanita* toxins has been developed (Block *et al.,* 1955), and treatment with an immune antiphalloidian serum is effective if given shortly after symptoms commence. Nonfatal cases persist about 30 days. During recovery all exertion must be avoided or fatal relapse may occur. Liver enlargement persists after convalescence. Pathology resembles that of phosphorus poisoning. It includes fatty degeneration, liver-kidney necrosis, endothelial-muscle-nerve cytolysis, and multiple hemorrhages, especially hemorrhagic enteritis.

In France, where mushroom shows are held annually, *Amanita* poisoning is so common that antiphalloidian serum is required by law to be stocked so as to be available to all physicians. Lewes (1948) reports poisoning of two German prisoners-of-war who collected *Amanita phalloides* in the woods, boiled them in water for three-quarters of an hour, then fried them. They used an unreliable traditional test ("if a silver coin does not blacken when placed in cooking mushrooms, they are safe to eat") and then consumed the mushrooms. Symptoms of poisoning began 9 and 12 hours later.

Buttenweiser and Bodenheimer (1924) report poisoning of a 17-day-old baby being nursed by its mother who had been poisoned by *Amanita phalloides*.

Amanita muscaria, fly agaric, has not been reported as causing human death since 1923. However, it is highly toxic and probably can cause death in the absence of proper treatment. It resembles *Amanita phalloides* but its cap is larger (3 to 8 inches across) and (except for white to brownish specimens taken in deep shade) is bright yellow to orange red. Usually the surface exhibits white or yellow warts. Some native tribes in Siberia concoct an intoxicating liquor from *Amanita muscaria,* or drink the urine of individuals who have eaten fly agaric, passage through the kidneys apparently enhancing the intoxicating properties (Taylor, 1949).

In 1898, the Count de Vecchi (attache of the Italian legation in Washington, D. C.), who was considered an expert in mycology, collected *Amanita muscaria* under the presumption that it was *Amanita caesaria* or *Amanita aurantiaca* and served it to himself and his physician. Despite treatment the Count died the following day. The physician, after energetic treatment with apomorphine and atropine, recovered (Prentiss, 1898).

2. Helvella (= Gyromitra)

The false morel or morchel, *Helvella esculenta* (Fig. 3), resembles closely, and is a close relative of, *Morchella esculenta,* the highly prized edible morel. Its name *(esculenta)* suggests that it is edible, and it is so regarded in some places. However, *Helvella esculenta* is another example of the variability of poisonous qualities of organisms. It is considered toxic in Germany but not in France. Ramsbottom (1953) regards it as safe when cooked, but Hendricks (1940) does not. Individual variation in response to *Helvella* is reported. Smith (1949) regards it as safe if no reaction occurs following the first ingestion, while Ramsbottom (1953) considers that it may be nontoxic on first ingestion but toxic on subsequent ingestions.

Helvella esculenta is widespread in the United States and has caused a large number of poisonings and fatalities (Cottingham, 1955). A toxic hemolysin, helvellic acid, has been known for many years, but since this

FIG. 3. False morel *(Helvella esculenta).*

acid is heat-labile and water-soluble, at least one other toxic fraction (probably also a hemolysin) must be present to account for poisonings caused by cooked *Helvella*. Six to eight hours after ingestion, vomiting, diarrhea, weakness, coma, and eventually death may occur. Postmortem examination reveals liver and kidney degeneration.

3. *Psilocybe*

Fatalities have been recorded from ingestion of *Psilocybe* in the north-western United States (Buck, 1961). They are small, relatively inconspicuous mushrooms of difficult positive identification (Fig. 4). The most dangerous species appears to be *Psilocybe baeocystis.*

Despite the obvious hazard, some people seem unable to resist wild "mushrooms." One avid mushroom hunter (Jacob, 1966) continues to taste new found species, commenting, "I think twice before I eat a wild mushroom — and then I eat it."

Many animals besides man feed regularly on the larger fungi: skunks, moles, shrews, squirrels, grouse, turtles, armadillos, and deer.

4. *Dryopteris (= Aspidium)*

Toxic fractions are found in a number of species of ferns *(Notholaena sinuata,* Jimmy fern; *Onoclea sensibilis,* sensitive fern; *Pteridium aquilinum,* bracken fern; and others). *Dryopteris felix-mas,* male fern, will be described as an example.

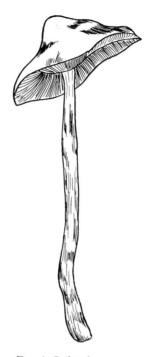

FIG. 4. *Psilocybe* sp.

Dryopteris felix-mas (Fig. 5) is a widely distributed species used locally as a vermifuge (rhizome and stipe consumed). It contains the enzyme thiaminase and perhaps other toxic elements. Thiaminase brings about destruction of thiamine (vitamin B) in the diet and produces avitaminosis. In 1914 a 23-year old Coloradoan died from an overdose of *Dryopteris* vermifuge (Hall, 1914).

At least five other species of *Dryopteris* and one other of *Pteridium* occur in the United States. The toxicity of these is unknown.

5. Equisetum

Horsetails (foxtails, rushes) of the genus *Equisetum* are small rushlike herbaceous perennials, growing anew each year from deeply buried rhizomes on waste ground. Four United States species *(Equisetum arvense, Equisetum palustre, Equisetum laevigatum,* and *Equisetum hiemale)* have caused livestock poisoning (Henderson *et al.,* 1952;

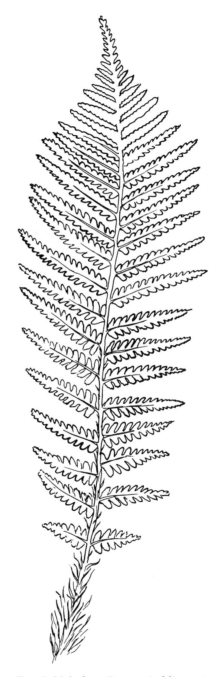

Fig. 5. Male fern *(Dryopteris felix-mas)*.

Sampson and Malmsten, 1942). Although no human poisonings are recorded, the horsetails are mentioned here as examples of toxic spore-bearing plants (Phylum Sphenophyta).

Although eight toxic principles have been isolated, the primary toxic fraction in *Equisetum* is undefined. The presence of a thiaminase is suspected.

6. *Taxus*

Several species of yew (also called ground hemlock) occur in the United States. *Taxus brevifolia,* the Pacific yew, ranges natively from southeastern Alaska to central California and western Montana. The Florida yew, *Taxus floridana,* occurs locally in northwestern Florida but is extremely rare. The Canadian yew, *Taxus canadensis,* occurs from Kentucky to Manitoba. The smooth cypress, *Cupressus glabra,* is sometimes improperly called "yew-wood." Although all *Taxus* species may have toxic qualities, it is the imported ornamental English yew, *Taxus baccata,* with which most reported human poisonings have been associated (Frohne and Pribilla, 1965). Canadian yew, *Taxus canadensis,* and Japanese yew, *Taxus cuspidata,* are also cultivated as ornamentals. Yews are hardy, slow-growing trees, which reach maturity in about 300 years.

English yew (sometimes called Irish yew) is an evergreen tree or shrub widely used as a hedge or decorative plant around homes. The alternate needlelike leaves are shiny dark green above and yellow-green below. The fruit is a red cuplike berry which looks good enough to eat. All parts of the yew are toxic, but the berries are most frequently involved in human poisoning. Toxicity seems to vary seasonally and geographically. O'Leary (1964) reports 20 cases of *Taxus* poisoning during 1959–1960 in Washington, D. C., Connecticut, Massachusetts, New Jersey, Ohio, Rhode Island, and Wisconsin. All cases were children (2½ months to 4 years old). *Taxus baccata* is considered more dangerous than any other poisonous tree or shrub in England (Garner, 1957).

The toxic fraction includes two alkaloids, taxine A and taxine B, which depress heart function and cause nervous system lesions. Onset of symptoms (nausea, abdominal pain, trembling, diastolic heart failure) occurs about 6 hours after ingestion. Treatment includes emesis, gastric lavage, and symptomatic treatment for circulatory failure and for alkaloid poisoning.

7. *Ricinus*

Ricinus communis, castor bean (castor-oil bean, palma christi, mole bean*), was introduced into North America from tropical Africa. The variegated white and black to brown beans which resemble engorged ticks are of major economic importance as the source of medicinal castor oil which is a widely used industrial lubricant and which can also be modified into paint drying oil, fabric and leather finishing oil, and turkey-red dye. World production is about 500,000 tons annually. The residue (pomace) left after extraction of the oil from the beans gives rise to dust which can produce marked dermatitis; eye, nose, and throat inflammation; and asthma.

Ricinus communis is cultivated in many commercial varieties as an ornamental and as shade for fowl. It also grows "wild" throughout much of the United States. Although an annual in most of the temperate zone, it develops into a perennial tree in the tropics and subtropics including some parts of the United States (Hawaii, California, Arizona, New Mexico, Texas, Florida, Puerto Rico, Virgin Islands).

Castor-oil bean leaves are only mildly toxic, but the beans are highly so. Ingestion of intact castor beans may cause only mild or even no poisoning. This is apparently due to the hard seed coat which resists digestion and slows absorption. If, on the other hand, the beans are "broken" prior to ingestion or by chewing, the contents of a single bean may cause death in a particularly sensitive person. Most cases of severe poisoning have involved two to seven beans, and eight beans are usually lethal.

The toxic fractions, ricin (a toxalbumin) and ricinin, are not oil-soluble and hence do not occur in the expressed and filtered castor oil. Ricin, with a minimum lethal dose of 0.0001 mg/kg by injection, is one of the most toxic substances known. Its oral toxicity is also high but variable (about 3% of injection toxicity). Ricin is apparently not readily absorbed through the wall of the digestive tract and may be partially destroyed by digestion. Nevertheless, it is absorbed through the intestinal wall much more readily than are most bacterial toxins and snake venoms. Ricin, but not ricinin, is heat-labile in water solution.

From a few minutes to 3 days after *Ricinus* ingestion the victim exhibits nausea, vomiting, diarrhea, abdominal pain, elevated temperature, drowsiness, disorientation, cyanosis, stupor, oliguria, and circulatory collapse. Laboratory findings include urine abnormalities (albumin, casts, red blood cells, hemoglobin), increased blood urea, and increased blood nonprotein nitrogen. In most fatal cases death occurs about 12 to 14 days

---

*Mole bean is also a common name for the poisonous caper spurge (garden spurge), *Euphorbia lathyrus*.

after ingestion. Mortality rate, based on limited data, is about 4%. Post-mortem pathology includes gastrointestinal hemorrhage and edema, and kidney hemolysis and degeneration.

Treatment includes emesis, gastric lavage, catharsis, maintenance of fluid and electrolyte balance, blood transfusions, urine alkalinization to minimize precipitations in the kidneys, and treatment for anuria.

O'Leary (1964) documents some 104 ingestions of castor beans in the United States in recent years, at least 98 of which were by children under 15 years of age. Ingestions were widespread geographically, occurring in Arizona, California, Colorado, Connecticut, Florida, Hawaii, Iowa, Kansas, Kentucky, Massachusetts, New Mexico, Ohio, Oregon, South Carolina, Tennessee, Texas, and Washington, D. C. Four of these ingestions resulted in fatalities—2 children in California and 2 adults in Hawaii. The 2 adults were soldiers who were chewing on castor beans in an attempt to obtain laxative effects.

8. *Phytolacca*

The perennial weed *Phytolacca americana* (pokeweed, scoke, garget, pokeberry, pigeonberry, American nightshade, inkberry,* red-ink) has all parts poisonous, most particularly the roots and unripe berries. As the berries ripen, their toxicity decreases but is not eliminated entirely.

Pokeweek is common in eastern North America, growing especially in disturbed rich soil of lowlands. It was used by the Indians and early settlers as a medicinal. It contains at least two toxins, a resin and a water-soluble saponin, neither of which is well-defined. It burns the mouth on being ingested.

Onset of symptoms occurs about 1 to 2 hours after ingestion: cramps, vomiting, diarrhea, impaired vision, salivation, perspiration, weak respiration, weak pulse, and (in terminal cases) coma and death. Recovery requires about 24 hours.

O'Leary (1964) documents some 118 *Phytolacca* ingestions, most by children under 12. At least 22 states have reported ingestions (Massachusetts to Kansas to New Mexico to Florida), and at least 2 deaths in children have been reported.

Poke, a commercial red ink, is extracted from the fruit, while the asparagus-like stems of the plant are used locally in eastern North America as cooked greens and were formerly used in a similar way by the American Indian.

---

*Inkberry is more properly used for *Ilex glabra, q.v.*

9. *Solanum*

This genus of nightshades contains some 1500 species throughout the world, many if not most of which are toxic. At least 15 species occurring in North America are known to be significantly poisonous (Kingsbury, 1964), of which eight have been clearly associated with human poisoning problems in the United States (Fruthaler, 1955).

Solanum americanum and *Solanum nigrum* are nearly identical morphologically and have the same common names (black nightshade, deadly nightshade,* common nightshade). They are annual herbs common in the eastern United States (New England to Florida to Nebraska), less common in the west. *Solanum americanum* is an indigenous species found primarily in established woods and grasslands. *Solanum nigrum,* an import from Europe, occupies disturbed areas and wastelands. Recent human ingestions of the toxic black berries are reported from Colorado, Florida, Indiana, Massachusetts, Michigan, Nebraska, Ohio and Washington (O'Leary, 1964).

Solanum dulcamara (European bittersweet, woody nightshade, climbing nightshade, marriage vine) is a climbing woody perennial found in woods and thickets throughout North America. Recent human ingestions of the toxic red berries are reported from Massachusetts, New York, Ohio, and Washington (O'Leary, 1964; Kingsbury, 1964). Its twigs and stem have been used to prepare a sedative, amodyne, and narcotic.

Solanum carolinense (horse nettle, bull nettle) is a weedy perennial occurring widely in cultivated, waste, and neglected areas of Eastern North America (Texas to Florida north to Canada). Scattered reports of poisonings by this species are difficult to evaluate. Kingsbury (1964) documents the death of a 6-year-old Pennsylvania boy from ingestion of the toxic yellow berries in 1963 (Green, 1958).

Solanum pseudocapsicum (Jerusalem cherry, natal cherry) is a common potted plant associated with Christmas. It is a native of the Old World. Recent ingestions of its toxic red berries are recorded in Connecticut, Massachusetts, Oregon, and Washington, D. C. (O'Leary, 1964), although documentation of resultant severe illness is lacking.

Solanum sodomeum (sodom apple, popolo) is a shrubby, spiny weed introduced from the Mediterranean area and now widespread in Hawaii where ingestion of its yellow fruit is reported to be a common cause of serious poisoning (Arnold, 1944).

Solanum tuberosum (potato) is an excellent example of the perplexity

---

*The common name "deadly nightshade" is more usually applied to *Atropa belladonna, q.v.*

associated with poisonous organisms. This vegetable, eaten safely by millions of persons daily, can be toxic. The tuber sprouts and sun-greened tissue of the tuber as well as the peelings sometimes contain enough of the toxin solanine to produce severe poisoning and even death. Boiling in water seems to remove part but not all of the toxin. The seeds, green stem, and leaves are also toxic (Hansen, 1925; Harris and Cockburn, 1918). The rarity of potato poisoning in man can be partially explained by variations in solanine content. Most potato tubers are about 0.009% solanine, whereas some of those associated with poisoning have been shown to contain 0.04% (Bamford, 1951).

Solanum triflorum (three-flower nightshade, cutleaf nightshade) is a widespread annual weed of North America (British Columbia to California east to Minnesota and Oklahoma, locally elsewhere) which occupies cultivated, disturbed, and waste areas. Although no human poisonings are documented, peas are condemned if this green to yellow berry is found among them. Separation from the peas is too difficult to be practical in most cases.

Solanum poisoning is similar for all species, involving ingestion of one or more of a series of glycoalkaloids referred to collectively as solanine. Solanine content varies with time, place, plant part, degree of maturity, and genetic makeup. Depending on severity, acute poisoning may result in burning in mouth and throat, weakness, nausea, dizziness, pupil dilation, and convulsions. Chronic poisoning is well known for domestic animals (Kingsbury, 1964) and can probably occur with man. Treatment includes emesis and treatment as for atropine poisoning.

In addition, *Solanum* spp. may sporadically develop toxic concentrations (>0.5%) of nitrates. Although nitrate poisoning from *Solanum* has clearly occurred in domestic animals (Whitehead and Moxon, 1952), no clear-cut case involving man is on record. Such poisoning, however, would be obscured by the toxic effects of solanine.

10. *Atropa*

Atropa belladonna (belladonna, deadly nightshade) is a perennial herb which has been imported from Europe as an ornamental. It is found wild where it has escaped from cultivation. Poisonings usually result from ingestion of the attractive black berries by children. As few as three berries can be fatal (Forsyth, 1954). The leaves are also poisonous, and poisoning can occur from handling the leaves as well as by eating plant parts.

Toxic fractions include the isomeric alkaloid *l*-hyoscyamine (from which atropine is made), scopolomine, hyoscine, hyoscyamine, apoatropine, belladonine, noratropine, tropacocaine, and meteloidine.

Toxicity increases with plant age. Cases of *Atropa belladonna* poisoning are confused with cases of *Amaryllis belladonna* poisoning, since both are commonly called belladonna, and with *Solanum americanum–Solanum nigrum* poisoning because all three species are called deadly nightshade. *Atropa* poisoning symptoms include dry mouth, pupil dilation, rapid heartbeat, and central nervous system depression. Emesis and thorough hand washing plus management for atropine poisoning make up the treatment.

In the lore of witchcraft, belladonna ointments were rubbed over a witch's body to enable her to fly through the air on broomsticks during celebrations of witches' Sabbath. Nonbelievers speculate that such witches flew only in the hallucinations associated with belladonna intoxication.

11. *Cicuta*

Species in the carrot family (Umbelliferae) resemble each other closely, and casual examination, even by a trained botanist, is often insufficient to distinguish species or even genera. The family contains many important food plants (including dill, celery, caraway, carrot, parsnip, parsley, and anise), but also some of the most toxic plants of North America including fool's parsley *(Aethusa cynapium),* bishop's weed *(Ammi visnaga),* and water parsnip *(Sium suave).* Two of the most toxic North American plant genera are also in this family — *Cicuta* (water hemlock), and *Conium* (poison hemlock).

At least nine species of *Cicuta* are known to be involved in poisonings in North America, but since the species are nearly indistinguishable morphologically, many *Cicuta* poisonings have never been pinned to a species. Their distribution can be used as evidence of the species involved in a particular poisoning:

Cicuta bolanderi	West central California
Cicuta bulbifera	Southern Canada, northern United States
Cicuta californica	Coastal California (36°–40° North Latitude)
Cicuta curtissi	Southeastern United States
Cicuta douglasi	High elevations, Arizona to Alaska
Cicuta mackenziana	Alaska and Adjoining Canada
Cicuta occidentalis	The Dakotas to Washington south to Nevada and New Mexico
Cicuta vegans	Northwestern United States and British Columbia

All are coarse perennial herbs occupying swampy or wet areas along streams or marshes.

The root of *Cicuta* spp. contains a highly toxic unsaturated alcohol, *(trans)*-heptadeca-8,10,12-triene-4,6-diyne-1,4-diol, also called cicutoxin (Anet *et al.,* 1953). Toxicity is not lost with age or from drying. Symptomatology is highly uniform. Some 15 to 60 minutes after ingestion, excessive salivation begins, followed quickly by tremors, alternating violent spasmodic convulsions and relaxation, abdominal pain, dilated pupils, elevated temperature, delirium, paralysis, and death due to respiratory failure (15 minutes to 8 hours after ingestion). Vomiting early in the syndrome is a good prognostic sign. Postmortem pathology is nonspecific.

Many human deaths have occurred, mostly in children who ate the attractive roots, possibly mistaking them for wild parsnip *(Pastinaca sativa)* or wild artichoke *(Helianthus annus),* both of which are edible (Haggerty and Conway, 1936).

In April 1963, two girls (ages 5 and 8) in East Meadow, New York, kindled a fire and roasted and nibbled at water hemlock (probably *Cicuta bulbifera)* root which they had found in the yard. The two became violently ill 2 hours later. They were rushed to a hospital where they recovered following treatment.

12. *Conium*

Conium maculatum (poison hemlock, spotted hemlock, California fern, Nebraska fern) closely resembles the water hemlocks (*Cicuta* spp.) and wild carrot *(Daucus carota).* The toxicity of *Cicuta* has been discussed. *Daucus carota,* one of the commonest weeds in North America, considered mildly toxic to nontoxic.

Conium maculatum was introduced from Europe and now is found throughout the United States and southern Canada as a luxuriant biennial (or perennial) weed of roadside ditches, farm-field edges, and waste areas. This is considered to be the "hemlock" used to put Socrates to death.

In May 1966, a married couple from Santa Fe, New Mexico, ate *Conium maculatum* while on a Sunday outing in the Sangre de Cristo Mountains. The wife, whose hobby was testing various wild plants as food, ate a substantial part of the root of one plant. The husband ate a little of the bulb. The woman went into convulsions within 2 hours and died the following day. The man, although ill, survived. At least 31 other such deaths have been recorded from *Conium maculatum* in the United States in recent years.

Conium maculatum contains at least five toxic alkaloids: coniine, N-methyl coniine, conhydrine, *l*-coniceine, and pseudoconhydrine. Coniine was the first alkaloid ever to be synthesized (by Ladenburg in 1886). In another demonstration of the variability of toxicity in poisonous organisms, *l*-coniceine predominates during vegetative growth, and coniine and N-methyl coniine predominate in the mature fruit. At certain times during plant growth as little as 1 square centimeter of plant tissue may produce fatal poisoning, whereas at other times consumption of as much as 4% of body weight may not produce death. Onset of symptoms usually occurs 12 to 60 minutes after ingestion. Symptoms include gradually increasing muscular weakness followed by paralysis with respiratory failure. Temporary albuminuria may occur, as may nausea, vomiting, and convulsions. The "mousey" odor of the plant may be detected in the breath of the victims. Postmortem pathology, involving mainly widespread congestion, has not been shown to be characteristic.

An extract of *Conium maculatum* has been used medicinally as a sedative.

13. *Aesculus*

The buckeyes and horsechestnut (*Aesculus* spp.) are trees or shrubs exhibiting large (1-inch) glossy brown seeds having a conspicuous scar or "buckeye." Some 26 species occur in the United States (all probably with toxic sprouts, young growth, and seeds), five of which have been significantly involved in poisoning.

Aesculus hipposcastanum (horsechestnut) was introduced as a shade tree from Europe and is now widely distributed in North America. *Aesculus pavia* (red buckeye, scarlet buckeye, woolly buckeye, firecracker plant) is a fertile-valley shrub or tree of the southeastern quarter of the United States. *Aesculus glabra* (Ohio buckeye, Texas buckeye, fetid buckeye) and *Aesculus octandra* (yellow buckeye, sweet buckeye) are rich-woods and riverbank trees of the eastern half ot the United States. *Aesculus californica* (California buckeye) is a low tree or shrub of dry hillsides and canyons of California.

Several toxic alkaloids, glycosides, and saponins occur in *Aesculus,* but the major toxic fraction appears to be aesculin, a hydroxycoumarin glycoside closely related to the anticoagulant rodenticides warfarin, fumarin, and tomorin (which were originally developed from spoiled sweetclover hay). Symptoms and treatment resemble those of anticoagulant rodenticide poisoning. O'Leary (1964) reports 34 *Aesculus* nut ingestions in recent years.

Flowers of *Aesculus californica* are reportedly poisonous to honey bees, and human poisoning from eating honey produced from the nectar of this plant has been reported (Anonymous, 1938). Bee larvae eating the honey often develop with missing legs, wings, or other deformities.

Seeds of *Aesculus* have been ground and used in breadstuffs in eastern North America (Fernald and Kinsey, 1943).

14. *Robinia*

The black locust, *Robinia pseudoacacia,* forms woods and thickets in eastern and central North America. It is also used as an ornamental. The bark and seeds (which are enclosed in a pealike pod) have caused human and domestic-animal poisoning, and the sprouts and leaves have caused poisoning of domestic animals. Recent pharmacologic interpretations are apparently absent. However, earlier reports (Power, 1901; Tasaki and Tanaka, 1918) specify a phytotoxin (robin) and a glycoside (robitin) as the major toxic fractions.

A few hours after poisoning, weak irregular pulse, anorexia, lassitude, weakness, nausea, vomiting (sometimes bloody diarrhea), coldness of extremities, marked pupil dilation, and stupor may develop. Fatality is rare.

Millspaugh (1892) reports several children poisoned from eating the seeds, and 32 boys at an orphans' home in Brooklyn were poisoned in 1887 after eating the inner bark of black-locust fence posts. Two of the orphan boys became stuporous, but all recovered (Emery, 1887).

Robinia is locally important as a honey source and was the original source of acacia perfume of commerce. It is a large but short-lived tree grown for timber, soil conservation, and shade as well as for ornamental purposes. It is an important honey plant.

15. *Ilex*

Holly (*Ilex* spp.) berries contain the toxin ilicin and on ingestion can cause vomiting, diarrhea, and mild narcosis. O'Leary (1964) documents 22 ingestions of holly berries during 1959–1961 in Colorado, Connecticut, Florida, Maryland, Massachusetts, Nevada, North Carolina, Ohio, Rhode Island, South Carolina, Tennessee, and Virginia. There is an unverified report of the death of an adult in Philadelphia from ingestion of holly-leaf tea.

At least eight species of holly are commonly used as ornamentals in the United States: *Ilex aquifolium* (English holly), *Ilex cornuta* (Chinese holly), *Ilex crenata* (Japanese holly), *Ilex glabra* (inkberry*), *Ilex laevigata* (smooth winterberry), *Ilex opaca* (American holly), *Ilex serrata* (Japanese winterberry) and *Ilex verticillata* (winterberry). All have bright red berries except *Ilex laevigata* (orange-red), *Ilex glabra* (black), and *Ilex crenata* (black).

The hollies and inkberry are evergreen while the winterberries are de-

---

*Phytolacca americana (pokeweed), *q.v.,* is also sometimes called inkberry.

ciduous. A number of nonornamental species, which are probably also toxic, occur in various parts of the country.

16. *Berberis*

Common barberry, *Berberis vulgaris,* has at times been cultivated for fruit used to make jellies and preserves. At least nine exotic species of *Berberis* are widely grown as ornamentals, and numerous indigenous species occur in the United States. A brilliant yellow dye is obtained from *Berberis* root. Cultivation of *Berberis vulgaris* has been discouraged because of its role as an alternate host for black stem rusts of cereals, especially pucciniasis graminis (black stem rust of wheat), and has now been largely eradicated in the northeastern United States.

A toxic fraction, berberine, has been isolated from the roots and bark of many *Berberis* species and perhaps occurs in all. Most reported ingestions have been of the nontoxic berries.

17. *Datura*

Members of the genus *Datura* are cosmopolitan weeds of world-wide distribution. The species are frequently confused with each other with regard to the many common names: thorn apple, jimson weed, Jamestown weed, apple of Peru, tolguacha, and others. About twelve species occur, of which four are common naturalized or cultivated in North America.

Datura stramonium (Fig. 6) is widely distributed across North America.

Datura metel is common in waste areas along the Coastal Plain from Florida to Texas.

Datura metaloides occupies plains, dry hills, and valleys from California to Florida north to Colorado and was used by the Indians as a narcotic.

Datura suaveolens is cultivated as an ornamental, especially in the southeastern United States where it does not form fruit.

Toxicity of *Datura* spp. is associated with a high content (0.25% to 0.7%) of solanaceous alkaloids (atropine, hyoscyamine, scopolamine, and others) which have been widely used as medicinals (stramonium, a "knockout drop," is also used to relieve asthma and bronchitis; scopolamine is used as a preanesthetic in childbirth, surgery, and ophthamology). In addition, *Datura* spp. may sporadically develop toxic concentrations ($>0.5\%$) of nitrates. Although nitrate poisoning from *Datura* has clearly occurred in domestic animals (Case, 1957), no clear-cut case involving man is on record. Such poisoning, however, would be obscured by the toxic effects of the solanaceous alkaloids present. In earlier times a tea made from *Datura* leaves was used to relieve asthma and produced

many cases of poisoning, and deliberate use of *Datura* for its hallucinogenic effects still occurs (Jacobziner and Raybin, 1960, 1961; Mitchell
and Mitchell, 1955; Rosen and Lechner, 1962).

FIG. 6. Jimson weed *(Datura stramonium).*

In 1676, Nathaniel Bacon of Jamestown, Virginia, raised a force to
attack the Indians without the consent of Governor Berkeley. Soldiers
dispatched to quell this "Bacon rebellion" were poisoned en route by
Datura leaves being used for salad giving rise to the common names
"Jamestown weed" and "jimson weed" (Jennings, 1935). British soldiers,
mistaking *Datura* for *Chenopodium album* (lamb's quarters), used it for
greens and suffered at least one fatality (Mitchell and Mitchell, 1955).

Children have been poisoned from sucking nectar from the corolla
tube of the flower (Morton, 1958) or from ingestion of the seeds. In addition, poisoning has occurred when *Datura* seeds have been mixed accidentally with beans and eaten (Goldberg, 1951; Stiles, 1951; Mitchell
and Mitchell, 1955). Small amounts can produce severe symptoms, and

larger amounts (for example, 4 to 5 gm of crude leaf or seed) death if treatment is not swift and effective. Contact with leaves or flowers produces a dermatitis in some individuals.

Symptoms vary with part and amount of plant ingested and with the physiologic idiosyncrasy of the poisoned individual. Symptoms may appear minutes to hours after ingestion: intense thirst, mydriatic visual disturbance, flushed skin, hyperirritability, incoherency, delirium, violence, reaching for imaginary objects, elevated temperature, rapid weak heartbeat, convulsions, and coma. Death may occur rapidly. If the victim recovers, major symptoms subside after 12 to 48 hours, with some (such as visual disturbance) persisting for about 2 weeks.

18. *Caesalpina (= Poinciana)*

Bird of paradise or poinciana *(Caesalpina gilliesi)* is an ill-scented ornamental outdoor perennial (southern half of the United States) or potted plant with large green poisonous seed pods. It has become naturalized in southern Texas, New Mexico, and Arizona.

The irritant pods may be eaten by children who usually recover after about 24 hours of nausea, vomiting, and diarrhea. Symptoms develop about 30 minutes after ingestion (Cann and Verhulst, 1958).

E. Other Inherently Toxic Plants

Numerous other toxic plants occur. Some of the more familiar will be mentioned briefly.

1. *Abrus*

Jequirity beans from *Abrus precatorius* (precatory bean, love bean, rosary pea, crabs-eye vine, weather plant, lucky bean, prayer bean, crabs-eye, and tiger-eye) are shiny, hard-shelled, and orange-red with a black tip. They were once used to make poison-tipped arrows and are now commonly used to make necklaces and costume decorations in the Caribbean Islands. Drilled or otherwise broken beads which are swallowed, chewed on, or brought in contact with open wounds can produce severe poisoning and even death from the hemorrhagic and cytolytic toxalbumin abrin. O'Leary (1964) reports five cases of jequirity bean poisoning in Florida and Massachusetts in 1959–1960. The plant is a woody vine of tropical regions grown for ornament and for its beans.

2. *Blighia*

The commonly eaten akee *(Blighia sapida)* of West Africa which has been introduced into Cuba, Jamaica, and Florida is highly toxic when unripe. The cotyledons are also poisonous. In Cuba and Jamaica about 50 cases occur annually, with some deaths. Most cases (85%) are in children. Only fully opened fruit should be picked. Fallen unripe fruit should be burned to prevent children from eating it. Water used in cooking akee should be discarded. Victims exhibiting only vomiting usually recover in a few days. Recovery after onset of convulsions or coma is rare.

3. *Convallaria*

The ornamental and medicinal lily-of-the-valley *(Convallaria majalis)* is considered toxic. It is a rhizomatous, perennial herb of Eurasia, widely grown in gardens and locally naturalized in eastern United States. Ingestions were reported in Indiana, Massachusetts, Michigan, and Ohio during 1959–1960. The rhizome and roots of this species yield a cardiac stimulant and diuretic similar to digitalis.

4. *Daphne*

Species of *Daphne,* especially *Daphne mezereum,* are considered to have toxic berries. Nine ingestions by children (19 to 42 months of age) in Washington and Oregon are reported by O'Leary (1964). *Daphne* spp. are ornamental shrubs indigenous to Eurasia.

5. ELEPHANT EARS

This name is applied to species of *Caladium, Colocasia, Dieffenbachia, Philodendron, Podophyllum,* and *Mandragora,* all of which contain a large amount of calcium oxalate and other toxins which produce burning and intoxication on ingestion. O'Leary (1964) records at least 33 ingestions of leaves, stems, or juice in the United States during 1959–1960 in children (9 months to 12 years of age). *Colocasia esculenta* is the widely eaten taro (dasheen) cultivated for its edible tubers. *Dieffenbachia picta* (also called dumb cane) was used by the Nazis in sterilization experiments. *Mandragora officinarum,* the legendary aphrodisiac "mandrake" of Europe, was considered (under the "doctrine of signatures") to resemble the human body, and when pulled from the ground was alleged to emit a horrible shriek sufficient to kill the collector. To collect it, dogs were tied to the plant and then called from a safe distance (Schery, 1952; Drach and Maloney, 1963). *Podophyllum peltatum,* may apple (Fig. 7), is used in treatment of paralysis and by rural folk of the eastern United States as an emetic and cathartic.

FIG. 7. May apple *(Podophyllum peltatum)*.

6. *Euphorbia*

The ornamental poinsettia, *Euphorbia pulcherrima,* has toxic leaves, and ingestion of a single leaf has been reported to have caused death in both children and adults.

At least five other species of *Euphorbia* are also dangerously toxic: *Euphorbia lathyrus* (caper garden spurge; can be fatal), *Euphorbia marginata* (snow on the mountain; juice used to brand cattle, will burn hands), *Euphorbia tirucalli* (milkbush), *Euphorbia mili* (crown-of-thorns), and *Euphorbia cyparissias* (cyprus spurge). At least 18 poisonings with 2 deaths from *Euphorbia* spp. have been recorded in recent years (O'Leary, 1964; Kingsbury, 1964).

7. HEATHERS

Various members of the family Ericaceae are toxic, including species of *Kalmia, Ledum, Leucothoe, Menziesia, Pieris* and *Rhododendron. Kalmia* spp. are evergreen shrubs or small trees cultivated for their showy flowers and used extensively for florist greens. *Ledum* spp. are small evergreen shrubs of cold bogs sometimes grown for ornament. *Leucothoe* spp. are evergreen or deciduous shrubs of acid swamps sometimes cultivated for ornament. *Menziesia* are deciduous shrubs, sometimes grown for ornament. *Pieris* are native or imported (from

Japan) evergreen shrubs. *Rhododendron* spp. are native and cultivated flowering shrubs (azaleas, rhododendron). The Delaware Indians used laurel *(Kalmia)* for suicide, and humans have been poisoned from ingesting honey made from *Kalmia* flowers (Howes, 1949). Powdered leaves of the Chinese *Rhododendron molle* are insecticidal.

8. Hedysarum

Sweet pea (sweet vetch), sometimes mistaken for related edible plants such as the high-mountain *Hedysarum alpinum,* has caused numerous poisonings. Illness from eating *Hedysarum mackenzi* (wild sweet pea) was suffered by members of the Richardson arctic expeditions between 1819 and 1848. Cultivated sweet pea *(Lathyrus odoratus)* is said to contain a paralytic alkaloid in its stems. Cases were reported in 1959 from Washington State.

9. Hippomane

The manchineel tree, *Hippomane mancinella* (also called manzanillo), with its crab apple-like fruit, is found in Central America, the West Indies, and Florida. Manchineel fatalities among early Spanish explorers led to its widespread destruction so that it is now largely restricted to the Everglades. A 1733 royal ordinance prescribed destruction of all manchineel trees at St. Barthelemy, Puerto Rico. The milky sap can cause severe dermatitis, and ingestion of the fruit can produce severe poisoning and death. Little and Wadsworth (1964) call it "the most poisonous tree of Puerto Rico and the Virgin Islands," In 1885, 54 Germans were poisoned by manchineel apples, and 5 died. In 1940, survivors of a sunken ship were similarly poisoned after landing their lifeboat on an uninhabited beach. Two persons were hospitalized after eating machineel apples at St. Thomas, Virgin Islands (Holdridge and Munoz-MacCormick, 1937; Little and Wadsworth, 1964).

10. Iris

Rootstocks of *Iris* spp. are said to be potentially fatal, and the leaves possibly toxic. Cases were reported in Ohio and Georgia in 1960. The rhizome of *Iris versicolor* is used to prepare a medicinal cathartic and diuretic.

11. JASMINES

Several species commonly called "jasmine" have toxic effects. *Jasminum nudiflorum,* a vinelike shrub of most of the United States, is one of the first plants to flower in the spring. *Cestrum nigrum* (night-blooming jasmine) is a common ornamental shrub or small tree of tropical America cultivated for its attractive flowers. On ingestion it produces atropine-like effects (hallucination, muscular and nervous irritability, tachycardia, elevated temperature, salivation, dyspnea, and paralysis). *Gardenia jasminoides* (cape jasmine gardenia), *Ervatamia coronaria* (crape jasmine), *Stephanotis* spp. (Spanish jasmine), and *Trachelospermum asminoides* (star jasmine, confederate jasmine) all have toxic qualities.

12. MISTLETOES

All mistletoes *(Loranthus* spp., *Phoradendron* spp., and *Viscum* spp.*)* are considered highly toxic. O'Leary (1964) documents 11 cases in children (8 months to 3 years of age) during 1959–1960 from ten different states. Fernald and Kinsey (1943) attribute the deaths of several children to eating the white berries of the American mistletoe *(Phoradendron flavescens),* also called false mistletoe.

13. MOONSEED

Moonseed poisoning has occurred in recent years in Hawaii *(Cocculus ferrandianus)* and the eastern United States *(Menispermum canadense; Cocculus carolinus,* Carolina moonseed, snailseed) (Fig. 8). *Cocculus*

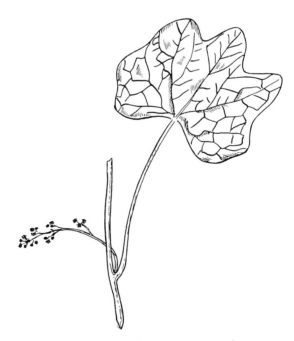

FIG. 8. Moonseed *(Menispermum canadense).*

rootstock and all parts of *Menispermum* are considered toxic. At least three human deaths have occurred from *Menispermum* poisoning in the United States in recent years. Moonseeds are twining shrubs or vines. *Cocculus* spp. are also called coral beads.

14. *Myristica*

Nutmeg *(Myristica fragrans,* a cultivated and naturalized West Indian tree) used sparingly as a condiment is safe, but in large quantities it can cause headache, dizziness, swelling and flushing of face, and abdominal pains. Several automobile accidents have been related to "nutmeg intoxication" following ingestion of large quantities of nutmeg on holiday eggnog. This effect may be more common than generally believed, owing to the difficulty of distinguishing nutmeg intoxication from alcoholic intoxication or a combination of the two. Mace is obtained from the aril surrounding the nutmeg seed. Intentional use of nutmeg for its hallucinatory effects is reported by O'Leary (1964). Besides its use as a condiment, nutmeg is also used to produce a valuable nondrying oil used to produce spicy flavor and odor in dentifrices, tobaccos, and perfumes.

15. *Nicotiana*

Tree tobacco *(Nicotiana glauca)* has produced at least two human deaths via ingestion poisoning in the United States in recent years. It is a South American shrub or small tree naturalized from southwestern Texas to southern California.

16. *Prunus*

Many species of *Prunus,* especially *Prunus serotina* (wild black cherry), are highly cyanogenic and can be dangerously poisonous. Fresh leaves of *Prunus serotina* (Fig. 9) contain an average of 212 mg of cyanide per 100 gm of leaves, more than ten times the amount considered dangerous. Although drying volatizes the cyanide, dried leaves may still be toxic. The fruit, especially the seed, is also toxic (Kingsbury, 1964).

17. *Rheum*

The common edible rhubarb *(Rheum rhaponticum)* (Fig. 10), indigenous to central Asia, has a toxic leaf, ingestion of which can be fatal. The stalk is the edible portion (Robb, 1919). It was used as a medicinal as early as 2700 B.C. in China.

18. *Thevetia*

Ingestion of yellow oleander, *Thevetia peruana,* has caused deaths in Hawaii and illness in Florida in recent years.

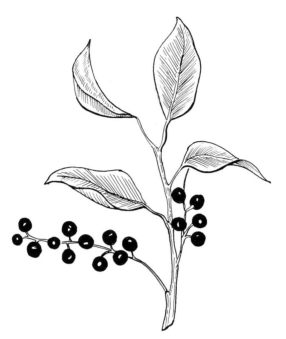

FIG. 9. Black cherry *(Prunus serotina)*.

FIG. 10. Rhubarb *(Rheum rhaponticum)*.

19. *Wistaria*

Several species of *Wistaria*, especially *Wistaria floribunda* (Japanese wistaria) and *Wistaria sinensis* (Chinese wistaria), are grown as ornamentals in the United States. At least 18 cases of children poisoned by consumption of seeds or pods were reported between 1958 and 1962 (Kingsbury, 1964; O'Leary, 1964). *Wistaria* spp. are woody vines or twining shrubs.

A systematic list of plants discussed is given in Table I.

TABLE I
SYSTEMATIC LIST OF DISCUSSED PLANTS

Kingdom – Plantae	Family – Polypodiaceae
Phylum – Cyanophyta	*Dryopteris*
Class – Myxophyceae	*Notholaena*
Order – Chroococcales	*Onoclea*
Family – Chroococcaceae	*Pteridium*
Microcystis	Phylum – Spenophyta
Order – Oscillatoriales	Class – Sphenopsida
Family – Nostocaceae	Order – Equisetales
Anabaena	Family – Equisetaceae
Aphanizomenon	*Equisetum*
Phylum – Phaeophyta	Phylum – Spermatophyta
Class – Phaeophyceae	Class – Conopsida
Order – Ectocarpales	Order – Pinales
Family – Desmarestiaceae	Family – Cupressaceae
Desmarestia	*Cupressus*
Phylum – Schizomycophyta	Order – Taxales
Class – Schizomycetes	Family – Taxaceae
Order – Eubacteriales	*Taxus*
Family – Enterobacteriaceae	Class – Angiospermae
Proteus	Sublcass – Dicotyledonae
Phylum – Eumycophyta	Order – Apocynales
Class – Ascomycetes	Family – Apocynaceae
Order – Helvellales	*Acocanthera*
Family – Helvellaceae	*Ervatamia*
Helvella	*Nerium*
Morchella	*Strophanthus*
Class – Basidiomycetes	*Thevetia*
Order – Agaricales	*Trachelospermum*
Family – Agaricaceae	Order – Asterales
Amanita	Family – Compositae
Psilocybe	*Aster*
Russula	*Eupatorium*
Phylum – Pteridophyta	*Grindelia*
Class – Pteropsida	*Gutierrezia*
Order – Filicales	*Helianthus*

TABLE I
SYSTEMATIC LIST OF DISCUSSED PLANTS (Continued)

Machaeranthera
Oonopsis
Sideranthus
Xylorrhiza
Order — Caryophyllales
Family — Chenopodiaceae
Atriplex
Beta
Chenopodium
Grayia
Family — Phytolaccaceae
Phytolacca
Order — Ebenales
Family — Ebenaceae
Diaospyros
Family — Polygonaceae
Rheum
Order — Ericales
Family — Ericaceae
Kalmia
Ledum
Leucothoe
Menziesia
Pieris
Rhododendron
Order — Euphorbiales
Family — Euphorbiaceae
Euphorbia
Hippomane
Ricinus
Order — Geraniales
Family — Linaceae
Linum
Order — Gentianales
Family — Loganiaceae
Strychnos
Order — Oleales
Family — Oleaceae
Jasminum
Order — Opuntiales
Family — Cactaceae
Lophophora
Order — Papaverales
Family — Cruciferae
Brassica
Stanleya

Order — Polemoniales
Family — Solanaceae
Atropa
Cestrum
Datura
Hyoscyamus
Mandragora
Nicotiana
Solanum
Order — Ranales
Family — Berberidaceae
Berberis
Podophyllum
Family — Lauraceae
Laurus
Family — Menispermaceae
Cocculus
Menispermum
Family — Myristicaceae
Myristica
Order — Rosales
Family — Leguminosae
Abrus
Astragulus
Caesalpina
Glycine
Hedysarum
Lathyrus
Robinia
Wistaria
Family — Rosaceae
Malus
Prunus
Order — Rubiales
Family — Rubiaceae
Gardenia
Order — Santales
Family — Loranthaceae
Loranthus
Phoradendron
Viscum
Family — Santalaceae
Comandra
Order — Sapindales
Family — Asclepiadaceae
Stephanotis

Family — Aquifoliaceae
  Ilex
Family — Hippocastanaceae
  Aesculus
Family — Sapindaceae
  Blighia
Order — Scrophulariales
  Family — Scrophulariaceae
    Castilleja
    Penstemon
Order — Thymelaeales
  Family — Thymelaeaceae
    Daphne
Order — Umbellales
  Family — Umbelliferae
    Aethusa
    Ammi
    Cicuta
    Conium

Daucus
Pastinaca
Sium
Order — Urticales
  Family — Moraceae
    Cannabis
Subclass — Monocotyledonae
  Order — Arales
    Family — Araceae
      Caladium
      Colocasia
      Dieffenbachia
      Philodendron
  Order — Liliales
    Family — Liliaceae
      Amaryllis
      Convallaria
      Iris
      Urginea

III. POISONOUS ANIMALS

Animal tissues may be rendered poisonous by bacterial and enzyme decomposition, but some are naturally toxic to man even when eaten as fresh as they can be prepared for the table. Animal toxins are powerful physiologic agents, and some may prove to be therapeutic drugs of great significance. For example, some shellfish (mollusc) toxins cause paralyzed muscles to contract, and some sea cucumber (echinoderm) toxins seem to cause tumors to dwindle. Mullet and goatfish toxins are hallucinogenic and may be of value in psychotherapy, while the red-beard sponge produces an antibiotic of potential therapeutic value.

A. "Rules of Thumb"

Many animal toxins are water-soluble, but most are heat-stable. Viscera and dark meat usually contain the highest toxic concentrations. Animals to be eaten should be eviscerated, the meat thoroughly washed and cooked, and the cooking broth discarded. Local eating customs should be followed (with due caution) and local quarantine regulations strictly adhered to.

Folk-tale tests for differentiating toxic from nontoxic forms are unreliable. Poisonous animals *cannot* be detected by their appearance or smell, by discoloration of metal, or by placing garlic in the cooking water.

B. Incidence

Throughout the world, poisonous animals apparently produce a greater morbidity and mortality than do poisonous plants, but ". . . it is impossible to say how many cases . . . occur annually owing to the unreliability or non-existence of health statistics in areas where poisoning occurs" (Walford, 1958; Fish and Cobb, 1954).

Table II illustrates the widespread and variable occurrence of poisoning by ingestion of animals.

TABLE II
SELECTED OUTBREAKS OF POISONING BY ANIMALS

Year(s)	Place	Vector	Cases	Deaths	Reference
1900–1963	Hawaii	Fish	433	7	Helfrich (1963)
1917–1958	California	Mussels, clams	479	33	Halstead (1965)
1946–1947	Fanning Island	Fish	95[a]	0	Ross (1947)
1949	Saipan	Eel	57	2	Khlentzos (1950)
1953–1960	Minamata, Japan	Fish	83	10	Kurland et al. (1960)
1955–1956	Indo-Pacific	Fish	40,000		Halstead (1958)
1957	Canada	Oyster, clam	94		Davies et al. (1958); Bond and Medcof (1958)
1962	Alaska	Little Neck clams	25	0	USPHS (1962)
1963–1964	Wake Island	Mullet	29	0	Banner and Helfrich (1964)

[a] In a population of 224 persons.

C. Inherently versus Secondarily Toxic Animals

Although many species of animals are inherently toxic, most human poisonings involve secondarily transvectored toxins. Inherent and secondary toxicity are closely intertwined taxonomically, and in some cases the origin of the toxin has not been satisfactorily demonstrated. Therefore, poisonous animals will be discussed systematically, with separate discussions of inherent versus secondary toxicity for each group.

D. Invertebrates

1. DINOFLAGELLATE PROTOZOA

Although transvection of toxins via mollusca and other marine animals is the usual dinoflagellate involvement with human poisoning, they can also cause respiratory irritation via wind blowing over protozoal blooms and can cause direct poisoning via ingestion. The species known to be toxic to man include *Gymnodinium brevis, Gonyaulax catenella, Gonyaulax tamarensis,* and *Pyrodinium phoneus.* Many others are suspected.

2. ANEMONES

In American Samoa and other tropical Pacific areas, sea anemones are regularly eaten, usually cooked. Most anemones seem never to be poisonous, although they possess dangerously venomous stinging cells (nematocysts) which may envenomize a person handling them or attempting to eat them raw. This envenomization system is destroyed by cooking. Species causing human poisoning after cooking include *Rhodactis howesi* (matelelei anemone), *Physobrachia douglasi,* and *Radianthus paumotensis* (matamala samasama anemone), all three of which occur generally from Polynesia to the Indian Ocean. The toxicity is apparently inherent.

Symptoms of anemone poisoning include acute gastritis, nausea, vomiting, abdominal pain, cyanosis, prostration, and an 8- to 36-hour stupor. Prolonged shock and death via pulmonary edema may result. Treatment is symptomatic (Farber and Lerke, 1963; Martin, 1960).

3. STARFISH

Starfish (Asteroidea) are apparently seldom ingested, and information on human poisonings is absent. However, toxicological studies indicate that many, if not most, starfish contain a powerful inherent heat-stable toxin.

4. SEA URCHINS

During their reproductive season, the ovaries of *Paracentrotus lividus* (sea urchin, Ireland to West Africa), *Tripneustes ventricosus* (white sea urchin, West Indies to Brazil to West Africa), and *Diadema antillarum* (sea needle, West Indies) may be highly poisonous (Halstead, 1965). Other poisonous species are reported to occur in Japan (Moru, 1943).

5. SEA CUCUMBERS

Poisoning of man by ingestion of sea cucumbers (Holothurioidea) is reported but unconfirmed. A high concentration of the toxin holothurin in the organs of Cuvier lends credence to these reports. Halstead (1965) lists 28 suspect species.

6. MOLLUSCA

A number of species of marine and freshwater mollusca are inherently toxic (for example, *Neptunea, Haliotis*) or secondarily toxic (for example,

Cardium, Mya) to man. Little is known of inherent molluscan toxins, but the secondary toxins, obtained when molluscs feed on certain dinoflagellate protozoa (*Gymnodinium brevis, Gonyaulax catenella, Gonyaulax tamarensis, Pyrodinium phoneus,* and probably others), have been widely studied.

Since some two million metric tons of marine molluscs are consumed annually in the world market, transvection of dinoflagellate toxins by mollusca is one of the most significant phenomena associated with poisonous animals. The molluscs are not harmed when they ingest the dinoflagellates. The involved toxins are stored in the digestive glands, gills, or siphons from where, when ingested, they can poison vertebrates. Since dinoflagellate abundance (directly dependent on available nutrients in the water) is seasonal, most mollusc poisoning in man is also seasonal. Specific toxins produced differ with the species of dinoflagellate involved. A single shellfish may contain a fatal toxic dose, and numerous cases and deaths have been reported from the United States.

Red tide, massive fish kills associated with red ocean water, is also associated with toxic dinoflagellates, the red color being produced by immense numbers of these protozoa in the water. Wind-borne spray from red tides of *Gymnocardium brevis* can cause respiratory irritation in man. Clinical symptoms of mollusc poisoning in man include tingling or burning of the lips, gums, tongue, and face which gradually spreads to other parts of the body until numbness develops in tingling areas. Weakness, dizziness, aching joints, excessive salivation, intense thirst, difficulty in swallowing, and difficult muscle movement are all common. Nausea, vomiting, diarrhea, and abdominal pain are uncommon. Muscular paralysis may increase, and death may result.

Toxins associated with shellfish poisoning are mostly (but not all) soluble heat-stable nitrogenous compounds.

Treatment includes removal of ingested shellfish by gastric lavage or emesis followed by catharsis. Alkaline fluids such as baking soda solution may be of value, since the toxin is rapidly destroyed in that medium. Other symptomatic treatment, including artificial respiration, may be indicated. No specific antidotes are available.

Prevention involves strict adherence to local shellfish quarantine regulations. Since heat does not destroy the toxin and boiling may not dissolve all of into the broth, cooking of shellfish cannot be depended on to render them nontoxic. Shellfish eating should be avoided during the warm months (North America, May to September, in some places March to November), and perpetually toxic species should not be used.

The following representative groups illustrate the widespread nature of inherently and secondarily poisonous molluscs.

a. Chiton. *Mopalia muscosa,* the mossy chiton, distributed from Alaska to Baja, California, transvects dinoflagellate poisons (Sommer and Meyer, 1937).

b. Abalone. The liver of two Japanese abalones *(Haliotis discus* and *Haliotis sieboldi)* have caused photosensitization poisoning in man (Hashimoto *et al.,* 1960), and it is probable that the same could occur with west American abalones *(Haliotis rufescens, Haliotis fulgens,* and *Haliotis kamtschatkana)* if the viscera were consumed. Americans ordinarily eat only the foot ("steak") of the abalone. Toxicity is considered secondary, originating with *Desmarestia* seaweed ingested by the abalone. Actual incidence of abalone poisoning is unknown but is probably low.

c. Murex and Dogwinkle. Several species of *Murex* are commonly eaten in Europe and poisonings from such ingestions are extremely rare if they ever occur. Nevertheless, poisoning reports appear in the literature (Plumert, 1902; 43 cases, 5 deaths), and researchers have detected some mildly toxic fractions (Whittaker, 1960). *Murex brandaris* (rockshell murex; Mediterranean Sea and West Africa), *Murex trunculus* (murex; Mediterranean Sea), and *Thais lapillus* (Atlantic dogwinkle; Atlantic shores, New York to Canada to Norway to Portugal) have all been reported as poisonous.

d. Top Shell. *Livona picta* (West Indian top shell), commonly eaten in southeastern Florida, the West Indies, and Central America, apparently can become seasonally poisonous and produce indigestion and nervous disorder. The poisoning may be due to a secondary toxin of algal origin (Arcisz, 1950).

e. Whelk and Triton. Whelk have caused human poisoning when the toxic salivary glands were ingested in whole shellfish raw, cooked, or canned. The toxin, tetramine, is apparently inherent. *Neptunea arthritica* and *Neptunea intersculpta* have caused whelk poisoning in Japan. Another Japanese whelk *(Buccinum leucostoma),* the red whelk *(Neptunea antigua)* of northern Europe, and the Oregon triton *(Argobuccinum oregonense)* have also been found to contain tetramine but have not been involved in reported human poisonings (Fange, 1960).

f. Limpet. The shield limpet *(Acmaea pelta)* of Pacific North America transvects dinoflagellate poisons (Sommer and Meyer, 1937).

g. Sea Hare. Records of sea hare poisoning go back at least to the first century, but toxicity of these molluscs to man is apparently unconfirmed. A toxic fraction, aplysin, has been detected. *Aplysia californica* (California sea hare, Monterey to San Diego), *Aplysia depilans* (sea hare, European North Atlantic and Mediterranean Sea), and *Aplysia punctata* (sea hare, Mediterranean Sea) have been reported as poisonous (Winkler, 1961).

h. Marine Mussel. Seven species of marine mussels have been shown capable of transvectoring dinoflagellate toxins: the California mussel (*Mytilus californianus,* west coast of North America), the bay mussel (*Mytilus edulis,* essentially world-wide), the meismuschel (*Mytilis galloprovincialis,* Mediterranean Sea), Australian mussels (*Mytilus planulatus* and *Modiolus areolatus),* ribbed mussel (*Modiolus demissus,* Virginia to Florida and introduced into San Francisco Bay), and northern horse mussel (*Modiolus modiolus,* western North America and Arctica).

Of these the bay mussel is by far the most important, having been involved in no less than 17 major poisoning outbreaks involving at least 197 cases and 126 deaths between 1793 and 1948 (Halstead, 1965).

The California mussel was involved in at least 116 cases and 8 deaths in California during 1917, 1918, and 1927 (Meyer *et al.,* 1928).

i. Scallop. The Atlantic deep sea scallop *(Placopecten magellanicus),* distributed from Labrador to North Carolina, has been reported to transvect dinoflagellate toxin in Nova Scotia and New Brunswick (Medcof and Gibbons, 1948; Gibbard and Naubert, 1948; Edwards, 1956).

j. Thorny Oyster. The spondylid or thorny oysters are distinct from the true oysters of the Family Ostreidae. Two species have been shown capable of transvectoring dinoflagellate toxins to man: the Atlantic thorny oyster *(Spondylus americanus),* distributed from North Carolina through the West Indies to Texas, and the Northern thorny oyster *(Spondylus ducalis)* of the Philippines, Indonesia, Micronesia, and New Guinea (O'Malley, 1962).

k. Clam. The common name clam is applied to at least six different molluscan families.

The mactrid clams include the solid surf clam *(Spisula solidissima),* found from Labrador to North Carolina, and the summer clam or gaper *(Schizothaerus nuttalli),* found from Alaska to Baja California and west to Japan. Both have been reported as transvectors of dinoflagellate toxins (Bond and Medcof, 1958; Pugsley, 1939).

The myacid clams include the soft-shell clam *(Mya arenaria)* distributed along the Atlantic and Pacific Coasts south to the Carolinas, California, Japan, and Great Britian (Fig. 11). Major poisoning outbreaks

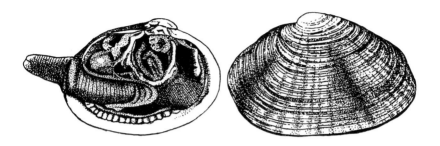

FIG. 11. Soft-shell clam *(Mya arenaria)*.

produced by this species transvectoring dinoflagellate toxins have occurred in Canada in 1945 (28 cases), 1954 (7 cases, 2 deaths), and 1957 (33 cases) (Bond and Medcof, 1958).

The sanguinolarid clams include the California jack-knife or razor clam *(Tagellus californicus),* distributed from Panama to California. It has been reported as a transvector of dinoflagellate toxin (Sommer and Meyer, 1937).

The solenid clams include the Atlantic jack-knife (or razor) clam *(Ensis directus)* found in eastern North America from the Gulf of St. Lawrence to Florida, and the northern razor clam *(Siliqua patula)* found from Alaska to California. Both transvector dinoflagellate toxins (Meyer, 1953; Davies *et al.,* 1958).

The tellinid clams include the bent-nose mud clam *(Macoma nasuta),* found from Alaska to California, and the white sand clam *(Macoma secta),* found from British Columbia to Baja California. Both transvect dinoflagellate toxins (Sommer and Meyer, 1937).

Two genera of the Family Veneridae *(Saxidomus* and *Tivela)* are also commonly called clams. These butter clams and pismo clams will be discussed under the Venerids.

1. Cockle. The heart cockle *(Cardium edule)* of European seas and the basket cockle *(Clinocardium nuttalli)* of the Pacific Ocean south to Japan and Baja California are both transvectors of dinoflagellate toxins (Charnot, 1945). The Japanese littleneck cockle and the red cockle (quahaug) are discussed under the Venerids.

m. Oyster. The common edible oyster *(Ostrea edulis)* and the Giant Pacific oyster *(Grassostrea gigas)* both occasionally transvect dinoflagellate toxins. In addition, the Giant Pacific oyster, distributed in Japan and along the Pacific Coast of North America, produces venerupin poisoning which has long been considered inherent but which has recently (Nakazima, 1963) been suspected as secondary poisoning associated with *Prorocentrum* dinoflagellates. Venerupin poisoning is also associated with *Tapes semidecussata* and *Dosinea japonica,* which will be discussed under the Venerids (Anderson, 1960; Engelsen, 1922).

n. Donax. Two donax are known to transvect dinoflagellate toxin. They are the common donax *(Donax denticulatus)* of the West Indies and the wedge-shell donax *(Donax serra)* of South Africa (Sapeika, 1958). *Heterodonax bimaculata* was reported to have caused poisoning in Northern California in 1939, but details are lacking (Halstead, 1965).

o. Piddock. The common (or flap-tipped) piddock *(Penitella penita),* found from Alaska to California, can transvect dinoflagellate toxin (Sommer and Meyer, 1937).

p. Freshwater Mussel. The unionid mussel *(Anodonta oregonensis)* occurs in freshwater streams from Alaska to Utah. It has been reported to transvector paralytic toxins, but the exact mode of this transvection in fresh water has not been elucidated (Humphreys and Gibbons, 1941).

q. Noah's Ark. Noah's Ark *(Arca noae)* of the Mediterranean Sea transvects dinoflagellate toxin (Engelsen, 1922).

r. Venerids. The Japanese callista *(Callista brevisiphonata)* causes inherent "callistin" poisoning during the spawning season (May to September) when its ovary contains a high concentrate of choline. Asano *et al.* (1953) report 119 cases near Mari, Hokkaido, Japan.

The Japanese dosinia *(Dosinia japonica)* and the Japanese littleneck cockle *(Tapes semidecussata),* plus the previously discussed Giant Pacific oyster *(Crassostrea gigas),* transvects *Prorocentrum* dinoflagellate toxin to produce a syndrome called venerupin poisoning. At least 542 cases and 185 deaths from venerupin poisoning have occurred in Japan in six outbreaks (1889, 1941, 1942, 1943, 1949, 1950) (Tarada, 1953).

The quahaug or red cockle *(Prothothaca steminea)* found from Alaska to California transvects dinoflagellate toxin (U. S. Public Health Service, 1962).

The Alaska butter clam *(Saxidomus giganteus)* found from Alaska to California, and the common Washington butter clam *(Saxidomus nuttalli)* found off California, both transvect dinoflagellate toxins which are concentrated in the siphon. Between 1929 and 1932, 97 poisonings by *Saxidomus nuttalli* with two deaths occurred in California (Sommer and

Meyer, 1937). *Saxidomus giganteus* can maintain toxicity year around. Studies made in 1943 on *Saxidomus* toxicity resulted in withholding of butter clam shipments and stopped shellfish operations in Alaska (Halstead, 1965).

The Pismo clam *(Tivela stultorum)* of California can transvect dinoflagellate toxin (Sommer and Meyer, 1937).

7. OCTOPUS, SQUID

Poisoning from ingestion of cephalopods is reported regularly from Japan, with 2874 persons affected in 1952–1955. Death occurred in 10 of these cases. The nature of the toxin is apparently unknown. The kinds of cephalopods involved are all commonly eaten, and no laboratory test has proved effective in detecting poisonous specimens. *Ommastrephes sloani pacificus* (squid, Pacific Ocean), *Octopus vulgaris* (octopus, nearly world-wide), and *Octopus dofleini* (Japanese octopus, Japan) are the species known to be involved (Kawabata and Halstead, 1957).

8. ARTHROPODS

Arthropods are the commonest animals on the Earth, with some 900,000 described species. Although many of these are venomous, few cause human poisoning, even though vast numbers (especially of Crustacea) are consumed annually by man.

a. Horseshoe Crabs. Three asiatic Merostomata *(Xiphosura)* cause human poisoning through ingestion of their seasonally toxic eggs, musculature, or viscera. Despite this periodic toxicity, the large masses of green unlaid eggs are widely eaten. Onset of poisoning symptoms usually occurs within 30 minutes: Dizziness, headache, nausea, slow pulse, decreased temperature, vomiting, abdominal cramps, diarrhea, cardiac palpitation, lip numbness, paresthesias of lower extremities, aphonia, "hot" mouth and throat, generalized muscular paralysis, trismus, hypersalivation, drowsiness, coma. The mortality rate is high. Death usually occurs within 16 hours. Treatment is symptomatic (Waterman, 1953).

Species involved include *Carcinoscorpius rotundicauda* (horseshoe crab, Philippines to India), *Taehypleus gigas* (Indo-Papuan king crab, India to Vietnam), and *Tachypleus tridentatus* (Moluccan crab, Japan to Vietnam).

b. Crustacea. Inherent toxicity of three species of crabs is reported to have caused human poisoning, but details are lacking. The species involved are *Angatea* sp. (white-shelled crab, Australasia), *Eriphia norfolcensis* (poison crab, Australia), and *Xanthodes reynaudi* (Indo-Pacific region) (Halstead, 1965).

E. Vertebrates

1. ELASMOBRANCHS

Poisoning as a result of eating sharks and rays is occasionally reported, usually due to ingestion of livers of tropical sharks. Warm-water species involved include the white shark *(Carcharodon carcharias)* cosmopolitan in tropical to warm-temperate oceans (Fig. 12); the black-tipped

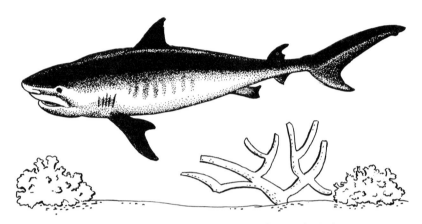

FIG. 12. Great white shark *(Carcharodon carcharias).*

shark *(Carcharhinus melanopterus)* of Indo-Pacific ocean waters from South Africa to Hawaii, the seven-gilled shark *(Heptranchias perlo)* of the Atlantic, Mediterranean, Indian, and Western Pacific oceans; the six-gilled shark *(Hexanchus grisseus)* of the Atlantic, Indian, and Pacific oceans; and the hammerhead shark *(Sphyrna zygaena)* of the tropical to warm-temperate Atlantic and Pacific Ocean areas.

The Greenland shark *(Somniosus microcephalus)* of the Arctic and North Atlantic is a cold-water species which has caused numerous human poisonings.

Although most shark poisonings have resulted from liver ingestion, other parts may also be toxic. The nature of the toxin involved is poorly known. Symptoms differ with degree of intoxication. Ingestion of toxic musculature usually results in gastrointestinal upset and diarrhea. Ingestion of toxic liver usually produces, within 30 minutes: nausea; vomiting; diarrhea; abdominal pain; headache; joint aches; mouth tingling; and burning of tongue, throat, and esophagus. Eventually muscular discoordination, breathing difficulty, and muscular paralysis can lead to coma and death.

Shark or ray viscera should not be eaten. Shark or ray musculature should be eaten only after investigation of the local situation and even then with due caution.

2. EELS

Moray eels *(Gymnothorax* spp.) and possibly other genera occupying tropical reefs have caused widespread poisoning (Fig. 13). These large

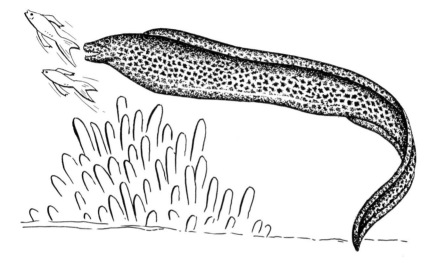

FIG. 13. Speckled moray *(Gymnothorax moringa).*

animals (10 feet or more long) have water-soluble toxins which render the flesh poisonous and make eel soup especially dangerous.

Symptoms of eel poisoning include tingling and numbness of lips, tongue, hand, and feet; leg heaviness; nausea; vomiting; diarrhea; abdominal pain; joint aches; metallic taste; difficulty in swallowing, foaming at the mouth; intense sweating; muscular discoordination; tetany convulsions; difficulty in breathing; and (in about 10% of cases) death. Recovery is slow.

Khlentzos (1950) details an outbreak of eel poisoning at Saipan associated with ingestion of a single *Gymnothorax flavimarginatus.* Of 57 persons intoxicated, 17 were hospitalized. Fourteen became comatose, and 2 died. Survivors were treated symptomatically.

Tropical moray eels should never be eaten even though natives of the area consider them safe.

Lampreys and hagfishes (Cyclostomata) often have toxin in the slime and flesh which produces gastrointestinal upset in man (Engelsen, 1922) (Fig. 14).

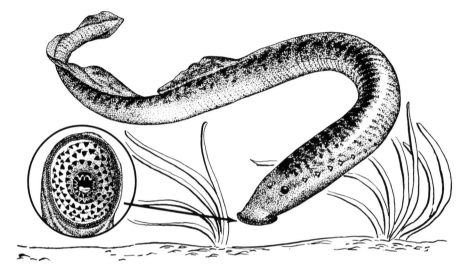

FIG. 14. Sea lamprey *(Petromyzon marinus)*.

3. FISHES

The flesh of many tropical fish apparently becomes poisonous by feeding on poisonous marine organisms. Some, such as puffers, triggerfish, and parrotfish, may be poisonous during most of the year. Others, such as and surgeonfish, moonfish, porcupine fish, filefish, and goatfish, may be poisonous only a part of the year. Still others including barracuda, pompano, mackerel, butterfly fish, snapper, sea bass, perch, and wrasse, are only sporadically poisonous and only in certain localities. In other places, apparently, they have always been safe to eat.

More than 300 species have been reported to cause fish poisoning. Of these, the puffers appear to have the most potent toxin, since mortality may reach 50%. In other types of fish poisoning, mortality ranges from less than 1% to about 10%, depending on physical condition of the individual, amount of fish eaten, potency of the toxin, and quality of medical care.

Morbidity may reach 50% of the population in limited areas of tropical countries where fish forms a large part of the diet. Fifty to 100 cases are reported yearly in Hawaii, with a mortality rate of less than 1%.

Fish poison exerts its primary effect on the peripheral nervous system.

The physiological effects produced by the toxins have not been pinpointed. Characteristic pathology has not been elucidated.

Although exact pinpointing of all poisonous fish by simple rules is not possible, the following generalized rules are used in survival training. Poisonous fish occur in all tropical waters, but chiefly in the Pacific, and may be caught comparatively close to shore. Most are characterized by a tough skin devoid of scales, and many are covered with rough thorn-like spines, long plates, or soft spines which give the appearance of hair. Many poisonous fish have jaws with an enamel-like covering without teeth which suggest a bird's bill. Their shapes are irregular, boxlike, or spherical. Other fish, such as certain herring, are poisonous especially during the spawning season. The roe (eggs) is the most toxic part. Poisonous fish in one locality may be harmless elsewhere.

Fish intoxication is of at least seven types: tetraodon, scombroid, ciguatera, hallucinary mullet, clupeoid, chimaeroid, and gempylid.

a. Tetraodon poisoning. Many puffer fishes (balloonfishes, swellfishes, fugu fishes, toadfishes, globefishes) (Fig. 15) and their relatives are

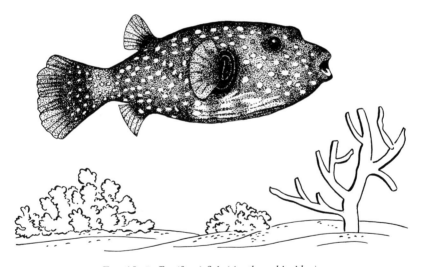

FIG. 15. Puffer (fugu) fish (*Arothron hispidus*).

inherently toxic (Yudkin, 1944; Mosher *et al.,* 1964). The emperors of ancient Japan canceled the inheritances of warriors indulging in the deadly fugu, which is still considered a delicacy in Japan. Alexander the Great prohibited his soldiers from eating puffer fishes. The viscera are especially toxic. The water-soluble endogenous toxin is found in puffers throughout the tropical Pacific, Atlantic, and Indian Oceans. Puffers have been used as a poison bait to destroy stray cats.

Species involved include death puffer *(Arothron hispidus,* Japan to tropical Pacific to South Africa to the Red Sea), white-spotted puffer *(Arothron meleagris,* Pacific Ocean from Central America to Indonesia), black-spotted puffer *(Arothron nigropunctatus,* Red Sea to Indo-Pacific oceans to Japan), Gulf puffer *(Sphaeroides annulatus,* Pacific coast of the Americas), and porcupine fish *(Diodon hystrix,* Arctic to north-temperate waters).

Symptoms of tetraodon poisoning include (about 10 to 45 minutes after ingestion) lip-tongue tingling, discoordination, numbness, salivation, weakness, nausea, vomiting, diarrhea, abdominal pain, twitching, paralysis, difficulty in swallowing, loss of voice, convulsions, and (more than 60% of cases) death.

b. Scombroid poisoning. Tuna, bonito, mackeral, skipjack, and othes scombroid fishes become secondarily toxic when exposure to warm temperature of freshly caught fish for several hours results in decomposition by *Proteus morganii* bacteria of histidine into highly toxic saurine (Kimata, 1961; Fig. 16). The toxins are not detectable by the usual smell of decomposition, although the flesh may have a sharp peppery taste.

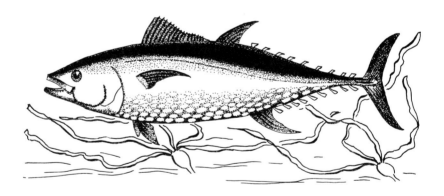

FIG. 16. Bluefin tuna *(Thunnus thynnus).*

Symptoms of scombroid poisoning develop within a few minutes after ingestion and include headache, dizziness, throbbing of large neck blood vessels, mouth dryness, heart palpitation, difficulty in swallowing, nausea, vomiting, diarrhea, and abdominal pain. Characteristic massive red welts develop. These itch intensely. Death from shock may occur. Otherwise recovery occurs within 8 to 24 hours.

Symptomatic treatments plus antihistamines are recommended. Since the fishes involved in scombroid intoxication are all commonly eaten without harm, this type of poisoning occurs sporadically and predictably. In general, scombroid fishes should be eaten soon after capture or should be frozen or canned immediately. Commercially canned scombroids are safe. Fish left in the sun more than 2 hours or showing gill pallor or off-odor should be discarded.

Ciguatera poisoning. Intoxication resulting from ingestion of numerous taxonomically unrelated fishes which are ecologically associated in narrow regions of coral reefs is called ciguatera (Banner and Helfrich, 1964). Peter Martyr (1555), in writing the history of the West Indies, refers to the danger of ciguatera poisoning. The West Indian theory that the toxin developed in fish which had eaten the poisonous manchineel berry (Halstead, 1959) still persists, but it has been shown to be untrue by modern scientists. Captain James Cook was poisoned (1776) by eating toxic fishes in New Caledonia, but recovered to continue his world voyage. During World War II (1941–1945) more than 400 Japanese soldiers and sailors died in Micronesia from fish poisoning, mostly ciguatera (Hiyama, 1950). Ciguatera is a secondary toxicity probably produced originally by a algae and passed up the food chain of small herbivorous fish to large carnivores, such as groupers and snappers. The thermostable toxin is not water-soluble and tends to concentrate in the viscera (Banner *et al.,* 1961; 1963*a,b*).

Symptoms of acute poisoning begin 30 minutes to 4 hours after ingestion and include numbness and tingling of the face and lips which spreads to fingers and toes. This is followed by nausea, vomiting, diarrhea, malaise, dizziness, abdominal pain, and muscular weakness. In severe poisoning, these symptoms progress to foaming at the mouth, muscular paralysis, difficulty in breathing, or convulsions. Death may occur from convulsions or respiratory arrest within 1 to 24 hours. If the patient recovers from the immediate symptoms, muscular weakness and tingling of the face, lips, and mouth may persist for weeks. A common characteristic is reversed temperature sensations, cold objects causing a searing pain ("electric shock") sensation, and hot things producing ᴀ sensation of cold.

More than 300 species of fish have been incriminated in ciguatera poisoning including those listed in Table III (see also Figs. 17 and 18).

d. Hallucinary mullet poisoning. Seasonal intoxication from certain mullet and surmullet can occur 10 minutes to 2 hours after ingestion.

TABLE III

SOME FISHES COMMONLY ASSOCIATED WITH CIGUATERA

Common Name[a]	Scientific name	Distribution
Surgeonfish	*Acanthurus glaucopareius*	Tropical Pacific
Surgeonfish	*Acanthurus triostegus*	Mid-Pacific
Ladyfish	*Albula vulpes*	All warm seas
Filefish	*Alutera scripta*	All warm seas
Snapper	*Aprion virescens*	Tropic Indo-Pacific
Triggerfish	*Balistoides niger*	Japan to Madagascar
Jack	*Caranx hippos*	Tropical Atlantic
Jack	*Caranx melampygus*	Japan to Tropic Pacific
Sea bass	*Cephalopholis argus*	Tropic Indo-Pacific
Herring	*Clupanodon thrissa*	Japan to India
Anchovy	*Engraulis japonicus*	Northeast Asia
Wrasse	*Epibulus insidiator*	Tropic Indo-Pacific
Sea bass	*Epinephelus fuscoguttatus*	Indo-Pacific
Snapper	*Gnathodentex aureolineatus*	Taumotu to E. Africa
Wrasse	*Julis gaimardi*	Tropic Indo-Pacific
Ocean bonito	*Katsuwonus pelamis*	Circumtropic
Trunkfish	*Lactophrys trigonus*	Eastern Tropic America to Cape Cod
Trunkfish	*Lactoria cornutus*	Tropical Pacific
Snapper	*Lethrinus miniatus*	Polynesia to E. Africa
Red snapper	*Lutjanus bohar*	Tropic Pacific to Red Sea
Red snapper	*Lutjanus gibbus*	Tropic Indo-Pacific
Snapper	*Lutjanus monostigma*	China, Polynesia to Red Sea
Red snapper	*Lutjanus vaigiensis*	Japan to East Africa
Snapper	*Monotaxis grandoculis*	Polynesia to E. Africa
Sea bass	*Mycteroperca venenosa*	Western Tropic Atlantic
Squirrelfish	*Myripristis murdjan*	Indo-Pacific
Porgie	*Pagellus erythrinus*	Black Sea to Scandanavia
Porgie	*Pagrus pagrus*	Western Europe, Mediterranean
Chinaman	*Paradicichthys venenatus*	Australia
Surmullet	*Parupeneus chryserydros*	Polynesia to East Africa
Sea bass	*Plectropomus oligacanthus*	Indonesia to Marshalls
Sea bass	*Plectropomus truncatus*	Micronesia, Indonesia, Philippines
Parrotfish	*Scarus caeruleus*	Florida to West Indies
Parrotfish	*Scarus microrhinos*	Indo-Pacific
Barracuda	*Sphyraena barracuda*	Hawaii to Red Sea; Brazil to Florida
Squaretail	*Tetragonurus cuvieri*	Temperate Zones
Surmullet	*Upeneus arge*	Polynesia, Micronesia
Sea bass	*Variola louti*	Tropic Indo-Pacific

[a] Common names of fishes are duplicated, variable, and generally unreliable.

Fig. 17. Red snapper *(Lutjanus apodus)*.

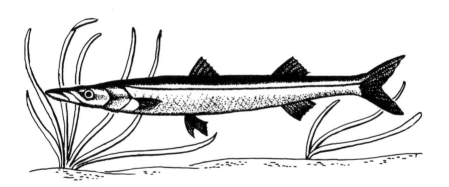

Fig. 18. Pacific barracuda *(Sphyraena argentea)*.

Helfrich, (1963) reports nine outbreaks in Hawaii, involving more than 50 persons. No fatalities were reported. Cases occurred in June, July, and August in restricted areas of Kauai and Molokai. The condition is strongly endemic and involves many species especially the mullets *(Mugil cephalus, Neomyxus chaptalii)*, the surmullet or goatfish *(Upeneus arge)*, and *Mulloidichthys samoensis*, and sometimes the rudderfish *(Kyphosus cinerascens)* and the surgeonfish *(Acanthurus sandvicensis)*. Symptoms include dizziness, loss of equilibrium, discoordination, and mental depression. If symptoms develop while the victim is asleep, he experiences terrible nightmares. Symptoms may include itching, throat-burning, muscular weakness and partial paralysis. While the heat-stable insoluble toxin is concentrated in the fish head, poisoning has resulted from ingestion of headless eviscerated fish.

e. Clupeoid poisoning. Ingestion of certain herring or sardine-like fishes has produced human poisoning in the Marshall Islands, New Caledonia, Fiji, the Society Islands, Indonesia, and Ceylon. Symptoms include labored respiration, cyanosis, cold sweat, painful cramps, dilated pupils, and (occasionally) death. The toxin is considered to be transvectored from dinoflagellates much as is done by shellfish (Randall, 1958) (Bartsch *et al.,* 1959).

f. Chimaeroid poisoning. The musculature and viscera of elephantfish *(Chimaera monstrosa;* North Atlantic, Mediterranean, South Africa) and ratfish *(Hydrolagus affinis,* North Atlantic; *Hydrolagus colliei,* American Pacific, Fig. 19) have a neurotoxin in their liver and roe

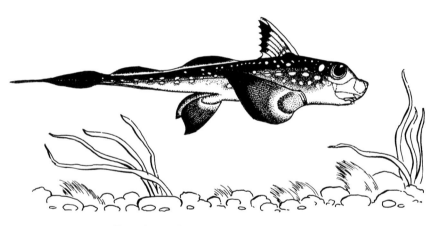

FIG. 19. Pacific ratfish *(Hydrolagus colliei).*

at certain times and places. Little is known of this phenomenon (Halstead, 1959).

g. Gempylid poisoning. Some escolars (Gempylidae) such as snake mackeral *(Lepidocybium flayobrunneur,* Pacific Ocean to Africa) and castor-oil fish *(Ruvettus praetiosus,* Atlantic Ocean to Indo-Pacific area, Fig. 20) contain large quantities of purgative oil which produces a rapidly developing diarrhea disassociated with cramping or pain (Cooper, 1964).

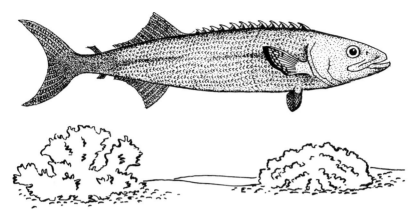

FIG. 20. Castor-oil fish (escolar) *(Ruvettus praetiosus)*.

h. Freshwater Fish Poisoning. The roe (ovaries) of certain fresh-water fish is toxic during the reproductive season. Ingestion is followed by headache, fever, vertigo, vomiting, abdominal cramps, and diarrhea. Species involved include sturgeons *(Acipenser huso,* Europe; *Acipenser sturio;* eastern America, Europe), catfishes *(Selenapsis herzbergi,* South America; *Pseudobagrus aurantiacus,* Japan), sculpins *(Scorpaenichthys marmoratus,* Pacific America), minnows *(Barbus barbus,* Europe; *Schizothorax intermedius,* Asia; *Tinca tinca,* Europe), and burbot *(Lota lota,* Europe) (Jensen, 1957).

4. TURTLES

Some marine turtles which can usually be eaten safely sometimes are found to be poisonous. The greatest concentration of toxin appears to be in the turtle liver. This phenomenon has not been explained but may be due to toxins passing along the food chain. Poisoning by marine turtles occurs most commonly in the Philippines, Indonesia, and Ceylon.

From a few hours to several days after ingestion, nausea, vomiting, diarrhea, abdominal pain, dizziness, "burning" lips (tongue, mouth, throat), difficult swallowing, excessive salivation, tongue ulcers, sleepiness, and death may occur. Death from kidney and liver damage occurs in about 44% of the cases.

Species involved include the green sea tutle *(Chelonia mydas),* the hawksbill turtle *(Eretmochelys imbricata),* and the leatherback turtle *(Dermochelys coriacea).* All three occupy tropical and subtropical seas throughout the world. *Dermochelys coriacea* is occasionally found in temperate waters (Cooper, 1964).

5. MAMMALS

Certain mammals, especially in colder climates, have been reported as poisonous to man. Much evidence points to the causative factor as the great concentrations of vitamin A in the livers and kidneys of these animals. In addition to dogs, seals, sea lions, and polar bears, toxic experiences are reported from several species of whales and dolphins.

a.   Canis.   Livers of the various breeds of sled dogs *(Canis familiaris),* such as the Eskimo (Fig. 21), chow, samoyede, and husky, sometimes

Fig. 21. Eskimo dog *(Canis familiaris* var. *eskimoensis).*

contain enough vitamin A to poison a man, as various arctic explorers have discovered much to their sorrow (Anonymous, 1942; Sutton, 1942).

b.   Erignathus.   The bearded seal, *Erignathus barbatus,* lives on ice edges along the sea throughout the Arctic. Its liver contains a high concentration of vitamin A and is considered highly toxic (Halstead, 1959).

c.   Neophoca.   The Australian sea lion, *Neophoca cinerea* (Fig. 22), occurs along the coast of South Australia. Its flesh is considered to be toxic (Rodahl and Moore, 1943).

d.   Thalarctos.   The polar bear, *Thalarctos maritimus* (Fig. 23), occurs throughout the Arctic. The numerous human poisonings caused by ingestion of the liver and kidneys of these bears is considered to be due to the high vitamin A concentrations in these organs. Symptoms in-

FIG. 22. Australian sea lion *(Neophoca cinerea)*.

FIG. 23. Polar bear *(Thalarctos maritimus)*.

clude severe headache, nausea, vomiting, diarrhea, abdominal pain, dizziness, drowsiness, irritability, collapse, light sensitivity, and convulsions. Death occurs only rarely (Rodahl and Moore, 1943).

TABLE IV
SYSTEMATIC LIST OF DISCUSSED ANIMALS

Kingdom – Animalia
  Phylum – Protozoa
    Class – Mastigophora
      Order – Dinoflagellata
        Family – Gymnodinidae
          Gymnodimium
        Family – Peridinidae
          Gonyaulax
          Pyrodinium
        Family – Prorocentridae
          Prorocentrum
  Phylum – Cnidaria
    Class – Anthozoa
      Order – Zoantharia
        Family – Actinidae
          Physobrachia
        Family – Actinodiscidae
          Rhodactis
        Family – Stoichactidae
          Radianthus
  Phylum – Echinodermata
    Class – Asteroidea
    Class – Echinoidea
      Order – Centrechinoida
        Family – Diadematidae
          Diadema
        Family – Echinidae
          Paracentrotus
        Family – Toxopneustidae
          Tripneustes
    Class – Holothurioidea
  Phylum – Mollusca
    Class – Amphineura
      Order – Polyplacophora
        Family – Mopalidae
          Mopalia
    Class – Gastropoda
      Order – Prosobranchia
        Family – Acmaeidae
          Acmaea
        Family – Buccinidae
          Buccinum
          Neptunea
        Family – Cymatidae
          Argobuccinum

Family – Haliotidae
  Haliotis
Family – Muricidae
  Murex
  Thais
Family – Trochidae
  Livona
Order – Opisthobranchia
  Family – Aplysidae
    Aplysia
Class – Pelecypoda
  Order – Filibranchia
    Family – Mytilidae
      Modiolus
      Mytilus
    Family – Pectinidae
      Placopecten
    Family – Spondylidae
      Spondylus
  Order – Eulamellibranchia
    Family – Arcidae
      Arca
    Family – Cardidae
      Cardium
      Clinocardium
    Family – Donacidae
      Donax
      Heterodonax
Family – Mactridae
  Schizothaerus
  Spisula
Family – Myacidae
  Mya
Family – Ostreidae
  Crassostrea
  Ostrea
Family – Pholodidae
  Penitella
Family – Sanguinolaridae
  Tagellus
Family – Solenidae
  Ensis
  Siliqua
Family – Tellinidae
  Macoma

Family — Unionidae
Anodonta
Family — Veneridae
Callista
Dosinea
Protothaca
Saxidomus
Tapes
Tivela
Class — Cephalopoda
Order — Dibranchiata
Family — Ommastrephidae
Ommastrephes
Family — Octopodidae
Octopus
Phylum — Arthropoda
Class — Merostomata
Order — Xiphosura
Family — Xiphosuridae
Carcinoscorpius
Tachypleus
Class — Crustacea
Order — Malacostraca
Family — Xanthidae
Angatea
Eriphia
Xanthodes
Phylum — Chordata
Class — Cyclostomata
Order — Hyperoartia
Family — Petromyzonidae
Petromyzon
Class — Chondrichthyes
Order — Selachii
Family — Carcharhinidae
Carcharhinus
Family — Hexanchidae
Hexanchus
Family — Heptranchidae
Heptranchias
Family — Isuridae
Carcharodon
Family — Sphyrnidae
Sphyrna
Family — Squalidae
Somniosus
Order — Holocephali
Family — Chimaeridae

Chimaera
Hydrolagus
Class — Osteichthyes
Order — Anguillida
Family — Muraenidae
Gymnothorax
Order — Acanthopterggii
Family — Gempylidae
Lepidocybium
Ruvettus
Order — Anacanthini
Family — Gadidae
Lota
Family — Tetraodontidae
Arothron
Sphaeroides
Order — Berycomorphida
Family — Holocentridae
Myripristis
Order — Chondrostei
Family — Acipenseridae
Acipenser
Family — Acanthuridae
Acanthurus
Family — Aluteridae
Alutera
Family — Carangidae
Caranx
Family — Kyphosidae
Kyphosus
Family — Lutjanidae
Lutjanus
Gnathodentex
Monotaxis
Aprion
Lethrinus
Paradicichthys
Family — Mugilidae
Neomyxus
Mugil
Family — Scaridae
Scarus
Family — Serranidae
Epinephelus
Variola
Plectropomus
Cephalopholis
Mycteroperca

TABLE IV
SYSTEMATIC LIST OF DISCUSSED ANIMALS (Continued)

Order – Isospondyli
Family – Albulidae
  Albula
Family – Clupeidae
  Clupanodon
Family – Engraulidae
  Engraulis
Order – Ostariophysi
Family – Aridae
  Selinapsis
  Pseudobagrus
Family – Cyprinidae
  Tinca
  Schizothorax
  Barbus
Order – Pecomorphida
Family – Sphyraenidae
  Sphyraena
Family – Mullidae
  Upeneus
  Parupeneus
  Mulloidichthys
Family – Tetragonuridae
  Tetragonurus
Family – Labridae
  Epibulus
  Julis
  Pagellus
  Pagrus
Family – Scombridae

  Katsuwonus
  Thunnus
Order – Plectoghathi
Family – Balistidae
  Balistoides
Family – Diodontidae
  Diodon
Family – Ostrachidae
  Lactophrys
  Lactoria
Order – Scleroparei
Family – Cottidae
  Scorpaenichthys
Class – Reptilia
Order – Testudinata
Family – Chelonidae
  Chelonia
  Eretmochelys
Family – Dermochelidae
  Dermochelys
Class – Mammalia
Order – Carnivora
Family – Canidae
  Canis
Family – Otaridae
  Neophoca
Family – Phocidae
  Erignathus
Family – Ursidae
  Thalarctos maritimus

IV. SOME OTHER NATURALLY OCCURRING TOXINS

Allergens. An allergen is any agent capable of producing an altered reaction to a specific substance. The commonest allergens are foods, pollens, fungi, hair, and drugs. Foods frequently causing allergy include egg, wheat, buckwheat, milk, peas, beans, nuts, fish, onion, potato, chocolate, mushrooms, and fruits (berries, cantaloupe) (Blumstein, 1935; Kingsbury, 1964).

Antimetabolites. An antimetabolite is a substance having a molecular structure similar to a substance essential for proper metabolism, but having a pharmacologic effect antagonistic to the essential substance. For example, vitamin K is antagonized by the antimetabolite dicumarol occurring in moldy hay; vitamins A and E are antagonized by rancid fats; and vitamin D is antagonized by tosamin from cereals (Best and Taylor, 1961).

Cyanogens. Substances, such as apple seeds, producing biologically significant quantities of cyanide are termed cyanogens (Kingsbury, 1964).

Estrogens. Certain natural substances possess the biologic activity of estrus-producing hormones. Estrogenic substances occur in the blood of mammals during pregnancy, and also in plants such as yeast and rape seed.

Goitrogens. A number of plants contain compounds which prevent normal inorganic iodide accumulation by the thyroid, thus inhibiting formation of thyroid hormone and causing a diseased condition known as goiter. Iodine-deficient diet also produces goiter, as do certain drugs. A number of species of *Brassica* (including cabbage, brussels sprouts, rutabaga, kale, broccoli, rape, and kohlrabi) contain goitrogens, as do chard *(Beta vulgaris cicla),* soybean *(Glycine max),* flax *(Linum usitatissimum),* and various other Cruciferae, Rosaceae, and Umbelliferae (Greer and Astwood, 1948).

Lathrogens. Toxins found in meal made from some of the vetches (e.g., *Lathyrus sativus* and *Lathyrus cicera)* produce a spastic paralysis (termed lupinosis) involving chiefly the legs. These toxins are termed lathrogens (Kingsbury, 1964).

REFERENCES

Anderson, L. S. (1960). *Am. J. Public Health* **50**, 71.
Anet, E. F., Lythgoe, B., Silk, M. H., and Trippett, S. (1953). *J. Chem. Soc.* p. 309.
Anonymous (1938). "Poisonous and Injurious Plants of Los Angeles County." County of Los Angeles, Livestock Department, California.
Anonymous (1942). "Is Polar Bear Liver Poisonous?" *J. Am. Med. Assoc.* **118**, 337.
Arcisz, W. (1950). *U. S. Fish Wildlife Serv. Spec. Sci. Rept.* **27**, 1.
Arnold, H. L. (1944). "Poisonous Plants of Hawaii." Tongg Publishing Co., Honolulu.
Asano, M., Takayanagi, F., and Kitamura, T. (1953). *Tohoku J. Agr. Res.* **3**, 321.
Bamford, F. (1951). "Poisons, Their Isolation and Identification." Blakiston, Philadelphia.
Banner, A. H., and Helfrich, P. (1964). *Hawaii Marine Lab. Univ. Hawaii, Tech. Rept.* **3**, iii + 48 pp.
Banner, A. H., Sasaki, S., Helfrich, P., Alender, C. B., and Scheuer, P. J. (1961). *Nature* **189**, 229.
Banner, A. H., Helfrich, P., Scheuer, P. J., and Yoshida, T. (1963a). *Proc. Gulf Caribbean Fisheries Inst.* **16**, 84.
Banner, A. H., Shaw, S. W., Alender, C. B., and Helfrich, P. (1963b). *South Pacific Comm. Tech. Paper* **141**. (Noumea, New Caledonia).
Bartsch, A. F., Drachman, R. H., and McFarren, E. F. (1959). *U. S. Public Health Serv. Rept.*

Best, C. H., and Taylor, N. B. (1961). "The Physiological Basis of Medical Practice." Williams and Wilkins, Baltimore.

Block, S. S., Stephens, R. L., Barreto, A., and Murrill, W. A. (1955). *Science* 121, 505.

Blumstein, G. L. (1935). *J. Allergy* 7, 74.

Bond, R. M., and Medcof, J. C. (1958). *Can. Med. Assoc. J.* 79, 19.

Borgstrom, G. (1961). "Fish as Food," Vol. I. Academic Press, New York.

Boylan, D. B., and Scheuer, P. J. (1967). *Science* 155, 52.

Buck, R. W. (1961). *New Engl. J. Med.* 265, 681.

Buttenweiser, S., and Bodenheimer, W. (1924). *Deut. Med. Wochschr.* 50, 607.

Cann, H. M., and Verhulst, H. L. (1958). "Poisonous Plants," p. 2. National Clearinghouse Poison Cont. Center.

Case, A. A. (1957). *J. Am. Vet. Med. Assoc.* 130, 323.

Charnot, A. (1945). *Mem. Soc. Sci. Nat. Maroc.* 47, 86, 90, 112.

Cooper, M. J. (1964). *Pacific Sci.* 18, 411.

Cottingham, J. O. (1955). *Proc. Indiana Acad. Sci.* 65, 210.

Davies, F. R., Edwards, H. I., Kitchen, W. L., and Tomlinson, H. O. (1958). *Can. J. Public Health* 49, 286.

Drach, G., and Maloney, W. H. (1963). *J. Am. Med. Assoc.* 184, 1047.

Dreisbach, R. H. (1966). "Handbook of Poisoning." Lange Medical Publishers, Los Altos, California.

Edwards, H. I. (1956). *J. Milk Food Technol.* 19, 331.

Emery, Z. T. (1887). *N. Y. Med. J.* 45, 92.

Engelsen, H. (1922). *Nord Hyg. Tidskr.* 3, 316.

Fange, R. (1960). *Ann. N. Y. Acad. Sci.* 90, 689.

Farber, L., and Lerke. P. (1963). *In* "Venomous and Poisonous Animals of the Far East" (H. L. Keegan and W. V. McFarlane, eds.), pp. 67–74. Pergamon, New York.

Fernald, M. L., and Kinsey, A. C. (1943). "Edible Wild Plants of Eastern North America," pp. 1847–1848. Idlewild Press, New York.

Fish, C. J., and Cobb, M. C. (1954). *U. S. Fish Wildlife Serv. Res. Rept.* 36.

Forsyth, A. A. (1954). *Min. Agr. Fisheries Food (London) Bull.* 161.

Frohne, D., and Pribilla, O. (1965). *Arch. Toxikol.* 21, 150.

Fruthaler, G. J. (1955). *Ochsner Clin. Rept.* July, 50–52.

Garner, R. J. (1957). "Veterinary Toxicology." Baillière, Tindall, and Cox, London.

Gibbard, J., and Naubert, J. (1948). *Am. J. Public Health* 38, 550.

Goldberg, R. E. (1951). *Today's Health* 29 (August), 38, 66.

Green, J. E. (1958). *Mod. Vet. Pract.* 39, (7), 60.

Greer, M. A., and Astwood, E. B. (1948). *Endocrinology* 43, 105.

Habekost, R. C., Fraser, I. M., and Halstead, B. W. (1955). *J. Wash. Acad. Sci.* 45, 101.

Haggerty, D. R., and Conway, J. A. (1936). *N. Y. State J. Med.* 36, 1511.

Hall, M. C. (1914). *J. Am. Med. Assoc.* 63, 242.

Halstead, B. W. (1958). *U. S. Public Health Rept.* 73, 302.

Halstead, B. W. (1959). "Dangerous Marine Animals." Cornell Maritime Press, Cambridge, Maryland.

Halstead, B. W. (1965–68). "Poisonous and Venomous Marine Animals of the World," Vols. I–III. U. S. Government Printing Office, Washington, D. C.

Hansen, A. A. (1925). *Science* 61, 340.

Harris, F. W., and Cockburn, F. (1918). *Am. J. Pharm.* 90, 722.

Hashimoto, Y., Naito, K., and Tsutsumi, J. (1960). *Bull. Japan. Soc. Sci. Fisheries* 26, 1216.

Helfrich, P. (1963). *Hawaii Med. J.* **22**, 361.

Henderson, J. A., Evans, E. V., and McIntosh, R. A. (1952). *J. Am. Vet. Med. Assoc.* **120**, 375.

Hendricks, H. V. (1940). *J. Am. Med. Assoc.* **114**, 1625.

Hiyama, Y. (1950). *U. S. Fish Wildlife Serv. Spec. Sci. Rept.* **25**.

Holdridge, L. R., and Munoz MacCormick, C. (1937). "Notes on Poisonous and Stinging-haired Plants of Puerto Rico." Caribbean National Forest, Rio Piedras, Puerto Rico.

Howes, F. N. (1949). *Kew Bull.* p. 167.

Humphreys, F. A., and Gibbons, R. J. (1941). *Can. J. Comp. Med. Vet. Sci.* **5**, 84.

Jacob, J. C. (1966). *Atlanta J. Constitution Mag.* pp. 30–32 (May 1, 1966).

Jacobson, M. (1958). *U. S. Dept. Agr. Handbook* **154**.

Jacobziner, H., and Raybin, H. W. (1960). *N. Y. State J. Med.* **60**, 3139.

Jacobziner, H., and Raybin, H. W. (1961). *N. Y. State J. Med.* **61**, 301.

Jennings, R. E. (1935). *J. Pediatr.* **6**, 657.

Jensen, E. T. (ed.) (1957). "Conference on Shellfish Toxicology." U. S. Public Health Service, Washington, D. C.

Kawabata, J., and Halstead, B. W. (1957). *Am. J. Trop. Med. Hyg.* **6**, 935.

Keegan, H. L., and MacFarlane, W. V. (1963). "Venomous and Poisonous Animals and Noxious Plants of the Pacific Region." Pergamon Press, New York.

Keegan, H. L., and Toshioka, S. (1958). Some Venomous Animals of the Far East. 406th Medical General Laboratory (U. S. Army), Camp Zama, Japan.

Khlentzos, C. T. (1950). *Am. J. Trop. Med.* **30**, 785.

Kimata, M. (1961). "Fish as Food" (G. Borgstrom, ed.), Vol. I, pp. 329–352. Academic Press, New York.

Kingsbury, J. M. (1964). "Poisonous Plants of the United States and Canada." Prentice-Hall, Englewood Cliffs, New Jersey.

Kurland, L. T., Faro, S. N., and Siedler, H. (1960). *World Neurol.* **1**, 370.

Lewes, D. (1948). *Brit. Med. J.* **2**, 383.

Link, K. P. (1959). *Circulation* **19**, 97.

Little, E. L. (1953). *U. S. Dept. Agr. Handbook* **41**.

Little, E. L., and Wadsworth, F. H. (1964). *U. S. Dept. Agr. Handbook* **249**.

Martin, E. J. (1960). *Pacific Sci.* **14**, 403.

Martyr, P. (1555). (The Eight Decades of Peter Martyr d'Anghera). English translation published by G. P. Putnam & Sons, 1912.

May, M. (1965). Okay To Eat Daisies—But Some Plants Kill. United Press International release, *Atlanta Journal,* March 19, 1965.

Medcof, J. C., and Gibbons, R. J. (1948). *Fisheries Res. Board Can. Progr. Rept. Atlantic Coast Sta.* **376**, 1.

Meyer, K. F. (1953). *N. Engl. J. Med.* **249**, 765, 804, 843.

Meyer, K. F., Sommer, H., and Schoenholz, P. (1928). *J. Prevent. Med.* **2**, 365.

Millspaugh, C. F. (1892). *Medicinal Plants,* Vol. I. John C. Yorston & Co., Philadelphia.

Mitchell, J. E., and Mitchell, F. N. *J. Pediatr.* **47**, 227.

Morton, J. F. (1958). *Proc. Florida State Hort. Soc.* **71**, 372.

Moru, J. (1943). Contribution a l'étude de la toxicité des animaux marins. Thesis, Faculty of Medicine, Paris.

Mosher, H. S., Fuhrman, F. A., Buchwald, H. D., and Fisher, H. G. (1964). *Science* **144**, 1100.

Nakazima, M. (1963). *In* "Poisonous and Venomous Marine Animals of the World" (B. W. Halstead, ed.), Vol. I, p. 766. U. S. Government Printing Office, Washington, D.C.

O'Leary, S. B. (1964). *Arch. Environ. Health* **9,** 216.

Olson, O. E., Dinkel, C. A., and Kamstra, L. D. (1954). *South Dakota Farm Home Res.* **6,** 12.

O'Malley, T. (1962). *In* "Poisonous and Venomous Marine Animals of the World" (B. W. Halstead, ed.), Vol. I, p. 235. U. S. Government Printing Office, Washington, D.C.

Plumert, A. (1902). *Arch. Schiffs Tropenhyg.* **6,** 15.

Power, F. B. (1901). *Pharmaceut. J.*

Prentiss, D. W. (1898). *Phila. Med. J.* **2,** 607.

Press, E. (1964). *Arch. Environ. Health* **9,** 142.

Pugsley, L. I. (1939). *Fisheries Res. Board Can. Progr. Rept. Pacific Coast Station* **40,** 11.

Ramsbottom, J. (1953). "Mushrooms and Toadstools." Collins, London.

Randall, J. E. (1958). *Bull. Marine Sci. Gulf Caribbean* **8,** 236.

Reynard, G. B., and Norton, J. B. S. (1942). *Univ. Maryland Agr. Expt. Sta. Tech. Bull.* **A10.**

Reynolds, W. A., and Lowe, F. H. *N. Engl. Med. Assoc.* **73,** 627.

Robb, H. F. (1919). *J. Am. Med. Assoc.* **73,** 627.

Rodahl, K., and Moore, T. (1943). *Biochem. J.* **37,** 166.

Rosen, C. S., and Lechner, M. (1962). *N. Engl. J. Med.* **267,** 448.

Ross, S. G. (1947). *Med. J. Australia* **2,** 617.

Sampson, A. W., and Malmsten, H. E. (1942). *Calif. Agr. Expt. Sta. Bull.* **593.**

Sapeika, N. (1958). *S. African Med. J.* **32,** 527.

Schery, R. W. (1952). "Plants for Man." Prentice-Hall, New York.

Schwimmer, M., and Schwimmer, D. (1955). "The Role of Algae and Plankton in Medicine." Grune and Stratton, New York.

Scott, H. G. (1963). *J. Environ. Health* **24,** 201.

Senior, V. E. (1960). *Can. J. Comp. Med.* **24,** 26.

Smith, A. H. (1949). "Mushrooms in Their Natural Habitats." Sawyer's Inc., Portland, Oregon.

Sommer, H., and Meyer, K. F. (1937). *Arch. Pathol.* **24,** 560.

Stiles, F. C. (1951). *J. Pediatr.* **39,** 354.

Sutton, R. L. (1942). *J. Am. Med. Assoc.* **118,** 1026.

Tarada, Y. (1953). *Mie Med. J.* **3,** 205.

Tasaki, B., and Tanaka, U. (1918). *J. Coll. Agr. Imperial Inst. Tokyo* **3,** 337.

Taylor, N. (1949). "Flight from Reality." Duell, Sloan and Pierce, New York.

Tucker, J. M., and Kimball, M. H. (1960). *Calif. Agr. Ext. Serv. Bull.* (July).

U. S. Public Health Service (1962). *Alaska's Health Welfare* **19,** 1.

Von Oettingen, W. F. (1958). "Poisoning." Saunders, Philadelphia.

Walford, L. A. (1958). "Living Resources of the Sea." Ronald Press, New York

Waterman, T. H. (1953). Reported in "Poisonous and Venomous Marine Animals of the World" (B. W. Halstead, ed.), Vol. I, p. 922. U. S. Government Printing Office, Washington, D. C.

Wittaker, V. P. (1960). *Ann. N. Y. Acad. Sci.* **90,** 695.

Whitehead, E. I., and Moxon, A. L. (1952). *S. Dakota Agr. Expt. Sta. Bull.* **424.**

Winkler, L. R. (1961). *Pacific Sci.* **15,** 211.

Youngken, H. W., and Karas, J. S. (1964). *U. S. Public Health Serv. Publ.* **1220.**

Yudkin, W. H. (1944). *Bull. Birmingham Oceanogr. Coll.* **9,** 1.

AUTHOR INDEX

Numbers in italics refer to pages on which the references are listed.

605

Jirgensons, B., 421, *446*
Joffe, A. Z., 424, 425, 426, *448*
Johansen, A., 336, *357*
Johanssen, A., 531, *537*
Johns, A. W., 136, *172*
Johnson, A. J., 397, *448*
Johnson, H. M., 386, *391*, 485, *486*
Johnson, K. M., 81, 83, 98, 100, 104, *108*, *110*
Johnston, H. H., 54, *70*
Jones, H. E. H., 418, *447*
Jordan, C. F., 249, *277*
Jordan, E. O., 227, 234, 235, *282*, 361, *391*
Jordan, R. T., 91, *110*
Jordan, W. S., Jr., 105, *110*
Joseph, P. R., 264, *282*
Joynt, M. F., 249, *277*
Jubb, G., 79, *110*
Juhasz, S., 411, 412, *448*
June, R. C., 240, *280*, *282*

K

Kabler, P. W., 97, *108*, 233, 249, *282*, *283*
Kadavy, J. L., 376, 378, 379, *390*, *392*, 521, *537*
Kagan, I. G., 180, 181, *220*, *221*
Kaiser, P., 305, *326*
Kaiser, R. L., 260, *282*
Kalter, S. S., 94, 99, 105, 106, *110*
Kampelmacher, E. H., 272, 273, *282*
Kamphans, S., 84, *110*
Kamstra, L. D., 551, *604*
Kanzawa, K., 334, 335, 336, 337, 341, 354, 355, *357*
Kao, C. T., 530, *537*
Kaplan, A. S., 88, *110*
Kaplan, H. S., 91, 92, *110*
Kaplan, M. T., 232, *282*
Karas, J. S., 549, *604*
Karashimada, T., 334, 335, 337, 341, 354, 355, *357*
Karlson, A. G., 254, *282*
Karzon, D. T., 104, 105, *108*
Kasel, J. A., 81, *110*
Kasza, L., 84, *108*, *110*
Kater, J. C., 209, *220*
Kato, T., 121, 122, *128*, *129*
Katz, S., 105, *110*

Kauffmann, F., 61, *70*, 468, *486*
Kaufman, H. E., 209, *221*
Kaufman, O. W., 303, 310, *324*, *325*
Kautter, D. A., 330, *357*, *486*, 529, *537*
Kawabata, J., 585, *603*
Kawabata, T., 237, *282*, 338, 355, *358*
Kean, B. H., 215, *220*
Keegan, H. L., 544, *603*
Kelen, A. E., 210, *220*
Kelly, C. B., 465, *486*
Kelly, S., 94, *110*
Kempe, H., 84, *112*
Kempe, J. D., 515, *539*
Kempe, L. L., 91, *110*, 309, 316, 317, 319, 320, *323*, *325*, 341, *357*, 483, *485*, 505, 507, 508, 515, 524, *535*, *537*, *538*
Kendereski, S., 236, *282*
Kennedy, J. C., 251, *280*
Kenner, B. A., 233, *282*
Kephart, R. E., 373, *390*
Kerr, D. E., 337, 338, 350, *356*
Kerr, K. B., 178, *222*
Kerrin, J. C., 243, *282*
Kessel, J. F., 210, *220*
Kessner, D. M., 239, *283*
Ketler, A., 93, *110*
Ketsawan, P., 184, *220*
Keuning, R., 404, 405, 406, *450*
Khlentzos, C. T., 578, 587, *603*
Kidder, R. W., 433, 434, *448*
Kiefer, E. D., 189, *221*
Kim, H. W., 100, 101, 105, 106, *108*, *113*
Kimata, M., 590, *603*
Kimball, M. H., 549, *604*
King, C. T., 254, *278*
King, G. J. G., 48, *72*
Kingsbury, J. M., 549, 560, 561, 570, 573, 575, 600, 601, *603*
Kinloch, J. P., 260, *283*
Kinosita, R., 428, 429, *448*
Kinsey, A. C., 564, 572, *602*
Kirby, A. C., 240, *283*
Kirschner, L., 274, 275, *283*
Kirsner, J. B., 133, *171*, 227, 230, 232, *278*
Kiss, M., 245, *284*
Kiss, P., 245, *284*
Kissling, R. E., 79, *110*

SUBJECT INDEX

A

Abalone *(Haliotis),* 598(T)
  human photosensitization poisoning by ingestion of liver of, 581
  toxin of, secondary, acquisition of, 581
Abattoirs
  spread of *Salmonella* contamination in, 48
Abrin, nature of, 568
Abrus, 568, 576 (T), *see also* Jequirity beans
  common names of, 568
Acetic acid, effect on microorganisms, 515
Achromobacter, inhibition of *S. aureus* by, 529
Achromobacter histaminium
  as synergist in food poisoning caused by marine products, 237
Acids
  effect on microorganisms, 515–516
    mechanism of, 515
Acocanthera spp., 545, 575 (T)
  toxin of, 545
Actinomyces
  as part of normal fecal flora, 225
  as possible etiological agent in food poisoning, 245–246
Adenoviruses
  animals infected by, 84
  human, gastrointestinal symptoms due to, 84
  identification by CF test, 105
    by HI test, 105
  immunology of, 84–85
  properties, 77
  stabilization by sodium ion, 89
  thermal effects on, 89, 90
Adults, food poisoning possibly caused by *E. coli* in, 239

Aerobacter aerogenes, type I, in food, origin of, 465, 466
Aeromonas, as part of normal fecal flora, 225
Aesculin, 564
  chemical nature of, 564
  human poisoning by, 564
  relation to anticoagulant rodenticides, 564
Aesculus californica
  flowers, toxicity to honey bee, 564
  human poisonings due to honey from, **564**
Aesculus spp., 564, 577 (T)
  common names of, 564
  human poisonings by, 551, 564
    number of species associated with, 564
  origin of, 564
  toxic alkaloids of, 564, *see also* Aesculin
  use of seeds, 564
Aflatoxicosis, 399–422, *see also* Aflatoxins
  outbreaks of, animals affected by, 399
  history, 399–400
Aflatoxin B, 404
Aflatoxin B$_1$
  distribution in tissues, 420
  excretion of, 420
  interaction with nucleic acids, 421
  molecular formula, 404, 405 (T)
  purified, carcinogenic potency of, 417–418
  toxic effects, in cell cultures, 411, 412 (T)
    assays based on, 414
  toxicity, 409
    comparative, 410 (T)
    to ducklings, 442
Aflatoxin B$_2$, 404
  molecular formula, 404, 405 (T)
Aflatoxin G$_1$, 404

sensitivity of, 389

Middle East

human infections with *Fasciola hepatica* in, 202

Milk, *see also* Dairy products

and milk products, as cause of botulism outbreaks in the U.S. 297 (T)

as vehicle for infectious hepatitis virus, 74, 79

as possible vehicle for avian leukosis agent, 74

as vehicle for food poisoning microorganisms, 247–258, *see also* individual organisms

contaminated, as possible source of listeriosis, 272

contamination of, modes of, 247

with tubercle bacilli, sources of, 252

excretion of toxic aflatoxin metabolites by cows into, 408, 418–419, 420

prevention of, 420

leptospirocidal effect, 274

pathogenic *E. coli* serotypes in, 239

raw, as vehicle for poliovirus, 79–80

as vehicle for tick-borne encephalitis viruses, 82

staphylococci in, 360

source of, 363

survival of cholera vibrios in, 265

transmission of *Brucella* by, 248, 249

of *C. diphtheriae* by, 257, 258

of *Salmonella* by, 50 (T), 54, 56 (T), 57

of shigellae by, 260

of septic sore throat and scarlet fever by, 255–256

of tuberculosis by, 252

"Milk sickness," 545

"Milk toxin"

nature of, 419, 420

production by lactating cattle and rats fed aflatoxin-containing diets, 419

Millet

as vehicle of ATA, 423, 424

isolation of fungal toxins from, 426

of *Fusarium sporotrichoides* from, 424

Mink

botulism type Cb in, 302 (T), 303

vehicles for, 302 (T)

susceptibility to botulinum toxins, 350–351

Mistletoes, toxicity of, 572

Moisture

limitation of, *see also* Curing and Drying

of, effect on growth of microorganisms, 516–523

of nonsporing bacteria, 517–518

of parasites, 522

of sporeforming bacteria, 518–522

of viruses and toxins, 522

Molds, *see also* Fungi

detection in spoiled food, media for, 467

losses of food or raw materials due to damage by, 396–397

spores, distribution of, 396

Moldy corn poisoning, 439–440

symptoms of, 440

Molluscs, 579–585, *see also* individual organisms

annual consumption of, 580

as transvectors of dinoflagellate toxins, 578, 580, 581, 582, 583, 584, 585

as vehicle for *Angiostrongylus cantonensis,* 183

for parasitic infections, 175–176

human poisoning by, prevention of, 580

seasonal, 580

symptoms of, 580

treatment of, 580

ingestion of toxic dinoflagellate by, 580

secondary toxins obtained by, 580

toxins, associated with shellfish poisoning, nature of, 580

effect on paralyzed muscles, 577

secondary, acquisition of, 580

storage of, 580

Monkeys, toxic effect of aflatoxins in, 410–411

Moonseed, 572

human poisoning by, 572–573

Morchella esculenta (edible morel), 553, 575 (T)

Mucin, gastric, activation of botulinum toxins by, 319

Mullets (Mullidae), 600 (T)

human poisoning, hallucinary, by, 578, 591

seasonal character of, 591, 593

species associated with, 593

symptoms of, 593–594

GENUS AND SPECIES INDEX